IE

TIME SERIES AND SYSTEM ANALYSIS WITH APPLICATIONS

TIME SERIES AND SYSTEM ANALYSIS WITH APPLICATIONS

SUDHAKAR M. PANDIT

Michigan Technological University

SHIEN-MING WU

University of Wisconsin, Madison

John Wiley and Sons

New York   Chichester   Brisbane   Toronto   Singapore

Library of Congress Cataloging in Publication Data:

Pandit, Sudhakar M. (Sudhakar Madhavrao)
  Times series and system analysis, with applications.

  Bibliography: p.
  Includes index.
  1. Time-series analysis.   2. System analysis.
I. Wu, Shien M.   II. Title.

QA280.P36   1983   519.5′5   88-15958
ISBN 0-471-86886-8

Printed in the United States of America

10   9   8   7   6   5   4   3   2   1

PREFACE

Engineers and scientists in system analysis use mathematical models, usually differential equations, developed from a conjectured physical mechanism. For complicated systems, an empirical approach, such as frequency analysis, is employed. On the other hand, statisticians and economists approximate their models, using difference equations, from plots of empirical autocorrelations and spectra. If time series and system analysis are brought together, it should be possible to avoid the considerable trial and error presently needed in both fields and vastly improve their applications. In the 1973 Ph.D. dissertation of S. M. Pandit, with S. M. Wu as advisor, a new philosophy of system analysis that bypasses this element of trial and error and provides models in the form of difference/differential equations directly from the observed data was outlined with the requisite mathematical foundation. The enthusiastic response of faculty and students to this philosophy and its extensive applications in diverse research investigations led to the need for this book.

An application of this new philosophy to time series modeling yields a sequential strategy, as in regression analysis. Once the time series is considered as the response of a system, it can be modeled with increasing degrees of freedom justified by the data. Successively higher order models are fitted by least squares until the improvement in the fit is statistically insignificant. The sequential modeling strategy can be conveniently carried out as a result of the ever-increasing capabilities of computers. The new modeling strategy can greatly reduce the tedious chore of searching for an appropriate model. We hope that this book will bring together time series and system analysis to provide system analysis specialists with a new tool and to make time series analysis useful to engineers and scientists.

The book is application-oriented. The new tool has been used for system identification, signature analysis, physical characterization, control, and even engineering design. Obviously, it is most useful for forecasting, which was the original purpose of the time series development. To enhance the ability for long-term forecasting, both stochastic and deterministic approaches are presented.

The first draft of the book grew out of lecture notes prepared for a one-semester course taught in 1973 at the University of Wisconsin, Madison, to graduate and senior undergraduate students. Revised drafts have been used in subsequent years for this course, as well as for an undergraduate-graduate two-course sequence at other schools. Students from disciplines such as civil and environmental, electrical, industrial, mechanical, mining, metallurgical, and nuclear engineering, as well as from economics and business have enrolled. Many of the examples in this book are based on the class projects undertaken by these students. We are grateful to the students for their gratifying response.

We especially thank Dr. Shiv. G. Kapoor and William Wittig for their contribution toward developing computer programs to fit models and for their help, along with Drs. T. Ungpiyakul and W. T. Tsai, in improving the final manuscript. We are indebted to Professor R. E. DeVor, University of Illinois, Urbana, and Professor W. R. DeVries, Rensselaer Polytechnic Institute, for their constructive comments.

Sudhakar M. Pandit
Shien-Ming Wu

We are grateful to the Literary Executor of the late Sir Ronald A. Fisher, F.R.S., to Dr. Frank Yates, F.R.S., and to Longman Group, Ltd., London, for permission to reproduce from Table III of their book *Statistical Tables for Biological, Agricultural and Medical Research* (6th Edition, 1974) and to Biometrika Trustees for permission to make use of some of the tables in *Biometrika Tables for Statisticians*, Volume I, by E. S. Pearson and H. O. Hartley.

CONTENTS

PREFACE   vii

1 INTRODUCTION   1

  1.1 TIME SERIES AND SYSTEM ANALYSIS   1

  1.2 CORRELATION, REGRESSION, DYNAMICS, AND
MEMORY   3

  1.3 STOCHASTIC DIFFERENCE/DIFFERENTIAL
EQUATIONS   5

  1.4 APPLICATIONS   6

  1.5 CHAPTERWISE OUTLINE AND SUMMARY   7

  1.6 GUIDELINES TO USER/INSTRUCTOR   11

2 AUTOREGRESSIVE MOVING AVERAGE MODELS   13

  2.1 LINEAR REGRESSION MODELS   16

    2.1.1 Simple Regression   17
    2.1.2 Multiple Regression   21

  2.2 FIRST ORDER AUTOREGRESSIVE MODEL   23

    2.2.1 Dependence of X_t on X_{t-1} 23
    2.2.2 Assumptions and Structure   24
    2.2.3 Estimation of ϕ_1 and σ_a^2 25
    2.2.4 Checks of Adequacy   27
    2.2.5 Prediction or Forecasting   29
    2.2.6 Random Walk as a Limit of AR(1)   30

  2.3 AUTOREGRESSIVE MOVING AVERAGE MODEL
ARMA(2,1)   33

    2.3.1 Dependence of a_t on X_{t-2} and a_{t-1} 34
    2.3.2 Special Case: AR(2) Model   36
    2.3.3 Checking the Adequacy of the AR(1) Model   37

2.3.4 Nonlinear Regression of the ARMA(2,1) Model   38
2.3.5 Other Special Cases: ARMA(1,1), MA(1)   40

2.4 ARMA($n,n-1$) MODELS AND MODELING
STRATEGY   40

2.4.1 ARMA(3,2) Model   41
2.4.2 ARMA($n,n-1$) Model   41
2.4.3 ARMA($n,n-1$) versus ARMA(n,m): ARIMA as
Special Cases   42
2.4.4 Adequacy of the ARMA($n,n-1$) Models   44

APPENDIX   **45**

A 2.1 LEAST SQUARE ESTIMATES IN SIMPLE
REGRESSION   45

A 2.1.1 Estimates of Y Intercept β_0 and Slope β_1 45
A 2.1.2 Mean, Variance, Covariance, and
Correlation of Random Variables and Their Linear
Forms   46
A 2.1.3 Mean and Variance of $\hat{\beta}_1$, $\hat{\phi}_1$, and $\hat{\rho}_k$ 50
A 2.1.4 Normal Distribution and Confidence
Intervals   54
A 2.1.5 Chi-Square and t-Distributions   59
A 2.1.6 Analysis of Variance and F-Distribution for
Model Adequacy Based on Residual Sum of
Squares   66

A 2.2 LEAST SQUARE ESTIMATES IN MULTIPLE
REGRESSION   69

A 2.2.1 Model and Least Square Estimates   69
A 2.2.2 Mean, Variance-Covariance Matrix of $\boldsymbol{\beta}$,
and Confidence Intervals   71
A 2.2.3 Variance-Covariance Matrix of $\hat{\phi}_1$, $\hat{\phi}_2$ for
AR(2) Model and Confidence Intervals   73

PROBLEMS   **74**

3 CHARACTERISTICS OF ARMA MODELS   **78**

3.1 GREEN'S FUNCTION AND STABILITY   79

3.1.1 Green's Function of the AR(1) System Using B
Operator   80
3.1.2 Physical Interaretation   80
3.1.3 Stability of the AR(1) System   85

3.1.4 Green's Function and the Orthogonal or Wold's
Decomposition   87
3.1.5 Green's Function of the ARMA(2,1) System:
Implicit Method   89
3.1.6 Green's Function of the ARMA(2,1) System:
Explicit Method   91
3.1.7 Green's Function of AR(2) and ARMA(1,1)
Systems   95
3.1.8 Why ARMA(2,1) Rather Than AR(2) After
AR(1)?   97
3.1.9 Stability of the ARMA(2,1) System   100
3.1.10 General Results   105

3.2   INVERSE FUNCTION AND INVERTIBILITY   106

3.2.1 AR(1) and MA(1) Models   107
3.2.2 ARMA(n,m) Models   108
3.2.3 Reasons for Invertibility   110

3.3   AUTOCOVARIANCE FUNCTION   111

3.3.1 Distribution Properties of a_t 112
3.3.2 Theoretical and Sample Autocovariance/
Autocorrelation Function   112
3.3.3 The AR(1) Model   115
3.3.4 The MA(1) Model   118
3.3.5 The ARMA(2,1) Model   119
3.3.6 Representation of Dynamics   125
3.3.7 Variance Decomposition   126
3.3.8 Relation Between the Green's Function and the
Autocovariance Function   128
3.3.9 The MA(2) Model   128
3.3.10 General Results   129

3.4   PARTIAL AUTOCORRELATION AND
AUTOSPECTRUM   130

3.4.1 Partial Autocorrelation   130
3.4.2 Autospectrum   132

APPENDIX   **132**

A 3.1   PARTIAL FRACTIONS   132

A 3.2   AN ALTERNATIVE POLAR FORM
REPRESENTATION OF THE GREEN'S FUNCTION
FOR THE ARMA(2,1) MODEL WITH COMPLEX
ROOTS   133

A 3.2.1 Trigonometric Form of a Complex Number   133
A 3.2.2 Polar Form Representation of the Green's
Function (Eq. 3.1.16)   134
A 3.3   STABILITY AND INVERTIBILITY CHECKS FROM
PARAMETERS   136

PROBLEMS   **138**

4   MODELING   **142**

4.1 THE SYSTEM APPROACH TO MODELING   143

4.1.1 Comparison With the ARMA(n,m) Approach   143
4.1.2 Consideration of Dynamics   144
4.1.3 Adequacy of the System Approach   147
4.1.4 Consideration of the Autocovariance Function   149
4.1.5 Reasons for the Moving Average Order $n-1$ 150

4.2 INCREMENT IN THE AUTOREGRESSIVE ORDER   151

4.3 ESTIMATION   153

4.3.1 AR Models   153
4.3.2 ARMA Models   154
4.3.3 Initial Guess Values Based on the Inverse
Function   155
4.3.4 An Example of Initial Values by the Inverse
Function   158

4.4 CHECKS OF ADEQUACY   159

4.4.1 Formulation as Testing of Hypothesis   160
4.4.2 Checking Criterion   161
4.4.3 Application of the F-Criterion   162
4.4.4 Checks on Residual (a_t) Autocorrelations   163

4.5 MODELING PROCEDURES WITH ILLUSTRATIVE
EXAMPLES   164

4.5.1 Sunspot Activity   167
4.5.2 IBM Stock Prices   169
4.5.3 Papermaking Data   170
4.5.4 Grinding Wheel Profile   172
4.5.5 Mechanical Vibratory Data   172

PROBLEMS   **174**

5   FORECASTING   **177**

5.1 A BRIEF HISTORICAL REVIEW   177

5.2 PREDICTION AS ORTHOGONAL PROJECTION 179

    5.2.1 Formulation 179
    5.2.2 Solution 180
    5.2.3 An Alternative Solution 182

5.3 FORECASTING BY CONDITIONAL EXPECTATION 182

    5.3.1 Conditional Expectation from Orthogonal
    Decomposition 183
    5.3.2 Rules for Conditional Expectation 184
    5.3.3 The AR(1) Model—Forecasts and Probability
    Limits 184
    5.3.4 The AR(1) Model—Correlation of Forecast
    Errors 186
    5.3.5 The AR(1) Model—Numerical Example 187
    5.3.6 General Results—ARMA(n,m) Model 188
    5.3.7 Illustrative Examples—ARMA Models 192
    5.3.8 Eventual Forecasts and Stability 197

5.4 UPDATING THE FORECASTS 198

5.5 EXPONENTIAL SMOOTHING 199

    5.5.1 The Concept of Exponentially Weighted
    Moving Average 200
    5.5.2 Interpretation and Advantages 201
    5.5.3 Relation with ARMA Models 204
    5.5.4 Genesis of ARMA Models from Exponential
    Smoothing 205
    5.5.5 Exponential Smoothing, ARMA Models, and
    Wiener-Kolmogorov Prediction Theory 206

PROBLEMS **207**

6 **UNIFORM SAMPLING OF CONTINUOUS SYSTEMS—
FIRST ORDER** **211**

6.1 FIRST ORDER DIFFERENTIAL EQUATION 213

    6.1.1 Solution 213
    6.1.2 Time Constant τ 214
    6.1.3 Stability 215

6.2 DIRAC DELTA FUNCTION AND ITS PROPERTIES 216

    6.2.1 Definition 217
    6.2.2 Relation with Unit Step Function 218

6.2.3 Properties   219

6.3   THE FIRST ORDER AUTOREGRESSIVE SYSTEM
       A(1)   219

6.3.1 The Stochastic Differential Equation and Its
Solution   220
6.3.2 Orthogonal Decomposition   221
6.3.3 The Green's Function   224
6.3.4 The Autocovariance Function   226
6.3.5 The Spectrum   228

6.4   UNIFORMLY SAMPLED FIRST ORDER
AUTOREGRESSIVE SYSTEM   228

6.4.1 The Sampled System Model   229
6.4.2 Expression for σ_a^2 230
6.4.3 Illustrative Example   230

6.5   LIMITING CASES—EFFECT OF SAMPLING
INTERVAL Δ AND THE PARAMETER α_0 231

6.5.1 Sampling Interval Δ 231
6.5.2 The Parameter α_0 232
6.5.3 Independent Increment Process   235
6.5.4 Physical Interpretation   237
6.5.5 How Large Is "Infinity" and How Small is
"Zero"?   238
6.5.6 Illustrative Example—IBM Stock Prices   239

APPENDIX   **239**

A 6.1 DIFFERENTIAL EQUATIONS AND THE
       EXPONENTIAL FUNCTION   239

A 6.2 PROPERTIES OF DIRAC DELTA FUNCTION   245

PROBLEMS   **246**

7 SECOND ORDER SYSTEM AND RANDOM VIBRATION   **248**

7.1   DIFFERENTIAL EQUATION FOR A DAMPED SPRING
MASS SYSTEM   249

7.1.1 Formulation of the Nonhomogeneous Equation   250
7.1.2 Solution of the Homogeneous Equation   250
7.1.3 Stability   251
7.1.4 An Experimental Example   253

7.2 SECOND ORDER AUTOREGRESSIVE SYSTEM A(2)   254

    7.2.1 The Green's Function of the A(2) System   255
    7.2.2 The Solution of the Nonhomogeneous Second
    Order Equation   259
    7.2.3 Orthogonal Decomposition and
    Autocovariance of the A(2) System   260
    7.2.4 Physical Examples of A(2) Systems   263
    7.2.5 Spectrum of the A(2) System   263

7.3 UNIFORMLY SAMPLED SECOND ORDER
AUTOREGRESSIVE SYSTEM   264

    7.3.1 Representation of the Sampled System   264
    7.3.2 Expression for θ_1 and σ_a^2 265
    7.3.3 Illustrative Examples   267

7.4 STABILITY REGION   268

    7.4.1 Additional Restrictions   269
    7.4.2 Reduced Stability Regions   269
    7.4.3 Static and Dynamic Stability   270

7.5 THE A(2) MODEL FROM DISCRETE DATA   272

    7.5.1 Nonuniqueness of the A(2) Model Parameters   272
    7.5.2 Multiplicity in Parameters ζ and ω_n 273
    7.5.3 Resolution of Multiplicity in b (or in $\alpha_0 = \omega_n^2$)   275
    7.5.4 Illustrations for Multiplicity   276
    7.5.5 Estimation   277
    7.5.6 Implications in Spectral Estimation   278

7.6 EFFECT OF SAMPLING INTERVAL, NATURAL
FREQUENCY, AND DAMPING RATIO   279

    7.6.1 Small $\omega_n\Delta$ 280
    7.6.2 Large $\omega_n\Delta$ 281
    7.6.3 Intermediate Values of $\omega_n\Delta$ 282
    7.6.4 Limiting Values of ζ 286
    7.6.5 An Illustrative Application of Limiting Cases—
    IBM Stock Prices   289

7.7 EXPERIMENTAL DEMONSTRATION AND
ILLUSTRATIVE APPLICATION   291

    7.7.1 Experimental Verification   292
    7.7.2 Application to Grinding Wheel Profile   292

PROBLEMS   **293**

8   AM(2,1) MODEL AND ITS APPLICATION TO EXPONENTIAL SMOOTHING   295

8.1   AM(2,1) MODEL AND ITS SAMPLED REPRESENTATION   296

8.2   DERIVATION OF EXPONENTIAL SMOOTHING   299

8.3   OPTIMAL VALUE OF λ AND ITS SENSITIVITY   300

8.4   AM(2,1) MODEL AS A RESULT OF FEEDBACK   301

8.5   OTHER LIMITING CASES AND INTERMEDIATE VALUES   305

8.6   AN ILLUSTRATIVE EXAMPLE—DIAMETER MEASUREMENTS OF A MACHINED PART   306

PROBLEMS   311

9   STOCHASTIC TRENDS AND SEASONALITY   313

9.1   ANALYSIS OF STOCHASTIC TRENDS AND SEASONALITY   314

9.1.1 Stochastic Trends   314
9.1.2 Stochastic Seasonality   319
9.1.3 Differencing or Seasonality Operators Before Modeling   323

9.2   EXAMPLES OF SERIES WITH STOCHASTIC TRENDS AND SEASONALITY   326

9.2.1 Modeling Investment and Money Market Rate   326
9.2.2 Modeling Hospital Patient Census   334
9.2.3 Modeling Consumer and Wholesale Price Indexes   341
9.2.4 Modeling Airline Passenger Ticket Sales   343

APPENDIX   352

A 9.1 CALCULATION OF THE STRENGTH g_k AND AMPLITUDE A_i CORRESPONDING TO MODE λ_k FOR THE ARMA(7,6) MODEL   352

PROBLEMS   353

10 DETERMINISTIC TRENDS AND SEASONALITY: NONSTATIONARY SERIES   355

10.1 LINEAR TRENDS   356

    10.1.1 Crack Propagation Data: Deterministic Part   357
    10.1.2 Stochastic Part   359
    10.1.3 Complete Model: Deterministic Plus Stochastic   361
    10.1.4 Physical Interpretation   362
    10.1.5 Calibration of Measuring Instruments   363

10.2 EXPONENTIAL TRENDS   363

    10.2.1 Basis Weight Response to Step Input   365
    10.2.2 First Order Dynamics   366
    10.2.3 Differential Equations   369
    10.2.4 Possibility of a Single Time Constant   370
    10.2.5 Chemical Relaxation—Estimation of Time Constant   371
    10.2.6 Another Differential Equation for Relaxation   374

10.3 PERIODIC TRENDS: SEASONALITY   375

    10.3.1 International Airline Passenger Data   376
    10.3.2 Exponential Growth Trend   378
    10.3.3 Addition of Periodic Trends   378
    10.3.4 Stochastic Part and the Combined Model   382
    10.3.5 Comparison with a Multiplicative Model   388

10.4 GENERAL NONSTATIONARY MODELS   392

    10.4.1 A General Model for Nonstationary Series   392
    10.4.2 Modeling as Decomposition   393
    10.4.3 Some Practical Aspects   395

10.5 MORE ILLUSTRATIVE EXAMPLES   396

    10.5.1 Consumer and Wholesale Price Index   396
    10.5.2 Wood Surface Profile   403

PROBLEMS   406

11 MULTIPLE SERIES: OPTIMAL CONTROL AND FORECASTING BY LEADING INDICATOR   412

11.1 TRANSFER FUNCTION AND ARMAV MODELS   413

    11.1.1 The Papermaking Process   413
    11.1.2 Transfer Function Model   415

11.1.3 Transfer Function versus State Variable
Approach 416
11.1.4 Large Delay or Dead Time 417
11.1.5 ARMAV Models 418
11.1.6 A Special Form of the ARMAV Models 421

11.2 MODELING AND ILLUSTRATIONS 424

11.2.1 Modeling Procedure 425
11.2.2 One Input–One Output Papermaking
System: Comparison with Deterministic Inputs 426
11.2.3 Two Inputs–One Output System With Dead
Time: Application to the Paper Pulping Digester 433

11.3 OPTIMAL CONTROL 443

11.3.1 Minimum Mean Squared Error Control
Strategy 444
11.3.2 Illustrative Example 1—First Order Model
with Lag 1 445
11.3.3 Illustrative Example 2—Second Order Model
with Lag 1 447
11.3.4 Illustrative Example 3—First Order Model
with Lag 2 448
11.3.5 Illustrative Example 4—Second Order Model
with Lag 2 449
11.3.6 Effect of Large Lag 451
11.3.7 General Results: One-Input One-Output
System 452
11.3.8 Improved Control by Additional Inputs 454
11.3.9 Optimal Control of a Two-Input One-Output
Paper Pulping Digester 458
11.3.10 General Results: Optimal Control of Multi-
Input One-Output Systems 459

11.4 FORECASTING BY LEADING INDICATOR 464

11.4.1 Forecasting Using Conditional Expectation 464
11.4.2 Conditional Expectation from Orthogonal
Decomposition 466
11.4.3 Forecast Errors and Probability Limits 467
11.4.4 An Illustrative Example 468
11.4.5 Usefulness of the Leading Indicator 469
11.4.6 General Results: Forecasting by a Single
Leading Indicator 471
11.4.7 General Results: Forecasting by Multiple
Leading Indicators 473

APPENDIX **475**

A 11.1 INITIAL VALUES FOR THE SPECIAL FORM OF
THE ARMAV MODELS 475

A 11.2 CONTINUOUS TO DISCRETE TRANSFER
FUNCTION 479

A 11.3 RELATION TO PROPORTIONAL PLUS INTEGRAL
PLUS DERIVATIVE (PID) CONTROL 481

PROBLEMS **482**

APPENDIX I: **LISTING OF DATA USED IN THE TEXT** **485**

APPENDIX II: **NORMAL, t, χ^2 AND F TABLES** **503**

 Table A: Normal Distribution 503
 Table B: χ^2 Distribution 504
 Table C: t-Distribution 506
 Table D: F-Distribution 508

APPENDIX III: **COMPUTER PROGRAMS** **514**

 Description of Program Input 515
 Example Runstream and Output for
 Chapter 2 518
 Example Runstream and Output for
 Chapter 4 521
 Example Runstream and Output for
 Chapter 10 526
 Example Runstream and Output for
 Chapter 11 532
 FORTRAN-77 Listing 541

REFERENCES **563**

APPLICATION BIBLIOGRAPHY **567**
 Iron and Steelmaking 567
 Manufacturing—General 568
   —Abrasive and Grinding 569
   —Machine Tool and Dynamics 570
   —Surface Characterization 571
   —Process Modeling 571
   —On-Line Monitoring 571
   —Tool Life 571
 Methodology and System Identification 571

Business, Economics, and Management 573
Nuclear Power Plant Surveillance 573
Solar Energy 574
Heat Transfer 574
Friction and Wear 574
Pulp and Papermaking 574
Vibrations 575
Signature Analysis 575
Biomedical Engineering 575
Miscellaneous 576

INDEX 577

TIME SERIES AND
SYSTEM ANALYSIS
WITH APPLICATIONS

1

INTRODUCTION

Practitioners and researchers in engineering, physical, biological, and social sciences are all concerned with the analysis of a sequence of observed data. The desirable goal of such a quantitative analysis of data is to develop a succinct but comprehensive characterization of the underlying system in the form of a mathematical model. This model can be used to analyze the system and predict its behavior under a changing environment. The information obtained from such an analysis can be further employed to alter the possible factors and variables in the system to achieve an optimal performance in some sense.

This book deals with system modeling, analysis, prediction, and control based on an ordered sequence of observed data. The system underlying the data may be in a form that can be readily visualized, such as for the data on a chemical process or a mechanical vibration, or it may be in an abstract form that is difficult to visualize such as for the data on stock prices. The methodology developed in this book correspondingly serves a dual purpose. It can provide models in the form of stochastic difference equations and differential equations from the data alone. The difference equations represent the abstract system as reflected in the data and can be used for prediction and control. On the other hand, the differential equations are an aid in understanding and visualizing the system underlying the data by analogy with physical systems; they can be used to characterize it, analyze its behavior such as stability, compare it with other systems, modify it by changing related constituents, and so on. Together with the available qualitative or quantitative knowledge of the system not contained in the data, the differential equations can be used for better system design, as is the common practice in physical sciences and engineering, where such equations are derived from known physical laws.

In this introductory chapter we have tried to outline the basic philosophy and the main concepts of the book. It is intended to provide a conceptual overview of the remaining chapters, link them together, and put them in proper perspective.

1.1 TIME SERIES AND SYSTEM ANALYSIS

A sequence of observed data, usually ordered in time, is called a *time series*, although time may be replaced by other variables such as space. In this book the term "data" will always mean the observations as a discrete sequence obtained at uniform intervals. The statistical methodology dealing with the analysis of such a sequence of data is called *time series analysis*. The feature of time series analysis that distinguishes it from other statistical analyses is the explicit recognition of the importance of the order in which the observations are made. Although in many other problems the observations are statistically independent, in almost all time series they are dependent.

The statistical dependence in the data is expressed by the correlation or *autocorrelation* between successive observations. Therefore, the existing methods of time series analysis are based on empirical or estimated autocorrelation or its Fourier transform autospectrum. The empirical autocorrelation is a poor estimator of the theoretical autocorrelation. This fact makes the techniques of time series analysis based on such estimates difficult and cumbersome, requiring heavy reliance on ad hoc trial and error procedures.

This difficulty can be avoided by consistently employing the methods of linear system analysis in both time series theory and inference to account for the dependence in the data. The time series is treated as one realization of the response of a stochastic system to uncorrelated or independent "white noise" input. The mathematical model for the dynamic system, either in continuous or discrete time, reduces the dependent or correlated time series output to the independent or uncorrelated input. The whole methodology can thus be summarized as finding such a model that accomplishes this reduction to independent data and then using standard statistical techniques for independent observations for estimation, prediction, and control.

The techniques of linear system analysis used for this purpose comprise the theory of difference/differential equations and the related transform techniques. No prior knowledge of these techniques will be assumed; they will be developed in a simple fashion by employing the respective delta functions. As will be seen, a systematic use of these techniques to model the dependence in the data, together with the standard statistical methods, makes the theory of time series and its inference far simpler and more straightforward than the existing methods. Additionally, the system analysis techniques are of great help in the physical interpretation of the results.

In this book we will be mainly concerned with the analysis of a single series of data or univariate time series. A simple extension of these methods to multiple time series that suffices in many practical cases is presented in Chapter 11. However, a comprehensive treatment of multivariate time series based on matrix methods is beyond the scope of this book.

1.2 CORRELATION, REGRESSION, DYNAMICS, AND MEMORY

In the statistical theory, the observed time series is considered as one possible realization of a *stochastic process*, which represents the mechanism generating it. The stochastic process or time series, for which the probability distribution and therefore all of its moments are independent of the origin, is called *strictly stationary*. On the other hand, when only the first two moments are considered and assumed to be independent of the origin, it is called *wide sense stationary* or simply, stationary.

For a stationary stochastic process the mean, which is the first moment, is constant and can be assumed to be zero without loss of generality. The second moment is the covariance, which divided by the variance (which is also constant) gives correlation. Therefore, the theory of a stationary stochastic process or time series is essentially the theory of its correlation. In this book we will mainly consider stationary or wide sense stationary time series. This includes certain series with infinite variance, such as random walk, that will be treated as limits of a stationary series. The nonstationarity caused by trends or seasonality will be considered in Chapters 9 and 10.

The theoretical correlation expresses the dependence of the time series observations on each other. This dependence can also be expressed by a regression model that represents the present observation as the sum of two independent, uncorrelated, or "orthogonal" parts: one dependent on the preceding ones and the other an independent sequence. For example, if X_t denotes the observation at time t, then the simplest such model is

$$X_t = \phi X_{t-1} + a_t \tag{1.2.1}$$

where ϕ is constant and a_t is a sequence of uncorrelated variables or shocks. Since this model expresses the dependence or regression of X_t on its own past values, it is called an *autoregressive* model; such models were first introduced by Yule (1927). At time $t-1$ this model is the usual linear regression model since the observation X_{t-1} is known, and it is therefore a fixed deterministic variable. Thus, the autoregressive model is a *conditional* regression model, in its static aspect at $t-1$, when X_{t-1} is fixed. A detailed consideration of this static conditional regression aspect of the model explained in Chapter 2 enables us to formulate a simple modeling strategy and provides the conditional least squares method of estimation.

When X_{t-1} is not fixed, Eq. (1.2.1) becomes a dynamic model. The observation X_{t-1} itself depends on X_{t-2} by the same model and recursively substituting, we have

$$X_t = \sum_{j=0}^{\infty} \phi^j a_{t-j}$$

If we write

$$G_j = \phi^j$$

then

$$X_t = \sum_{j=0}^{\infty} G_j a_{t-j} \qquad (1.2.2)$$

Equation (1.2.2) explicitly expresses the dynamic structure of the "AR(1)" model (1.2.1). It shows how the past shocks a_{t-j} affect the present observation X_t or how a_{t-j} are "remembered." The function G_j that summarizes this dependence or memory is called the *Green's function* and will play a fundamental role in our entire treatment of system modeling, prediction, and analysis.

The difference equation (1.2.1) or its "solution" (1.2.2) express the regression, dynamics, or memory in discrete time. The dynamics or memory can also be expressed in continuous time by the differential equation

$$\frac{dX(t)}{dt} + \alpha_0 X(t) = Z(t) \qquad (1.2.3)$$

for which the solution can be shown to be

$$X(t) = \int_0^{\infty} G(v)Z(t-v)\,dv \qquad (1.2.4)$$

which correspond to Eqs. (1.2.1) and (1.2.2), respectively. Here, $Z(t)$ and $G(v)$ are continuous time analogues of a_t and G_j. The Green's function $G(v)$ is also called the impulse response function; we will use the name *Green's function* to emphasize the basic theory of difference and differential equations. It should be noted that the Green's function expresses the dynamics or memory in terms of the shocks a_t and not the observations X_t. Because of these shocks, the system cannot be exactly predicted and the stronger the dynamics or memory expressed by the larger values of the Green's function, the more unpredictable the system's behavior is at future times.

In this manner, the dependence or correlation in the data will be treated in this book, not by the empirical estimate of the correlation, but by the conditional regression and dynamic aspects of models such as Eqs. (1.2.1) and (1.2.3). The conditional regression aspect will enable us to use standard statistical methods based on independent data for time series inference. The dynamic aspect will enable us to use the methods of system analysis based on the elementary theory of linear difference/differential equations for prediction, control, and interpretation of the underlying system from the data.

1.3 STOCHASTIC DIFFERENCE/DIFFERENTIAL EQUATIONS

The first order models, (1.2.1) or (1.2.3), assume that the dependence or dynamics in the data is of a very simple nature, which may not be the actual case. However, we will show that an arbitrary stationary time series or stochastic system can be approximated as closely as we want by a model with successively more complex dynamics.

$$X_t - \phi_1 X_{t-1} - \phi_2 X_{t-2} - \ldots - \phi_n X_{t-n}$$
$$= a_t - \theta_1 a_{t-1} - \theta_2 a_{t-2} - \ldots - \theta_{n-1} a_{t-n+1} \quad (1.3.1)$$

for increasing orders of n. This linear stochastic difference equation is called a discrete Autoregressive Moving Average model denoted by ARMA($n, n-1$). The solution of this stochastic difference equation can again be expressed in the form (1.2.2). A detailed consideration of the static and dynamic aspects of this model will lead to a simple modeling strategy that consists of fitting such models for successively large orders of n till an adequate representation is obtained.

Similarly, more complex dynamics in continuous time can be represented by a linear stochastic differential equation

$$\frac{d^n X(t)}{dt^n} + \alpha_{n-1} \frac{d^{n-1} X(t)}{dt^{n-1}} + \ldots + \alpha_0 X(t)$$
$$= b_{n-1} \frac{d^{n-1} Z(t)}{dt^{n-1}} + \ldots + b_1 \frac{dZ(t)}{dt} + Z(t) \quad (1.3.2)$$

for increasing orders of n. This stochastic differential equation is called a continuous Autoregressive Moving Average model, denoted by AM($n, n-1$). The solution of this stochastic differential equation can be expressed in the form (1.2.4).

We will show how to obtain *both* the difference and differential equations of the type (1.3.1) and (1.3.2) from the data to represent the underlying dynamic system and how to use them for prediction, control, characterization, and other applications. In a sense this book may be considered as the theory and application of linear stochastic difference/differential equations emphasizing system analysis from time series data. As an important part of this theory, we will show that when a stochastic system governed by a differential equation such as (1.3.2) is sampled at uniform intervals, the resultant discrete system has a representation of the same form as (1.3.1). This result together with the parametric relationships developed in Chapters 6 to 8 enables us to fit both models simultaneously to the data by the method of conditional least squares and to use either of the models depending upon the application.

The deterministic difference equations can be solved by means of recursive methods, which are ideally suited for digital computers. We will make full use of these recursive techniques in dealing with stochastic difference equations. However, we will avoid the undue stress on recursive methods at the expense of analytical solutions. In fact, analytical solutions provide considerable insight into the problems leading to simpler procedures, and therefore they are equally emphasized. The Green's function represents the crux of such a solution and is extensively used. Moreover, the theory of stochastic differential equations becomes much easier to follow once the analytical solutions of the difference equations are grasped properly. The formal similarity between the solutions of stochastic difference/differential equations and their interrelationships introduce considerable simplicity and elegance in the treatment.

Although the differential/difference equations used in this book are linear, this does not mean that they cannot be used to model nonlinear systems in the usual sense in science, engineering, and econometrics. The dependence in the observed data even from such nonlinear systems can still be represented by models presented in this book. In a probabilistic sense, the representation may be exact up to the first two moments—mean and autocovariance—and approximate for the higher moments. In a deterministic sense, the representation may be a "linearization" over the given interval, which may provide some clues to the possible nonlinear representation. Moreover, genuine nonlinear/nonstationary trends in the data can be modeled by the methods of Chapters 9 and 10.

1.4 APPLICATIONS

The methodology developed in this book is strongly application-oriented. In fact, the methodology can be effectively learned and its significance and importance thoroughly grasped only by employing it in the analysis of real-life data. To aid this learning process, real data from different disciplines is used throughout to introduce, explain, illustrate, and illuminate the theoretical treatment and the computational procedures. This is the reason why we have avoided, as much as possible, the use of a convenient set of simulated data concocted to illustrate a point.

A good insight into the system underlying the data requires a fairly comprehensive grasp of its fundamentals. This involves a considerably specialized knowledge of the field from which the data is taken. On the other hand, the scope of application of the methodology developed in this book is not restricted to any specific field. Therefore, to avoid burdening the reader with the details of the specific system underlying the data, we have chosen data from a few typical fields, which we hope most of the readers will easily comprehend. Only a brief description of the underlying system is provided in such cases. Similarly, the interpretation of the differential equations obtained from the data has been limited to that which can be understood and appreciated by the general reader, although these equations are capable of providing a rather deep understanding of the system by means of specialized information about the system.

The real-life data has been chosen from the following fields:

CHAPTERS

(A) Physical Sciences and Engineering

		CHAPTERS
(i)	Mechanical Vibrations	3, 4, 7
(ii)	Chemical Processes	10
(iii)	Papermaking	2, 4, 5, 10, 11
(iv)	Metal Cutting-Grinding	4, 7
(v)	Sunspot Activity	2, 4
(vi)	Mechanics-Crack Propagation	10

(B) Business, Industry, and Management

(i)	Stock Market	2, 4, 7
(ii)	Banking	9
(iii)	Quality Control	8
(iv)	Aviation	9, 10

(C) Social and Behavioral Sciences

(i)	Consumer and Wholesale Prices	9, 10
(ii)	Hospital Utilization	9

(D) Economics

(i)	Gross National Product	10
(ii)	Expenditure on Producers' Durables	10

A more detailed listing of references, describing the application of the methodology to many specialized areas, is given at the end following "References." Readers with specialized interests may supplement the material in the book by additional reading on the references cited in the "Application Bibliography."

1.5 CHAPTERWISE OUTLINE AND SUMMARY

Chapter 2

The Autoregressive Moving Average (ARMA) models are introduced in Chapter 2 to express the dependence of the present observation on the past ones. With five sets of real data it is shown how these models naturally arise as an extension of linear regression models expressing the dependence of one set of variables on another. The ARMA models are, in fact, conditional regression models, because when the past observations are known and therefore fixed, they reduce to the usual linear regression models. A detailed analysis of this static or conditional regression aspect leads to the least squares method of estimation, forecasting or prediction, and to a simple procedure of modeling successively complex dependence by ARMA($n,n-1$) models for increasing orders of n.

Chapter 3

The unconditional regression or the dynamic aspect of the ARMA models is studied in Chapter 3. Three characteristics of the ARMA systems which describe the dynamics in different ways are derived and discussed. The first is the Green's function, which expresses the influence of past a_t's on X_t. The second is the "inverse function," which expresses the influence of past X_t's on the present X_t. The third is the autocovariance/autocorrelation function, which is a statistical characterization of the dependence between X_t and X_{t-1}, X_{t-2}, ... Certain restrictions on the parameters of the ARMA models are imposed by the dynamics, namely stability and invertibility, which are considered in detail.

Chapter 4

The static and dynamic characteristics developed in the preceding chapters are employed in Chapter 4 to evolve a simple and rational procedure of modeling ARMA systems from the time series data. It is shown that this new procedure consists of approximating the data successively by ARMA($n,n-1$) models, increasing n in steps of two, until adequate approximation is obtained. The estimation procedure, including the methods of obtaining initial guess values for nonlinear regression, is discussed. Simple checks of adequacy are provided. The complete modeling procedure is illustrated by means of five sets of real data introduced in Chapter 2.

Chapter 5

The ARMA models so obtained are used in Chapter 5 for forecasting and prediction. It is shown how the forecasts at arbitrary lead times can be computed by conditional expectation and how the probability limits on the forecasts can be obtained under the normality assumption of a_t's. (In addition to the checks of adequacy in Chapter 4, this is the only place where the normal distribution of a_t's is used; the rest of the book only assumes the independence or uncorrelatedness of a_t's.) Chapter 5 also shows how the forecasting by ARMA models connects the profound prediction theory initiated by Kolmogorov and Wiener in a very abstract framework, with empirical methods of forecasting starting with the simple concept of averaging and leading to exponential smoothing. The relation between exponential smoothing and the ARMA models is examined in detail.

Chapter 6

Chapter 6 begins the consideration of the continuous systems by developing the theory and applications of a stochastic system governed by a first order differential equation. The physical meaning and stability characteristics of a first order system are studied and then a continuous autoregressive model of first order, A(1), is introduced in a simple fashion by employing the Dirac delta function. The Green's function and the autocovariance function of the A(1) system are derived, and it is shown that when the system is sampled at uniform intervals, it can be represented by an AR(1) model. The parametric relationships between the A(1) and AR(1) models are derived and used to obtain the A(1) model from discrete data. Some limiting cases of the A(1) model, including random walk, are discussed.

Chapter 7

Chapter 7 treats the most important and widely applicable continuous stochastic system governed by a second order differential equation. For a better intuitive appreciation, this system is formulated as a random vibration of a mass-spring-dashpot system, which provides a physical meaning to the abstract model. After analyzing the Green's function and autocovariance function, the discrete representation for the uniformly sampled second order system is derived. The parametric relationships between the continuous and discrete models are developed and used to obtain the continuous model from the discrete data. The effects of sampling interval and natural frequency are studied. An experimental demonstration of the theory and an illustrative example are presented.

Chapter 8

A more general second order system, AM(2,1), is analyzed in Chapter 8. It is shown that the well-known technique of exponential smoothing is a special case of the uniformly sampled AM(2,1) model. The AM(2,1) model is interpreted as a result of a first order system with a first order feedback. This interpretation explains why the AM(2,1) model would be basic for many of the economic, business, or quality control systems in which exponential smoothing has been successful. Other limiting cases of the AM(2,1) model and illustrative examples are also discussed.

Chapter 9

It is shown in Chapter 9 that a large number of practical time series with trend and seasonality can be modeled and analyzed by the methods of Chapter 4. An analysis of Green's function together with simulated time series are given to show how the Green's function determines the seasonality and trends. Data from hospital census, consumer and wholesale price indices, and investment and money market rates are used to illustrate the modeling of trend and seasonality.

Chapter 10

This chapter extends the methodology to the data which have nonstationary and seasonal trends. Linear, exponential, and sinusoidal trends are treated by means of real data obtained from different disciplines. A new approach of modeling, based on decomposing the data into deterministic and stochastic parts, is developed. Its advantages are illustrated in comparison with other approaches, such as differencing, to take care of the nonstationarity.

Chapter 11

This chapter deals with modeling, forecasting, and control of multiple time series. By employing a representation of ARMA models similar to state space techniques, it is possible to extend the modeling procedure of Chapter 4 in a simple fashion to multiple series under certain broad assumptions. By means of two or three sets of data used at a time, this chapter illustrates how to model them and how to use these models for deriving optimal control equations and forecasting by leading indicators.

Emphasis on First and Second Order Systems

The first and second order systems have been emphasized and dealt with in detail throughout. The first order systems enable the reader to easily grasp the basic concepts without being overpowered by the mathematics. The second order systems have some additional features and practical importance worthy of detailed treatment. The expressions for the higher order systems are given, but their derivation is omitted or referred to outside literature. These expressions can easily be incorporated in the computer programs, so that the methodology can also be applied to higher order systems. Illustrative examples dealing with such higher order systems are given whenever necessary.

Notation

This book does not deal with the measure theory or the distribution theory of time series; in fact, one of the main advantages of our approach is that the probabilistic concepts can be made much simpler by the techniques of system analysis. Hence, in many cases, it is neither necessary nor convenient to distinguish between population and sample, random variables and observation, estimated and theoretical values, by using different letters or hats. Therefore, unless absolutely necessary, such a distinction has not been made. In particular, the estimated values of the parameters in a time series model are taken as "true" values in forecasting, the computed a_t's from the data are not distinguished by $\hat{a}_t$, both the random variable and its observed value are denoted by X_t, and so on.

Exercises

The exercises at the end of each chapter are an essential part of learning. They are meant not only for practice and to clarify the concepts discussed in the text, but also to illuminate additional aspects not discussed in the text. Many of them are forerunners of the concepts in the succeeding chapters. Time series analysis is best learned by analyzing data of one's own interest, and such analysis should supplement the exercises whenever possible.

Starred Sections

Some of the relatively advanced sections that might require a little more sophisticated background than the rest have been starred. These sections may be omitted without interrupting the main flow. Some starred sections contain general results and expressions, outside references for a deeper understanding or comparison, and a brief discussion of partial autocorrelations, spectrum, and other characteristics not directly necessary in the development of the methodology.

1.6 GUIDELINES TO USER/INSTRUCTOR

This book has been written as a text for a formal course for senior and introductory-level graduate students. It can also be used for self-study and reference. In this section we will try to acquaint the reader with the mathematical level of the contents to provide some guidelines as to how it may be adopted to individual needs.

Chapter 2, Chapter 4 (Section 4.5), and Chapters 9 and 10 excluding Appendixes do not involve any mathematical derivations. If the formulae used therein are taken for granted, one can read them independently and understand the basic approach and the detailed procedure for modeling discrete systems. On the basis of this material alone, it is possible to use the available computer programs based on the text to obtain the required models from stationary or nonstationary series.

The simplest level of mathematics is used in Chapter 5 on forecasting. The idea of conditional expectation is explained as a geometric concept of orthogonal projection in Section 5.2. However, even this section may be omitted by starting from Section 5.3.2 and taking for granted the rules for conditional expectation. If the computation of the Green's function by the method of comparing the coefficients in Section 3.1.5 is briefly explained, then obtaining forecasts and their variances follows immediately. This material is also adequate to obtain the optimal control equation and the control error variance in Chapter 11. Thus, the *mechanics* of modeling, forecasting, and control can be learned and used without much mathematical background.

The algebra in Chapter 3 is based on the summation of a geometric series, explained simply by long division, and to some extent simple partial fraction decomposition. These two concepts are developed ab-initio and extensively used in this and subsequent chapters. This development is necessary to understand the dynamics of ARMA models, which is very important in physical interpretation and intuitive appreciation.

The calculus of ordinary linear differential equations of the first and second order is used in Chapters 6 and 7. It is developed from the beginning by introducing the Dirac delta function. However, the reader is expected to know or take for granted the results about differentiation and integration of the exponential function (reviewed in Appendix A 6.1) that is also used in Chapter 8.

As a text for a one-semester course, this book may be used at the discretion of the instructors. At the University of Wisconsin, Madison, we have found that it is possible to cover Chapters 2 to 11 by discussing Chapters 2 to 5 in detail and dealing with only a few selected examples (subject to the students' field of interest) from Chapters 9 to 11 to explain the underlying concept of modeling, analysis, and physical interpretation. The theory of continuous AM models in Chapters 6 to 8 can be learned without proof, but the basic analogy between the continuous systems and the discrete systems should be emphasized.

A division for a two-quarter course could be Chapters 1 to 6 and then Chapters 7 to 11. Alternatively: (I) Discrete stationary systems and forecasting—Chapters 1 to 5 and 11; (II) Continuous systems and nonstationary series—Chapters 6 to 10. In any case, the students should always be encouraged to do term projects based on real data from their fields of specialization to fully appreciate the use of modeling.

2

AUTOREGRESSIVE MOVING AVERAGE MODELS

The enclosed Figures 2.1 to 2.5 show plots of observed data from five different fields. The first figure shows the observed values of adjustments in the gate opening in a papermaking process taken every twenty minutes. The second one is a record of the activity in the sun in terms of the yearly sunspot numbers from the years 1749 through 1924. The third is a plot of IBM stock prices (daily) from May 17, 1961, through November 2, 1962. The fourth shows the response of a mechanical vibratory system of mass, spring, and dashpot. The fifth depicts the profile of a grinding wheel. All data are listed in Appendix I at the end.

A common characteristic of all the plots in the figures is that they do not seem to have a strict trend such as a straight line or a sine wave. All of them seem to fluctuate more or less slowly around a fixed level, although the level seems to change in Figures 1 and 3. These characteristics allow us to assume that all the sets are "stationary"; in other words, the probabilistic structure of each set does not change from point to point, and in particular, does not depend on the time origin.

The elementary statistical techniques are primarily based on the assumption that the data are statistically independent or uncorrelated. These are not applicable directly to the data under consideration which are *not* independent. In fact, the most important and useful characteristic of each set of data is the dependence or correlation between the observations. It is the dependence which characterizes the dynamics or the "memory" of the underlying system. This dependence/dynamics/memory of the system enables us to predict the future values of the system from the past values once we quantify the dependence. The nature of the dependence or dynamics distinguishes one system from another. It is therefore important to express this dependence by some mathematical model.

The purpose of this chapter is to introduce the Autoregressive Moving Average (ARMA) models in the form of stochastic linear difference equations to express this dependence. In Section 2.1 we first review the linear regression models for expressing the dependence of one variable on another. This method is extended in Section 2.2 to express the interdependence of the observations on a single variable leading to a first order Autoregressive model AR(1). The regression structure of this model, its estimation by conditional least squares, and one-step ahead forecasts are

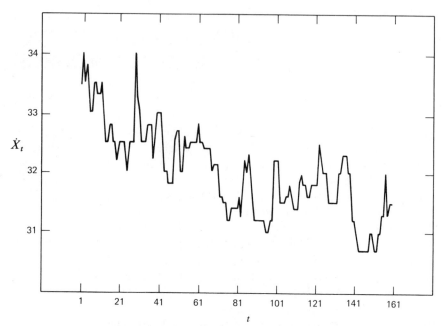

Figure 2.1   Papermaking process—input gate opening.

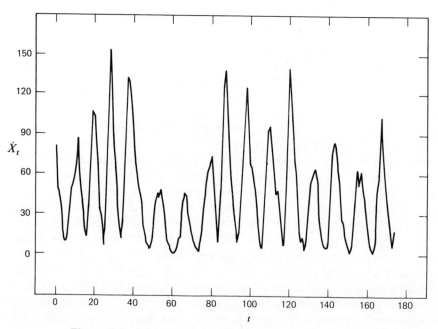

Figure 2.2   Wolfer's sunspot numbers: yearly (1749–1924).

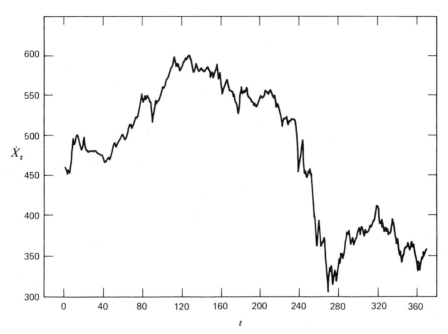

Figure 2.3   IBM stock prices: daily (May 17, 1961 to November 2, 1962).

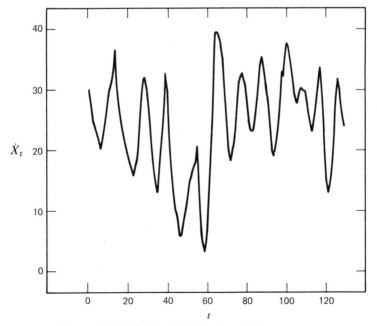

Figure 2.4   Mechanical vibratory displacement.

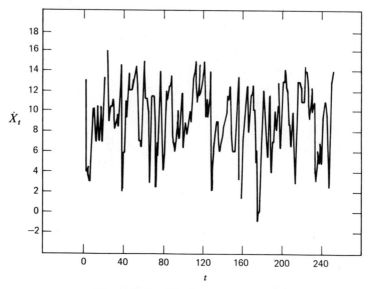

Figure 2.5   Grinding wheel profile.

examined. Section 2.3 shows how the ARMA(2,1) model naturally arises when the AR(1) model is inadequate to represent the dependence. The special cases of this model and its differences with the AR(1) model are discussed. The inadequacy of the ARMA(2,1) model leads to ARMA($n,n-1$) models in Section 2.4. This development gradually provides the procedures for modeling, estimation, and checking, which will be further elaborated in Chapter 4.

The reader is advised to study this chapter carefully. Although simple and elementary, it overshadows Chapters 3 to 5; all the basic concepts from these chapters are presented here. A good grasp of this chapter will considerably facilitate the understanding of the next three chapters. The treatment in this chapter is rather elementary; this makes it easier to grasp the concepts without getting bogged down in mathematics.

2.1 LINEAR REGRESSION MODELS

The main question in this chapter is: How can we account for the dependence in the data? If we use mathematical models to express it, what kind of models should we employ? For example, in the case of IBM stock price series, we need a model that expresses the dependence of today's price on yesterday's. In general, we want to know how an observation at time t, say $\dot{X}_t$, depends on previous observations $\dot{X}_{t-1}$, $\dot{X}_{t-2}$, etc. (It is convenient to use average subtracted data, $X_t = \dot{X}_t - \overline{X}$, for modeling. The dot will represent original data, *before* subtracting the average.)

2.1.1 Simple Regression

To answer this question let us first review the well-known method of linear regression that expresses the dependence of one set of observations $\dot{y}_t$ on another set $\dot{x}_t$ under the assumption that $\dot{y}_t$'s are independent or uncorrelated. For example, suppose we observed the pressure of a gas $\dot{y}_t$ at different temperatures $\dot{x}_t$ for $t = 1, 2, \ldots, N$ observations; we can plot $\dot{y}_t$ versus $\dot{x}_t$ as shown in Fig. 2.6. Then, we can express the dependence of $\dot{y}_t$ on $\dot{x}_t$ by a linear regression of first order

$$\dot{y}_t = \beta_0 + \beta_1 \dot{x}_t + \varepsilon_t, \qquad t = 1, 2, \ldots, N$$

which represents a straight line with intercept β_0 and slope β_1. It can be shown that the "best fit" in the sense of minimizing the sum of squares of the "residuals" ε_t's is obtained if we set

$$\hat{\beta}_0 = \bar{y} - \hat{\beta}_1 \bar{x}$$

and

$$\hat{\beta}_1 = \frac{\sum\limits_{t=1}^{N}(\dot{y}_t - \bar{y})(\dot{x}_t - \bar{x})}{\sum\limits_{t=1}^{N}(\dot{x}_t - \bar{x})^2}$$

where

$$\bar{x} = \frac{1}{N}\sum_{t=1}^{N}\dot{x}_t, \qquad \bar{y} = \frac{1}{N}\sum_{t=1}^{N}\dot{y}_t$$

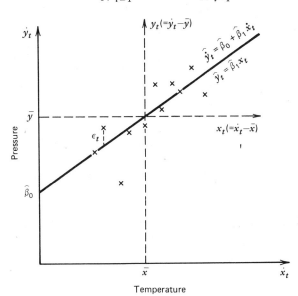

Figure 2.6 Simple linear regression.

Derivation of these expressions is given in Appendix A 2.1.1. $\hat{\beta}_0$ and $\hat{\beta}_1$ so obtained are called least squares estimates of β_0 and β_1 since they minimize the sum of squares of ε_t. In the regression theory, hats on the variables such as $\hat{\beta}$, are used to denote the sample or estimated values in order to distinguish them from the population or "true" values. When such a distinction is not required in our subsequent work, we will omit the hats on the estimated values for the sake of simplicity.

If we adjust the data by subtracting the mean or average value from each observation, that is, replace $\dot{y}_t$ by $\dot{y}_t - \bar{y}$ and $\dot{x}_t$ by $\dot{x}_t - \bar{x}$, and denote the new variables by y_t and x_t, then the first order regression may be represented by

$$y_t = \beta_1 x_t + \varepsilon_t, \qquad t = 1, 2, \ldots, N, \qquad \varepsilon_t \sim \text{NID}(0, \sigma_\varepsilon^2) \quad (2.1.1)$$

where NID denotes Normally Independently Distributed. Now the least squares estimates of β_1 and σ_ε^2 are given by

$$\hat{\beta}_1 = \frac{\displaystyle\sum_{t=1}^{N} y_t x_t}{\displaystyle\sum_{t=1}^{N} x_t^2} \qquad (2.1.2)$$

and

$$\hat{\sigma}_\varepsilon^2 = \frac{1}{N} \sum_{t=1}^{N} (y_t - \hat{\beta}_1 x_t)^2 \qquad (2.1.3)$$

$$= \frac{\text{Residual Sum of Squares}}{\text{Number of Residuals}} \dagger$$

Note how the regression model (2.1.1) expresses the dependence of y_t on x_t. It decomposes y_t into two parts: one completely dependent on x_t given by $\beta_1 x_t$, and another independent of x_t given by ε_t. The first part is known completely as soon as we know x_t; the second part is not known unless we know both x_t and y_t. If both x_t and y_t are known, then ε_t can be computed simply by

$$\varepsilon_t = y_t - \beta_1 x_t \qquad (2.1.4)$$

When y_t is not known, ε_t is a random variable with mean zero and variance σ_ε^2. It is assumed that for different values of t, ε_t are normally independently distributed.

Regression equation (2.1.1), which expresses the dependence of y_t on x_t, can be used to predict the unknown value of y_t from x_t, if we assume

---

† For an unbiased estimate, number of estimated parameters should be subtracted from the denominator, see Eqs. (A 2.1.69) and (A 2.1.70).

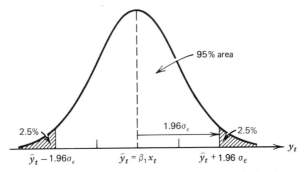

Figure 2.7   Normal distribution of the observation y_t with the forecast $\hat{y}_t$ as the mean.

that there is causation between y_t and x_t. If the predicted value is denoted by $\hat{y}_t$, then clearly (for known β_1)

$$\hat{y}_t = \beta_1 x_t \qquad (2.1.5)$$

which is also obvious from the pressure versus temperature plot in Fig. 2.6.

However, the observed value of y_t will not be $\beta_1 x_t$, but will differ from this predicted value by ε_t which is unknown. Although we cannot state the exact value of y_t, we can make a probabilistic statement about it using the predicted value and the fact that $\varepsilon_t \sim \text{NID}(0, \sigma_\varepsilon^2)$. Since the observed value is

$$\text{Observation} = \text{Prediction} + \text{Error}$$

$$y_t = \beta_1 x_t + \varepsilon_t$$

$$= \hat{y}_t + \varepsilon_t$$

its distribution may be represented by Fig. 2.7.

For example, 95% probability limits for y_t, if we assume known values of β_1 and σ_ε, are

$$\hat{y}_t \pm 1.96\sigma_\varepsilon \qquad (2.1.6)$$

that is,

$$\beta_1 x_t \pm 1.96\sigma_\varepsilon$$

In other words, we are 95% confident that the (unknown) random variable y_t will lie between $\hat{y}_t - 1.96\sigma_\varepsilon$ and $\hat{y}_t + 1.96\sigma_\varepsilon$.†

† These are based on normal distribution because β_1 and σ_ε are assumed known. Otherwise, estimation error in these parameters enters the forecast error and inflates the probability limits, which need to be calculated using t-distribution (see Appendix A 2.1.5). For large N the two procedures give almost the same results. Since N is usually large in time series, the simpler procedure based on normal distribution is used for illustrative purposes, assuming known β_1 and σ_ε.

To illustrate the computations involved in the simple regression, let us consider the following data:

$$t = 1 \quad 2 \quad 3 \quad 4 \quad 5 \qquad (N=5)$$

$$\dot{x}_t = 5 \quad 6 \quad 3 \quad 2 \quad 5$$

$$\dot{y}_t = 7 \quad 6 \quad 5 \quad 4 \quad 6$$

$$\bar{x} = \frac{1}{N} \sum_t \dot{x}_t = \frac{1}{5}(5+6+3+2+5) = 4.2$$

$$\bar{y} = \frac{1}{N} \sum_t \dot{y}_t = \frac{1}{5}(7+6+5+4+6) = 5.6$$

Subtracting the mean from each observation and replacing $\dot{x}_t$ by x_t and $\dot{y}_t$ by y_t, we have

$$t = 1 \qquad 2 \qquad 3 \qquad 4 \qquad 5$$
$$x_t = 0.8 \qquad 1.8 \quad -1.2 \quad -2.2 \qquad 0.8$$
$$y_t = 1.4 \qquad 0.4 \quad -0.6 \quad -1.6 \qquad 0.4$$

Then, by Eqs. (2.1.2) and (2.1.3) we obtain

$$\hat{\beta}_1 = \frac{\sum\limits_{t=1}^{5} y_t x_t}{\sum\limits_{t=1}^{5} x_t^2} = \frac{6.4}{10.8} = 0.59$$

$$\hat{\sigma}_\varepsilon^2 = \frac{1}{5} \sum_{t=1}^{5} (y_t - 0.59\, x_t)^2$$

$$= \frac{1}{5}[(0.928)^2 + (-0.662)^2 + (0.108)^2 + (-0.302)^2 + (-0.072)^2]$$

$$= 0.281$$

and if we assume that these estimated values are true values

$$y_t = \beta_1 x_t + \varepsilon_t \qquad\qquad t=1,2,3,4,5$$
$$= 0.59\, x_t + \varepsilon_t \qquad\qquad \varepsilon_t \sim \text{NID}(0,0.281)$$

When both y_t and x_t are known, ε_t can be computed by

$$\varepsilon_t = y_t - 0.59\, x_t$$

The predicted value $\hat{y}_t$ is given by

$$\hat{y}_t = \beta_1 x_t = 0.59\, x_t$$

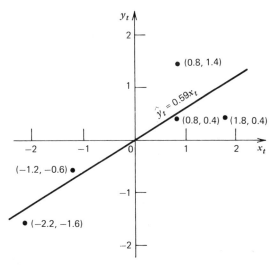

Figure 2.8   Plot of the regression equation with observed data.

The 95% probability limits for the observation y_t are

$$\hat{y}_t \pm 1.96\sigma_\varepsilon = \beta_1 x_t \pm 1.96\ \sigma_\varepsilon$$

$$= 0.59\ x_t \pm 1.96 \times \sqrt{0.281}$$

$$= 0.59\ x_t \pm 1.04$$

y_t versus x_t is plotted in Fig. 2.8.

Strictly speaking, the use of normal distribution in providing probability limits for predictions from a regression equation is not justified for small sample sizes. t-distribution should be used for small samples. However, usual sample sizes in time series are so large that the t-distribution tends to normal, which is then valid for computing probability limits. Hence, normal distribution has been used above for a sample size of 5 only for illustrative purposes.

2.1.2 Multiple Regression

In the simple first order regression model considered above, there was only one dependent variable $\dot{x}_t$, that is, the pressure was assumed to depend only on temperature. Suppose it also depends on concentration, specific volume, and so on. In fact, there could be n dependent variables, say $\dot{x}_{t1}$, $\dot{x}_{t2}, \ldots, \dot{x}_{tn}$. Then, after subtracting the averages from the respective observations, the regression equation corresponding to Eq. (2.1.1) may be written as

$$y_t = \beta_1 x_{t1} + \beta_2 x_{t2} + \ldots + \beta_n x_{tn} + \varepsilon_t, \qquad t = 1, 2, \ldots, N \quad (2.1.7)$$

$$\varepsilon_t \sim \text{NID}(0, \sigma_\varepsilon^2)$$

If N sets of observations $(y_t, x_{t1}, x_{t2}, \ldots, x_{tn})$, $t = 1, 2, \ldots, N$, are known, then the least squares estimates of $\beta_1, \beta_2, \ldots, \beta_n$ and σ_ε^2 are best written by means of vector notation. Given

$$\mathbf{Y} = \begin{bmatrix} y_1 \\ y_2 \\ \vdots \\ y_N \end{bmatrix}, \quad \mathbf{X} = \begin{bmatrix} x_{11} & x_{12} & \cdots & x_{1n} \\ x_{21} & x_{22} & \cdots & x_{2n} \\ \vdots & \vdots & & \vdots \\ x_{N1} & x_{N2} & \cdots & x_{Nn} \end{bmatrix}, \quad \boldsymbol{\beta} = \begin{bmatrix} \beta_1 \\ \beta_2 \\ \vdots \\ \beta_n \end{bmatrix} \quad (2.1.8)$$

then the least squares estimates are (see their derivation in Appendix A 2.2)

$$\hat{\boldsymbol{\beta}} = (\mathbf{X}'\mathbf{X})^{-1} \mathbf{X}'\mathbf{Y} \quad (2.1.9)$$

and

$$\hat{\sigma}_\varepsilon^2 = \frac{1}{N} \sum_{t=1}^{N} (y_t - \hat{\beta}_1 x_{t1} - \ldots - \hat{\beta}_n x_{tn})^2 \quad (2.1.10)$$

Equation (2.1.7) decomposes y_t into n parts $\beta_i x_{ti}$ dependent on x_{ti}, and one part ε_t that is independent of all the x_{ti}, $i = 1, 2, \ldots, n$. If $(y_t, x_{t1}, x_{t2}, \ldots, x_{tn})$, $t = 1, 2, \ldots, N$, are given and the least squares estimates obtained above are taken as true values, ε_t can be computed by

$$\varepsilon_t = y_t - \beta_1 x_{t1} - \beta_2 x_{t2} - \ldots - \beta_n x_{tn} \quad (2.1.11)$$

The prediction of y_t is (with known β_i's and σ_ε^2)

$$\hat{y}_t = \beta_1 x_{t1} + \beta_2 x_{t2} + \ldots + \beta_n x_{tn}$$

and the 95% probability limits for the true value are

$$\hat{y}_t \pm 1.96\sigma_\varepsilon \quad (2.1.12)$$

or

$$\beta_1 x_{t1} + \beta_2 x_{t2} + \ldots + \beta_n x_{tn} \pm 1.96\sigma_\varepsilon$$

Note that for the special case $n = 2$

$$\mathbf{X} = \begin{bmatrix} x_{11} & x_{12} \\ x_{21} & x_{22} \\ \vdots & \vdots \\ x_{N1} & x_{N2} \end{bmatrix}, \quad \mathbf{X}'\mathbf{X} = \begin{bmatrix} \sum_{t=1}^{N} x_{t1}^2 & \sum_{t=1}^{N} x_{t1} x_{t2} \\ \sum_{t=1}^{N} x_{t1} x_{t2} & \sum_{t=1}^{N} x_{t2}^2 \end{bmatrix} \quad (2.1.13)$$

and

$$\mathbf{X'Y} = \begin{bmatrix} \sum_{t=1}^{N} x_{t1}y_t \\ \sum_{t=1}^{N} x_{t2}y_t \end{bmatrix}$$

(2.1.13)

2.2 FIRST ORDER AUTOREGRESSIVE MODEL

Now consider the papermaking data shown in Fig. 2.1. Since this is a plot of papermaking input X_t (or $X_t = X_t - \overline{X}$, after subtracting the average $\overline{X}$) versus time t, it may appear that the data can be represented by a regression model like

$$X_t = \beta_0 + \beta_1 t + \varepsilon_t$$

However, this model requires that ε_t's and therefore X_t's be independent. But the plot of X_t in Fig. 2.1 shows that if X_{t-1} is large, X_t in general tends to be large; if X_{t-1} is small, X_t tends to be small. Therefore, the above model is clearly inappropriate for the data since X_t may depend on X_{t-1}, X_{t-1} on X_{t-2}, and so on. In fact, we want a model that expresses this dependence of X_t on X_{t-1} or X_{t-1} on X_{t-2}, rather than that of X_t on t.

2.2.1 Dependence of X_t on X_{t-1}

A close look at the regression models in the last section indicates that we can use them in a modified form in the present case. The first order model expresses the dependence between y_t and x_t in the pair (y_t, x_t) and relates y_t with x_t, y_{t-1} with x_{t-1}, and so on. The same model can be used in the present case to express the dependence between X_t and X_{t-1} in the pair (X_t, X_{t-1}), and to thus relate X_t with X_{t-1}, X_{t-1} with X_{t-2}, and so on. The plot of X_t versus X_{t-1} for $t = 2, 3, \ldots, 160$ is shown in Fig. 2.9. The points are clearly scattered around a straight line through the origin. This straight line trend also shows that X_t does depend on X_{t-1} (X_{t-1} on X_{t-2}, X_{t-2} on X_{t-3}, and so on) as we had intuitively suspected. Therefore, we can write the model relating X_t and X_{t-1} as

$$X_t = \phi_1 X_{t-1} + a_t$$

(2.2.1)

We have written ϕ_1 in place of β_1 and a_t in place of ε_t to distinguish this model from the first order regression model (2.1.1). Whereas Eq. (2.1.1) expresses the dependence of one variable on another at the same time, the model (2.2.1) expresses the dependence of the variable on itself at different times. In other words, the variables X_t regress on themselves; hence, the model (2.2.1) is called an Autoregressive model of order one

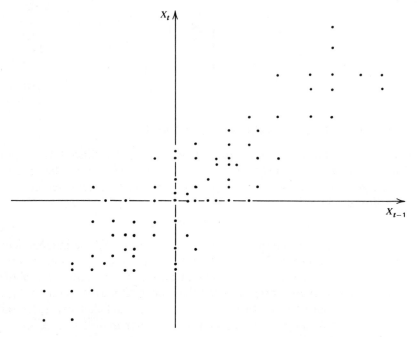

Figure 2.9 Papermaking data: X_t versus X_{t-1}.

and denoted by AR(1). Note that just like ordinary regression, t is only a nominal index, that is,

$$X_{t+1} = \phi_1 X_t + a_{t+1}$$
$$X_{t+2} = \phi_1 X_{t+1} + a_{t+2}$$

and

$$X_{t-1} = \phi_1 X_{t-2} + a_{t-1}$$
$$X_{t-2} = \phi_1 X_{t-3} + a_{t-2}$$

and so on.

2.2.2 Assumptions and Structure

What are the assumptions in the AR(1) model (2.2.1)? The crucial assumption in the regression model (2.1.1) was that ε_t at different values of t are independent. Similarly in the AR(1) model (2.2.1), a_t's at different t are independent, that is, a_t is independent of a_{t-1}, a_{t-2}, etc. For convenience, we will also assume the distribution of a_t to be normal, so that just like ε_t,

$$a_t \sim \text{NID}(0, \sigma_a^2) \tag{2.2.2}$$

Note that the AR(1) model is completely specified only when σ_a^2 is given, in addition to ϕ_1.

The second assumption implicit in the regression model (2.1.1) pertains to the number of variables. By restricting to simple regression rather than multiple, we are implicitly assuming that ε_t does not depend on some other variables like x_{t2}, x_{t3}, etc., as clearly seen by comparing Eq. (2.1.1) with Eq. (2.1.7). Similarly, since Eq. (2.2.1) is self-regressive or autoregressive of order one, it implicitly assumes that a_t does not depend on X_{t-2}, X_{t-3}, etc. (In fact, it will be shown in Chapter 3 that if a_t is independent of $a_{t-1}, a_{t-2}, a_{t-3} \ldots$, then a_t is also independent of X_{t-2}, X_{t-3}, ...)

Like the regression model (2.1.1), the AR(1) model (2.2.1) can also be interpreted as an orthogonal decomposition of X_t into two parts: one completely dependent on X_{t-1} given by $\phi_1 X_{t-1}$, and another independent of X_{t-1} given by a_t. At time $t-1$, when X_{t-1} is observed and known or fixed, the AR(1) model (2.2.1) is the same as a regression model. Therefore, a regression of this kind is called "conditional regression." At time $t-1$, since X_t is an unknown random variable, so is a_t with its distribution property given by Eq. (2.2.2). As soon as X_t is observed and known at time t, a_t no longer remains a random variable but becomes a fixed number, which can then be computed by [compare with Eq. (2.1.4)]

$$a_t = X_t - \phi_1 X_{t-1} \tag{2.2.3}$$

Equation (2.2.3), when simply viewed as a different form of Eq. (2.2.1), has an interesting interpretation. Since X_t is a dependent series and a_t is an independent one, the AR(1) model is seen to be a device to transform or reduce dependent data to independent data. As seen in Eq. (2.2.3), this is accomplished by removing from X_t the part that depends on X_{t-1}.

2.2.3 Estimation of ϕ_1 and σ_a^2

If we are given a set of data, how can we "fit" an AR(1) model to it and obtain the estimates of ϕ_1 and σ_a^2? Since the AR(1) model is just a conditional regression, we can get the conditional least squares estimates by minimizing the sum of squares of a_t's. Therefore, following Eqs. (2.1.1) to (2.1.3), we have

$$\hat{\phi}_1 = \frac{\sum_{t=2}^{N} X_t X_{t-1}}{\sum_{t=2}^{N} X_{t-1}^2} \tag{2.2.4a}$$

$$\hat{\sigma}_a^2 = \frac{1}{N-1} \sum_{t=2}^{N} (X_t - \hat{\phi}_1 X_{t-1})^2 \tag{2.2.5a}$$

$$= \frac{1}{N-1} \sum_{t=2}^{N} a_t^2 = \frac{\text{Residual Sum of Squares}}{\text{Number of Residuals}} \dagger \qquad (2.2.5)$$

It is assumed in the expressions above that the average $\overline{X}$ has been subtracted from the data. When this is *not* the case, the estimate of ϕ_1 takes the form

$$\hat{\phi}_1 = \frac{\sum_{t=2}^{N} (\dot{X}_t - \overline{X})(\dot{X}_{t-1} - \overline{X})}{\sum_{t=2}^{N} (\dot{X}_{t-1} - \overline{X})^2} \qquad (2.2.4)$$

The parameter ϕ_1 in the AR(1) model, or its estimate given by Eqs. (2.2.4) and (2.2.4a), measures the extent of the dependence of X_t on X_{t-1}. If this dependence or relation between X_t and X_{t-1} is strong, ϕ_1 will be large in magnitude, and if it is weak, ϕ_1 will be small in magnitude. In fact, if ϕ_1 is zero, the AR(1) model becomes

$$X_t = a_t$$

so that X_t is really an independent or uncorrelated series. In this case the plot of X_t versus X_{t-1} will look random and scattered everywhere; no strong trend like the one in the plot for the papermaking data in Fig. 2.9 will be present.

On the other hand, if $\phi_1 > 1$ or $\phi_1 < -1$, it can be seen from Eq. (2.2.1) that X_t may increase or decrease without bound, because a_t's have a fixed finite variance and cannot continually increase in magnitude to keep X_t within bound (see Problem 2.1j). Such an "exploding" X_t is a nonstationary or unstable time series. For a stationary or stable series, X_t remains bounded (in the sense that it has finite variance) and we need

$$|\phi_1| < 1 \qquad (2.2.6)$$

[See Problems 2.9, 2.10, and Eq. (A 2.1.14).]

Since the expression (2.2.4) gives an estimate of the dependence or the relation between the values of X_t one "lag" apart, it is called the estimated autocorrelation at lag one and denoted by $\hat{\rho}_1$. Similarly, the estimated autocorrelation $\hat{\rho}_k$ at k lags is given by

$$\hat{\rho}_k = \frac{\sum_{t=k+1}^{N} (\dot{X}_t - \overline{X})(\dot{X}_{t-k} - \overline{X})}{\sum_{t=k+1}^{N} (\dot{X}_{t-k} - \overline{X})^2} = \frac{\sum_{t=k+1}^{N} X_t X_{t-k}}{\sum_{t=k+1}^{N} X_{t-k}^2}$$

---

† For an unbiased estimate, number of estimated parameters should be subtracted from the denominator.

which is one of several forms that can be used (see Chapter 3). The theoretical autocorrelation ρ_k is the limit of $\hat{\rho}_k$ as $N \to \infty$. Autocorrelation function ρ_k represents an important characteristic of the linear stochastic models and will be discussed in detail in Chapter 3. For a stationary time series the mean, variance, and covariance (or correlation) remains the same for all t. (See Section 1.2 and Appendix A 2.1.2 for definitions.)

If Eqs. (2.2.4) and (2.2.5) are used for the papermaking data, we get

$$\hat{\phi}_1 = 0.8983, \qquad \hat{\sigma}_a^2 = 0.094$$

Since the highest value of ϕ_1 is 1.0, the value $\hat{\phi}_1 = 0.8983$ mathematically confirms the strong dependence of X_t on X_{t-1} for the papermaking data, as indicated by Fig. 2.9.

2.2.4 Checks of Adequacy

To check whether the AR(1) model with these parameters is appropriate and adequate, we have to see whether there is any evidence against the basic assumptions discussed in Section 2.2.2. For example, the plots of a_t versus a_{t-1} and a_t versus X_{t-2} are shown in Figs. 2.10 and 2.11, to check the dependence of a_t on a_{t-1} and a_t on X_{t-2}, respectively. It is clear from the plots that there is no visible evidence against the assumption of independence; the points are scattered all around and there is no trend indicating dependence.

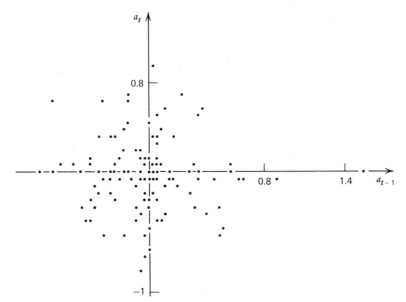

Figure 2.10 Plot of a_t versus a_{t-1}: papermaking data, AR(1) model.

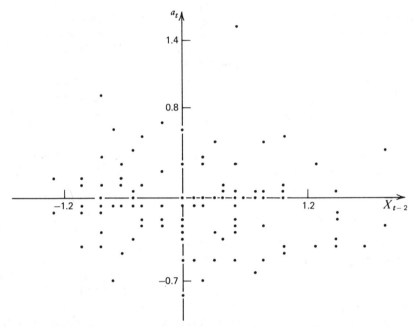

Figure 2.11 Plot of a_t versus X_{t-2}: papermaking data, AR(1) model.

This can be further confirmed by computing the correlation between a_t and a_{t-1}, and a_t and X_{t-2} as

$$\hat{\rho}(a_t, a_{t-1}) = \frac{\sum\limits_{t=3}^{N} a_t a_{t-1}}{\sum\limits_{t=3}^{N} a_{t-1}^2} = -0.065$$

$$\hat{\rho}(a_t, X_{t-2}) = \frac{\sum\limits_{t=3}^{N} a_t X_{t-2}}{\sqrt{\left(\sum\limits_{t=3}^{N} a_t^2\right)\left(\sum\limits_{t=3}^{N} X_{t-2}^2\right)}} = 0.0006$$

In fact, it can be easily verified that if indeed a_t is independent of a_{t-1}, $a_{t-2}, \ldots$, then a_t is also independent of $X_{t-1}, X_{t-2}, \ldots$. This statement is applicable to all ARMA models and not only to AR(1). A simpler and more illuminating procedure of checking will become clear in Section 2.3.3. An upper limit for the absolute value of the estimated autocorrelation below which the data may be considered to be independent is approximately $\dfrac{2}{\sqrt{N}}$ (see Appendix A 2.1.3 to A 2.1.4 and Section 4.4.4).

2.2.5 Prediction or Forecasting

The model for the papermaking data may be written as

$$X_t = 0.8983 X_{t-1} + a_t \qquad a_t \sim \text{NID}(0, 0.094)$$

Again exploiting the analogy with the regression model (2.1.1), we can use the AR(1) model for prediction and forecasting. For example, when we are at time $t-1$ so that X_{t-1} is known, the model

$$X_t = \phi_1 X_{t-1} + a_t$$

is just a regression model and therefore, as shown in Eq. (2.1.5), the prediction of X_t at time $t-1$, or "one step ahead prediction at time $t-1$" denoted as $\hat{X}_{t-1}(1)$, is

$$\hat{X}_{t-1}(1) = \phi_1 X_{t-1} \qquad (2.2.7)$$

The "one step ahead prediction error," which is the difference between observation and prediction, is

$$e_{t-1}(1) = X_t - \hat{X}_{t-1}(1) = a_t \qquad (2.2.8)$$

Equation (2.2.8) provides another interpretation of a_t's: a_t's are one step ahead prediction errors. This fact is very important and is true for all models. It also shows that the least square estimate of the parameter ϕ_1, which minimizes the sum of squares of a_t's, is also the estimate that minimizes the mean squared error of one step ahead prediction. At time $t-1$, X_t is a random variable and so is the prediction error $e_{t-1}(1)$, which has the distribution

$$e_{t-1}(1) \sim \text{NID}(0, \sigma_a^2) \qquad (2.2.9)$$

Since the observation will be equal to the predicted value plus the prediction error, its conditional distribution is

$$(X_t | X_{t-1}) \sim \text{NID}(\phi_1 X_{t-1}, \sigma_a^2)$$

as shown in Fig. 2.12 (compare with Fig. 2.7).

For example, the 95% probability limits [see Eq. (2.1.6)] for the predictions or forecasts are

$$\hat{X}_{t-1}(1) \pm 1.96 \sigma_a \qquad (2.2.10)$$

that is,

$$\phi_1 X_{t-1} \pm 1.96 \sigma_a$$

For the papermaking data, suppose we observed the input data at time $t-1 = 9$ as 33.5 and wanted to predict the input data at $t = 10$. Then (notice that $\overline{X} = 32.02$)

$$X_{t-1} = \dot{X}_9 - \overline{X} = 33.5 - 32.02 = 1.48$$
$$\hat{X}_{t-1}(1) = \hat{X}_9(1) = 0.8983 \, (1.48) = 1.33$$

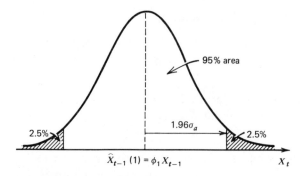

Figure 2.12   Conditional distribution of X_t given X_{t-1}.

so that the predicted value of $\dot{X}_t$ at $t = 10$ is

$$32.02 + 1.33 = 33.35$$

and the 95% probability limits are

$$33.35 \pm 1.96 \times \sqrt{0.094}$$

that is,

$$33.35 \pm 0.601$$

The prediction error $e_{t-1}(1)$ has the distribution

$$e_{t-1}(1) \sim \text{NID}(0,\sigma_a^2) \equiv \text{NID}(0,0.094)$$

and the conditional distribution of X_t at $t = 10$, given X_{t-1}, $t-1 = 9$, is

$$(X_{10}|X_9) \sim N(\phi_1 X_9, \sigma_a^2)$$
$$\sim N(0.8983 \times 1.48, 0.094)$$
$$\sim N(1.33, 0.094)$$

At time $t = 10$ the input data X_{10} is observed to be 33.3, and therefore the actual one step ahead prediction error is

$$33.3 - 33.35 = -0.05$$

2.2.6 Random Walk as a Limit of AR(1)

Obviously, the AR(1) model would not be adequate for every stationary stochastic system, as it was for the papermaking data. However, it does provide a good approximation for many systems characterized by inertia. This can be clearly seen by considering the IBM stock prices data. If we use expressions (2.2.4) and (2.2.5) to compute the parameters of an AR(1) model, from this data we get

$$\hat{\phi}_1 = 0.999 \qquad \hat{\sigma}_a^2 = 52.61$$

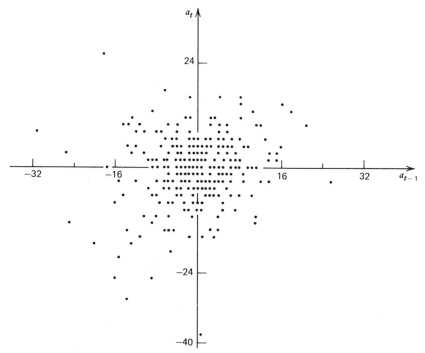

Figure 2.13 Plot of a_t versus a_{t-1}: AR(1) model IBM data.

The plots of a_t versus a_{t-1} and a_t versus X_{t-2}, illustrated in Figs. 2.13 and 2.14, do not show any visible evidence against the basic assumptions.

The corresponding correlations are found to be

$$\hat{\rho}(a_t, a_{t-1}) = +0.087, \qquad \hat{\rho}(a_t, X_{t-2}) = -0.0078$$

Thus, the AR(1) model (with the rounded-off parameters)

$$X_t = 0.999 X_{t-1} + a_t \qquad a_t \sim \text{NID}(0, 52.61)$$

is seen to be adequate for the IBM data. A slightly different form of the model can also be shown to be adequate, as will be seen in Chapter 4.

This AR(1) model can be very nearly represented by

$$X_t = X_{t-1} + a_t \tag{2.2.11}$$

that is,

$$X_t - X_{t-1} = a_t$$

or

$$\nabla X_t = a_t$$

where ∇ denotes the difference operator. This is an interesting limiting form of the AR(1) model as ϕ_1 approaches unity. First of all, it shows that

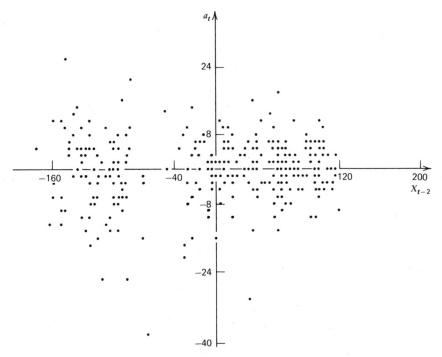

Figure 2.14 Plot of a_t versus X_{t-2}: AR(1) model IBM data.

the system is characterized by extremely high inertia, that is, strong dependence or memory. Its value or response remains unchanged as it moves from $t-1$ to t, except for the random independent increment a_t. But for this increment a_t, the system would stay in the same position indefinitely.

An immediate consequence of the limiting AR(1) model (2.2.11) is that the prediction of X_t based on X_{t-1} is simply

$$\hat{X}_{t-1}(1) = X_{t-1}$$

Thus, the best prediction or forecast of tomorrow's stock price is today's stock price, at least for the IBM stock price data under consideration.

Since t is only a nominal index, therefore,

$$X_{t-1} = X_{t-2} + a_{t-1}$$

and if we substitute this in Eq. (2.2.11), then

$$X_t = X_{t-2} + a_t + a_{t-1}$$

Continuing this way, we find that

$$X_t = a_t + a_{t-1} + a_{t-2} + \ldots\ldots$$

$$= \sum_{j=0}^{\infty} a_{t-j} \qquad (2.2.12)$$

which is a special case of the expansion considered in detail in Chapter 3. Thus, X_t is simply a sum of independent random variables. In other words, the series is generated every time by adding a random step to the previous position. For this reason, model (2.2.12) and therefore (2.2.11) is called a "random walk" model. The result that stock prices behave like random walk and therefore the best forecast of tomorrow's price is today's price was conjectured by Bachelier as far back as 1900. It is now known as Bachelier's hypothesis. (See Problem 2.18 for deciding how close to 1 should $\hat{\phi}_1$ be to justify a random walk model.)

2.3 AUTOREGRESSIVE MOVING AVERAGE MODEL ARMA(2,1)

Now consider the Wolfer's sunspot numbers data shown in Fig. 2.2. If we use expressions (2.2.4) and (2.2.5) to estimate the parameters of an AR(1) model from this data, we get

$$\hat{\phi}_1 = 0.81, \qquad \hat{\sigma}_a^2 = 409.08$$

To check the basic assumptions (Section 2.2.2) the plots of a_t versus a_{t-1} and a_t versus X_{t-2} are shown in Figs. 2.15 and 2.16. There is ample evidence against the independence assumptions; in fact, the plot in Fig. 2.16 gives a clear indication of the negative dependence of a_t on X_{t-2}, since

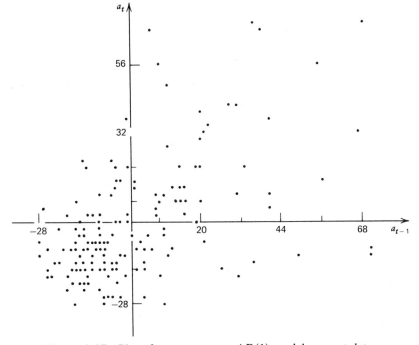

Figure 2.15   Plot of a_t versus a_{t-1}—AR(1) model sunspot data.

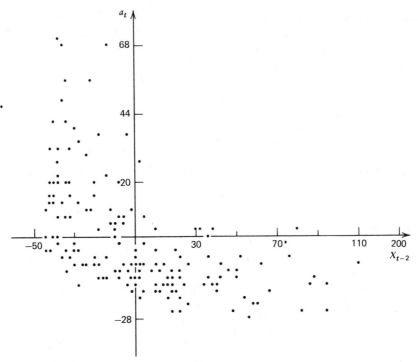

Figure 2.16   Plot of a_t versus X_{t-2}: AR(1) model sunspot data.

the points are scattered around a line with negative slope. Furthermore, this is also confirmed by the estimates

$$\hat{\rho}(a_t, a_{t-1}) = 0.53, \qquad \hat{\rho}(a_t, X_{t-2}) = -0.38$$

This raises the question: What can be done, in general, when the basic assumption no longer holds?

2.3.1 Dependence of a_t on X_{t-2} and a_{t-1}

To answer this question, recall how we obtained the AR(1) model at the beginning of Section 2.2.1. There, the problem was to express the dependence of X_t on X_{t-1}; it was solved by using regression equation (2.2.1) to decompose X_t into two parts: one dependent on X_{t-1} written as $\phi_1 X_{t-1}$, and another independent of X_{t-1}, written as a_t.

In the present case, since an AR(1) model is not adequate and appropriate, let us replace a_t by a_t' and write the model as

$$X_t = \phi_1 X_{t-1} + a_t' \qquad (2.3.1)$$

Now we have the original series X_t as well as the residual series a_t and, since a'_t may depend on X_{t-2} and/or a_{t-1}, we can similarly use regression equation (2.1.7) to decompose a'_t into *three* parts: one dependent on X_{t-2}, one dependent on a_{t-1}, and one independent of these two. Thus,

$$a'_t = \phi_2 X_{t-2} - \theta_1 a_{t-1} + a_t \qquad (2.3.2)$$

and substituting in Eq. (2.3.1), we have a new model

$$X_t = \phi_1 X_{t-1} + \phi_2 X_{t-2} + a_t - \theta_1 a_{t-1} \qquad (2.3.3a)$$

or

$$X_t - \phi_1 X_{t-1} - \phi_2 X_{t-2} = a_t - \theta_1 a_{t-1} \qquad (2.3.3)$$

This model expresses the dependence of X_t on its two preceding values, X_{t-1} and X_{t-2}, and has an "autoregressive dependence" of order two. It also includes the dependence on preceding a_t values of order one. Therefore, this model is called Autoregressive Moving Average model of order two and one respectively, denoted by ARMA(2,1). The left-hand side of Eq. (2.3.3) is the autoregressive part, whereas the right-hand side is the moving average part of the model. (The term "moving average" is somewhat of a misnomer, but we adopt it because it has become too common.)

The distribution of a_t is again the same as given by (2.2.2). The assumptions for the AR(1) model are now valid for the ARMA(2,1) model. However, since the model form itself now removes the dependence of a_t on X_{t-2} and a_{t-1} by the terms $\phi_2 X_{t-2}$ and $\theta_1 a_{t-1}$, we can write the basic assumptions for the ARMA(2,1) model as

a_t is independent of $a_{t-2}, a_{t-3}, \ldots$

which also implies that

a_t is independent of $X_{t-3}, X_{t-4}, \ldots$

The ARMA(2,1) model is also a conditional regression model since at time $t-1$, X_{t-1}, X_{t-2} and a_{t-1} are known and it becomes a regression model. When X_t is known, a_t can be computed by

$$a_t = X_t - \phi_1 X_{t-1} - \phi_2 X_{t-2} + \theta_1 a_{t-1} \qquad (2.3.4)$$

This equation, viewed again as a different form of Eq. (2.3.3), shows that the ARMA(2,1) model is also devised to transform or reduce the dependent series X_t into the independent series a_t; compare Eq. (2.2.3).

However, there is a fundamental difference between Eq. (2.3.4) and the corresponding Eq. (2.2.3) for the AR(1) model, in addition to the simple increase in the number of terms due to the higher order of the ARMA(2,1) model. Unlike Eq. (2.2.3), Eq. (2.3.4) requires preceding a_t's to compute the present one. Therefore, the a_t's for an ARMA(2,1) model have to be computed recursively, *starting from the beginning*. The initial value of a_t at $t = 0$ is generally taken as zero, which is the mean of a_t.

This difference, as we will see, makes the estimation of the ARMA(2,1) model much more difficult compared to the AR(1) model.

2.3.2 Special Case: AR(2) Model

A special case of the ARMA(2,1) model, for which the estimation can be done along the same lines as the AR(1) model, is obtained by setting $\theta_1 = 0$. Then, the ARMA(2,1) model (2.3.3) takes the form

$$X_t = \phi_1 X_{t-1} + \phi_2 X_{t-2} + a_t \qquad (2.3.5a)$$

or

$$X_t - \phi_1 X_{t-1} - \phi_2 X_{t-2} = a_t \qquad (2.3.5)$$

Since the moving average part is absent, this is just an autoregressive model of order two, denoted by AR(2). For this model X_t depends only on X_{t-1} and X_{t-2}, so that a_t can be computed from X_t, X_{t-1}, and X_{t-2} alone, without having to go to the beginning for computing previous a_t's. Therefore, the conditional least square estimates of ϕ_1, ϕ_2, and σ_a^2 can be obtained by using the ordinary regression formulae (2.1.9) and (2.1.10), as in the case of the AR(1) model. Using the special case of Eq. (2.1.7) for $n=2$, given by Eq. (2.1.13), we have

$$\mathbf{Y} = \begin{bmatrix} X_3 \\ X_4 \\ \vdots \\ X_N \end{bmatrix} \qquad \mathbf{X} = \begin{bmatrix} X_2 & X_1 \\ X_3 & X_2 \\ \vdots & \vdots \\ X_{N-1} & X_{N-2} \end{bmatrix}$$

$$\mathbf{X'X} = \begin{bmatrix} \sum_{t=2}^{N-1} X_t^2 & \sum_{t=2}^{N-1} X_t X_{t-1} \\ \sum_{t=2}^{N-1} X_t X_{t-1} & \sum_{t=1}^{N-2} X_t^2 \end{bmatrix}$$

$$\mathbf{X'Y} = \begin{bmatrix} \sum_{t=3}^{N} X_t X_{t-1} \\ \sum_{t=3}^{N} X_t X_{t-2} \end{bmatrix}$$

$$\begin{bmatrix} \hat{\phi}_1 \\ \hat{\phi}_2 \end{bmatrix} = (\mathbf{X'X})^{-1}\mathbf{X'Y}$$

$$\hat{\sigma}_a^2 = \frac{1}{N-2} \sum_{t=3}^{N} (X_t - \hat{\phi}_1 X_{t-1} - \hat{\phi}_2 X_{t-2})^2 \qquad (2.3.6)$$

For the Wolfer's sunspot series it was noted earlier that the a_t's of the fitted AR(1) series showed a strong negative dependence on X_{t-2}, as seen in Fig. 2.16. This indicates the possibility of an AR(2) model. If the expressions (2.3.6) are used to estimate ϕ_1, ϕ_2, and σ_a^2 for this data, we get

$$\hat{\phi}_1 = 1.34, \qquad \hat{\phi}_2 = -0.65, \qquad \hat{\sigma}_a^2 = 236.85$$

thus giving an AR(2) model for the sunspot series

$$X_t = 1.34 X_{t-1} - 0.65 X_{t-2} + a_t$$

Observe that the value of ϕ_2, which measures the dependence of X_t on X_{t-2}, is negative as expected. The reason why ϕ_2 is most often negative will be further studied in Chapter 7.

2.3.3 Checking the Adequacy of the AR(1) Model

A comparison of this AR(2) model for the sunspot series with the AR(1) fitted at the beginning of this section brings out some interesting points. First of all, the value of ϕ_2 expressing the dependence of X_t on X_{t-2} turns out to be fairly large. The large value of ϕ_2 shows the importance of including X_{t-2}, and therefore the AR(1) model, limiting this dependence to only X_{t-1}, is indeed inadequate. Second, there is a drastic reduction in the residual sum of squares, Σa_t^2 as indicated by its "mean or average" value σ_a^2 dropping from 409.08 to 236.85 when we go from the AR(1) to the ARMA(2,1) model (with $\theta_1 = 0$). In other words, the ARMA(2,1) model accounts for a much larger portion of the total variance of the X_t series through the dependence of X_t on X_{t-1} and X_{t-2} than does the AR(1) model. Thus, the significant reduction in the residual sum of squares also points out the inadequacy of the AR(1) model and the need for an AR(2) or ARMA(2,1) model. Just as between two sets of parameter estimates for a given model, the one that gives a significantly smaller sum of squares of a_t's is better; similarly, between two models, the one that gives a significantly smaller sum of squares of a_t's is better.

We had already guessed the inadequacy of the AR(1) model from the plots of a_t versus a_{t-1} and a_t versus X_{t-2} in Figs. 2.15 and 2.16, indicating the violation of the basic assumptions. The two points discussed above show that rather than using plots or estimated correlations, a more reliable way of checking the adequacy of an AR(1) model is by fitting an ARMA(2,1) model. If the additional parameters ϕ_2, θ_1, and/or the reduction in the residual sum of squares are significantly large, then the AR(1) model is inadequate and an ARMA(2,1) model is needed; otherwise, the AR(1) model is adequate.

Observe that this was also the kind of argument that was implicitly used in fitting the AR(1) models in the last section. We suspected the dependence of X_t on X_{t-1} from the trend in the plot of X_t versus X_{t-1}, such as the one in Fig. 2.9. However, a better way of ascertaining the adequacy of the independence assumption, that is, the adequacy of the "AR(0)" model

$$X_t = a_t$$

was to fit an AR(1) [same as ARMA(1,0)] model. For the AR(0) model, σ_a^2 is the same as the variance of the series, which is 0.5503 for the papermaking series, 7093.0 for the IBM series, and 1208.0 for the sunspot series. When we increase the order and fit AR(1) models to these sets of data, we find that ϕ_1 values are all quite large, 0.8983, 0.999, and 0.81, respectively, while there is a drastic reduction in the sum of squares as shown by the σ_a^2 of 0.094, 52.61, and 409.08, respectively. Therefore, in all these cases the assumption of independence or the AR(0) model is inadequate or inappropriate, and an AR(1) model is definitely better (also see Appendix A 2.1.6).

A question now arises: How large should the values of the estimated additional parameters (such as ϕ_2, θ_1) or the reduction in the sum of squares be, to justify going from an AR(1) to ARMA(2,1) model? For the Wolfer's sunspot series the estimate of ϕ_2 is as large as -0.65 and the "mean sum of squares" σ_a^2 is reduced from 409.08 to 236.85; thus, the need for an AR(2) model is obvious. However, if the values of ϕ_2 and θ_1 of an ARMA(2,1) model are small, say 0.1 each, and the reduction in sum of squares is only 1%, should we still consider the AR(1) model inadequate and decide that an ARMA(2,1) model is required?

Whether the value of a parameter is significantly different from zero or not can be determined from the confidence interval on its estimates provided by the estimation computer routine. In the above case, when the estimates of ϕ_2 and θ_1 are 0.1, if, say, a 95% confidence interval on each is ± 0.2, then the estimates can be considered insignificantly small. A statistical test for the significance of the reduction in residual sum of squares will be given in Chapter 4.

2.3.4 Nonlinear Regression of the ARMA(2,1) Model

To find out whether the AR(1) model is adequate for the sunspot data we should have fitted an ARMA(2,1) model rather than the AR(2) model, as we did earlier. The AR(2) model was fitted for illustrative purposes primarily because the estimation of parameters for autoregressive models is quite simple and can be accomplished by the method of *linear* least squares appropriate for linear regression. However, to fit an ARMA(2,1) model, the method of nonlinear least squares is required, which, in essence, reduces to iterative least squares.

Although the conditional regression or dependence is linear for both AR(2) and ARMA(2,1) models, and the unconditional regression is also linear for the AR(2) model, the unconditional regression for the ARMA(2,1) model is nonlinear. To see this, note that for the model

$$X_t = \phi_1 X_{t-1} + \phi_2 X_{t-2} - \theta_1 a_{t-1} + a_t$$

when all the observations up to $t-1$ are known, so that a_{t-1} is also known, the dependence of X_t on X_{t-1}, X_{t-2}, as well as on a_{t-1}, is clearly linear. Thus, the conditional regression, given observations up to X_{t-1}, is linear whether θ_1 is zero or not, since the right-hand side of the equation is linear in the unknown parameters.

Now consider the unconditional regression or dependence. For an AR(2) model, the equation

$$X_t = \phi_1 X_{t-1} + \phi_2 X_{t-2} + a_t$$

expressing the regression of X_t on its past values is linear. However, for the ARMA(2,1) model, to express the regression of X_t on its past values, we have to recursively use

$$a_{t-1} = X_{t-1} - \phi_1 X_{t-2} - \phi_2 X_{t-3} + \theta_1 a_{t-2}$$

in

$$X_t = \phi_1 X_{t-1} + \phi_2 X_{t-2} - \theta_1 a_{t-1} + a_t$$

Thus,

$$X_t = \phi_1 X_{t-1} + \phi_2 X_{t-2} - \theta_1(X_{t-1} - \phi_1 X_{t-2} - \phi_2 X_{t-3} + \theta_1 a_{t-2}) + a_t$$
$$= (\phi_1 - \theta_1)X_{t-1} + (\phi_2 + \theta_1\phi_1)X_{t-2} + \theta_1\phi_2 X_{t-3} - \theta_1^2 a_{t-2} + a_t$$

We still have a term involving a_{t-2} that has to be expressed in terms of X_{t-2}, X_{t-3}, and a_{t-3}, and so on. However, we can see at this step itself that when the dependence of X_t is expressed in terms of the past X_t, the resultant equation will be nonlinear in the unknown parameters ϕ_1, ϕ_2, and θ_1, since their products and squares are involved. Thus, the regression becomes nonlinear and requires a nonlinear least squares method for estimation.

The nonlinear least squares method achieves the minimization of the sum of squares of a_t's in a stepwise fashion. It starts with some initial values of the parameters, computes a_t's recursively, and obtains the sum of squares. The "direction" along which the sum of squares will be smaller is then found by methods such as linearization and/or steepest descent. The new values of parameters with this smaller sum of squares are then taken as initial values and the routine continues until the minimum sum of squares is obtained. The method also provides confidence intervals on the parameter estimates and other relevant information together with the estimate of σ_a^2. The method can also be used for linear regression if desired.

Suppose the nonlinear least squares routine is used to estimate the parameters ϕ_1, ϕ_2, and θ_1 of an ARMA(2,1) model. If the estimated value of θ_1 is small (as judged by the fact that its confidence interval at, say, the 95% level includes zero), then there is reason to believe that an AR(2) model may also provide an adequate fit. This belief can be verified by actually fitting an AR(2) model when desired. Since the AR(2) model has fewer parameters, it is naturally expected that its residual sum of squares would be larger than that of an ARMA(2,1) model. If this increase is not significant (using the statistical test in Chapter 4), then we conclude that the AR(2) model is also adequate.

2.3.5 Other Special Cases: ARMA(1,1), MA(1)

Other special cases of the ARMA(2,1) model can be obtained by setting $\phi_2 = 0$. When $\phi_2 = 0$, we have a model

$$X_t = \phi_1 X_{t-1} - \theta_1 a_{t-1} + a_t \tag{2.3.7}$$

which is an ARMA(1,1) model. Further, if we set $\phi_1 = 0$ in the ARMA(1,1) model, that is, $\phi_1 = \phi_2 = 0$ in the ARMA(2,1) model, we have

$$X_t = a_t - \theta_1 a_{t-1} \tag{2.3.8}$$

Since this model involves only the moving average part of order one, it is denoted by MA(1). To see whether these special cases are appropriate and adequate for the given data, we can proceed along the same lines as the AR(2) case discussed above. If the estimated values of ϕ_1 or/and ϕ_2 are small, then the ARMA(1,1)/MA(1) model may be appropriate. The adequacy can then be checked by considering the significance of increase in the sum of squares.

Note that the AR(1) model itself is a special case of ARMA(2,1) with $\phi_2 = \theta_1 = 0$. It is therefore more advantageous to first fit an ARMA(2,1) model, see whether the AR(1) model seems appropriate because the values of ϕ_2 and θ_1 are close to zero, and then fit the AR(1) model to check its adequacy. If, however, the values of ϕ_2 and θ_1 are not small, for example, their 95% confidence intervals do not include zero, then clearly an ARMA(2,1) model *is* required and an AR(1) model would not be adequate. To check the adequacy of the ARMA(2,1) model we have to go to higher order models, which we will now discuss.

2.4 ARMA(n,n−1) MODELS AND MODELING STRATEGY

Suppose an ARMA(2,1) model is inadequate; then clearly the source of inadequacy is in the basic assumption given in Section 2.3.1. When the assumption is violated, then a_t of the inadequate ARMA(2,1) model depends on either $X_{t-3}, X_{t-4}, \ldots$, and/or $a_{t-2}, a_{t-3}, \ldots$. The next possible candidates for dependence are X_{t-3} and a_{t-2}.

2.4.1 ARMA(3,2) Model

If we follow the same argument as the one used for going from the AR(1) to the ARMA(2,1) model in Eqs. (2.3.1) to (2.3.3), we can write the inadequate ARMA(2,1) model as

$$X_t = \phi_1 X_{t-1} + \phi_2 X_{t-2} + a_t'$$

decompose the a_t' into four parts: one dependent on X_{t-3}, one dependent on a_{t-1}, one dependent on a_{t-2}, and one independent of these three

$$a_t' = \phi_3 X_{t-3} - \theta_1 a_{t-1} - \theta_2 a_{t-2} + a_t$$

and write the new model as

$$X_t = \phi_1 X_{t-1} + \phi_2 X_{t-2} + \phi_3 X_{t-3} + a_t - \theta_1 a_{t-1} - \theta_2 a_{t-2} \quad (2.4.1a)$$

or

$$X_t - \phi_1 X_{t-1} - \phi_2 X_{t-2} - \phi_3 X_{t-3} = a_t - \theta_1 a_{t-1} - \theta_2 a_{t-2} \quad (2.4.1)$$

This model has an autoregressive dependence of order 3 and a moving average dependence of order 2. Therefore, this is an ARMA(3,2) model and is the next logical choice after the ARMA(2,1) model.

The distribution of a_t is again the same as given by Eq. (2.2.2). Similar to the AR(1) and ARMA(2,1) models, the basic assumption for the ARMA(3,2) model is written as

$$a_t \text{ is independent of } a_{t-3}, a_{t-4}, \ldots$$

which implies

$$a_t \text{ is independent of } X_{t-4}, X_{t-5}, \ldots$$

2.4.2 ARMA(n,n−1) Model

Proceeding in this way, we arrive at the ARMA(n,n−1) model

$$X_t - \phi_1 X_{t-1} - \phi_2 X_{t-2} - \ldots - \phi_n X_{t-n} = a_t - \theta_1 a_{t-1}$$
$$- \theta_2 a_{t-2} - \ldots - \theta_{n-1} a_{t-n+1}, \qquad a_t \sim \text{NID}(0, \sigma_a^2) \quad (2.4.2)$$

The form of the model removes the dependence of a_t on $X_{t-1}, X_{t-2}, \ldots,$ X_{t-n} and $a_{t-1}, a_{t-2}, \ldots, a_{t-n+1}$ by the autoregressive part of order n and the moving average part of order $n-1$, respectively. Hence, the basic assumption, valid for this model, may now be written in the form

$$a_t \text{ is independent of } a_{t-n}, a_{t-n-1}, \ldots$$

which also implies that

$$a_t \text{ is independent of } X_{t-n-1}, X_{t-n-2}, \ldots$$

The ARMA(n,n−1) is a conditional regression model since at time $t-1$, $X_{t-1}, X_{t-2}, \ldots, X_{t-n}$ and $a_{t-1}, a_{t-2}, \ldots, a_{t-n+1}$ are known and it becomes a regression model. When X_t is known, a_t can be computed by

$$a_t = X_t - \phi_1 X_{t-1} - \phi_2 X_{t-2} - \ldots - \phi_n X_{t-n}$$
$$+ \theta_1 a_{t-1} + \theta_2 a_{t-2} + \ldots + \theta_{n-1} a_{t-n+1} \quad (2.4.3)$$

which, considered as a different form of Eq. (2.4.2), shows that the ARMA$(n, n-1)$ model reduces the dependent series X_t to the independent series a_t, by removing the autoregressive and the moving average dependence of X_t on preceding n X_t's and $n-1$ a_t's.

The computation of a_t's by Eq. (2.4.3) requires the preceding a_t's to compute the present one as long as at least one of the θ's is nonzero. Therefore, the a_t's have to be computed recursively, starting from the beginning, when the moving average part is present. In that case, the unconditional regression is nonlinear. The special case of the AR(n) model can, of course, be estimated by the explicit linear least squares method; however, if the formulae of a nonlinear least squares method are used, the nonlinear routine will converge in the first step.

The evolution of ARMA$(n, n-1)$ models starting from the independence assumption, or AR(0) model, is schematically presented in Fig. 2.17. It shows how ARMA$(n, n-1)$ models of increasing order arise, when the basic assumption is successively relaxed, by including the dependence of a_t on X_{t-n-1} and a_{t-n} in the succeeding model. All the special cases, including the pure autoregressive and pure moving average models, are shown. The chart covers all the possible ARMA models, and it is easy to see that as n increases, the number of models multiply very rapidly.

2.4.3 ARMA(n,n−1) versus ARMA(n,m): ARIMA as Special Cases

Our approach of concentrating on the ARMA$(n, n-1)$ models and treating other models as their special cases is radically different from the approaches available in the time series literature, in which all the ARMA(n, m) models for arbitrary values of n and m are treated on par. A large part of the literature deals with the estimation of parameters when the orders n and m are known. A comprehensive account of the time series theory and inference by spectral methods may be found in Hannan (1970). However, the question of determining the values of n and m *before fitting a model* is very difficult and has been solved with rigorous statistical tests only when $m = 0$ [see Anderson (1971)]. Box and Jenkins (1970) have provided some empirical guidelines for the determination of n and m when one of them is zero.

Contrary to this approach, we do not assume that the data has been generated by an ARMA(n, m) model for some fixed values of n and m and then try to find n and m by trial and error. Our modeling strategy is to

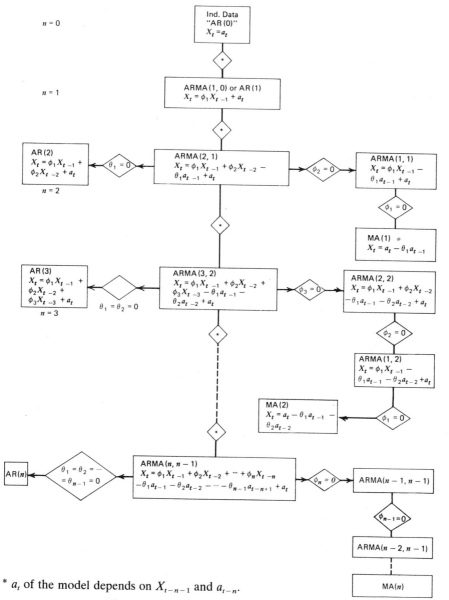

* a_t of the model depends on X_{t-n-1} and a_{t-n}.

Figure 2.17   Genesis of ARMA models.

(*Note:* All the models in the preceding stages are special cases of the
ARMA($n,n − 1$) in the succeeding stages.)

approximate the dependence in the data more and more closely by an increasing sequence of ARMA$(n,n-1)$ models. We stop at the value of n beyond which we do not get significant improvement in the approximation, as judged by the reduction in the residual sum of squares. As we will show in Chapter 4, this strategy enables us to devise a simple modeling procedure that can be completely executed on a digital computer. In other words, once the data is supplied to the modeling routine, it can output all the adequate models pointing out the most adequate one; no empirical examination of plots of autocorrelations, spectra, residuals, etc. is required.

If an ARMA(n,m) model with $m \neq n - 1$ is appropriate or desired for some reason, ARMA$(n,n-1)$ strategy is still the easiest way of arriving at them, as discussed in detail in Chapter 4. It should also be emphasized that the so-called integrated ARMA (ARIMA) models can also be arrived at as special cases of ARMA$(n,n-1)$. Thus, for example, the random walk $\nabla X_t = a_t$ is AR(1) with $\phi_1 = 1$; integrated random walk $\nabla^2 X_t = a_t$ is ARMA(2,1) with $\phi_1 = 2$, $\phi_2 = -1$, and $\theta_1 = 0$, and so forth.

2.4.4 Adequacy of the ARMA$(n,n-1)$ Models

A question now arises: Is it always possible to represent or approximate a set of data by an ARMA$(n,n-1)$ model? In other words, can every stochastic system be represented well by an ARMA$(n,n-1)$ model? This question can be answered in the affirmative for stationary stochastic systems, including the limiting cases such as random walk. By employing the elementary theory of linear operators on Hilbert space, it can be shown that any stationary stochastic system can be approximated as closely as we want by an autoregressive moving average model of order $(n,n-1)$ for discrete, continuous, scalar, as well as vector cases. A simplified version of this proof will be given in Chapter 4 (Section 4.1.3).

In the next Chapter 3, we will study two important characterizations of the ARMA models: Green's function and autocovariance. Then in Chapter 4, we will examine these two characterizations and clearly show why the moving average order in the sequence is always $n-1$ in general, rather than n or some arbitrary number m. This will mathematically confirm our approach of treating other ARMA models as special cases of ARMA$(n,n-1)$ models for appropriate n.

The theory developed in Chapter 3 can also be used to show that a sum of two variables will in general have an ARMA(2,1) model if each has an AR(1) model. (See Problems 3.21 to 3.23.) Thus a sum of n AR(1) variables will have in general an ARMA$(n,n-1)$ model. Since many time series are aggregates of individual effects, they can be readily modeled by the ARMA$(n,n-1)$ strategy.

Additional sustenance to the $\text{ARMA}(n, n-1)$ modeling strategy comes from the fact that if a continuous autoregressive moving average process, in the form of a linear differential equation of autoregressive order n and *arbitrary* moving average order, is sampled at uniform intervals, the resultant sampled process is $\text{ARMA}(n, n-1)$. A simple case for $n = 2$ will be considered in Chapters 7 and 8.

Although there is no theoretical limit on how large n should be, in practice the value of n is usually small; in fact, $n = 2$ suffices in many cases. A large variety of stochastic systems in practice can be adequately modeled by the $\text{ARMA}(2,1)$ model. The importance of the $\text{ARMA}(2,1)$ model will become especially clear when we study continuous systems in Chapters 6 to 8.

APPENDIX A 2.1

LEAST SQUARES ESTIMATES IN SIMPLE REGRESSION

A 2.1.1 Estimates of Y Intercept β_0 and Slope β_1

Consider the simple regression model using the notation of Section 2.1.1.

$$\dot{y}_t = \beta_0 + \beta_1 \dot{x}_t + \varepsilon_t \qquad \text{(A 2.1.1)}$$

Least squares estimates of the Y intercept β_0 and slope β_1 are obtained by minimizing the sum of squares of the errors ε_t:

$$\Sigma \varepsilon_t^2 = \Sigma (\dot{y}_t - \beta_0 - \beta_1 \dot{x}_t)^2$$

where Σ in this section denotes summation over t from 1 to N. If we differentiate with respect to β_0, β_1 and equate to zero, the estimates $\hat{\beta}_0$ and $\hat{\beta}_1$ must satisfy

$$\frac{\partial \Sigma \varepsilon_t^2}{\partial \beta_0} = -2\Sigma (\dot{y}_t - \beta_0 - \beta_1 \dot{x}_t) = 0 \qquad \text{(A 2.1.2)}$$

$$\frac{\partial \Sigma \varepsilon_t^2}{\partial \beta_1} = -2\Sigma \dot{x}_t (\dot{y}_t - \beta_0 - \beta_1 \dot{x}_t) = 0 \qquad \text{(A 2.1.3)}$$

Dividing equation (A 2.1.2) by $-2N$ gives

$$\hat{\beta}_0 = \bar{y} - \hat{\beta}_1 \bar{x} \qquad \text{(A 2.1.4)}$$

Since

$$\Sigma \bar{x}(\dot{y}_t - \bar{y}) = \bar{x}(\Sigma \dot{y}_t - \Sigma \bar{y}) = 0$$

and

$$\Sigma \bar{x}(\dot{x}_t - \bar{x}) = \bar{x}(\Sigma \dot{x}_t - \Sigma \bar{x}) = 0$$

by definition of the averages $\bar{x}$ and $\bar{y}$, substituting for β_0 in (A 2.1.3) and dividing by -2 gives

$$\Sigma \dot{x}_t[(\dot{y}_t - \bar{y}) - \beta_1(\dot{x}_t - \bar{x})] = \Sigma \dot{x}_t[(\dot{y}_t - \bar{y}) - \beta_1(\dot{x}_t - \bar{x})]$$
$$- \Sigma \bar{x}[(\dot{y}_t - \bar{y}) - \beta_1(\dot{x}_t - \bar{x})] = 0$$

that is

$$\Sigma(\dot{x}_t - \bar{x})[(\dot{y}_t - \bar{y}) - \beta_1(\dot{x}_t - \bar{x})] = 0$$

or

$$\hat{\beta}_1 = \frac{\Sigma(\dot{y}_t - \bar{y})(\dot{x}_t - \bar{x})}{\Sigma(\dot{x}_t - \bar{x})^2} = \frac{\Sigma y_t x_t}{\Sigma x_t^2} \qquad \text{(A 2.1.5)}$$

A 2.1.2 Mean, Variance, Covariance, and Correlation of Random Variables and Their Linear Forms

Although the least square estimates can be obtained without considering the random nature of y_t, to know how good these estimates are, one has to know their statistical properties. For this, one needs some basic results from statistics, briefly summarized in this section.

Two primary characteristics of a random variable are its expected value (average over infinite number of observations) or the *mean*, and the variance, which is the expected value of the *square* of its deviation from the mean. The mean denotes the center of its probability distribution and the variance denotes the spread to indicate how far it can deviate from the center or the mean. The square root of the variance is called *standard error* or *standard deviation*. For two random variables an additional characteristic is their *covariance*, which is the expected value of the product of their deviations from the respective means. The covariance expresses the extent of dependence between a pair of random variables and is zero when they are *independent*.

If X and Y are two random variables with means μ_x and μ_y, variances σ_x^2 and σ_y^2, and covariance σ_{xy} and we denote the expected value by E, then

$$\text{Mean of } X \quad = \mu_x = E(X) \qquad \text{(A 2.1.6)}$$
$$\text{Mean of } Y \quad = \mu_y = E(Y) \qquad \text{(A 2.1.7)}$$
$$\text{Variance of } X \quad = \sigma_x^2 = E(X - \mu_x)^2 \qquad \text{(A 2.1.8)}$$
$$\text{Variance of } Y \quad = \sigma_y^2 = E(Y - \mu_y)^2 \qquad \text{(A 2.1.9)}$$
$$\text{Covariance } (X, Y) = \gamma_{xy} = E(X - \mu_x)(Y - \mu_y) \qquad \text{(A 2.1.10)}$$

Although the covariance is a good indicator in the sense that the stronger the dependence the larger its value, its practical defect is that it is affected by the scales of X and Y. For example, if both X and Y are multiplied by 2, their dependence is unchanged, yet the covariance gets multiplied by 4 as can be easily seen from (A 2.1.6), (A 2.1.7), and (A 2.1.10). This scaling effect can be eliminated by dividing the covariance by the square roots of variances, or standard deviations, and this "scaleless" covariance is called *correlation*, denoted by ρ_{xy}. Thus,

$$\text{Correlation } (X,Y) = \frac{\text{Covariance } (X,Y)}{\sqrt{\text{Var}(X)\ \text{Var}(Y)}} \qquad \text{(A 2.1.11)}$$

$$= \frac{E(X-\mu_x)(Y-\mu_y)}{\sqrt{E(X-\mu_x)^2\ E(Y-\mu_y)^2}} \qquad \text{(A 2.1.11a)}$$

that is,

$$\rho_{xy} = \frac{\gamma_{xy}}{\sigma_x\ \sigma_y} \qquad \text{(A 2.1.11b)}$$

Not only is the correlation scaleless, but it can also be shown to be always less than or equal to one in absolute value. To show this, note that although a random variable can be positive or negative, its square cannot be negative and therefore the expected value of its square or variance also cannot be negative. Now consider the random variable $aX + bY$.

$$\text{Mean of } (aX + bY) = \mu_{ax+by} = E(aX+bY)$$
$$= aE(X) + bE(Y)$$
$$= a\mu_x + b\mu_y \qquad \text{(A 2.1.12)}$$

$$\text{Variance of } (aX+bY) = \sigma^2_{ax+by} = E(aX+bY - \mu_{ax+by})^2$$
$$= E[a(X-\mu_x) \quad + b(Y-\mu_y)]^2$$
$$= E[a^2(X-\mu_x)^2 \quad + 2ab(X-\mu_x)(Y-\mu_y)$$
$$+ b^2(Y-\mu_y)^2]$$
$$= a^2E(X-\mu_x)^2 \quad + 2abE(X-\mu_x)(Y-\mu_y)$$
$$+ b^2E(Y-\mu_y)^2$$
$$= a^2\ \text{Var}(X) \quad + 2ab\ \text{Cov}(X,Y) + b^2\ \text{Var}(Y)$$
$$\text{(A 2.1.13)}$$

(A 2.1.13) is an extremely important result that will be used in many different ways in the book. First let us use it to show that the correlation is always less than or equal to one in absolute value. Let

$$a = 1 \quad \text{and} \quad b^2 = \frac{\text{Var}(X)}{\text{Var}(Y)} \quad \text{or} \quad b = \pm \sqrt{\frac{\text{Var}(X)}{\text{Var}(Y)}}$$

Then (A 2.1.13) gives

$$\text{Var}(X) \pm 2\sqrt{\frac{\text{Var}(X)}{\text{Var}(Y)}} \, \text{Cov}(X,Y) + \text{Var}(X) \geq 0$$

or

$$\pm \, \text{Cov}(X,Y) \leq \sqrt{\text{Var}(X) \, \text{Var}(Y)} \qquad \text{(A 2.1.14a)}$$

that is,

$$|\rho_{xy}| \leq 1 \qquad \text{(A 2.1.14)}$$

Second, if X and Y are independent, so that $\text{Cov}(X,Y) = 0$ or the correlation $\rho_{xy} = 0$, then (A 2.1.13) reduces to

$$\text{Var}(aX+bY) = a^2 \, \text{Var}(X) + b^2 \, \text{Var}(Y) \qquad \text{(A 2.1.15)}$$

In particular, (A 2.1.15) says that the variance of the sum or difference of *independent* random variables is the sum of their variances ($a = 1, b = \pm 1$).

Results (A 2.1.12) and (A 2.1.15) can now be generalized to any number of random variables. Thus, the mean of a linear combination of random variables is the same linear combination of the means. The variance of a linear combination of *independent* random variables is a linear combination of the variances with *coefficients squared*. Note that the independence is *not* required for the mean but *is* required for the variance. Also, the coefficients get squared and therefore the variances always get added, which is intuitively clear since the variability is always increased whether you add or subtract random variables.

The expectation operation used in the definitions (A 2.1.6) to (A 2.1.10) is a theoretical operation on the "population" or an infinite sample. In practice, one needs to estimate these quantities from a *finite* sample of N observations, say $X_1, X_2, \ldots, X_N$ and $Y_1, Y_2, \ldots, Y_N$. The easiest way to find the estimates of these quantities is by replacing the expectation operation over infinite sample in (A 2.1.6) to (A 2.1.11) by the *averaging* operation over the available finite sample. This gives a set of estimates of the mean, the variance, the covariance, and the correlation as

$$\overline{X} = \frac{1}{N} \Sigma \, X_i \qquad \text{(A 2.1.16)}$$

$$\hat{\sigma}_x^2 = \frac{1}{N} \Sigma \, (X_i - \overline{X})^2 \qquad \text{(A 2.1.17)}$$

$$\hat{\gamma}_{xy} = \frac{1}{N} \Sigma \, (X_i - \overline{X})(Y_i - \overline{Y}) \qquad \text{(A 2.1.18)}$$

$$\hat{\rho}_{xy} = \frac{\Sigma \, (X_i - \overline{X})(Y_i - \overline{Y})}{\sqrt{\Sigma \, (X_i - \overline{X})^2 \, \Sigma \, (Y_i - \overline{Y})^2}} \qquad \text{(A 2.1.19)}$$

corresponding to (A 2.1.6), (A 2.1.8), (A 2.1.10), and (A 2.1.11a), respectively, where Σ stands for summation over i from 1 to N as before.

If we substitute the observed values from a sample in (A 2.1.16) to (A 2.1.19), we will get fixed numbers as estimates. However, these numbers will be different for different sets of samples, since the *estimators* defined by (A 2.1.16) to (A 2.1.19) are themselves random variables.[†] It is therefore important to know how well they estimate the quantity they are estimating. An important desirable feature of an estimator is its unbiasedness. An estimator is said to be *unbiased* if its mean or expected value is the parameter it is estimating. For example, $\overline{X}$ is an unbiased estimator of μ_x because it is easy to show that the mean of $\overline{X}$ is μ_x by using (A 2.1.12):

$$E(\overline{X}) = E\left(\frac{1}{N} \Sigma X_i\right)$$

$$= E\left(\frac{1}{N}X_1 + \frac{1}{N}X_2 + \cdots + \frac{1}{N}X_N\right)$$

$$= \frac{1}{N}E(X_1) + \frac{1}{N}E(X_2) + \cdots + \frac{1}{N}E(X_n)$$

$$= \frac{\mu_x}{N} + \frac{\mu_x}{N} + \cdots + \frac{\mu_x}{N} = \mu_x \qquad \text{(A 2.1.20)}$$

since each X_i is a sample from the same population with mean μ_x.

Thus, the estimator $\overline{X}$ is centered around μ_x, the parameter it is estimating. However, for a given sample the computed average will not be equal to μ_x. How far such computed averages will deviate from μ_x for different sets of samples is indicated by the variance of $\overline{X}$. Just as (A 2.1.12) was used to find the mean of $\overline{X}$, we should use (A 2.1.13) to find the variance of $\overline{X}$. However, to use (A 2.1.13), we must know the covariance of all possible pairs from $X_1, X_2, \ldots, X_N$, which is generally quite difficult in practice. This difficulty can be overcome if the observations are *independent*, so that the covariances are all zero and we can use the simpler

---

[†] Therefore an estimator, as a random variable, is often denoted by capital letters such as $\overline{X}$, whereas its value, the estimate, for a given sample, is denoted by a small letter such as $\overline{x}$. To avoid the extra notation, we do not make this distinction for other parameters. Properties such as mean, variance and distribution refer to an estimator, whereas computed values refer to an estimate.

expression (A 2.1.15), as is normally assumed in elementary statistics. Thus, for independent observations $X_1, X_2, \ldots, X_N$

$$
\begin{aligned}
\mathrm{Var}(\overline{X}) &= \mathrm{Var}\left[\frac{1}{N}X_1 + \frac{1}{N}X_2 + \ldots + \frac{1}{N}X_N\right] \\
&= \frac{1}{N^2}\mathrm{Var}(X_1) + \frac{1}{N^2}\mathrm{Var}(X_2) + \ldots + \frac{1}{N^2}\mathrm{Var}(X_N) \\
&= \frac{N\,\mathrm{Var}(X)}{N^2} \\
&= \frac{1}{N}\mathrm{Var}(X)
\end{aligned}
\tag{A 2.1.21}
$$

that is,

$$
\sigma_{\bar{x}}^2 = E[\overline{X} - E(\overline{X})]^2 = E(\overline{X} - \mu_x)^2 = \frac{1}{N}\sigma_x^2 \tag{A 2.1.21a}
$$

or

$$
\sigma_{\bar{x}} = \frac{1}{\sqrt{N}}\sigma_x \tag{A 2.1.21b}
$$

Thus, (A 2.1.21a) and (A 2.1.21b) show that the standard deviation of $\overline{X}$, which measures the "distance" between the estimator $\overline{X}$ and the parameter μ_x it is estimating, reduces by a scale of $1/\sqrt{N}$ as the number of observations N is increased. In particular, $\sigma_{\bar{x}} \to 0$ as $N \to \infty$, which means that the larger the number of observations, the closer the estimator $\overline{X}$ is to the parameter it is estimating. This is another desirable property of an estimator called *consistency*.

However, in practice, we have a finite N. How "close" $\overline{X}$ is from μ_x for finite N may be specified by giving the probability of an interval called "confidence interval" around $\bar{x}$ in which μ lies. For this we need the distribution of $\overline{X}$ discussed in Section A 2.1.4.

A 2.1.3 Mean and Variance of $\hat{\beta}_1$, $\hat{\phi}_1$, and $\hat{\rho}_k$

In the model (A 2.1.1), ε_t have zero mean, are independent at different t, that is,

$$
E(\varepsilon_t) = 0
$$

$$
\begin{aligned}
\mathrm{Cov}(\varepsilon_t, \varepsilon_{t-k}) &= E\{[\varepsilon_t - E(\varepsilon_t)][\varepsilon_{t-k} - E(\varepsilon_{t-k})]\} \\
&= E(\varepsilon_t\varepsilon_{t-k}) \\
&= 0 \quad \text{for } k \neq 0
\end{aligned}
$$

and have variance σ_ε^2, that is,

$$
\begin{aligned}
\mathrm{Var}(\varepsilon_t) &= E[\varepsilon_t - E(\varepsilon_t)]^2 \\
&= E\varepsilon_t^2 \\
&= \sigma_\varepsilon^2
\end{aligned}
$$

Note that for zero mean variables the variance and covariance expressions simplify to $E(\varepsilon_t^2)$ and $E(\varepsilon_t \varepsilon_{t-k})$. The terms β_0 and $\beta_1 \dot{x}_t$ are deterministic (nonrandom) variables and therefore their expected values are the same. In fact, the usual deterministic variables can be thought of as "degenerate" random variables with the same mean and *zero variance*, and they will be "independent" of any other random variable. Thus, for the model (A 2.1.1), if we use (A 2.1.12)

$$
\begin{aligned}
\text{Mean of } \dot{y}_t = E(\dot{y}_t) &= E(\beta_0 + \beta_1 \dot{x}_t + \varepsilon_t) \\
&= E(\beta_0 + \beta_1 \dot{x}_t) + E\varepsilon_t \\
&= E(\beta_0) + E(\beta_1 \dot{x}_t) + 0 \\
&= \beta_0 + \beta_1 \dot{x}_t \quad\quad\quad\text{(A 2.1.22)}
\end{aligned}
$$

and by (A 2.1.15)

$$
\begin{aligned}
\text{Var}(\dot{y}_t) &= \text{Var}(\beta_0 + \beta_1 \dot{x}_t + \varepsilon_t) \\
&= \text{Var}(\beta_0 + \beta_1 \dot{x}_t) + \text{Var}(\varepsilon_t) \\
&= 0 + E(\varepsilon_t^2) \\
&= \sigma_\varepsilon^2 \quad\quad\quad\quad\quad\quad\text{(A 2.1.23)}
\end{aligned}
$$

or

$$
\text{Var}(\dot{y}_t) = E(\dot{y}_t - E\dot{y}_t)^2 = E(\dot{y}_t - \beta_0 - \beta_1 \dot{x}_t)^2 = E(\varepsilon_t^2) = \sigma_\varepsilon^2
$$

Also, if we take the average over $t=1$ to N in (A 2.1.1)

$$
\begin{aligned}
\bar{y} &= \beta_0 + \beta_1 \bar{x} + \bar{\varepsilon} \\
E\bar{y} &= E\beta_0 + E\beta_1 \bar{x} + E\bar{\varepsilon} \\
&= \beta_0 + \beta_1 \bar{x}
\end{aligned}
$$

Now considering the estimator $\hat{\beta}_1$ given by (A 2.1.5), we see that

$$
\begin{aligned}
Ey_t &= E(\dot{y}_t - \bar{y}) \\
&= E\dot{y}_t - E\bar{y} \\
&= \beta_0 + \beta_1 \dot{x}_t - \beta_0 - \beta_1 \bar{x} \\
&= \beta_1(\dot{x}_t - \bar{x}) \\
&= \beta_1 x_t
\end{aligned}
$$

and since the only random variable in the right-hand side of (A 2.1.5) is

$y_t = (\dot{y}_t - \bar{y})$, x_t being constants, we can apply (A 2.1.12) to (A 2.1.5) and get

Mean of $\hat{\beta}_1 = E(\hat{\beta}_1)$

$$= E\left(\frac{\Sigma y_t x_t}{\Sigma x_t^2}\right)$$

$$= \frac{1}{\Sigma x_t^2} E(y_1 x_1 + y_2 x_2 + \ldots + y_N x_N)$$

$$= \frac{1}{\Sigma x_t^2} (x_1 E y_1 + x_2 E y_2 + \ldots + x_n E y_N)$$

$$= \frac{1}{\Sigma x_t^2} (x_1 \beta_1 x_1 + x_2 \beta_1 x_2 + \ldots x_N \beta_1 x_N) = \beta_1 \quad \text{(A 2.1.24)}$$

so that $\hat{\beta}_1$ is an *unbiased* estimator of β_1.

Similarly, $\dot{y}_t$'s are independent because ε_t's are independent, and therefore if we apply (A 2.1.15) and note that $\Sigma \bar{y} x_t = 0$

$$\text{Var}(\hat{\beta}_1) = \text{Var}\left[\left(\frac{x_1}{\Sigma x_t^2}\right)\dot{y}_1 + \left(\frac{x_2}{\Sigma x_t^2}\right)\dot{y}_2 + \ldots + \left(\frac{x_N}{\Sigma x_t^2}\right)\dot{y}_N\right]$$

$$= \left(\frac{x_1}{\Sigma x_t^2}\right)^2 \text{Var}(\dot{y}_1) + \left(\frac{x_2}{\Sigma x_t^2}\right)^2 \text{Var}(\dot{y}_2) + \ldots$$

$$+ \left(\frac{x_N}{\Sigma x_t^2}\right)^2 \text{Var}(\dot{y}_N)$$

$$= \frac{\Sigma x_t^2}{(\Sigma x_t^2)^2} \text{Var}(\dot{y}_t)$$

$$= \frac{1}{\Sigma x_t^2} \text{Var}(\dot{y}_t) = \frac{\sigma_\varepsilon^2}{\Sigma x_t^2} \quad \text{(A 2.1.25)}$$

As $\Sigma x_t^2 \to \infty$ as $N \to \infty$, (A 2.1.25) shows that $\text{Var}(\hat{\beta}_1) \to 0$ as $N \to \infty$. Since the mean of $\hat{\beta}_1$ is the "true" value β_1, this shows that $\hat{\beta}_1$ gets "closer" to the true value with an increasing number of observations; in other words, the estimator is consistent. Similarly, since $E\bar{y} = \beta_0 + \beta_1 \bar{x}$ and $E\hat{\beta}_1 = \beta_1$, it follows from (A 2.1.4) that

$$E(\hat{\beta}_0) = \beta_0 \quad \text{(A 2.1.26)}$$

and by (A 2.1.15), since $\bar{y}$ can be shown to be independent of $\hat{\beta}_1$,

$$\text{Var}(\hat{\beta}_0) = \text{Var } \bar{y} + \bar{x}^2 \text{Var}(\hat{\beta}_1) = \sigma_\varepsilon^2\left(\frac{1}{N} + \frac{\bar{x}^2}{\Sigma x_t^2}\right) \quad \text{(A 2.1.27)}$$

The results of the simple regression model

$$y_t = \beta_1 x_t + \varepsilon_t$$

are not exactly applicable to the AR(1) model

$$X_t = \phi_1 X_{t-1} + a_t$$

even though, like ε_t's, a_t's are independent, because X_{t-1} is a random variable unlike the deterministic variable x_t. However, at time $t-1$, the observation X_{t-1} is known or fixed and deterministic, and then the AR(1) model can be considered as a *conditional* linear regression model and (A 2.1.5) can be used to get the estimator

$$\hat{\phi}_1 = \frac{\Sigma X_t X_{t-1}}{\Sigma X_{t-1}^2}$$

if we assume X_0 is available; otherwise, the sums can be truncated as in (2.2.4) and (2.2.4a). Under this conditional least squares set-up, $\hat{\phi}_1$ is unbiased and consistent, that is, the mean of $\hat{\phi}_1$ is ϕ_1 and the variance tends to zero as $N \to \infty$.

However, under the unconditional least squares set-up, when X_{t-1} is considered as a random variable and $t-1$ has no special consideration, these results are only approximately true. Similar approximation can be used for large N to compute the approximate variance of $\hat{\phi}_1$ by replacing ΣX_{t-1}^2 by its expected value $N \text{Var}(X_{t-1}) = N \text{Var}(X_t) = N\sigma_a^2/(1-\phi_1^2)$ (see Problem 2.10) in using (A 2.1.25):

$$\text{Var}(\hat{\phi}_1) \simeq \frac{\sigma_a^2}{\Sigma X_{t-1}^2} \simeq \frac{\sigma_a^2}{N \text{Var}(X_t)} = \frac{(1-\phi_1^2)}{N} \qquad \text{(A 2.1.28)}$$

and

$$\text{Var}(\hat{\phi}_1) \simeq \frac{\sigma_a^2}{N\sigma_a^2} \qquad \text{when} \qquad \phi_1 = 0$$

$$= \frac{1}{N} \qquad \qquad\qquad\qquad\qquad \text{(A 2.1.29)}$$

A similar argument can be used for the sample autocorrelation estimator

$$\hat{\rho}_k = \frac{\Sigma X_t X_{t-k}}{\Sigma X_{t-k}^2}$$

to find its approximate variance when $\rho_k \equiv 0$ as

$$\text{Var}(\hat{\rho}_k) \simeq \frac{1}{N} \qquad \text{(A 2.1.30)}$$

Note that since $\hat{\rho}_1 = \hat{\phi}_1$, (A 2.1.29) is really a special case of (A 2.1.30) and, incidentally, both these variances are the same as $\mathrm{Var}(\overline{X})/\mathrm{Var}(X)$, as seen from (A 2.1.21). The approximate standard deviation or standard error $1/\sqrt{N}$ of the estimated autocorrelation is particularly useful in determining how small $\hat{\phi}_1$ or $\hat{\rho}_k$ should be, so that ϕ_1 or ρ_k can be considered to be zero and the data independent. If we assume normal distribution as discussed in the next section, the absolute value of $\hat{\phi}_1$ or $\hat{\rho}_k$ should be less than $2/\sqrt{N}$, as a quick approximate check on the independence of data. An approximation better than (A 2.1.30) for this purpose will be given in Section 4.4.4. It should also be noted that for small and moderate N, the estimators $\hat{\phi}_1$ and $\hat{\rho}_k$ are biased. Moreover, $\hat{\rho}_k$ are highly correlated with each other, which makes them rather poor estimators of ρ_k. Therefore, we will not use $\hat{\rho}_k$ in this book, except as a secondary check on the independence of the residuals from a fitted model, discussed in Section 4.4.4. When AR(1) model is adequate, the approximation (A 2.1.28) can be used for $\mathrm{Var}(\hat{\phi}_1)$ to find a confidence interval for ϕ_1.

A 2.1.4 Normal Distribution and Confidence Intervals

A random variable is completely defined by its probability distribution, of which the mean and the variance are primary characteristics. Results derived so far in Sections A 2.1.1 to A 2.1.3 did not require the assumption of any distribution. However, to make specific probability statements regarding the least square estimates and predictions or forecasts based on them, indicating how much they could differ from their actual values, it is necessary to assume a distribution for the data. The most commonly assumed distribution is the normal distribution, in which the probability mass is distributed around the mean by a bell-shaped curve, such as the ones shown in Figures 2.7 and 2.12, with its spread determined by the variance.

If a random variable X has normal distribution with mean μ_x and variance σ_x^2, written in short as $X \sim N(\mu_x, \sigma_x^2)$, then the probability density of the values x of X is given by the expression

$$f(x) = \frac{1}{\sqrt{2\pi}\,\sigma_x} e^{-[(x-\mu_x)^2/2\sigma_x^2]}, \quad -\infty < x < \infty \quad \text{(A 2.1.31)}$$

For any random variable X with probability density $f(x)$, the probability that X will have values between a and b, denoted by $P(a \leqslant X \leqslant b)$, is given by

$$P(a \leqslant X \leqslant b) = \int_a^b f(x)\, dx \quad \text{(A 2.1.32)}$$

Obviously, X must be between $-\infty$ and $+\infty$, so that

$$P(-\infty < X < \infty) = \int_{-\infty}^{\infty} f(x)\, dx = 1 \quad \text{(A 2.1.33)}$$

and the expected value of any function of the random variable X, say $g(X)$, can be defined by

$$E[g(X)] = \int_{-\infty}^{\infty} g(x) f(x) \, dx \qquad \text{(A 2.1.34)}$$

In particular, taking $g(X) = X$, we get the definition of mean μ_x

$$\mu_x = E(X) = \int_{-\infty}^{\infty} x f(x) \, dx \qquad \text{(A 2.1.35)}$$

Taking $g(X) = (X - \mu_x)^2$, we get the definition of variance

$$\sigma_x^2 = E(X - \mu_x)^2 = \int_{-\infty}^{\infty} (x - \mu_x)^2 f(x) \, dx \qquad \text{(A 2.1.36)}$$

and taking $g(X, Y) = (X - \mu_x)(Y - \mu_y)$, we get the definition of covariance

$$\gamma_{x,y} = E(X - \mu_x)(Y - \mu_y) \qquad \text{(A 2.1.37)}$$
$$= \int_{-\infty}^{\infty} \int_{-\infty}^{\infty} (x - \mu_x)(y - \mu_y) f(x,y) \, dx dy$$

where $f(x,y)$ is the joint probability density of random variables X, Y. When X and Y are independent, the joint density is of the product form, say $f(x)f(y)$, and then their covariance (and therefore correlation) is zero:

$$\gamma_{x,y} = E(X - \mu_x)(Y - \mu_y) \qquad \text{(A 2.1.38)}$$
$$= \int_{-\infty}^{\infty} (x - \mu_x)f(x) \, dx \int_{-\infty}^{\infty} (y - \mu_y)f(y) \, dy = 0$$

since each integral is zero by (A 2.1.33) and (A 2.1.35).

The actual expressions for $f(x)$ or $f(x,y)$ will not be required in this book because we deal with their first two "moments" defined by (A 2.1.35) to (A 2.1.37). The probability computation such as (A 2.1.32) is done with the help of tables as explained below. In particular, the normal distribution is completely defined by these two moments. Note that (A 2.1.36) is a special case of (A 2.1.37) when Y is replaced by X.

The most useful property of the normal distribution is that a linear combination of normally distributed variables is again normally distributed, with mean and variance given by (A 2.1.12) and (A 2.1.13). Thus, if

$$X \sim N(\mu_x, \sigma_x^2), \quad Y \sim N(\mu_y, \sigma_y^2)$$

then,

$$aX + bY \sim N[(a\mu_x + b\mu_y), (a^2\sigma_x^2 + 2ab\sigma_{xy} + b^2\sigma_y^2)] \quad \text{(A 2.1.39a)}$$

In particular, if X and Y are independent, so that $\text{Cov}(X, Y) = \gamma_{xy} = \sigma_{xy} = 0$, then

$$aX + bY \sim N[(a\mu_x + b\mu_y), (a^2\sigma_x^2 + b^2\sigma_y^2)] \qquad \text{(A 2.1.39)}$$

and these results can be extended to any number of variables.

These fundamental results can be used to facilitate the Normal probability computation by a single table. Consider the following transformation of X:

$$Z = \frac{X - \mu_x}{\sigma_x} = \frac{\text{Variable } - \text{ Mean}}{\text{Standard Deviation}} \qquad \text{(A 2.1.40)}$$

Now

$$Z = \frac{1}{\sigma_x}X - \frac{\mu_x}{\sigma_x}$$

is a linear combination of two independent normally distributed random variables: X with mean μ_x, variance σ_x^2, and the degenerate random variable, say μ_x/σ_x, with mean μ_x/σ_x and variance zero. If we take $a = 1/\sigma_x$ and $b = -1$ and apply (A 2.1.39)

$$\mu_z = E(Z) = \frac{\mu_x}{\sigma_x} - \frac{\mu_x}{\sigma_x} = 0 \qquad \text{(A 2.1.41}a\text{)}$$

$$\sigma_z^2 = E(Z - \mu_z)^2 = E(Z^2)$$

$$= \frac{\sigma_x^2}{\sigma_x^2} + (-1)^2 \times 0 = 1 \qquad \text{(A 2.1.41}b\text{)}$$

so that

$$Z \sim N(0,1) \qquad \text{(A 2.1.41}c\text{)}$$

Thus, Z has a normal distribution with mean zero and variance one, known as a standard normal distribution. The cumulative probabilities $\Phi(z) = P(-\infty < Z \leqslant z)$ for various values of z up to two decimal places are given in Table A (Appendix II). The values most used in practice are

$$\Phi(1.65) = 0.95053,$$
$$\Phi(1.96) = 0.975, \qquad \text{(A 2.1.42)}$$
$$\Phi(2.58) = 0.99506$$

Considering the symmetry of the standard normal distribution, we can see that

$$P(-1.65 \leqslant Z \leqslant 1.65) = \Phi(1.65) - \Phi(-1.65)$$
$$= \Phi(1.65) - [1 - \Phi(1.65)] \qquad \text{(A 2.1.43)}$$
$$\approx 0.90 = 90\%$$

Similarly,

$$P(-1.96 \leqslant Z \leqslant 1.96) = 0.95 = 95\% \qquad \text{(A 2.1.44)}$$
$$P(-2.58 \leqslant Z \leqslant 2.58) = 0.99 = 99\% \qquad \text{(A 2.1.45)}$$

In other words, 90%, 95%, and 99% probability limits on Z are ± 1.65,

± 1.96 and ± 2.58, respectively. Since by (A 2.1.40)

$$X = \mu_x + Z\sigma_x \qquad \text{(A 2.1.46)}$$
$$= \text{Mean} + Z \times \text{Standard Deviation}$$

90%, 95%, and 99% probability limits on X are

$$\mu_x \pm 1.65\sigma_x, \qquad \mu_x \pm 1.96\sigma_x, \qquad \text{and} \quad \mu_x \pm 2.58\sigma_x \quad \text{(A 2.1.47)}$$

respectively. In general, $p \times 100\%$ probability limits are

$$\mu_x \pm z_{\frac{p+1}{2}}\sigma_x \qquad \text{(A 2.1.48)}$$

where

$$\Phi(z_{\frac{p+1}{2}}) = p + \frac{1-p}{2} = \frac{p+1}{2} \qquad \text{(A 2.1.49)}$$

Thus, if we know the mean μ_x and variance σ_x^2 of a normally distributed variable, arbitrary probability limits can be found from (A 2.1.48) using Table A to find $z_{(p+1)/2}$ defined by (A 2.1.49). For example, 80% probability limits can be found by noting from Table A that

$$\Phi(z_{\frac{p+1}{2}}) = \frac{0.8 + 1}{2} = 0.9$$

gives $z_{(p+1)/2} = 1.28$ so that 80% probability limits are

$$\mu_x \pm 1.28\sigma_x$$

It is now clear [see (A 2.1.22) to (A 2.1.23)] that the simple regression model

$$\dot{y}_t = \beta_0 + \beta_1\dot{x}_t + \varepsilon_t, \qquad \varepsilon_t \sim \text{NID}(0, \sigma_\varepsilon^2)$$

is equivalent to

$$\dot{y}_t \sim \text{NID}(\beta_0 + \beta_1\dot{x}_t, \sigma_\varepsilon^2) \qquad \text{(A 2.1.50)}$$

and the simpler

$$y_t = \beta_1 x_t + \varepsilon_t, \qquad \varepsilon_t \sim \text{NID}(0, \sigma_\varepsilon^2)$$

is equivalent to

$$y_t \sim \text{NID}(\beta_1 x_t, \sigma_\varepsilon^2) \qquad \text{(A 2.1.51)}$$

If we assume that the parameters β_0, β_1, and σ_ε^2 are known, the forecasts or predictions in both cases are the means $\beta_0 + \beta_1\dot{x}_t$ or $\beta_1 x_t$ and, say, 95% probability limits on the forecasts are given by

$$\beta_0 + \beta_1\dot{x}_t \pm 1.96\,\sigma_\varepsilon \qquad \text{and} \qquad \beta_1 x_t \pm 1.96\,\sigma_\varepsilon$$

which was used in (2.1.6) of the text.

However, in practice neither the mean nor the variance are known. Suppose σ_x^2 is known but not μ_x; then, the average $\overline{X}$ was shown to be a good estimator of the mean μ_x. Now we can make a probability statement about how far $\overline{X}$ is from μ_x. Since $\overline{X}$ is a linear combination of normally distributed independent variables $X_1, X_2, \ldots, X_N$, it follows from (A 2.1.20), (A 2.1.21a), and (A 2.1.39) that

$$\overline{X} \sim N\left(\mu_x, \frac{\sigma_x^2}{N}\right) \tag{A 2.1.52}$$

and if we take 95% for illustration

$$P(\mu_x - 1.96\ \sigma_x/\sqrt{N} \leqslant \overline{X} \leqslant \mu_x + 1.96\ \sigma_x/\sqrt{N}) = 0.95 = 95\%$$

that is,

$$P(\overline{X} - 1.96\ \sigma_x/\sqrt{N} \leqslant \mu_x \leqslant \overline{X} + 1.96\ \sigma_x/\sqrt{N}) = 0.95 = 95\%$$

This is implied in the statement that the 95% confidence interval on the mean μ_x is

$$\overline{x} \pm 1.96\ \sigma_x/\sqrt{N}$$

and similarly, 90% and 99% confidence intervals are

$$\overline{x} \pm 1.65\ \sigma_x/\sqrt{N}, \qquad \overline{x} \pm 2.58\ \sigma_x/\sqrt{N}$$

Similarly, the least square estimators $\hat{\beta}_0$ and $\hat{\beta}_1$ are linear combinations of independent normal variables $\dot{y}_1, \dot{y}_2, \ldots, \dot{y}_N$ and therefore, if we use (A 2.1.24) to (A 2.1.27)

$$\hat{\beta}_0 \sim N\left[\beta_0, \sigma_\varepsilon^2\left(\frac{1}{N} + \frac{\overline{x}^2}{\Sigma x_t^2}\right)\right] \tag{A 2.1.53}$$

$$\hat{\beta}_1 \sim N\left(\beta_1, \frac{\sigma_\varepsilon^2}{\Sigma x_t^2}\right) \tag{A 2.1.54}$$

Therefore, 95% confidence intervals for β_0 and β_1 are, respectively,

$$\hat{\beta}_0 \pm 1.96\ \sigma_\varepsilon \sqrt{\frac{1}{N} + \frac{\overline{x}^2}{\Sigma x_t^2}} \tag{A 2.1.55}$$

$$\hat{\beta}_1 \pm 1.96\ \frac{\sigma_\varepsilon}{\sqrt{\Sigma x_t^2}} \tag{A 2.1.56}$$

Another important property of the normal distribution is the central limit tendency, by which an average or weighted average of a large number of random variables tends to be normally distributed even when the variables are not. In other words, even when X or y_t have a distribution different from normal but satisfying some mild restrictions, it can be shown that $\overline{X}$ or estimators such as $\hat{\beta}_0$ and $\hat{\beta}_1$ tend to be normally distributed.

Using such methods, it can be shown that as $N \to \infty$, $\hat{\phi}_1$ or generally $\hat{\rho}_k$ has normal distribution with mean ϕ_1 or ρ_k. Therefore, for finite N, the approximate 95% confidence intervals may be taken as, by (A 2.1.28) for an adequate AR(1) model,

$$\hat{\phi}_1 \pm \frac{1.96 \, \sigma_a}{\sqrt{\Sigma X_{t-1}^2}} \qquad \text{(A 2.1.57)}$$

or

$$\hat{\phi}_1 \pm 1.96 \sqrt{(1 - \hat{\phi}_1^2)/N}$$

and when $\phi_1 = 0$ or $\rho_k = 0$, by (A 2.1.29) and (A 2.1.30),

$$\pm \frac{1.96}{\sqrt{N}}$$

with the values 1.65 for 90%, 2.58 for 99%, and in general $z_{(p+1)/2}$ defined by (A 2.1.49) for $p \times 100\%$. A rigorous proof of these results and the exact distribution for small and moderate N are somewhat tedious and the reader is referred to statistical texts such as Anderson (1971).

These results are particularly helpful as an approximate check on the hypothesis that $\phi_1 = 0$ or $\rho_k = 0$. The hypothesis is rejected at a 5% level of significance (that is, there is a 5% risk that such rejection is wrong or ϕ_1/ρ_k is actually zero but we wrongly conclude that it is not) if $|\hat{\phi}_1|$ or $|\hat{\rho}_k|$ is greater than $1.96/\sqrt{N}$. This provides a quick check on the independence of the data or on the residuals from a fitted model when used to compute $\hat{\rho}_k$. This is, however, a check on individual ρ_k. For independent data *all* ρ_k should be zero, and a check for this is based on Chi-square distribution which we now discuss.

A 2.1.5 Chi-Square and *t*-Distributions

The Chi-square (χ^2) distribution is a name given to the distribution of the sum of squares of independent standard normal variables. If $Z_1, Z_2, \ldots$, Z_v are *independent* standard normal variables, then χ_v^2 denotes their sum of squares, that is,

$$\chi_v^2 \equiv Z_1^2 + Z_2^2 + \ldots + Z_v^2, \qquad Z_i \sim \text{NID}(0,1) \qquad \text{(A 2.1.58)}$$

where Z is defined by (A 2.1.40). It follows from the definition of Z that χ_v^2 may also be more generally defined by

$$\chi_v^2 \equiv \frac{(X_1 - \mu_1)^2}{\sigma_1^2} + \frac{(X_2 - \mu_2)^2}{\sigma_2^2} + \ldots + \frac{(X_v - \mu_v)^2}{\sigma_v^2},$$
$$X_i \sim \text{NID}(\mu_i, \sigma_i^2) \qquad \text{(A 2.1.58a)}$$

It follows immediately from (A 2.1.41b) that the mean of χ_ν^2 is

$$E(\chi_\nu^2) = E(Z_1^2) + E(Z_2^2) + \ldots + E(Z_\nu^2) = \nu \qquad \text{(A 2.1.59)}$$

The integer ν, which is the number of *independent* standard normal variables required for χ_ν^2, and which is also the mean of χ_ν^2, is called the *"degrees of freedom"* of χ_ν^2. If we take $\sigma_i^2 = \sigma^2$ in (A 2.1.58a), it follows that the sum of squares of ν independent normal variables with mean zero and variance σ^2 is distributed as $\sigma^2\chi_\nu^2$, that is,

$$\sum_{i=1}^{\nu} Y_i^2 \sim \sigma^2\chi_\nu^2, \qquad Y_i \sim \text{NID}(0,\sigma^2) \qquad \text{(A 2.1.60)}$$

$$E\sum_{i=1}^{\nu} Y_i^2 = E(\sigma^2\chi_\nu^2) = \nu\sigma^2 \qquad \text{(A 2.1.61)}$$

a result that is often useful in practice. Although the probability density function of χ_ν^2 can be found from (A 2.1.31) using the definition (A 2.1.58), it will not be required in this book. All the probability computation can be done by using Table B of cumulative χ_ν^2 probabilities given in Appendix II.

The Chi-square distribution is needed when the variance is unknown. Using χ^2, we can show that the estimator $\hat{\sigma}_x^2$ defined by (A 2.1.17), simply replacing the expectation by averaging, turns out to be biased (unlike $\overline{X}$), and we can modify it to make it unbiased.

$$\hat{\sigma}_x^2 = \frac{1}{N}\Sigma(X_i - \overline{X}), \qquad N\hat{\sigma}_x^2 = \Sigma(X_i - \overline{X})^2$$

Thus, $N\hat{\sigma}_x^2$ is a sum of squares of zero mean normal variables $X_1 - \overline{X}$, $X_2 - \overline{X}, \ldots, X_N - \overline{X}$, of which only $N-1$ are independent because of the restriction

$$\Sigma(X_i - \overline{X}) = \Sigma X_i - N\overline{X} = 0$$

from the definition of $\overline{X}$. Therefore, by (A 2.1.60)

$$N\hat{\sigma}_x^2 \sim \sigma_x^2\chi_{N-1}^2, \qquad \text{or} \qquad \frac{N\hat{\sigma}_x^2}{\sigma_x^2} \sim \chi_{N-1}^2 \qquad \text{(A 2.1.62)}$$

where

$$\nu = N-1$$

that is,

$$\frac{\text{Degrees of}}{\text{Freedom}} = \frac{\text{\# of}}{\text{Independent}} - \frac{\text{\# of}}{\text{Parameter}} \qquad \text{(A 2.1.63)}$$
$$\text{Observations} \qquad \text{Estimators}$$

In other words, we lost 1 degree of freedom because we used one estimator $\overline{X}$ in place of the parameter μ_x. Note that by (A 2.1.60)

$$\Sigma(X_i - \mu_x)^2 \sim \sigma_x^2 \chi_N^2$$

which has the full N degrees of freedom because known μ_x is used.

By (A 2.1.61)

$$E(N\hat{\sigma}_x^2) = E\Sigma(X_i - \overline{X})^2 = (N-1)\sigma_x^2 \qquad \text{(A 2.1.64)}$$

or

$$E(\hat{\sigma}_x^2) = \frac{N-1}{N}\sigma_x^2 \qquad \text{(A 2.1.64a)}$$

showing that $\hat{\sigma}_x^2$ is biased and *underestimates* σ_x^2. We can correct the bias by using a new estimator

$$s^2 = \frac{N}{N-1}\hat{\sigma}_x^2 = \frac{1}{N-1}\Sigma(X_i - \overline{X})^2 \qquad \text{(A 2.1.65)}$$

for which

$$Es^2 = \sigma_x^2$$

so that it is an unbiased estimator of σ_x^2. Also,

$$\Sigma(X_i - \overline{X})^2 = (N-1)s^2 \sim \sigma_x^2\chi_{N-1}^2,$$

or

$$\frac{(N-1)s^2}{\sigma_x^2} \sim \chi_{N-1}^2 \qquad \text{(A 2.1.66)}$$

A similar argument can be used when the variance σ_ε^2 of $\dot{y}_t$ in a simple regression model is unknown. Just like $\hat{\sigma}_x^2$, the least square estimator $\hat{\sigma}_\varepsilon^2$ defined by (2.1.3) in the text turns out to be biased. If we rewrite it in terms of original data $\dot{y}_t$ using (A 2.1.4)

$$\hat{\sigma}_\varepsilon^2 = \frac{1}{N}\Sigma(y_t - \hat{\beta}_1 x_t)^2$$

$$= \frac{1}{N}\Sigma(\dot{y}_t - \overline{y} - \hat{\beta}_1\dot{x}_t + \beta_1\overline{x})^2$$

$$= \frac{1}{N}\Sigma(\dot{y}_t - \hat{\beta}_0 - \hat{\beta}_1\dot{x}_t)^2 \qquad \text{(A 2.1.67)}$$

Thus, $N\hat{\sigma}_\varepsilon^2$ is a sum of squares normally distributed zero mean random variables from N independent observations with *two* parameters β_0 and β_1 replaced by their estimates. Therefore, it has $N-2$ degrees of freedom by (A 2.1.63) and

$$N\hat{\sigma}_\varepsilon^2 \sim \sigma_\varepsilon^2\chi_{N-2}^2 \qquad \text{(A 2.1.68)}$$

so that by (2.1.61)

$$EN\hat{\sigma}_\varepsilon^2 = E\Sigma(\dot{y}_t - \hat{\beta}_0 - \hat{\beta}_1\dot{x}_t)^2 = (N-2)\sigma_\varepsilon^2$$

or

$$E\hat{\sigma}_\varepsilon^2 = \frac{N-2}{N}\sigma_\varepsilon^2 \qquad (A\ 2.1.69)$$

showing that it is biased, underestimating σ_ε^2. An unbiased estimator would be

$$s_R^2 = \frac{N}{N-2}\hat{\sigma}_\varepsilon^2 = \frac{1}{N-2}\Sigma(\dot{y}_t - \hat{\beta}_0 - \hat{\beta}_1\dot{x}_t)^2 \qquad (A\ 2.1.70)$$

and

$$Es_R^2 = \sigma_\varepsilon^2$$

with

$$\Sigma(\dot{y}_t - \hat{\beta}_0 - \hat{\beta}_1\dot{x}_t)^2 = (N-2)s_R^2 \sim \sigma_\varepsilon^2\chi_{N-2}^2,$$

or

$$\frac{(N-2)s_R^2}{\sigma_\varepsilon^2} \sim \chi_{N-2}^2 \qquad (A\ 2.1.71)$$

Thus, s_R^2 should be used rather than $\hat{\sigma}_\varepsilon^2$ if an unbiased estimate is desired.

The Chi-square distribution can be used to provide an approximate overall check on many $\hat{\rho}_k$'s simultaneously, to see whether the residuals from which they are estimated are independent and the fitted model is adequate. As indicated in the last section, if the data are independent so that $\rho_k \equiv 0$, then for large N, approximately,

$$\hat{\rho}_k \sim N\left(0, \frac{1}{N}\right)$$

If $\hat{\rho}_k$'s are independent for different k, the degrees of freedom for the residuals from an ARMA (n,m) model can be estimated by applying (A 2.1.58a) to $\hat{\rho}_k$ and using (A 2.1.63),

$$\sum_{k=1}^{K}\frac{\hat{\rho}_k^2}{\text{Var}(\hat{\rho}_k)} = \sum_{k=1}^{K}\frac{\hat{\rho}_k^2}{1/N} \qquad (A\ 2.1.72)$$

$$= N\sum_{k=1}^{K}\hat{\rho}_k^2 \sim \chi_{K-n-m}^2$$

for the residuals from an ARMA(n,m) model. Actually, the distribution of $\hat{\rho}_k$ used above is approximate and $\hat{\rho}_k$'s are not independent for different k. However, checks based on (A 2.1.72) may be used as a quick supplementary criterion for checking independence of a_t's, as discussed in Section 4.4.4.

When both μ_x and σ_x^2 are unknown, we cannot use the 95% limits $\bar{x} \pm 1.96\sigma_x/\sqrt{N}$ for μ_x. It would seem natural to replace σ_x by s, the unbiased estimator. However, note that the number 1.96 is the 97.5% point of Z distribution and the limits $\bar{x} \pm 1.96\sigma_x/\sqrt{N}$ result from [see (A 2.1.52)]

$$Z = \frac{\overline{X} - \mu_x}{\sigma_x/\sqrt{N}} \sim N(0,1)$$

which is no longer true when σ_x is replaced by s, a random variable. Thus, we need the distribution of $(\overline{X} - \mu_x)/s/\sqrt{N}$, which can be expressed in terms of known distributions by

$$\frac{\overline{X} - \mu_x}{s/\sqrt{N}} \sim \frac{(\overline{X} - \mu_x)}{\sigma_x/\sqrt{N}} \Big/ \frac{s}{\sigma_x} \sim \frac{Z}{\sqrt{\dfrac{(N-1)s^2}{\sigma_x^2}} \Big/ (N-1)}$$

$$\sim \frac{\text{Standard Normal Variable}}{\sqrt{\chi_{N-1}^2/(N-1)}} \sim t_{N-1}$$

$$(A\ 2.1.73)$$

where it can be shown that $\overline{X}$ is independent of s^2. In general, the ratio of two independent random variables, a standard normal with the square root of a Chi-square divided by its degrees of freedom, is called a t-distribution, that is,

$$t_\nu = \frac{Z}{\sqrt{\chi_\nu^2/\nu}}, \quad Z \text{ independent of } \chi_\nu^2 \qquad (A\ 2.1.74)$$

Thus, the confidence intervals are now obtained using (A 2.1.73) by replacing $1.96 = z_{0.975}$ with $t_{N-1,0.975}$, the 0.975 cumulative probability point from a t-table, and using s in place of σ_x, that is,

$$\bar{x} \pm t_{N-1,0.975} s/\sqrt{N}$$

Similarly, in the simple linear regression model, the forecast for a given $x = (\dot{x} - \bar{x})$ is

$$\hat{y} = \hat{\beta}_0 + \hat{\beta}_1\dot{x} = \bar{y} + \hat{\beta}_1 x$$

and its variance is defined as the variance of the forecast error $\dot{y} - \hat{y}$. Note that $\dot{y}$, being an unknown observation (not contained in the data used for estimating $\hat{\beta}_0$ and $\hat{\beta}_1$ in $\hat{y}$) is independent of $\hat{y}$. Therefore, when

σ_ε^2 is known

$$
\begin{aligned}
\mathrm{Var}(\dot{y} - \hat{y}) &= \mathrm{Var}(\dot{y}) + \mathrm{Var}(\hat{y}) \\
&= \sigma_\varepsilon^2 + \mathrm{Var}(\bar{y} + \hat{\beta}_1 x) \\
&= \sigma_\varepsilon^2 + \mathrm{Var}(\bar{y}) + \mathrm{Var}(\hat{\beta}_1 x) \\
&= \sigma_\varepsilon^2 \left(1 + \frac{1}{N} + \frac{x^2}{\Sigma x_t^2} \right)
\end{aligned}
\qquad (A\ 2.1.75)
$$

since it can be shown that $\bar{y}$ is independent of $\hat{\beta}_1$. Therefore, when σ_ε^2 is known but β_0 and β_1 are unknown so that their estimated values $\hat{\beta}_0$ and $\hat{\beta}_1$ are used in forecasting, 95% probability limits for the unknown observation are

$$
\hat{y} \pm 1.96\sigma_\varepsilon \sqrt{1 + \frac{1}{N} + \frac{x^2}{\Sigma x_t^2}}
\qquad (A\ 2.1.76)
$$

When σ_ε^2 is unknown, we can use s_R, defined by (A 2.1.70) and (A 2.1.71), in place of σ_ε, replace $z_{0.975} = 1.96$ by $t_{N-2,0.975}$, and write the 95% probability limits as

$$
\hat{y} \pm t_{N-2,0.975}\, s_R \sqrt{1 + \frac{1}{N} + \frac{x^2}{\Sigma x_t^2}}
\qquad (A\ 2.1.77a)
$$

or

$$
\hat{\beta}_0 + \hat{\beta}_1 \dot{x} + t_{N-2,0.975}\, \hat{\sigma}_\varepsilon \sqrt{\frac{N}{N-2}} \sqrt{1 + \frac{1}{N} + \frac{x^2}{\Sigma x_t^2}}
\qquad (A\ 2.1.77)
$$

or

$$
\bar{y} + \hat{\beta}_1 x + t_{N-2,0.975}\, \hat{\sigma}_\varepsilon \sqrt{\frac{N}{N-2}} \sqrt{1 + \frac{1}{N} + \frac{x^2}{\Sigma x_t^2}}
\qquad (A\ 2.1.77b)
$$

As the number of observations $N \to \infty$, $\Sigma x_t^2 \to \infty$, $N/(N-2) \to 1$, $t_{N-2,0.975} \to z_{0.975} = 1.96$, and (A 2.1.77) becomes

$$
\hat{\beta}_0 + \hat{\beta}_1 \dot{x} \pm 1.96\hat{\sigma}_\varepsilon \equiv \bar{y} + \hat{\beta}_1 x \pm 1.96\hat{\sigma}_\varepsilon
$$

which is the same as assuming estimated $\hat{\beta}_0$, $\hat{\beta}_1$, and $\hat{\sigma}_\varepsilon$ to be true values and using standard normal Table A for finding probability limits as discussed in Section A 2.1.4; compare (A 2.1.50) and (A 2.1.51). For example, it is seen from Table C at the end of the book that for $N = 122$,

$t_{N-2,0.975} = t_{120,0.975} = 1.98$, and

$$1.98 \times \sqrt{\frac{122}{120}} \sqrt{1 + \frac{1}{122}} = 2.005$$

Thus, the assumption of known parameters underestimates the probability limit by about 2%. This justifies, via the conditional regression set-up for the AR(1) model, using the 95% probability limits for the forecast of $\dot{X}_{t+1}$ at t as

$$\overline{X} + \hat{\phi}_1 X_t \pm 1.96\hat{\sigma}_a$$

in Section 2.2.5 of the text, rather than possibly using

$$\overline{X} + \hat{\phi}_1 X_t + t_{N-2,0.975}\, \hat{\sigma}_a \sqrt{\frac{N}{N-2}} \sqrt{1 + \frac{1}{N} + \frac{X_t^2}{\Sigma X_t^2}}$$

For small and moderate N, of course, use of t-Table C rather than normal Table A is desirable.

It should be pointed out that, although similar in appearance, (A 2.1.76) to (A 2.1.77b) are *not* confidence intervals but probability limits. Confidence intervals are intervals on a *parameter* such as μ_x, whereas probability limits in (A 2.1.76) to (A 2.1.77b) are intervals on a *random variable* $\dot{y}$ not yet observed. For 90% and 99% probability limits, we use $z_{0.95} = 1.65$ or $t_{N-2,0.95}$ and $z_{0.995} = 2.58$ or $t_{N-2,0.995}$ instead of $z_{0.975} = 1.96$ or $t_{N-2,0.975}$ used above for illustration. Note that

$$t_{N-2,0.975}\, \hat{\sigma}_\varepsilon \sqrt{\frac{N}{N-2}} \sqrt{1 + \frac{1}{N} + \frac{x^2}{\Sigma x_t^2}} > 1.96\hat{\sigma}_\varepsilon$$

shows that for the same 95% confidence, we pay the penalty of a larger interval (less accurate) for forecasts in using estimated values for unknown parameters, compared to known parameters.

In fact, for a given $\dot{x}$, a 95% *confidence interval* about the regression line for the *expected value of* $\dot{y}$, given by $\beta_0 + \beta_1 \dot{x}$, can be found using the variance of its estimator $\hat{y} = \hat{\beta}_0 + \hat{\beta}_1 \dot{x}$ derived in (A 2.1.75) as

$$\hat{\beta}_0 + \hat{\beta}_1 \dot{x} \pm 1.96\sigma_\varepsilon \sqrt{\frac{1}{N} + \frac{x^2}{\Sigma x_t^2}} \qquad \text{(A 2.1.78)}$$

when σ_ε^2 is known, and

$$\hat\beta_0 + \hat\beta_1 \dot{x} \pm t_{N-2,0.975}\, \hat\sigma_\varepsilon \sqrt{\frac{N}{N-2}} \sqrt{\frac{1}{N} + \frac{x^2}{\Sigma x_t^2}} \qquad \text{(A 2.1.79)}$$

when σ_ε^2 is unknown.

A 2.1.6 Analysis of Variance and *F*-Distribution for Model Adequacy Based on Residual Sum of Squares

Whether $\dot{y}_t$ depends on $\dot{x}_t$ significantly or is independent of $\dot{x}_t$ is most effectively determined by a statistical test of hypothesis $\beta_1 = 0$, by means of a technique called analysis of variance. This refers to the decomposition of the term $\Sigma(\dot{y}_t - \bar{y})^2$ in the variance estimator:

$$
\begin{aligned}
\dot{y}_t - \bar{y} &= \dot{y}_t - \hat\beta_0 - \hat\beta_1 \bar{x}, && \text{by (A 2.1.4)} \\
&= (\dot{y}_t - \hat\beta_0 - \hat\beta_1 \dot{x}_t) + \hat\beta_1 x_t \\
&= \text{Residual } (\hat\varepsilon_t) + \text{Slope term} && \text{(A 2.1.80)}
\end{aligned}
$$

It was shown in (A 2.1.68) to (A 2.1.70) that $s_R^2 = \Sigma(\dot{y}_t - \hat\beta_0 - \hat\beta_1 \dot{x}_t)^2 / (N-2)$ provides an unbiased estimator of residual variance σ_ε^2, *irrespective of whether* β_1 *is zero or not*. On the other hand, since $\hat\beta_1 \sim N(\beta_1, \sigma_\varepsilon^2 / \Sigma x_t^2)$ by (A 2.1.54)

$$
\begin{aligned}
E\Sigma(\hat\beta_1 x_t)^2 &= (\Sigma x_t^2) E\hat\beta_1^2 \\
&= (\Sigma x_t^2) E[(\hat\beta_1 - \beta_1) + \beta_1]^2 \\
&= (\Sigma x_t^2)[E(\hat\beta_1 - \beta_1)^2 + E\beta_1^2] \\
&= \sigma_\varepsilon^2 + \beta_1^2 \Sigma x_t^2 \qquad \text{(A 2.1.81)}
\end{aligned}
$$

Therefore, $\Sigma(\hat\beta_1 x_t)^2$ would tend to be larger when β_1 is not zero and smaller—of the same order as σ_ε^2—when β_1 is zero. In other words, the ratio $\Sigma(\hat\beta_1 x_t)^2 / s_R^2$ would be larger when $\beta_1 \neq 0$ and can be used to test the hypothesis that $\beta_1 = 0$. To determine how large the computed ratio should be, so that we can safely reject the hypothesis with a small risk such as 5% (called the level of significance), we need the probability distribution of the ratio, which is a random variable.

The distribution used for this purpose is the *F*-distribution, which is the distribution of a ratio of two independent Chi-squares divided by their respective degrees of freedom. Thus, the random variable F_{ν_1,ν_2} denotes

$$F_{\nu_1,\nu_2} \equiv \frac{\chi^2_{\nu_1}/\nu_1}{\chi^2_{\nu_2}/\nu_2}, \quad \chi^2_{\nu_1} \text{ independent of } \chi^2_{\nu_2} \qquad \text{(A 2.1.82)}$$

The values $F_{\nu_1,\nu_2,P}$ of F_{ν_1,ν_2}, for which the cumulative probability $P(0 < F_{\nu_1,\nu_2} \le F_{\nu_1,\nu_2,P}) = P = 0.95, 0.99$ and 0.999 are given in Table D in Appendix II. When the computed F value exceeds $F_{\nu_1,\nu_2,P}$, we reject the hypothesis, with a $(1-P) \times 100\%$ risk of being wrong; that is, there is a $(1-P) \times 100\%$ probability that the hypothesis is true but we reject it because $F > F_{\nu_1,\nu_2,P}$.

To develop the F-distribution for testing $\beta_1 = 0$, consider the sum of squares on both sides of (A 2.1.80). The cross product sum on the right-hand side vanishes because if we use (A 2.1.4)

$$2\Sigma(\dot{y}_t - \hat{\beta}_0 - \hat{\beta}_1\dot{x}_t)\hat{\beta}_1 x_t = 2\hat{\beta}_1\Sigma(\dot{y}_t - \bar{y} - \hat{\beta}_1 x_t)x_t$$
$$= 2\hat{\beta}_1\Sigma(y_t - \hat{\beta}_1 x_t)x_t = 0$$

by the definition of $\hat{\beta}_1 = \Sigma y_t x_t/\Sigma x_t^2$. Therefore, the analysis of variance and the distribution of each term under the "null" hypothesis $\beta_1 = 0$ are

$$\Sigma(\dot{y}_t - \bar{y})^2 = \Sigma(\dot{y}_t - \hat{\beta}_0 - \hat{\beta}_1\dot{x}_t)^2 + \hat{\beta}_1^2\Sigma x_t^2$$

$$\sigma_\varepsilon^2\chi^2_{N-1} \qquad \equiv \sigma_\varepsilon^2\chi_{N-2} \qquad\qquad + \sigma_\varepsilon^2\chi^2_1 \qquad \text{(A 2.1.83)}$$

$$A_1 \qquad\qquad = A_0 \qquad\qquad\qquad + (A_1 - A_0)$$

The result $\Sigma(\dot{y}_t - \hat{\beta}_0 - \hat{\beta}_1 \dot{x}_t)^2/\sigma_\varepsilon^2 \sim \chi^2_{N-2}$ was proved in (A 2.1.71). Since $\dot{y}_t \sim \text{NID}(\beta_0 + \beta_1\dot{x}_t, \sigma_\varepsilon^2)$, when $\beta_1 = 0$, $\dot{y}_t \sim \text{NID}(\beta_0, \sigma_\varepsilon^2)$ and therefore by (A 2.1.66), $\Sigma(\dot{y}_t - \bar{y})^2 \sim \sigma_\varepsilon^2\chi^2_{N-1}$. Also, if we divide (A 2.1.83) by σ_ε^2, since the left-hand side and the first term on the right-hand side are distributed as the sum of squares of $N-1$ and $N-2$ independent standard normal variables, it follows that $\hat{\beta}_1^2\Sigma x_t^2/\sigma_\varepsilon^2$ must be distributed as the sum of squares of $(N-2) - (N-1) = 1$ standard normal variables, that is, χ^2_1 and independent of the first term. Thus, the ratio

$$F = \frac{\hat{\beta}_1^2\Sigma x_t^2}{\Sigma(\dot{y}_t - \hat{\beta}_0 - \hat{\beta}_1\dot{x}_t)^2/N-2} \sim F(1, N-2) \qquad \text{(A 2.1.84)}$$

by the definition of F-distribution (A 2.1.82), where $F(\nu_1,\nu_2)$ denotes an F-distribution with ν_1 and ν_2 degrees of freedom.

The above testing of hypothesis $\beta_1 = 0$ can now be described in a slightly different way, so that it can be generalized to provide a check on model adequacy for the ARMA$(n, n-1)$ modeling strategy evolved in Sections 2.3 and 2.4 of the text. This test of hypothesis is equivalent to the question whether the two-parameter adequate model

$$\dot{y}_t = \beta_0 + \beta_1 \dot{x}_t + \varepsilon_t$$

is required or the one-parameter model, obtained by restricting $\beta_1 = 0$ in this model,

$$\dot{y}_t = \beta_0 + \varepsilon_t$$

suffices. For this restricted model $\hat{\beta}_0 = \bar{y}$, and therefore the term $A_1 = \Sigma(\dot{y}_t - \bar{y})^2$ in (A 2.1.83), is the residual sum of squares of this restricted, "lower order" model. The term $A_0 = \Sigma(\dot{y}_t - \hat{\beta}_0 - \hat{\beta}_1 \dot{x}_t)^2$ is the residual sum of squares of the unrestricted, two parameter, "higher order" model. The ratio (A 2.1.84) can be alternatively written as

$$F = \frac{A_1 - A_0}{1} \div \frac{A_0}{N-2} \sim F(1, N-2) \qquad \text{(A 2.1.84a)}$$

This can also be thought of as the test criterion for answering the question, for the average subtracted data, whether the model

$$y_t = \beta_1 x_t + \varepsilon_t$$

is required or the model obtained by restricting $\beta_1 = 0$

$$y_t = \varepsilon_t$$

suffices.

Now under the conditional least squares set-up, this criterion can be applied for the question whether, for the average subtracted data, the adequate AR(1) model

$$X_t = \phi_1 X_{t-1} + a_t$$

is required so that the data is dependent, or the model obtained by restricting $\phi_1 = 0$

$$X_t = a_t$$

suffices so that the data is independent. In this case, $A_1 = \Sigma(X_t - \bar{X})^2$ and $A_0 = \Sigma a_t^2$ of the fitted AR(1) model. If the computed value F from (A 2.1.84a) is greater than, say, $F_{1, N-2, 0.95}$, then we reject the hypothesis that $\phi_1 = 0$ and conclude that the data is dependent, at a 5% level of significance.

As an illustration, consider results for the papermaking data in Section 2.3.3. $A_1/(N-1) \simeq$ estimated variance of the series $= 0.5503$, and $A_0/$

$(N-1)$ = estimated variance of the residuals of the AR(1) = 0.094. Therefore, since $N = 160$, if we apply (2.1.84a)

$$F = \frac{0.5503 - 0.094}{0.094} \times (160 - 2) = 768.65$$

whereas from Table D at the end of the book

$$F_{1,\infty,0.95} = 3.84 \qquad \text{and even} \qquad F_{1,\infty,0.999} = 10.83$$

Since $F > F_{1,\infty,0.999}$, the hypothesis $\phi_1 = 0$ is rejected and the data is clearly dependent.

Note that the F-test given by (A 2.1.84) and (A 2.1.84a) can be applied only when the two-parameter regression model (or the AR(1) model) *is* adequate, since the entire derivation of F-distribution is based on this model as the correct model. Thus, for example, it is not strictly applicable for the sunspot data for which the AR(1) model is not adequate, and a higher order model, at least AR(2), is required. However, in such cases, the F value for the lower order models, before the adequate model is reached, is usually so large that the preceding models are rejected as desired. Thus, a reasonable procedure, with some modifications elaborated in Chapter 4, is to continue increasing the order of the model until *all* the three criteria for the residual a_t's are satisfied: F-test, individual $|\hat{\rho}_k| < 2/\sqrt{N}$, and $N\Sigma\rho_k^2 < \chi^2_{K-n-m}$ discussed earlier.

The F-test in the form (A 2.1.84a) can be readily generalized to more parameters. Thus, to test whether an adequate r-parameter model is required or the lower order model with $(r-s)$ parameters obtained by restricting s parameters to zero suffices, we can use (A 2.1.84a) with 1 replaced by s, 2 replaced by r, and A_1 and A_0 the residual sum of squares of the lower and higher order models with $(r-s)$ and r parameters, respectively. Its application for general ARMA models in a variety of ways is fully discussed in Section 4.4.

APPENDIX A 2.2

LEAST SQUARE ESTIMATES IN MULTIPLE REGRESSION

A 2.2.1 Model and Least Square Estimates

Results in Appendix A 2.1 can be generalized to the multiple regression model conveniently by the matrix notation used in Section 2.1.2 of the text. If we use the notation of (2.1.8), the multiple regression model

$$y_t = \beta_1 x_{t1} + \beta_2 x_{t2} + \ldots \beta_n x_{tn} + \varepsilon_t,$$

$$t = 1, 2, \ldots, N$$

$$\varepsilon_t \sim \text{NID}(0, \sigma_\varepsilon^2)$$

can be succinctly written as

$$\mathbf{Y} = \mathbf{X}\boldsymbol{\beta} + \boldsymbol{\varepsilon} \qquad \text{(A 2.2.1)}$$

where $\mathbf{Y}$, $\boldsymbol{\beta}$, and $\boldsymbol{\varepsilon}$ are column vectors of observations $(y_1, y_2, \ldots, y_N)'$, parameters $(\beta_1, \beta_2, \ldots, \beta_n)'$, and errors $(\varepsilon_1, \varepsilon_2, \ldots, \varepsilon_N)'$, with $'$ denoting the transpose of a matrix/vector and $\mathbf{X}$ is an $N \times n$ matrix with a typical element x_{ij}, $i = 1, 2, \ldots, N$, $j = 1, 2, \ldots, n$. Although this model was introduced in the text for the average subtracted data for convenience, it is general enough to represent a model for the original data as well. Thus, the simple regression model (A 2.1.1) is a special case of this model if we identify $\dot{y}_t$ with y_t, β_1 with β_0, β_2 with β_1, x_{t2} with $\dot{x}_t$ and $x_{t1} \equiv 1$.

Least square estimates are obtained by minimizing

$$\Sigma \varepsilon_t^2 = \boldsymbol{\varepsilon}'\boldsymbol{\varepsilon} = (\mathbf{Y} - \mathbf{X}\boldsymbol{\beta})'(\mathbf{Y} - \mathbf{X}\boldsymbol{\beta})$$

If $\partial f / \partial \boldsymbol{\beta}$ denotes a column vector $(\partial f / \partial \beta_1, \partial f / \partial \beta_2, \ldots, \partial f / \partial \beta_n)'$, then the least square estimates must satisfy

$$\frac{\partial \Sigma \varepsilon_t^2}{\partial \boldsymbol{\beta}} = \frac{\partial}{\partial \boldsymbol{\beta}} [\mathbf{Y}'\mathbf{Y} - \mathbf{Y}'\mathbf{X}\boldsymbol{\beta} - \boldsymbol{\beta}'\mathbf{X}'\mathbf{Y} + \boldsymbol{\beta}'\mathbf{X}'\mathbf{X}\boldsymbol{\beta}]$$

$$= -2\mathbf{X}'\mathbf{Y} + 2\mathbf{X}'\mathbf{X}\boldsymbol{\beta} = 0$$

which gives the normal equations

$$\mathbf{X}'\mathbf{X}\boldsymbol{\beta} = \mathbf{X}'\mathbf{Y} \qquad \text{(A 2.2.2)}$$

and therefore the least square estimates

$$\hat{\boldsymbol{\beta}} = (\mathbf{X}'\mathbf{X})^{-1} \mathbf{X}'\mathbf{Y} \qquad \text{(A 2.2.3)}$$

where the vector derivatives used above can be verified by writing the matrices in full.

To illustrate, consider the special case (A 2.1.1), in a slightly modified form, with $\beta_0 = \beta_0' - \beta_1 \bar{x}$,

$$\dot{y}_t = \beta_0' + \beta_1(\dot{x}_t - \bar{x}) + \varepsilon_t$$

$$= \beta_0' + \beta_1 x_t + \varepsilon_t, \qquad \Sigma x_t = \Sigma(\dot{x}_t - \bar{x}) = 0$$

If we apply the above results with $x_{t1} \equiv 1$, $x_{t2} \to x_t$, and $y_t \to \dot{y}$ [see (2.1.13)]

$$\mathbf{X}'\mathbf{X} = \begin{bmatrix} \Sigma x_{t1}^2 & \Sigma x_{t1} x_{t2} \\ \Sigma x_{t1} x_{t2} & \Sigma x_{t2}^2 \end{bmatrix} = \begin{bmatrix} N & \Sigma x_t \\ \Sigma x_t & \Sigma x_t^2 \end{bmatrix} = \begin{bmatrix} N & 0 \\ 0 & \Sigma x_t^2 \end{bmatrix}$$

$$\mathbf{X}'\mathbf{Y} = \begin{bmatrix} \Sigma x_{t1} y_t \\ \Sigma x_{t2} y_t \end{bmatrix} = \begin{bmatrix} \Sigma \dot{y}_t \\ \Sigma x_t \dot{y}_t \end{bmatrix}$$

so that

$$\begin{bmatrix} \hat{\beta}_0' \\ \hat{\beta}_1 \end{bmatrix} = \begin{bmatrix} N & 0 \\ 0 & \Sigma x_t^2 \end{bmatrix}^{-1} \begin{bmatrix} \Sigma \dot{y}_t \\ \Sigma x_t \dot{y}_t \end{bmatrix}$$

$$= \begin{bmatrix} 1/N & 0 \\ 0 & 1/\Sigma x_t^2 \end{bmatrix} \begin{bmatrix} \Sigma \dot{y}_t \\ \Sigma x_t \dot{y}_t \end{bmatrix} \tag{A 2.2.4}$$

which gives

$$\hat{\beta}_0' = \frac{1}{N} \Sigma \dot{y}_t = \bar{y} \qquad \text{that is,} \qquad \hat{\beta}_0 = \bar{y} - \hat{\beta}_1 \bar{x}$$

and

$$\hat{\beta}_1 = \frac{\Sigma x_t \dot{y}_t}{\Sigma x_t^2} = \frac{\Sigma x_t (\dot{y}_t - \bar{y})}{\Sigma x_t^2} = \frac{\Sigma x_t y_t}{\Sigma x_t^2}$$

as before in Section A 2.1.1.

A 2.2.2 Mean, Variance-Covariance Matrix of $\hat{\beta}$, and Confidence Intervals

By the model (A 2.2.1) $EY = X\beta$ and substituting in (A 2.2.3)

$$\begin{aligned} E(\hat{\beta}) &= E(X'X)^{-1} X'Y \\ &= (X'X)^{-1} X'EY \\ &= (X'X)^{-1} X'X\beta \\ &= \beta \end{aligned} \tag{A 2.2.5}$$

showing that the least square estimators are unbiased.

The variance-covariance matrix of a vector random variable is defined as the matrix with variances of each element along the diagonal and covariances as proper off-diagonal elements. Thus,

$$V(\hat{\beta}) = E[(\hat{\beta} - E(\hat{\beta}))(\hat{\beta} - E(\hat{\beta}))'] \tag{A 2.2.6}$$

$$= \begin{bmatrix} \text{Var}(\hat{\beta}_1) & \text{Cov}(\hat{\beta}_1, \hat{\beta}_2) & \cdots & \text{Cov}(\hat{\beta}_1, \hat{\beta}_n) \\ \text{Cov}(\hat{\beta}_2, \hat{\beta}_1) & \text{Var}(\hat{\beta}_2) & \cdots & \text{Cov}(\hat{\beta}_2, \hat{\beta}_n) \\ \cdots & & \cdots & \cdots \\ \text{Cov}(\hat{\beta}_n, \hat{\beta}_1) & \cdots & & \cdots & \text{Var}(\hat{\beta}_n) \end{bmatrix}$$

In particular, each element of the error vector ε has the same variance σ_ε^2 and all the elements are independent, so that their covariance is zero. Hence,

$$V(\varepsilon) = \sigma_\varepsilon^2 I$$

where I stands for the identity matrix. Also,

$$V(Y) = V(\varepsilon) = \sigma_\varepsilon^2 I$$

and it can be verified from the definition of the variance-covariance matrix given above that for any $m \times N$ matrix $\mathbf{A}$

$$V(\mathbf{AY}) = \mathbf{A}V(\mathbf{Y})\mathbf{A}'$$

If we apply these results to the least square estimator given by (A 2.2.3)

$$
\begin{aligned}
V(\hat{\boldsymbol{\beta}}) &= V[(\mathbf{X'X})^{-1}\mathbf{X'Y}] \\
&= [(\mathbf{X'X})^{-1}\mathbf{X'}]V(\mathbf{Y})[(\mathbf{X'X})^{-1}\mathbf{X'}]' \\
&= (\mathbf{X'X})^{-1}\mathbf{X'}\sigma_\varepsilon^2\mathbf{IX}(\mathbf{X'X})^{-1} \\
&= (\mathbf{X'X})^{-1}\sigma_\varepsilon^2 \qquad\qquad\qquad\qquad\text{(A 2.2.7)}
\end{aligned}
$$

where $\mathbf{X'X}$ as well as $(\mathbf{X'X})^{-1}$ are symmetric matrices, which is also clear from (A 2.2.6).

In view of the covariances between, say, $\hat{\beta}_1$ and $\hat{\beta}_2$, etc., one should consider a confidence region (n-dimensional) for the estimates $\hat{\beta}_1$, $\hat{\beta}_2$, . . . , $\hat{\beta}_n$. This is complicated. However, an easier procedure is to get confidence intervals on individual parameters β_i, using $\text{Var}(\hat{\beta}_i)$ along the diagonal of $V(\hat{\boldsymbol{\beta}})$, which do provide some idea of how close each $\hat{\beta}_i$ is to its true value β_i. If we assume σ_ε^2 is known and consider the normal distribution of ε_t, say, 95% confidence intervals are

$$\hat{\beta}_i \pm 1.96\sqrt{\text{Var}(\hat{\beta}_i)}, \qquad i = 1, 2, \ldots, n \qquad\text{(A 2.2.8)}$$

where $\hat{\beta}_i$ is the ith element of (A 2.2.3) and $\text{Var}(\hat{\beta}_i)$ is the ith diagonal element of the matrix $(\mathbf{X'X})^{-1}\sigma_\varepsilon^2$ by (A 2.2.6) and (A 2.2.7).

To illustrate, again consider the modified simple regression model, for which $(\mathbf{X'X})^{-1}$ matrix was obtained in (A 2.2.4). Therefore, if we use (A 2.2.7)

$$
V\begin{bmatrix}\hat{\beta}_0' \\ \hat{\beta}_1\end{bmatrix} = V\begin{bmatrix}\bar{y} \\ \hat{\beta}_1\end{bmatrix} = \begin{bmatrix}1/N & 0 \\ 0 & 1/\Sigma x_t^2\end{bmatrix}\sigma_\varepsilon^2
$$

Thus,

$$\text{Var}(\hat{\beta}_0') = \text{Var}(\bar{y}) = \sigma_\varepsilon^2/N$$

$$\text{Var}(\hat{\beta}_1) = \sigma_\varepsilon^2/\Sigma x_t^2$$

which were found earlier. However, in addition

$$\text{Cov}(\hat{\beta}_0', \hat{\beta}_1) = \text{Cov}(\bar{y}, \hat{\beta}_1) = 0$$

so that under normal assumption, $\bar{y}$ is independent of $\hat{\beta}_1$, a result which was repeatedly used earlier, for example, the derivation of (A 2.1.27) or (A 2.1.75), etc.

Results (A 2.2.7) and (A 2.2.8) can also be used to find confidence intervals on the parameters of an ARMA(n,m) model. Although the regression in this case is nonlinear, the model can be linearized around some initial guess values of parameters, new parameters can be estimated by (A 2.2.3), and iterations continued until convergence occurs. $(\mathbf{X}'\mathbf{X})^{-1}\hat{\sigma}_a^2$ matrix of the last iteration then provides the confidence intervals by (A 2.2.8).

A 2.2.3 Variance-Covariance Matrix of $\hat{\phi}_1$, $\hat{\phi}_2$ for AR(2) Model and Confidence Intervals

To illustrate the use of these results for ARMA models, let us consider the AR(2) model. It follows from (A 2.2.7) and the results for the AR(2) model given in Section 2.3.2 that for large N the variance-covariance matrix of $\hat{\phi}_1$, $\hat{\phi}_2$ is given by

$$V\begin{bmatrix} \hat{\phi}_1 \\ \hat{\phi}_2 \end{bmatrix} = (\mathbf{X}'\mathbf{X})^{-1}\sigma_a^2 = \begin{bmatrix} \Sigma X_t^2 & \Sigma X_t X_{t-1} \\ \Sigma X_t X_{t-1} & \Sigma X_t^2 \end{bmatrix}^{-1} \sigma_a^2 \quad \text{(A 2.2.9)}$$

This can be expressed in a simple form in terms of parameters, by arguments similar to those used in deriving (A 2.1.28), by replacing ΣX_t^2 and $\Sigma X_t X_{t-1}$ by their expected values: $N\text{Var}(X_t) = N\gamma_0$ and $N\text{Cov}(X_t, X_{t-1}) = N\gamma_1$, respectively. It can be shown [see Problem (3.17)] that for the AR(2) model

$$\gamma_0 = \frac{(1-\phi_2)\sigma_a^2}{(1+\phi_2)[(1-\phi_2)^2 - \phi_1^2]} \quad \text{(A 2.2.10)}$$

$$\gamma_1 = \frac{\phi_1}{(1-\phi_2)}\gamma_0 \quad \text{(A 2.2.11)}$$

Therefore,

$$V\begin{bmatrix} \hat{\phi}_1 \\ \hat{\phi}_2 \end{bmatrix} \simeq \begin{bmatrix} N\gamma_0 & N\gamma_1 \\ N\gamma_1 & N\gamma_0 \end{bmatrix}^{-1} \sigma_a^2$$

$$= \frac{\sigma_a^2}{N\gamma_0} \begin{bmatrix} 1 & \dfrac{\phi_1}{(1-\phi_2)} \\ \dfrac{\phi_1}{(1-\phi_2)} & 1 \end{bmatrix}^{-1}$$

$$= \frac{(1+\phi_2)[(1-\phi_2)^2 - \phi_1^2]}{N(1-\phi_2)} \frac{(1-\phi_2)^2}{[(1-\phi_2)^2 - \phi_1^2]}$$

$$\cdot \begin{bmatrix} 1 & \dfrac{-\phi_1}{(1-\phi_2)} \\ \dfrac{-\phi_1}{(1-\phi_2)} & 1 \end{bmatrix}$$

$$= \frac{1}{N}\begin{bmatrix} 1-\phi_2^2 & -\phi_1(1+\phi_2) \\ -\phi_1(1+\phi_2) & 1-\phi_2^2 \end{bmatrix} \quad \text{(A 2.2.12)}$$

from which the estimated variance-covariance matrix can be found by substituting the estimated $\hat{\phi}_1$, $\hat{\phi}_2$.

If we use the variances given by the diagonal elements of (A 2.2.12), say, 95% confidence intervals on ϕ_1 and ϕ_2 can be written as

$$\hat{\phi}_1 \pm 1.96 \sqrt{\frac{1-\phi_2^2}{N}}, \qquad \hat{\phi}_2 \pm 1.96 \sqrt{\frac{1-\phi_2^2}{N}}$$

Again, the hypothesis $\phi_2 = 0$ will be rejected at a 5% level of significance if $|\hat{\phi}_2| > 1.96/\sqrt{N}$.

It will be shown in Chapter 9 that seasonal and/or polynomial trends in the series lead to $\phi_2 \rightarrow -1$ for an AR(2) model, and $\phi_1 \rightarrow 1$ for an AR(1) model. Expressions (A 2.2.12) and (A 2.1.28) show that the variances of the parameter estimates (and therefore confidence intervals) tend to zero in such cases and the estimation situation is quite "favorable." Therefore, from the point of view of statistical estimation also, we do not need any modification in the ARMA($n,n-1$) modeling strategy when integrated ARMA models such as $\nabla X_t = a_t$ or $\nabla^2 X_t = a_t$ are appropriate, as stressed at the end of Section 2.4.3. Rather than try to guess the presence of such trends (requiring special operators to remove them) before modeling, it is quite straightforward and safe to let the estimated model parameters tell us about their presence or absence after modeling. This will be illustrated by many examples in Chapter 9.

PROBLEMS

2.1 The following data

$t =$	1	2	3	4	5	6	7	8
$\dot{X}_t =$	3.8	4.5	3.4	4.6	5.8	7.1	5.8	5.0

is fitted to an AR(1) model $[X_t = \phi_1 X_{t-1} + a_t, a_t \sim \text{NID}(0,\sigma_a^2)]$
(a) Plot X_t versus t.
(b) Find the estimate of ϕ_1.
(c) Compute the values of a_t, $t = 2, 3, \ldots, 8$.
(d) Plot X_t versus X_{t-1} and show the line $\hat{X}_t = \phi_1 X_{t-1}$ on the same figure.
(e) Compute $\hat{\sigma}_a^2$.
(f) Plot a_t versus a_{t-1}.
(g) Plot a_t versus X_{t-2}.
(h) Compute $\hat{\rho}(a_t, a_{t-1})$.
(i) Compute $\hat{\rho}(a_t, X_{t-2})$.

(j) Starting with $X_1 = \dot{X}_1 - \overline{X}$, generate two new series of data using $\phi_1 = 1.3$ and $\phi_1 = -1.3$ in the AR(1) model with the a_t's found in (c) and plot them on the graph of (a). How do these two new time series differ from the original one? (Do not forget to subtract the average.)

2.2 For a series with 300 observations, given that $\hat{\sigma}_a^2 = 10$, find the residual sum of squares, if the adequate model is (i) AR(1) and (ii) AR(2).

2.3 Using the results of Problem (2.1)
(a) Find the one step ahead prediction of $\dot{X}_9$ at time $t = 8$.
(b) Find the 95% probability limits for the one step ahead prediction of $\dot{X}_9$ at time $t = 8$.
(c) What is the conditional distribution of $\dot{X}_9$, given $\dot{X}_8 = 5.0$?
(d) If at time $t = 9$, the observed $\dot{X}_9$ is 2.7, find the one step ahead prediction error of $\dot{X}_9$ at time $t = 8$.

2.4 The model for IBM stock prices has been found to be AR(1) with
$$X_t = 0.999 X_{t-1} + a_t, \qquad a_t \sim NID(0, 52.61)$$
Suppose $\overline{X} = 478$
(a) If yesterday's stock price was 487, what is the best prediction of today's stock price?
(b) What are the 95% and 90% probability limits on this prediction?
(c) If today's stock price turns out to be 491, what is the prediction error?

2.5 For an AR(1) model, $X_t = 25$ and the 95% probability limits on $\hat{X}_t(1)$ are (15.5, 9.5). Find the values of ϕ_1 and σ_a^2 of the model.

2.6 For the following sets of data listed in Appendix I, assuming the AR(1) model, write a computer program and use the computer to
(a) Obtain estimates of ϕ_1 and σ_a^2.
(b) Plot a_t versus a_{t-1}.
(c) Plot a_t versus X_{t-2}.
      (1) Mechanical Vibratory Displacement Data
      (2) Grinding Wheel Profile

2.7 Fit an AR(2) model to the two sets of data in Problem (2.6).

2.8 For each of the two sets of data, state whether the AR(1) model is adequate in your opinion or the AR(2) model may be required. State your reasons.

2.9 Prove that for large $N \to \infty$
$$|\hat{\phi}_1| \leq 1$$

provided the series is stationary, that is, it has the same mean, variance, and covariance at all t. [*Hint:* $\Sigma(X_t \pm X_{t-1})^2 \geq 0$ or $\Sigma X_t^2 + \Sigma X_{t-1}^2 \pm 2\Sigma X_t X_{t-1} \geq 0$.]

2.10 (a) For the AR(1) model show that

$$\text{Var}(X_t) = \gamma_0 = \frac{\sigma_a^2}{(1 - \phi_1^2)}$$

[*Hint:* See (A 2.1.15).]

(b) Where did you use stationarity in the above proof?

(c) What is the variance of X_t for a random walk model?

2.11 A series with 200 observations has variance $\hat{\gamma}_0 = 10$. An AR(1) model with $\hat{\phi}_1 = 0.8$ has been fitted to this series. Assuming the model to be adequate, find its (i) $\hat{\sigma}_a^2$ and (ii) residual sum of squares.

2.12 Using the basic definitions of Appendix A 2.1.2, show that for a stationary time series with zero mean, the definition of *auto*correlation ρ_k between X_t and X_{t-k} takes the form

$$\text{Autocorrelation } (X_t, X_{t-k}) = \rho_k$$

$$= \frac{\text{Cov}(X_t, X_{t-k})}{\text{Var}(X_t)}$$

$$= \frac{\gamma_k}{\gamma_0} = \frac{EX_t X_{t-k}}{EX_t^2}$$

2.13 Using the data given for the illustrative example in Section 2.1.1, find

(a) $\hat{\beta}_0$ and an unbiased estimate of σ_ε^2.

(b) The variances of $\hat{\beta}_0$ and $\hat{\beta}_1$.

(c) The 95% confidence intervals on β_0, β_1 assuming (i) known σ_ε^2 and (ii) unknown σ_ε^2.

(d) The 95% *probability limits* for forecast $\hat{y}$ when $\dot{x} = 6$ assuming (i) known σ_ε^2 and (ii) unknown σ_ε^2.

(e) The 95% *confidence interval* about the regression line for the expected value of $\dot{y}$ assuming (i) known σ_ε^2 and (ii) unknown σ_ε^2 when $\dot{x} = 6$.

2.14 Repeat Problem (2.13c) to (2.13e) using 90% levels.

2.15 Assuming AR(1) to be an adequate model for (i) papermaking and (ii) IBM Stock Prices data, find the approximate

(a) variance of $\hat{\phi}_1$.

(b) 95% confidence interval on ϕ_1.

2.16 Find approximate 95% probability limits for one step ahead predictions of $\dot{X}_9$ at time $t=8$, using the results of Problem (2.1), but taking into account that $\hat{\phi}_1$ and $\hat{\sigma}_\varepsilon^2$ are estimated values with finite N (that is, do not assume them to be known). Compare with Problem (2.3b).

2.17 Assuming the AR(2) model to be adequate for the sunspot series, find
   (a) The variance-covariance matrix of $\hat{\phi}_1$, $\hat{\phi}_2$.
   (b) The 95% confidence intervals (individual) on ϕ_1 and ϕ_2.
   (c) The correlation of $\hat{\phi}_1$, $\hat{\phi}_2$.

2.18 (a) If inclusion of 1 in a $p \times 100\%$ confidence interval on ϕ_1 is taken as the criterion for treating a limiting AR(1) model as a random walk, show that $\hat{\phi}_1$ should be in the interval

$$\frac{N - Z_{(p+1)/2}^2}{N + Z_{(p+1)/2}^2} < \hat{\phi}_1 < 1$$

where N is the number of observations and $Z_{(p+1)/2}$ is the $(p+1)/2$ cumulative probability point of the standard normal distribution. When $\hat{\phi}_1$ lies in this interval let us say that the random walk model is justified at $(1 - p) \times 100\%$ "level of significance."
   (b) Is the random walk model for the IBM data in the text justified at 5% level of significance?
   (c) What is the smallest value of $\hat{\phi}_1$ at which the random walk is still justified at 5% level for the IBM data?
   (d) What is the highest level of significance at which the given (rounded off) value of $\hat{\phi}_1$ for the IBM data justifies the random walk model?
   (e) What other criterion and definition of the level of significance can you suggest for this purpose?

3

CHARACTERISTICS OF ARMA MODELS

In Chapter 2 the ARMA models were derived as a simple extension of the ordinary regression between two variables to the conditional regression expressing the dependence of one variable on its own past values. This extension was exploited to obtain the conditional least squares estimates of the parameters and one step ahead forecasts. It was also pointed out in that chapter that in their conditional regression aspects the ARMA models are similar to the ordinary regression models. However, this does not mean that the structure of the underlying systems is similar for these two kinds of models. We will see in this chapter that the ARMA systems have a rather deep structure compared to the ordinary regression models.

The principal difference stems from the fact that the regression systems are static whereas the ARMA systems are dynamic. A disturbance ε_t entering a regression system at time t affects only y_t but not y_{t+1}. By the time the system response proceeds from t to $t + 1$, the disturbance ε_t is "forgotten"; thus, the system has no "memory" or dynamics. Hence, the regression system has only a static dependence of one variable on another.

This static dependence is shared by the ARMA models in their conditional regression form. However, in their unconditional form the ARMA models also have dynamics or memory. A disturbance a_t affecting the system is "remembered" and continues to affect the system at subsequent times. It is this memory or dynamics which gives rise to the dependence in the data and is represented by an ARMA model. In this chapter we will discuss the characteristics of the system that describe the dynamics in different ways and therefore characterize the ARMA systems.

The first is the "Green's function," which describes the dynamics or the memory in terms of the past a_t's; it explains how the a_t's affect or influence the response X_t by expressing the response as a linear combination of a_t's. The Green's function imposes certain restrictions on the parameters of ARMA models called stability conditions. The second is the "inverse function," which describes the dynamics or memory in terms of the influence of past X_t's on the present X_t by decomposing X_t as a linear combination of past X_t's. The third is the autocovariance function or autocorrelation function, which is a statistical characterization of the dependence between the sequence of random variables $X_t, X_{t-1}, X_{t-2}, \ldots$. We will treat these three characteristics in detail in the first three sections since they will be repeatedly used in the sequel. In the fourth section we will briefly touch upon two more characteristics, namely the partial autocorrelation and spectrum, only for the sake of completeness; we will not need them further.

3.1 GREEN'S FUNCTION AND STABILITY

Consider the AR(1) model for the papermaking data in Chapter 2 that we will simply represent as

$$X_t = 0.9X_{t-1} + a_t$$

If we recursively substitute for X_{t-1}, X_{t-2}, etc., we have

$$
\begin{aligned}
X_t &= 0.9(0.9X_{t-2} + a_{t-1}) + a_t \\
&= (0.9)^2(0.9X_{t-3} + a_{t-2}) + 0.9a_{t-1} + a_t \\
&= (0.9)^3(0.9X_{t-4} + a_{t-3}) + (0.9)^2a_{t-2} + 0.9a_{t-1} + a_t \\
& \quad \cdots \\
&= a_t + 0.9a_{t-1} + (0.9)^2a_{t-2} + (0.9)^3a_{t-3} + \cdots
\end{aligned}
$$

For a general AR(1) model

$$X_t - \phi_1 X_{t-1} = a_t \tag{3.1.1}$$

we similarly have

$$
\begin{aligned}
X_t &= \phi_1(\phi_1 X_{t-2} + a_{t-1}) + a_t \\
&= a_t + \phi_1 a_{t-1} + \phi_1^2 a_{t-2} + \cdots \\
&= \sum_{j=0}^{\infty} \phi_1^j a_{t-j}
\end{aligned}
\tag{3.1.2}
$$

Equation (3.1.1) is a difference equation, with the left-hand side representing the homogeneous part whereas the right-hand side represents the "forcing function," to use the terminology of differential equations.

The solution of the first order difference equation (3.1.1) is given by Eq. (3.1.2), which expresses the solution as a linear combination of the values of the forcing function. The coefficient function ϕ_1^j in this expansion, which really represents the essential part of the solution, will be called the "Green's function" [see Miller (1968)]. If we denote the Green's function by G_j, then for an AR(1) model

$$G_j = \phi_1^j \tag{3.1.3}$$

and the solution (3.1.2) can be written as

$$X_t = \sum_{j=0}^{\infty} G_j a_{t-j} = \sum_{j=-\infty}^{t} G_{t-j} a_j \tag{3.1.4}$$

Note that

$$G_0 = 1.$$

An arbitrary ARMA model can be expressed in the form (3.1.4). The equivalence of the two forms of summations in Eq. (3.1.4) should be carefully noted. Either of these forms will be used in the sequel as per convenience. A numerical illustration of this equivalence will be given in Section 3.1.2. Note that the Green's function form (3.1.4) of an ARMA model is essentially an infinite order moving average model MA(∞).

3.1.1 Green's Function of the AR(1) System Using B Operator

Just as with differential equations, the manipulation of difference equations becomes much simpler by using operator methods. Following these methods, we will use the "backshift operator" B defined by

$$BX_t = X_{t-1}$$

and, in general,

$$B^j X_t = X_{t-j} \qquad (3.1.5)$$

The AR(1) model (3.1.1) can then be written as

$$(1 - \phi_1 B)X_t = a_t \qquad (3.1.1a)$$

The solution (3.1.2) and the Green's function can be readily obtained from this form (3.1.1a). It is easy to see, either by long division or by the sum of a geometric series, that

$$\begin{aligned}
X_t &= \frac{a_t}{(1 - \phi_1 B)} \\
&= (1 + \phi_1 B + \phi_1^2 B^2 + \phi_1^3 B^3 + \ldots)a_t \\
&= a_t + \phi_1 a_{t-1} + \phi_1^2 a_{t-2} + \ldots \\
&= \sum_{j=0}^{\infty} \phi_1^j a_{t-j} \\
&= \sum_{j=0}^{\infty} G_j a_{t-j}
\end{aligned}$$

3.1.2 Physical Interpretation

As a characteristic of the system dynamics, the Green's function can be interpreted in two ways. First, it is clear from the above expansion that G_j is the weight, given in the present response, to the shock or disturbance a_t, which entered the system j time units back. The quantity G_j indicates how well the system remembers the shocks a_{t-j}. The larger the value of

ϕ_1 in the AR(1) system, the more clearly is the shock a_{t-j} remembered for a fixed value of j, and also the more distant past is included in the memory. For the papermaking model considered above, the weight given to a_{t-6} is still as high as $(0.9)^6 = 0.532$, and the memory of a_{t-6} is quite strong. Whereas if the value of ϕ_1 was, say, 0.4, the weight G_6 would be only $(0.4)^6 = 0.004$ and a_{t-6} would be almost forgotten.

A second interpretation of the Green's function G_j is that it characterizes how slow or fast the dynamic response of the system to any particular a_t decays. In other words, if a single a_t is injected into the system, the Green's function determines how quickly the system will return to its equilibrium or mean position, which we always take to be zero. If the value of ϕ_1 is small, then the response decays very fast and the system quickly returns to its equilibrium position if undisturbed by other a_j's.

Figure 3.1 shows the generation of the response for an AR(1) model with $\phi_1 = 0.5$, which is also numerically illustrated in Table 3.1. The values of the Green's function given by ϕ_1^i are shown in the third row of Table 3.1. Values of a_t given in the second row of Table 3.1 are plotted in Fig. 3.1(a). $G_{t-j}a_j$ for $j=1, 2, \ldots, 14$ are shown from the 4th to 17th row in Table 3.1, with only $G_{t-j}a_j$ for $j=1, 2, \ldots, 6$ plotted in Fig. 3.1(b) to (g). By the second equality in (3.1.4) we have (assuming $a_t = 0$ for $t \le 0$):

$$X_t = \sum_{j=0}^{t} G_{t-j}a_j$$

$$= G_{t-0}a_0 + G_{t-1}a_1 + G_{t-2}a_2 + \ldots + G_0a_t$$

Hence, the sum of the values in each column from the 4th to 17th row, which is shown at the bottom of Table 3.1, is the value of X_t at time t, $t=1, 2, \ldots, 14$.

Graphically, the value of X_t, plotted in Fig. 3.1(h) is the sum of the values of $G_{t-j}a_j$, $j=0, 1, 2, \ldots, t$. For instance, for $t = 3$, the value of X_3 is obtained in Table 3.1 or Fig. 3.1 as follows: $(a_0 = 0)$

$$X_3 = [G_{t-1}a_1 + G_{t-2}a_2 + G_{t-3}a_3]_{t=3}$$

$$= G_2a_1 + G_1a_2 + G_0a_3$$

$$= 0 - 0.5 + 2$$

$$= 1.5$$

For $t = 4$

$$X_4 = [G_{t-1}a_1 + G_{t-2}a_2 + G_{t-3}a_3 + G_{t-4}a_4]_{t=4}$$

$$= G_3a_1 + G_2a_2 + G_1a_3 + G_0a_4$$

$$= 0 - 0.25 + 1 - 2$$

$$= -1.25$$

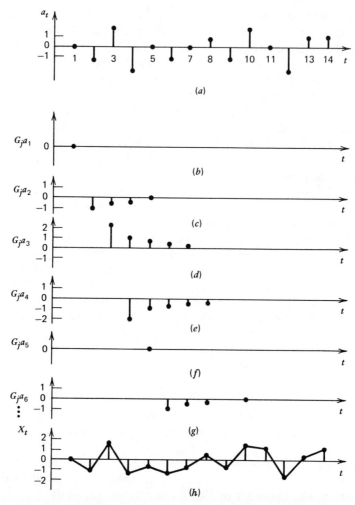

Figure 3.1   Graphical illustration of response generation by Green's function, AR(1) system: $\phi_1 = 0.5$.

which are the same as those directly obtained from the difference form of the model:

$$X_3 = 0.5X_2 + a_3 = 0.5(-1) + 2$$

$$= 1.5$$

$$X_4 = 0.5X_3 + a_4 = 0.5(1.5) + (-2)$$

$$= -1.25$$

Table 3.1 Numerical Illustration of Response Generation with $a_t = 0$ for $t \leqslant 0$: Second Equality of (3.1.4)

t		0	1	2	3	4	5	6	7	8	9	10	11	12	13	14
a_t		0	0	−1	2	−2	0	−1	0	1	−1	2	0	−2	1	1
G_t		1	.5	.25	.125	.0625	.0313	.0156	.0078	.0039	.0020	.0010	.0005	.0002	.0001	.0001
j																
0	$G_{t-0}a_0$	0	0	0	0	0	0	0	0	0	0	0	0	0	0	0
1	$G_{t-1}a_1$		0	0	0	0	0	0	0	0	0	0	0	0	0	0
2	$G_{t-2}a_2$			−1	−.5	−.25	−.125	−.0625	−.0313	−.0156	−.0078	−.0039	−.0020	−.0010	−.0005	−.0002
3	$G_{t-3}a_3$				2	1	.5	.25	.125	.0625	.0313	.0156	.0078	.0039	.0020	.0010
4	$G_{t-4}a_4$					−2	−1	−.5	−.25	−.125	−.0625	−.0313	−.0156	−.0078	−.0039	−.0020
5	$G_{t-5}a_5$						0	0	0	0	0	0	0	0	0	0
6	$G_{t-6}a_6$							−1	−.5	−.25	−.125	−.0625	−.0313	−.0156	−.0078	−.0039
7	$G_{t-7}a_7$								0	0	0	0	0	0	0	0
8	$G_{t-8}a_8$									1	.5	.25	.125	.0625	.0313	.0156
9	$G_{t-9}a_9$										−1	−.5	−.25	−.125	−.0625	−.0313
10	$G_{t-10}a_{10}$											2	1	.5	.25	.125
11	$G_{t-11}a_{11}$												0	0	0	0
12	$G_{t-12}a_{12}$													−2	−1	−.5
13	$G_{t-13}a_{13}$														1	.5
14	$G_{t-14}a_{14}$															1
	$X_t = \displaystyle\sum_{j=0}^{t} G_{t-j}a_j$	0	0	−1	1.5	−1.25	−.625	−1.3125	−.656	.672	−.664	1.668	.834	−1.583	.2085	1.104

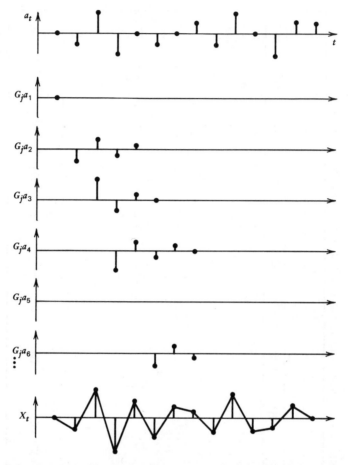

Figure 3.2 Graphical illustration of response generation by Green's function, AR(1) system: $\phi_1 = -0.5$.

By the first equality of (3.1.4), the generation of the response X_t can also be expressed as (since $a_t = 0$ for $t \leq 0$)

$$X_t = \sum_{j=0}^{t} G_j a_{t-j}$$

$$= G_0 a_{t-0} + G_1 a_{t-1} + G_2 a_{t-2} + \ldots + G_t a_0$$

This is illustrated in Table 3.2 for the AR(1) model with $\phi_1 = 0.5$. The Tables 3.1 and 3.2 also numerically explain the equality of the two infinite sums in Eq. (3.1.4).

The generation of the response X_t by the difference equation (3.1.1)

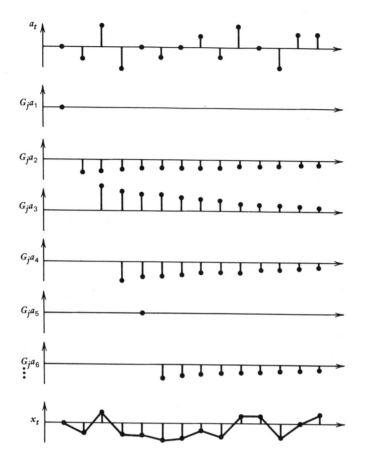

Figure 3.3   Graphical illustration of response generation by Green's function, AR(1) system: $\phi_1 = 0.9$.

or by the expansion with the Green's function given by Eq. (3.1.2) for $\phi_1 = -0.5$ and $\phi_1 = 0.9$ is similarly illustrated in Figs. 3.2 and 3.3. Note that the response fluctuates rapidly for negative values of ϕ_1 and becomes smooth as the value becomes positive, and for large positive values, there is a tendency toward trends and long runs.

3.1.3 Stability of the AR(1) System

So far we have considered the values of ϕ_1 less than one in absolute value. For these values, G_j asymptotically goes to zero at a rate depending upon the magnitude of ϕ_1. Thus, for values of ϕ_1 less than one in absolute value,

Table 3.2 Numerical Illustration of Response Generation with $a_t = 0$ for $t \leq 0$: First Equality of (3.1.4)

$j=$	t	0	1	2	3	4	5	6	7	8	9	10	11	12	13	14
	a_t	0	0	−1	2	−2	0	−1	0	1	−1	2	0	−2	1	1
	G_t	1	.5	.25	.125	.0625	.0313	.0156	.0078	.0039	.0020	.0010	.0005	.0002	.0001	.0001
0	G_0a_{t-0}	0	0	−1	2	−2	0	−1	0	1	−1	2	0	−2	1	1
1	G_1a_{t-1}		0	0	−.5	1	−1	0	−.5	0	.5	−.5	1	0	−1	.5
2	G_2a_{t-2}			0	0	−.25	.5	−.5	0	−.25	0	.25	−.25	.5	0	−.5
3	G_3a_{t-3}				0	0	−.125	.25	−.25	0	−.125	0	.125	−.125	.25	0
4	G_4a_{t-4}					0	0	−.0625	.125	−.125	0	−.0625	0	.0625	−.0625	.125
5	G_5a_{t-5}						0	0	−.0313	.0625	−.0625	0	−.0313	0	.0313	−.0313
6	G_6a_{t-6}							0	0	−.0156	.0313	−.0313	0	−.0156	0	.0156
7	G_7a_{t-7}								0	0	−.0078	.0156	−.0156	0	−.0078	0
8	G_8a_{t-8}									0	0	−.0039	.0078	−.0078	0	−.0039
9	G_9a_{t-9}										0	0	−.0020	.0039	−.0039	0
10	$G_{10}a_{t-10}$											0	0	−.0010	.0020	−.0020
11	$G_{11}a_{t-11}$												0	0	−.0005	.0010
12	$G_{12}a_{t-12}$													0	0	−.0002
13	$G_{13}a_{t-13}$														0	0
14	$G_{14}a_{t-14}$															0
	$X_t = \sum\limits_{j=0}^{t} G_j a_{t-j}$	0	0	−1	1.5	−1.25	−.625	−1.3125	−.656	.672	−.664	1.668	.834	−1.583	.2085	1.104

given sufficient time, the system asymptotically returns to its equilibrium position if only one a_t is injected. Therefore, under the condition

$$|\phi_1| < 1 \qquad (3.1.6)$$

an AR(1) system is said to be "asymptotically stable."

For the random walk model

$$X_t = X_{t-1} + a_t$$

since $\phi_1 = 1$, we have

$$G_j \equiv 1$$

For this limiting AR(1) system, if a single a_t is injected, the system remains in the position a_t indefinitely and therefore it is clearly *not* asymptotically stable. However, the response remains bounded, that is, at subsequent values of time it will not exceed a_t. Thus, the system is stable. Therefore, the condition of stability of an AR(1) system becomes

$$|\phi_1| \leq 1 \qquad (3.1.7)$$

Note that if the system is asymptotically stable, it is stable but not vice versa, as is seen from the random walk example.

On the other hand, if the absolute value of ϕ_1 is greater than one, that is, if the stability condition (3.1.7) is not satisfied, then

$$G_j \to \infty, \text{ as } j \to \infty$$

In other words, if a single a_t of any small value is injected in the system, then, given sufficient time, the system response can exceed any bound, however large in magnitude. Therefore, in this case the AR(1) system is unstable.

It should be emphasized that the stability condition is completely determined by the "homogeneous" or the autoregressive part, which is the left-hand side of Eq. (3.1.1). It is independent of the right-hand side; in fact, even if a_t were deterministic and not stochastic (random) as in our case, the stability condition would still be (3.1.7). For example, the stability condition (3.1.7) is the same for AR(1) and ARMA(1,1) models. Although the stability condition strictly refers to a system, we will use it interchangeably for system, model, process, and time series.

3.1.4 Green's Function and the Orthogonal or Wold's Decomposition

The expansion of X_t with the help of the Green's function, given by Eq. (3.1.2) for an AR(1) system, can also be considered as an "orthogonal decomposition" of X_t. For random variables the notion of independence or uncorrelatedness is the same as orthogonality. The a_t's may be considered as "axes" and the Green's function G_j the "coordinate" of X_t corresponding to the axis a_{t-j}. Thus, the decomposition (3.1.2) expresses X_t as the sum of perpendicular vectors $G_j a_{t-j}$ in an infinite dimensional space.

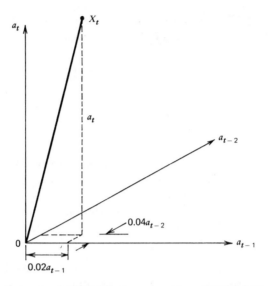

Figure 3.4 Orthogonal or Wold's decomposition of X_t, AR(1) model with
$\phi_1 = 0.2$: $X_t \simeq a_t + 0.2a_{t-1} + 0.04a_{t-2}$.

Although the space is theoretically infinite dimensional, the asymptotic
stability condition (3.1.6), when satisfied, guarantees that for all practical
purposes one has to add only a finite number of vectors to get X_t. This is
graphically illustrated in Fig. 3.4.

The concept of an orthogonal or canonical decomposition of a stationary stochastic process or time series X_t into the sum of uncorrelated
random variables is very important and was introduced in the time series
literature by Wold (1938). Hence, the expansion such as (3.1.2) is known
as Wold's decomposition. It can be used to derive all the statistical properties of the time series. For example, suppose we want to find the variance
γ_0 of the AR(1) process X_t. Then, since the variance of the sum of independent or uncorrelated random variables is the sum of their variances
[see (A 2.1.15)]

$$\gamma_0 = \mathrm{Var}(X_t) = \mathrm{Var}\left(\sum_{j=0}^{\infty} \phi_1^j a_{t-j}\right)$$

$$= \sum_{j=0}^{\infty} \mathrm{Var}(\phi_1^j a_{t-j})$$

$$= \left(\sum_{j=0}^{\infty} \phi_1^{2j}\right)\sigma_a^2$$

$$= \frac{\sigma_a^2}{1 - \phi_1^2} \tag{3.1.8}$$

if we assume that the condition of asymptotic stability (3.1.6) is satisfied. Note that as ϕ_1 tends to 1, that is, the system approaches the boundary of stability, the variance γ_0 tends to ∞. Thus, the random walk process has infinite variance. As in the case of the IBM series in the last chapter, we treat the random walk as a limiting case of AR(1) with ϕ_1 close to one and variance very large. Other properties such as covariances and correlations can also be obtained using Wold's decomposition, as we will see in section three of this chapter and in Chapter 6.

3.1.5 Green's Function of the ARMA(2,1) System: Implicit Method

We now turn our attention to the ARMA(2,1) model and derive its Green's function and stability conditions:

$$X_t - \phi_1 X_{t-1} - \phi_2 X_{t-2} = a_t - \theta_1 a_{t-1} \qquad (3.1.9)$$

or

$$(1 - \phi_1 B - \phi_2 B^2)X_t = (1 - \theta_1 B)a_t \qquad (3.1.9a)$$

The simple method of substitution used for the AR(1) model becomes very cumbersome for this model, and hence we use the method of comparing the coefficients. Suppose

$$X_t = \sum_{j=0}^{\infty} G_j a_{t-j} = \left(\sum_{j=0}^{\infty} G_j B^j \right) a_t$$

Substituting this in Eq. (3.1.9a), we have

$$(1 - \phi_1 B - \phi_2 B^2)\left(\sum_{j=0}^{\infty} G_j B^j \right) a_t = (1 - \theta_1 B)a_t$$

Since a_t's are orthogonal, this gives the operator identity

$$(1 - \phi_1 B - \phi_2 B^2)(G_0 + G_1 B + G_2 B^2 + G_3 B^3 + \dots) \equiv (1 - \theta_1 B)$$

If we equate the coefficients of equal powers of B

$$0 : G_0 = 1$$
$$1 : G_1 - \phi_1 = -\theta_1 \Rightarrow G_1 = \phi_1 - \theta_1$$
$$2 : G_2 - \phi_1 G_1 - \phi_2 = 0 \Rightarrow G_2 = \phi_1^2 - \phi_1 \theta_1 + \phi_2 \qquad (3.1.10)$$

and

$$G_j = \phi_1 G_{j-1} + \phi_2 G_{j-2}, \qquad j \geq 3$$

that is,

$$(1 - \phi_1 B - \phi_2 B^2)G_j = 0, \qquad j \geq 2 \qquad (3.1.11)$$

The use of Eqs. (3.1.10) and (3.1.11) is very convenient when only a first few G_j's are needed. For example, it will be seen in Chapter 5 that if one needs, say, four step ahead forecasts, then the probability limits depend on G_1, G_2, and G_3, which are best calculated by (3.1.10) and (3.1.11). However, if the entire form of G_j and its behavior for large j is needed, one should use the explicit form developed in the subsequent Section 3.1.6.

For example, suppose $\phi_1 = 1.3$, $\phi_2 = -0.4$, and $\theta_1 = 0.4$. Then,

$$
\begin{aligned}
G_0 &= 1 \\
G_1 &= \phi_1 - \theta_1 = 1.3 - 0.4 = 0.9 \\
G_2 &= \phi_1 G_1 + \phi_2 G_0 \\
&= 1.3 \times 0.9 - 0.4 \\
&= 0.77 \\
G_3 &= \phi_1 G_2 + \phi_2 G_1 = 1 - 0.36 \\
&= 0.64
\end{aligned}
$$

and so on. It should be noted that the Green's function G_j satisfies the homogeneous or autoregressive Eq. (3.1.11), which is the left-hand side of Eq. (3.1.9), for $j \geq 2$. This result is true in general, and following the above method, we can easily verify that for an ARMA(n,m) model

$$(1 - \phi_1 B - \phi_2 B^2 - \ldots - \phi_n B^n) G_j = 0, \quad j \geq \max(n, m+1) \quad (3.1.12)$$

In particular, for an ARMA($n, n-1$) model, the Green's function may be recursively computed from the operator identity

$$
\begin{aligned}
(1 - \phi_1 B - \phi_2 B^2 - \ldots - \phi_n B^n)&(G_0 + G_1 B + G_2 B^2 + \ldots) \\
&\equiv (1 - \theta_1 B - \theta_2 B^2 - \ldots - \theta_{n-1} B^{n-1})
\end{aligned}
$$

that gives, on equating the coefficients of equal powers of B,

$$
\begin{aligned}
0 &: G_0 = 1 \\
1 &: G_1 - \phi_1 G_0 = -\theta_1 \\
2 &: G_2 - \phi_1 G_1 - \phi_2 G_0 = -\theta_2 \\
3 &: G_3 - \phi_1 G_2 - \phi_2 G_1 - \phi_3 G_0 = -\theta_3 \quad\quad (3.1.13)
\end{aligned}
$$

$$\ldots$$

$$
\begin{aligned}
n-1 &: G_{n-1} - \phi_1 G_{n-2} - \ldots - \phi_{n-1} G_0 = -\theta_{n-1} \\
n &: G_n - \phi_1 G_{n-1} - \ldots - \phi_n G_0 = 0
\end{aligned}
$$

that is,

$$(1 - \phi_1 B - \phi_2 B^2 - \ldots - \phi_n B^n) G_j = 0 \quad j \geq n \quad (3.1.14)$$

Eqs. (3.1.10) and (3.1.11) are special cases of Eqs. (3.1.13) and (3.1.14) for $n=2$, and Eq. (3.1.14) is a special case of Eq. (3.1.12) for $m = n-1$.

3.1.6 Green's Function of the ARMA(2,1) System: Explicit Method

The Green's function G_j for the ARMA(2,1) model is implicitly given by Eqs. (3.1.10) and (3.1.11), from which it can be computed recursively. To obtain an explicit expression for G_j, we will follow the method of inverting the autoregressive operator as in the AR(1) case. Suppose we factor the autoregressive part of the ARMA(2,1) model as

$$(1 - \phi_1 B - \phi_2 B^2) = (1 - \lambda_1 B)(1 - \lambda_2 B) \qquad (3.1.15)$$

that is,

$$\left.\begin{array}{r} \lambda_1 + \lambda_2 = \phi_1 \\ \lambda_1\lambda_2 = -\phi_2 \end{array}\right\} \qquad (3.1.15a)$$

where λ_1 and λ_2 are the characteristic roots (of the second order linear difference equation) given by

$$\lambda^2 - \phi_1\lambda - \phi_2 = 0$$

so that

$$\lambda_1, \lambda_2 = \frac{1}{2}\left(\phi_1 \pm \sqrt{\phi_1^2 + 4\phi_2}\right) \qquad (3.1.15b)$$

The Wold's decomposition of the ARMA(2,1) process X_t is given by

$$X_t = \frac{(1 - \theta_1 B)a_t}{(1 - \phi_1 B - \phi_2 B^2)} = \frac{(1 - \theta_1 B)a_t}{(1 - \lambda_1 B)(1 - \lambda_2 B)}$$

This expansion can be evaluated as in the AR(1) case if we use partial fractions. If we assume that λ_1, λ_2 are distinct (for which the formulae are given in Appendix A 3.1),

$$
\begin{aligned}
X_t &= \frac{(1 - \theta_1 B)}{(1 - \lambda_1 B)(1 - \lambda_2 B)}a_t \\[2mm]
&= \left[\frac{\left(1 - \dfrac{\theta_1}{\lambda_1}\right)}{\left(1 - \dfrac{\lambda_2}{\lambda_1}\right)} \cdot \frac{1}{(1 - \lambda_1 B)} + \frac{\left(1 - \dfrac{\theta_1}{\lambda_2}\right)}{\left(1 - \dfrac{\lambda_1}{\lambda_2}\right)} \cdot \frac{1}{(1 - \lambda_2 B)}\right]a_t \\[2mm]
&= \left[\frac{(\lambda_1 - \theta_1)}{(\lambda_1 - \lambda_2)} \cdot \frac{1}{(1 - \lambda_1 B)} + \frac{(\lambda_2 - \theta_1)}{(\lambda_2 - \lambda_1)} \cdot \frac{1}{(1 - \lambda_2 B)}\right]a_t \\[2mm]
&= \sum_{j=0}^{\infty}\left[\left(\frac{\lambda_1 - \theta_1}{\lambda_1 - \lambda_2}\right)\lambda_1^j + \left(\frac{\lambda_2 - \theta_1}{\lambda_2 - \lambda_1}\right)\lambda_2^j\right]a_{t-j}
\end{aligned}
$$

which gives the Green's function as

$$G_j = \left(\frac{\lambda_1 - \theta_1}{\lambda_1 - \lambda_2}\right) \lambda_1^j + \left(\frac{\lambda_2 - \theta_1}{\lambda_2 - \lambda_1}\right) \lambda_2^j \qquad (3.1.16)$$

An Alternative Derivation

G_j derived above can also be obtained as a solution of the homogeneous difference Eq. (3.1.11) with initial condition (3.1.10). Just as the solution of an nth order differential equation is a linear combination of exponential functions, similarly the solution of an nth order difference equation is a linear combination of terms

$$\lambda^j$$

where λ's are the characteristic roots. Since the characteristic roots of (3.1.11) are λ_1 and λ_2, we have

$$G_j = g_1\lambda_1^j + g_2\lambda_2^j \qquad (3.1.16a)$$

where the two consonants g_1, g_2 are to be evaluated from the initial conditions (3.1.10). Thus,

$$G_0 = g_1 + g_2 = 1$$
$$G_1 = g_1\lambda_1 + g_2\lambda_2 = \phi_1 - \theta_1 = \lambda_1 + \lambda_2 - \theta_1$$

Solving these, we have

$$g_1 = \frac{\lambda_1 - \theta_1}{\lambda_1 - \lambda_2}$$

$$g_2 = \frac{\lambda_2 - \theta_1}{\lambda_2 - \lambda_1} \qquad (3.1.17)$$

which again gives Eq. (3.1.16).

As an example, for the case $\phi_1 = 1.3$, $\phi_2 = -0.4$, and $\theta_1 = 0.4$ considered above, the characteristic roots by Eq. (3.1.15b) are

$$\lambda_1, \lambda_2 = \frac{1}{2}(1.3 \pm \sqrt{1.69 - 1.6})$$

$$= 0.8, 0.5$$

Hence, by Eq. (3.1.16) the Green's function takes the form

$$G_j = \left(\frac{0.8 - 0.4}{0.8 - 0.5}\right)(0.8)^j + \left(\frac{0.5 - 0.4}{0.5 - 0.8}\right)(0.5)^j$$

$$= \frac{1}{0.3}[0.4(0.8)^j - 0.1(0.5)^j]$$

$$G_0 = 1$$

$$G_1 = \frac{1}{0.3}[0.32 - 0.05] = 0.9$$

$$G_2 = \frac{1}{0.3}[0.256 - 0.025] = 0.77$$

etc., as in Section 3.1.5.

For this example $\phi_1^2 + 4\phi_2 = 0.09$, which is greater than zero, and therefore the roots λ_1 and λ_2 are real. It is clear from Eq. (3.1.15b) that whenever $\phi_1^2 + 4\phi_2 \geq 0$, the roots are real and the Green's function G_j, given by Eq. (3.1.16), is a sum of exponentials, as the above example shows. The form of the Green's function and the way it generates the data X_t from a_i's is shown in Fig. 3.5 for this example.

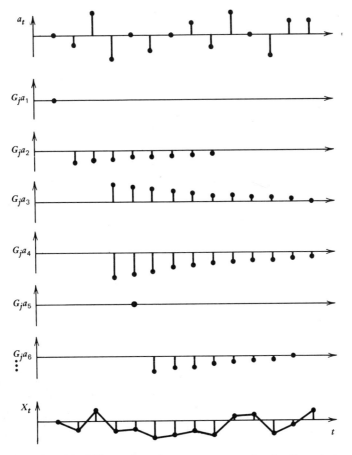

Figure 3.5   Graphical illustration of response generation by Green's function, ARMA(2,1) model: $\phi_1 = 1.3$, $\phi_2 = -0.4$, $\theta_1 = 0.4$.

Complex Roots (See Appendix A 3.2)

When $\phi_1^2 + 4\phi_2 < 0$, the roots λ_1, λ_2 are complex conjugate and so are g_1, g_2. Let

$$\lambda_1, \lambda_2 = \frac{1}{2}(\phi_1 \pm \sqrt{\phi_1^2 + 4\phi_2}) = re^{\pm i\omega}, i = \sqrt{-1}$$

where

$$r = |\lambda_1| = |\lambda_2| = \sqrt{-\phi_2} \qquad (3.1.15c)$$

$$\omega = \cos^{-1}\frac{\phi_1}{2\sqrt{-\phi_2}} = \cos^{-1}\frac{(\lambda_1 + \lambda_2)}{2\sqrt{\lambda_1 \lambda_2}} \qquad (3.1.15d)$$

and

$$g_1, g_2 = ge^{\pm i\beta} \qquad (3.1.17a)$$

where

$$g = |g_1| = |g_2|$$

$$= \frac{1}{2}\sqrt{1 + \left[\frac{\phi_1 - 2\theta_1}{\sqrt{-(\phi_1^2 + 4\phi_2)}}\right]^2} = \frac{A}{2} \qquad (3.1.17b)$$

$$\beta = \tan^{-1}\left[\frac{-\phi_1 + 2\theta_1}{\sqrt{-(\phi_1^2 + 4\phi_2)}}\right]$$

$$= \tan^{-1}\left[\frac{\text{Imaginary Part of } g_1}{\text{Real Part of } g_1}\right]\dagger \qquad (3.1.17c)$$

then, substituting in (3.1.16a) gives (also see Appendix A 3.2)

$$G_j = ge^{i\beta}(re^{i\omega})^j + ge^{-i\beta}(re^{-i\omega})^j$$

$$= gr^j [e^{i(j\omega + \beta)} + e^{-i(j\omega + \beta)}]$$

$$= 2gr^j \cos(j\omega + \beta) = r^j A \cos(j\omega + \beta) \qquad (3.1.16b)$$

which is a damped cosine wave with amplitude $2g$ given by (3.1.17b), phase β given by (3.1.17c), the damping governed by ϕ_2 via (3.1.15c), and the frequency ω given by (3.1.15d). Note that when $\phi_2 = -1$, there is no damping and G_j is an undamped cosine wave; the damping increases as the value of ϕ_2 increases from -1 toward zero.

As an example, consider the mechanical vibration system data in Fig. 2.4, which clearly exhibits a pseudo-periodic behavior. It will be shown in

---

† Although this expression readily generalizes to higher order models, care should be taken to use the proper quadrant for the angle from the signs of real and imaginary parts.

2.36
−0.40

Chapter 4 that the ARMA(2,1) model for this data has $\phi_1 = 1.43$, $\phi_2 = -0.61$, and $\theta_1 = -0.54$ so that $\phi_1^2 + 4\phi_2 = -0.4 < 0$.

$$r = \sqrt{-\phi_2} = \sqrt{0.61} = 0.78$$

$$\omega = \cos^{-1}\left(\frac{1.43}{2 \times 0.77}\right) = 23.56° = 0.41 \text{ radians}$$

$$g = \frac{1}{2}\sqrt{1 + \frac{[1.43 - 2(-0.54)]^2}{0.4}}$$

$$= \frac{1}{2}\sqrt{1 + \frac{2.51^2}{0.4}} = 2.05$$

$$\beta = \tan^{-1}\frac{-2.51}{\sqrt{0.4}} = -75.9° = -1.32 \text{ radians}$$

so that the Green's function is

$$G_j = (0.77)^j \, 4.08 \cos (0.42j - 1.32)$$

A plot of generation of X_t by this Green's function is shown in Fig. 3.6.

3.1.7 Green's Functions of AR(2) and ARMA(1,1) Systems

The Green's function for the AR(2) system is obtained by putting $\theta_1 = 0$ in Eq. (3.1.16) as:

$$G_j = \frac{1}{(\lambda_1 - \lambda_2)} [\lambda_1^{j+1} - \lambda_2^{j+1}] \tag{3.1.18}$$

Figure 3.7 illustrates this Green's function of AR(2) as a sum of exponentials for $\phi_1 = 1.3$, $\phi_2 = -0.4$. Note that in reality the plot of the Green's function consists of discrete points. However, for the sake of illustration, the points are joined by a continuous curve.

When ϕ_2 is very small, we have seen in Chapter 2 that the ARMA(2,1) model almost degenerates to ARMA(1,1). For example if $\phi_1 = 0.85$, $\phi_2 = -0.04$, then

$$\lambda_1 = 0.8, \, \lambda_2 = 0.05$$

and the model and its Green's functions are practically of ARMA(1,1) nature as shown in Fig. 3.8. The generation of response is shown in Fig. 3.9.

An expression for the Green's function of an ARMA(1,1) model can be found by specializing Eq. (3.1.16). Since the ARMA(1,1) model is the

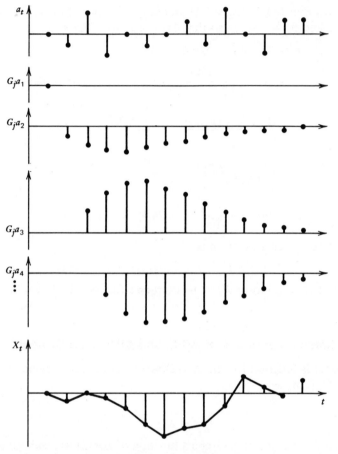

Figure 3.6 Generation of response by Green's function, ARMA(2,1) model: $\phi_1 = 1.4$, $\phi_2 = -0.59$, $\theta_1 = -0.55$.

same as ARMA(2,1) with

$$\phi_2 = -\lambda_1\lambda_2 = 0$$

and

$$\phi_1 = \lambda_1 + \lambda_2$$

we can get the desired expression by substituting $\lambda_2 = 0$ and $\lambda_1 = \phi_1$ in Eq. (3.1.16):

$$G_j = \frac{\lambda_1 - \theta_1}{\lambda_1}(\lambda_1^j) + \frac{0 - \theta_1}{0 - \lambda_1}(0^j)$$

$$= \frac{\lambda_1 - \theta_1}{\lambda_1}(\lambda_1^j) + \frac{\theta_1}{\lambda_1}(0^j) \qquad (3.1.19a)$$

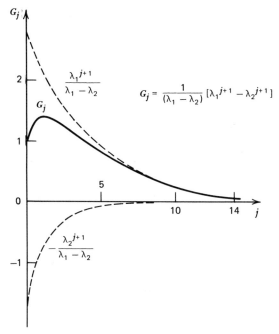

Figure 3.7 Green's functions of AR(2) model as a sum of two exponentials, $\phi_1 = 1.3$, $\phi_2 = -0.4$.

The expression $0^j = 0$ for $j \geq 1$, but for $j = 0$ we must define $0^0 = 1$ in order to get

$$G_0 = \frac{\lambda_1 - \theta_1}{\lambda_1} + \frac{\theta_1}{\lambda_1} = 1$$

Since we always have $G_0 = 1$, the Green's function of an ARMA(1,1) model may be more conveniently written as

$$G_j = \frac{\lambda_1 - \theta_1}{\lambda_1} (\lambda_1^j)$$

$$= (\lambda_1 - \theta_1)\lambda_1^{j-1}$$

$$= (\phi_1 - \theta_1) \phi_1^{j-1}, \qquad j \geq 1 \qquad (3.1.19)$$

Note that (3.1.19) is valid only for $j \geq 1$ and would give an incorrect result if used for $j = 0$.

3.1.8 Why ARMA(2,1) Rather Than AR(2) After AR(1)?

When both ϕ_2 and θ_1 are zero (that is, $\lambda_2 = \theta_1 = 0$), the Green's function (3.1.16) reduces to that of an AR(1) model with $\lambda_1 = \phi_1$. The transition

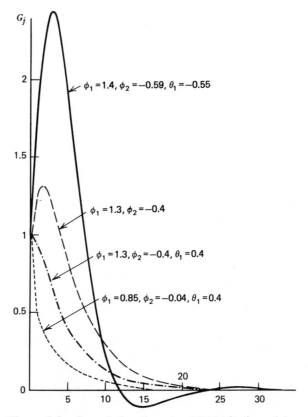

Figure 3.8 Green's functions for ARMA(2,1) models.

from AR(1) to ARMA(2,1) is a transition from simple first order dynamics, characterized by a single exponential λ_1^j, to slightly complex dynamics requiring two exponentials λ_1^j and λ_2^j for characterization. It is seen that this transition from AR(1) to ARMA(2,1) is accomplished by adding a term λ_2^j with suitable coefficients to the Green's function of an AR(1) system, that is,

$$\text{AR(1)} \quad : \quad G_j = \lambda_1^j$$

$$\text{ARMA(2,1):} \quad G_j = g_1\lambda_1^j + g_2\lambda_2^j$$

The two coefficients g_1 and g_2 are always restricted by

$$G_0 = g_1 + g_2 = 1$$

Thus, only one coefficient can be free, say g_2, and the value of g_1 must be

$$g_1 = 1 - g_2$$

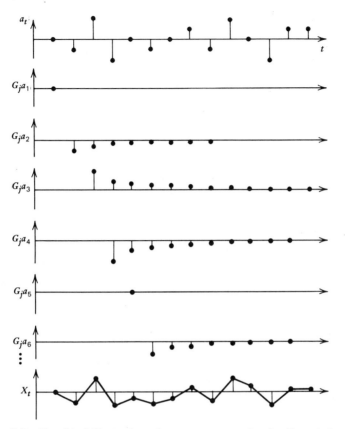

Figure 3.9   Graphical illustration of response generation by Green's function, ARMA(2,1) model: $\phi_1 = 0.85$, $\phi_2 = -0.04$, $\theta_1 = 0.4$.

If the free coefficient g_2 happens to take the value that satisfies [see Eq. (3.1.10)]

$$G_1 = g_1\lambda_1 + g_2\lambda_2 = \lambda_1 + \lambda_2 = \phi_1$$

that is,

$$(1 - g_2)\lambda_1 + g_2\lambda_2 = \lambda_1 + \lambda_2, \qquad \text{or} \qquad g_2 = \frac{\lambda_2}{\lambda_2 - \lambda_1}$$

then we get an AR(2) model. However, if the free coefficient does not satisfy this condition and takes any other value, then we have an ARMA(2,1) model with

$$(1 - g_2)\lambda_1 + g_2\lambda_2 = \phi_1 - \theta_1 = \lambda_1 + \lambda_2 - \theta_1$$

or

$$\theta_1 = \lambda_2 + g_2(\lambda_1 - \lambda_2)$$

This is one of the reasons why we treat the AR(2) model as a special case of the ARMA(2,1) model, which will be further elaborated in Chapter 4. It is the one free coefficient that introduces the moving average parameter θ_1 and determines the relative weights for λ_1 and λ_2 in the dynamics. (Also see Problem 3.21)

3.1.9 Stability of the ARMA(2,1) System

The interpretation of the Green's function in terms of the memory of the past a_t's and their influence on the future response given for the AR(1) system also holds for the ARMA(2,1) system. If we follow the same reasoning, it is easy to see from the expression for G_j given by Eq. (3.1.16) that the ARMA(2,1) system is asymptotically stable if

$$|\lambda_1| < 1, \qquad |\lambda_2| < 1 \tag{3.1.20}$$

These conditions of asymptotic stability can be expressed directly in terms of the autoregressive parameters by the following manipulations using Eqs. (3.1.20) and (3.1.15a).

Since

$$\phi_2 = -\lambda_1\lambda_2$$

(3.1.20) implies

$$|\phi_2| < 1$$

that is,

$$-1 < \phi_2 < 1$$

Also,

$$\lambda_1(1-\lambda_2) < (1-\lambda_2)$$

or

$$\lambda_1 + \lambda_2 - \lambda_1\lambda_2 < 1$$

that is,

$$\phi_1 + \phi_2 < 1$$

and

$$-(1+\lambda_2) < \lambda_1(1+\lambda_2)$$

or

$$-\lambda_1\lambda_2 - (\lambda_1+\lambda_2) < 1$$

that is,

$$\phi_2 - \phi_1 < 1$$

Thus, the asymptotic stability conditions for the ARMA(2,1) system are given by

$$\phi_1 + \phi_2 < 1, \qquad \phi_2 - \phi_1 < 1, \qquad |\phi_2| < 1 \tag{3.1.21}$$

which are equivalent to (3.1.20). The stability region for an ARMA(2,m) system is shown in Fig. 3.10. Those familiar with linear system analysis can readily derive Eq. (3.1.21) by applying the well-known Routh's criteria, as discussed in Appendix A 3.3.

For all the examples of ARMA(2,1) models considered earlier, the systems are asymptotically stable. For example,

$$\phi_1 = 1.3, \qquad \phi_2 = -0.4$$

$$\phi_1 + \phi_2 = 1.3 - 0.4 = 0.9 < 1$$

$$\phi_2 - \phi_1 = -1.7 < 1$$

$$|\phi_2| = |-0.4| < 1$$

showing that conditions (3.1.21) are satisfied. For this case the roots are $\lambda_1 = 0.8$, $\lambda_2 = 0.5$, so that (3.1.20) is satisfied. Similarly, for the case $\phi_1 = 0.85$ as above, but $\phi_2 = -0.04$, since $\lambda_1 = 0.8$, $\lambda_2 = 0.05$, it is easy to see that both Eqs. (3.1.20) and (3.1.21) are satisfied. Figures 3.5, 3.8, and 3.9 show that the Green's functions in the case of these two corresponding systems asymptotically tend to zero in an exponential way. For the case $\phi_1 = 1.4$, $\phi_2 = -0.59$

$$\phi_1 + \phi_2 = 1.4 - 0.59 = 0.81 < 1$$

$$\phi_2 - \phi_1 = -0.59 - 1.4 = -1.99 < 1$$

$$|\phi_2| = |-0.59| < 1$$

and

$$\lambda_1, \lambda_2 = 0.7 \pm 0.316 \, i$$

showing that the conditions (3.1.21) or (3.1.20) are satisfied. Figure 3.8 shows that in this case the Green's function asymptotically tends to zero in a damped sine wave fashion. Note that since the conditions (3.1.20) and

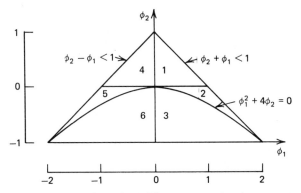

Figure 3.10 Stability region of AR(2) or ARMA(2,m) system (invertibility region of ARMA(n,2): replace ϕ by θ).

(3.1.21) are equivalent, only one of them needs to be checked to ascertain asymptotic stability of the system.

As in the case of the AR(1) system, the asymptotic stability implies that the memory of an a_t from sufficiently distant past is zero. That is, given sufficient time, the influence of a single a_t injected in the system will eventually decay, and the system will return to its equilibrium or mean position. The rate and the type of decay will depend upon the values of the roots λ_1 and λ_2.

Now consider an ARMA (2,1) system with $\phi_1 = 1.5$, $\phi_2 = -0.5$, and $\theta_1 = 0.3$, that is,

$$(1 - 1.5B + 0.5B^2)X_t = (1 - 0.3B)a_t$$

that is,

$$(1 - B)(1 - 0.5B)X_t = (1 - 0.3B)a_t$$

Since one of the roots is one, it is clear that the system is *not* asymptotically stable. However, is it stable? This question can be answered by seeing whether the Green's function remains bounded. Since the two roots $\lambda_1 = 1$, $\lambda_2 = 0.5$ are distinct, we can use the expression (3.1.16) to get

$$G_j = \left(\frac{0.7}{0.5}\right)(1)^j - \frac{0.2}{0.5}(0.5)^j$$

From this form it is clear that

$$G_j \to \frac{0.7}{0.5} \quad \text{as} \quad j \to \infty$$

and therefore G_j remains bounded and the system is stable. The Green's function for this case and the response for the same set of a_t's as those used in the preceding examples are shown in Fig. 3.11.

Hence, if both the roots are less than one in absolute value, the ARMA(2,1) system is asymptotically stable. If one of the roots is one in absolute value, it is stable but not asymptotically stable. If one or both of the two roots are greater than one in absolute value, then obviously the system is unstable, since $G_j \to \infty$ as $j \to \infty$. The only question that remains is: What happens if *both* the roots are one in absolute value? If the roots are opposite in sign, then the system is still stable since by Eq. (3.1.16) the Green's function is, with $\lambda_1 = 1$, $\lambda_2 = -1$,

$$G_j = \left(\frac{1 - \theta_1}{2}\right)(1)^j + \left(\frac{1 + \theta_1}{2}\right)(-1)^j$$

so that

$$G_j \to 1 \quad \text{or} \quad -\theta_1 \quad \text{as} \quad j \to \infty$$

which is thus bounded and the system is stable but not asymptotically stable.

However, if both the roots are one in absolute value and of the same sign then, as we now show, the system is unstable. Suppose $\lambda_1 = \lambda_2 = 1$.

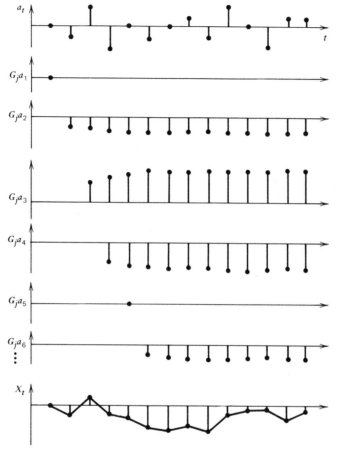

Figure 3.11 Green's function and response for ARMA(2,1) model: $\phi_1 = 1.5$, $\phi_2 = -0.5$, $\theta_1 = 0.3$.

Since the roots are equal, Eq. (3.1.16) cannot be used, as it is based on distinct roots. However, G_j can be obtained directly in this case

$$
\begin{aligned}
X_t &= \frac{(1 - \theta_1 B)}{(1 - B)(1 - B)} a_t \\
&= (1 + B + B^2 + \ldots)(1 + B + B^2 + \ldots)(1 - \theta_1 B) a_t \\
&= (1 + 2B + 3B^2 + 4B^3 + \ldots)(1 - \theta_1 B) a_t \\
&= \sum_{j=0}^{\infty} [(j+1) - \theta_1 j] a_{t-j} \\
&= \sum_{j=0}^{\infty} [(1 - \theta_1) j + 1] a_{t-j}
\end{aligned}
$$

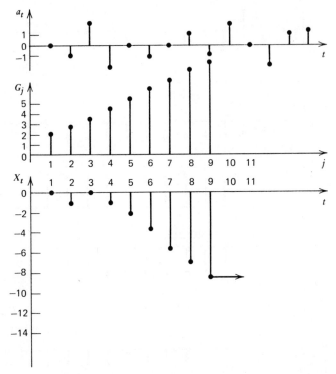

Figure 3.12 Response for the integrated random walk $(1 - B)^2 X_t = a_t$.

Therefore,

$$G_j = (1 - \theta_1)j + 1 \to +\infty \qquad \text{as} \qquad j \to \infty$$

Figure 3.12 shows the response for such a system with $\theta_1 = 0$, that is, for the limiting AR(2) model

$$(1 - 2B + B^2)X_t = a_t \tag{3.1.22}$$

or

$$(1 - B)^2 X_t = a_t$$

that is,

$$\nabla^2 X_t = a_t$$

where ∇ is the difference operator $(1 - B)$. Note that since

$$(1 - B)X_t = a_t$$

that is,

$$X_t = \sum_{j=0}^{\infty} a_{t-j}$$

is a random walk, the limiting AR(2) model (3.1.22), which can also be written as

$$X_t = \sum_{i=0}^{\infty} B^i \left(\sum_{j=0}^{\infty} a_{t-j} \right)$$

is called integrated random walk. Similarly, models such as

$(1-B)^d(1-\phi_1 B - \phi_2 B^2 - \ldots - \phi_n B^n)X_t$
$$= (1-\theta_1 B - \theta_2 B^2 - \ldots - \theta_m B^m)a_t$$

are sometimes called integrated or accumulated ARMA (ARIMA) models. We will treat them simply as special limiting cases of ARMA models.

*3.1.10 General Results

From the above discussion the stability conditions for an ARMA(2,1) or AR(2) system may be written as

$$|\lambda_1| \le 1, \qquad |\lambda_2| \le 1 \tag{3.1.23}$$

and if $\lambda_1 = \lambda_2$, then $|\lambda_1| = |\lambda_2| < 1$.

The explicit form of the Green's function for an ARMA(n,m) system, in general, or an ARMA ($n,n-1$) system, in particular, and their stability condition can be obtained in the same way as the ARMA(2,1) case discussed above. If we repeat the partial fraction expansion for distinct roots for the ARMA($n,n-1$) model, it can be shown that the Green's function has the form:

$$G_j = g_1\lambda_1^j + g_2\lambda_2^j + \ldots + g_n\lambda_n^j \tag{3.1.24}$$

where $\lambda_1, \lambda_2, \ldots, \lambda_n$ are the roots of

$$\lambda^n - \phi_1\lambda^{n-1} - \phi_2\lambda^{n-2} - \ldots - \phi_n = 0 \tag{3.1.25}$$

and

$$g_i = \frac{(\lambda_i^{n-1} - \theta_1\lambda_i^{n-2} - \ldots - \theta_{n-1})}{(\lambda_i - \lambda_1)(\lambda_i - \lambda_2) \ldots (\lambda_i - \lambda_{i-1})(\lambda_i - \lambda_{i+1}) \ldots (\lambda_i - \lambda_n)} \tag{3.1.26}$$

$$i = 1, 2, \ldots, n$$

where the denominator is the product of all terms $(\lambda_i - \lambda_j)$ for $j = 1, 2,$ $\ldots, n$ excluding the zero term $(\lambda_i - \lambda_i)$. Each real root λ_i in Eq. (3.1.24) provides an exponential dynamic mode like an AR(1) model, whereas a complex conjugate pair of roots gives an exponentially decaying sinusoidal mode, whose frequency and decay rate can be obtained as illustrated in Section 3.1.6.

Results for ARMA(n,m), $m < n-1$ follow by setting the last $(n-1-m)\theta_i$'s zero. It can be verified that Eq. (3.1.16) is obtained from these expressions by putting $n = 2$. General expressions for ARMA(n,m) are given in Pandit (1973).

As in the ARMA(2,1) case, Eq. (3.1.24) shows that the conditions of asymptotic stability for an ARMA (n,m) system are [see (3.1.20)]

$$|\lambda_k| < 1, \qquad k = 1, 2, \ldots, n \tag{3.1.27}$$

whereas stability conditions are [see (3.1.23)]

$$|\lambda_k| \leq 1, \qquad k = 1, 2, \ldots, n \tag{3.1.28}$$

and if $\lambda_i = \lambda_j$, $i \neq j$, then $|\lambda_i| = |\lambda_j| < 1$, $i,j = 1, 2, \ldots, n$. From now on, unless we want to specifically distinguish asymptotic stability, we will refer to the conditions (3.1.27) and their special cases (3.1.20) and (3.1.21) as stability conditions (see also Appendix A 3.3).

What we have called Green's function in this chapter is also referred to in the literature as weighting function [see Wiener (1949) and Pugachev (1957)] or as ψ weights [see Box and Jenkins (1970)] etc. The latter refer to the asymptotic stability conditions as stationarity conditions and treat ARMA processes with unit roots as nonstationary processes.

3.2 INVERSE FUNCTION AND INVERTIBILITY

The Green's function represents the memory or dynamics of a system in terms of the past a_t's and shows how they affect the response X_t. It was obtained by expressing X_t as a linear combination of the a_t's. The boundedness of the Green's function imposes stability conditions on the autoregressive parameters of the ARMA models.

The dynamics of an ARMA system can also be represented by expressing X_t as a linear combination of the past X_t's. The coefficient function in this expansion is called the "inverse function," since it is obtained by inverting the operator that yields the Green's function. The inverse function will be denoted by I_j; its boundedness imposes "invertibility" restrictions on the moving average parameters of the ARMA models. As we will see, these are exactly similar to the stability conditions discussed in the last section. Thus, I_j is defined by

$$X_t = \sum_{j=1}^{\infty} I_j X_{t-j} + a_t \tag{3.2.1}$$

or

$$a_t = (1 - I_1 B - I_2 B^2 - \ldots)X_t \tag{3.2.1a}$$

which is essentially an infinite order autoregressive model AR (∞).

3.2.1 AR(1) and MA(1) Models

For the AR(1) model, since

$$X_t = \phi_1 X_{t-1} + a_t$$

it immediately follows that

$$I_1 = \phi_1, \qquad I_j = 0, \qquad j > 1$$

Note that the Green's function G_j for the AR(1) model was obtained by the operator

$$\frac{1}{(1 - \phi_1 B)}$$

whereas the operator that gives the inverse function I_j is

$$(1 - \phi_1 B)$$

Since there is no moving average parameter in the AR(1) model, there are no invertibility restrictions; in fact, the inverse function always remains bounded.

For the MA(1) model

$$X_t = (1 - \theta_1 B) a_t \tag{3.2.2}$$

$$a_t = \frac{1}{(1 - \theta_1 B)} X_t$$

$$= (1 + \theta_1 B + \theta_1^2 B^2 + \ldots) X_t$$

$$\therefore \ X_t = \sum_{j=1}^{\infty} -\theta_1^j X_{t-j} + a_t \tag{3.2.3}$$

thus,

$$I_j = -\theta_1^j \tag{3.2.4}$$

and the invertibility condition is

$$|\theta_1| < 1 \tag{3.2.5}$$

For the MA(1) model

$$G_0 = 1$$

$$G_1 = -\theta_1$$

$$G_j = 0 \qquad j > 1$$

and the operator that gives G_j is

$$(1 - \theta_1 B)$$

whereas the operator that gives the inverse function I_j as

$$\frac{1}{(1 - \theta_1 B)}$$

3.2.2 ARMA(n,m) Models

It is seen that the inverse functions $I_j (j \geq 1)$ for the AR(1) model is of the same form as the Green's function of the MA(1) model; whereas the inverse function for the MA(1) model is of the same form as the Green's function for the AR(1) model. Although we have always been considering ARMA($n, n-1$) models, to see this duality between I_j and G_j let us consider the ARMA(1,2) model

$$(1 - \phi_1 B)X_t = (1 - \theta_1 B - \theta_2 B^2)a_t \qquad (3.2.6)$$

If we use the method of comparing coefficients and substitute Eq. (3.2.1a) for a_t, we get the operator identity

$$(1 - \phi_1 B) \equiv (1 - \theta_1 B - \theta_2 B^2)(1 - I_1 B - I_2 B^2 - \ldots)$$

and the recursive equations now follow from Eqs. (3.1.10) and (3.1.11) by interchanging ϕ and θ

$$-I_1 - \theta_1 = -\phi_1 \Rightarrow I_1 = \phi_1 - \theta_1 \qquad (3.2.7)$$

$$I_j = \theta_1 I_{j-1} + \theta_2 I_{j-2} \; j \geq 2, \; I_0 = -1$$

that is,

$$(1 - \theta_1 B - \theta_2 B^2)I_j = 0, j \geq 2 \qquad (3.2.8)$$

A similar recursive relation can be obtained for ARMA(n,m) models by interchanging ϕ and θ in Eqs. (3.1.12) to (3.1.14) and replacing G_j by $-I_j$. This duality between the Green's function and the inverse function is shown by the block diagram in Fig. 3.13.

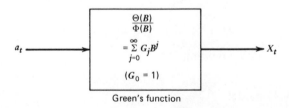

Green's function

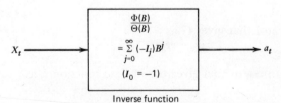

Inverse function

Figure 3.13   Duality between the Green's function and the inverse function.

The explicit expression for the inverse function I_j can be obtained by duplicating Eqs. (3.1.15) and (3.1.16) and by interchanging the role of autoregressive and moving average parameters. Let v_1, v_2 be the characteristic roots of the moving average operator on the right-hand side of Eq. (3.2.6), that is,

$$(1 - \theta_1 B - \theta_2 B^2) = (1 - v_1 B)(1 - v_2 B) \tag{3.2.9}$$

$$v_1 + v_2 = \theta_1, \qquad v_1 v_2 = -\theta_2 \tag{3.2.9a}$$

and v_1, v_2 can be computed as the roots of

$$v^2 - \theta_1 v - \theta_2 = 0$$

so that

$$v_1, v_2 = \frac{1}{2} (\theta_1 \pm \sqrt{\theta_1^2 + 4\theta_2}) \tag{3.2.9b}$$

Then, it follows immediately from Eq. (3.1.16) by interchanging ϕ's and θ's, λ's and v's and by change of sign that when the roots are distinct,

$$I_j = -\left(\frac{v_1 - \phi_1}{v_1 - v_2}\right) v_1^j - \left(\frac{v_2 - \phi_1}{v_2 - v_1}\right) v_2^j \tag{3.2.10}$$

When $\theta_1^2 + 4\theta_2 \geq 0$, the roots are real and I_j is a sum of exponentials. When $\theta_1^2 + 4\theta_2 < 0$, the roots are complex and I_j is a damped sine wave. The examples of G_j in Section 3.1 illustrate I_j when ϕ's are replaced by θ's.

The invertibility conditions for the ARMA(1,2) model or, in general, for an ARMA(n,2) model are

$$|v_1| < 1, \qquad |v_2| < 1 \tag{3.2.11}$$

or

$$\theta_1 + \theta_2 < 1, \qquad \theta_2 - \theta_1 < 1, \qquad |\theta_2| < 1 \tag{3.2.12}$$

The invertibility region for the ARMA(n,2) model can thus be seen from Fig. 3.10 if we replace ϕ's by θ's.

The inverse function I_j for the MA(1), MA(2) models can be obtained from Eq. (3.2.10) by putting $\phi_1 = \theta_2 = 0$ and $\phi_1 = 0$, respectively. The inverse function I_j for an ARMA(1,1) model, obtained by putting $\theta_2 = v_2 = 0$ and $\theta_1 = v_1$, is

$$I_j = \frac{\theta_1 - \phi_1}{-\theta_1} \theta_1^j = (\phi_1 - \theta_1)\theta_1^{j-1} \tag{3.2.13}$$

A special case of this for $\phi_1 = 1$, that is, the inverse function for the model

$$(1 - B)X_t = (1 - \theta_1 B)a_t \tag{3.2.14}$$

given by

$$I_j = (1 - \theta_1)\theta_1^{j-1} \qquad (3.2.15)$$

will be useful later in dealing with exponential smoothing.

The explicit form of the inverse function for an ARMA(n,m) system, in general, similarly follows from Eqs. (3.1.24) to (3.1.26) by replacing G_j by $-I_j$, λ_k by v_k, and interchanging n,m, and ϕ,θ. The invertibility conditions for the general model are

$$|v_k| < 1, \; k = 1, 2, \ldots, m \qquad (3.2.16)$$

where v_k are the roots

$$v^m - \theta_1 v^m\text{-}1 - \ldots - \theta_m = 0$$

If the ARMA(n,m) model is symbolically denoted by

$$\Phi(B)X_t = \Theta(B)a_t \qquad (3.2.17)$$

that is,

$$(1 - \phi_1 B - \phi_2 B^2 - \ldots - \phi_n B^n)X_t$$
$$= (1 - \theta_1 B - \theta_2 B^2 - \ldots - \theta_m B^m)a_t$$

or

$$(1 - \lambda_1 B)(1 - \lambda_2 B) \ldots (1 - \lambda_n B)X_t$$
$$= (1 - v_1 B)(1 - v_2 B) \ldots (1 - v_m B)a_t$$

then the stability and invertibility conditions may be summarized in the following table:

Table 3.3 Stability and Invertibility Conditions of ARMA(n,m) Systems

System	$\Phi(B)$ $\lambda_1, \lambda_2, \ldots, \lambda_n$	$\Theta(B)$ $v_1 \; v_2, \ldots, v_m$
Stable	$|\lambda_k| < 1$	—
Invertible	—	$|v_k| < 1$
Stable & Invertible	$|\lambda_k| < 1$	$|v_k| < 1$

3.2.3 Reasons for Invertibility

If the roots v_k are greater than one in absolute value, then the inverse function increases without bound. This means that the more distant we go in the past, the greater the influence of the past X_t's on the present one. Such a situation is physically meaningless, and therefore the invertibility condition is usually imposed on the ARMA models.

There is also an important practical reason for requiring invertibility in connection with the nonlinear least squares routine. In estimating the parameters by this routine, a_t's are recursively computed at each iteration for computing the sum of squares of a_t's. The recursive relation that gives a_t's for an ARMA(n,m) model is

$$a_t = X_t - \phi_1 X_{t-1} - \phi_2 X_{t-2} - \ldots - \phi_n X_{t-n}$$
$$+ \theta_1 a_{t-1} + \theta_2 a_{t-2} + \ldots + \theta_m a_{t-m}$$

This computation is equivalent to computing a_t's by the inverse function, since by substituting successively for $a_{t-1}, a_{t-2}, \ldots$ above, we get

$$a_t = X_t - \sum_{j=1}^{\infty} I_j X_{t-j}$$

If the moving average parameters θ_i's do not satisfy invertibility conditions, then the I_j and therefore computed a_t's get so large that they exceed the allowable limit in the computer. This error can be corrected by giving initial values of θ_i well within the invertibility limits.

3.3 AUTOCOVARIANCE FUNCTION

The Green's function and the inverse function characterize the ARMA models from a systems point of view; they would remain the same even if the random variable a_t was replaced by some deterministic "input" or forcing function. In fact, the operator

$$\frac{\Theta(B)}{\Phi(B)} = \frac{(1 - \theta_1 B - \theta_2 B^2 - \ldots - \theta_m B^m)}{(1 - \phi_1 B - \phi_2 B^2 - \ldots - \phi_n B^n)}$$

which yields the Green's function, may be considered as a transfer function that transforms the input a_t into the output X_t by

$$X_t = \frac{\Theta(B)}{\Phi(B)} a_t$$

irrespective of whether the input a_t is deterministic or stochastic. Similarly, the inverse function is a transfer function that transforms the input X_t into the output a_t by

$$a_t = \frac{\Phi(B)}{\Theta(B)} X_t$$

and the operator $\Phi(B)/\Theta(B)$ yields the inverse function. The stability and invertibility conditions ensure that in either representation a finite input will always produce finite output. The Green's function and the inverse function are thus the characteristics of the system represented by the operators $\Theta(B)/\Phi(B)$ and $\Phi(B)/\Theta(B)$, respectively; they are independent of the nature of the input or the output.

3.3.1 Distribution Properties of a_t

When a_t is a random variable for a fixed t or a stochastic process for different values of t, its probabilistic behavior is characterized by the distribution property

$$a_t \sim \text{NID}(0, \sigma_a^2)$$

This property states that a set of, say, N a_t's $\{a_1, a_2, \ldots, a_N\}$ has multivariate normal distribution, with each a_t having fixed mean zero, variance σ_a^2, and covariance between any two a_t's is zero; all these properties do not involve the value of t. Since the covariance involved here is that of a stochastic process with itself at different values of t, it is called "autocovariance." In general, the autocovariance at lag k would be given by

$$\text{Cov}(a_t, a_{t-k}) = E(a_t - \mu)(a_{t-k} - \mu)$$

where E denotes the "expected value" and μ is the mean of a_t, and the autocovariance at lag zero is the variance. In the present case, since $\mu = 0$, the autocovariance of a_t may be specified by

$$\begin{aligned} E(a_t a_{t-k}) &= \sigma_a^2, & k = 0 \\ &= 0, & k \neq 0 \end{aligned}$$

or more succinctly by

$$E(a_t) = 0, \qquad E(a_t a_{t-k}) = \delta_k \sigma_a^2 \tag{3.3.1}$$

where δ_k is the Kronecker delta function that is zero for all k, except at $k = 0$ when it is one. Since a multivariate normal distribution function involves only means, variances, and covariances, the stochastic process a_t is completely specified by the first two moments given by Eq. (3.3.1). Without the normality assumption, the first two moments (3.3.1) unchanged over t make a_t (and therefore X_t, as will be seen in the next section) a wide sense stationary or simply stationary stochastic process. With the additional assumption of normal distribution, determined by these two moments, a_t (and X_t) also become stationary in the strict sense. (See Section 1.2 for definitions.)

3.3.2 Theoretical and Sample Autocovariance/Autocorrelation Function

Since X_t is a linear combination of a_t's with the Green's function as coefficients, it will also have multivariate normal distribution with mean zero. This distribution is therefore also completely characterized by the autocovariance function of X_t defined by

$$\gamma_k = E(X_t X_{t-k})$$

which is the autocovariance at lag k, and for $k = 0$ gives the variance of X_t. Knowing the autocovariance function of a_t given by Eq. (3.3.1) and the Green's function of an ARMA system, the autocovariance function γ_k of the output X_t can be determined. In this section we will consider this probabilistic characteristic that characterizes the stochastic process X_t. When the autocovariance γ_k is divided by variance, we get the autocorrelation function ρ_k, at lag k

$$\rho_k = \frac{\gamma_k}{\gamma_0}$$

Note that

$$\rho_0 = 1$$

The autocovariance function γ_k or the autocorrelation function ρ_k expresses the dependence of X_t on X_{t-k}. Their estimates can be obtained directly from the data, before fitting the model, by

$$\hat{\gamma}_k = \frac{1}{N} \sum_{t=k+1}^{N} X_t \cdot X_{t-k} = \frac{1}{N} \sum_{t=k+1}^{N} (\dot{X}_t - \overline{X})(\dot{X}_{t-k} - \overline{X}) \quad (3.3.2)$$

where X_t is observed data after subtracting the average, $\overline{X}$, and

$$\hat{\rho}_k = \frac{\hat{\gamma}_k}{\hat{\gamma}_0} \quad (3.3.3)$$

These are called sample autocovariance and autocorrelation functions to distinguish them from the theoretical values γ_k and ρ_k. It was shown in Chapter 2 that for the AR(1) model $\hat{\rho}_1 = \hat{\phi}_1$.

Just like Green's/inverse function, the *theoretical* autocovariance function completely characterizes an ARMA model. That is, given a model, we can find any of these functions, and given a form of these functions, we can find the ARMA model corresponding to it. However, unlike the Green's/inverse function, the autocovariance function can be estimated from a sample of observations by Eqs. (3.3.2) and (3.3.3), without knowing the model. A large part of the existing time series literature therefore deals with the use of this estimated sample autocorrelation for the modeling and estimation of time series.

Such use of the sample autocorrelation function would be appropriate if it was a good estimate of the theoretical autocorrelation. Unfortunately, this is not the case. In fact, sample autocorrelations are very poor estimates; they often have large variances and can be highly correlated with each other, presenting a distorted version of the true autocorrelations. These discouraging properties of the estimated sample autocorrelation function and the consequent dangers in the modeling procedures essentially based on them have been pointed out by Kendall (1945).

As we will see in this section, for pure autoregressive and pure moving average models of low order ($m,n \le 2$), the theoretical autocorrelation function has rather simple forms. Under fortunate circumstances, it is sometimes possible to discern these forms from their distorted version given by the estimated sample autocorrelation function. It is then possible to make use of a plot of autocorrelations and partial autocorrelations (defined in Section 3.4) to guess the form of the model. However, even in these simple cases, a modeling procedure based on sample autocorrelation ends up with trial and error when the guess is found to be wrong after fitting. The only systematic method of modeling that remains is to try all the possible ARMA models of order ($m,n < k$) for $k = 1, 2, 3, \ldots$, etc. As indicated in Chapter 2, a simpler and more straightforward modeling procedure is possible based on the ARMA ($n,n-1$) strategy, which does not involve the examination of sample autocorrelations.

Similar remarks apply to checking the independence of a_t's for model adequacy by residual autocorrelations, that is, the autocorrelation estimates obtained from Eqs. (3.3.2) and (3.3.3) replacing X_t's by the residuals a_t's obtained after fitting the model. If a_t's were an actual uncorrelated or white noise series, their estimated autocorrelations would be distributed approximately normally with mean zero and variance N^{-1} (see Appendix A 2.1.4). Therefore, values smaller than $2/\sqrt{N}$ may be considered insignificant in testing whether a given series is white noise or independent. However, it is pointed out in Durbin (1970) that such a test is no longer appropriate for testing whether a_t's are uncorrelated or not, since the a_t's are *estimated* residuals obtained by using estimated parameter values. For this reason we will make use of the residual autocorrelation only as a supplementary criterion in checking. The main checking criterion will be based on reductions in the sum of squares of a_t's, as illustrated in Section 2.3.3. Hence, neither the sample nor the residual autocorrelations play a significant role in our approach.

However, as a statistical characterization, the theoretical autocovariance function γ_k is important and we will derive and discuss it in detail for the various ARMA models. In this development we will need the covariances between $a_{t-\ell}$ and X_{t-k} that can be obtained from Eq. (3.3.1) and the Green's function expansion, true for any ARMA model

$$X_t = \sum_{j=0}^{\infty} G_j a_{t-j}$$

It immediately follows that since $G_0 = 1$

$$E(a_t X_t) = \sigma_a^2$$

and

$$E(a_t X_{t-k}) = 0, \qquad k > 0$$

from Eq. (3.3.1). These relations may also be written as

$$E(a_t X_{t-k}) = \delta_k \sigma_a^2 \tag{3.3.4}$$

Similarly, in general,

$$\begin{aligned}
E(a_{t-\ell} X_{t-k}) &= E\left[a_{t-\ell} \left(\sum_{j=0}^{\infty} G_j a_{t-k-j} \right) \right] \\
&= \sum_{j=0}^{\infty} G_j E(a_{t-\ell} a_{t-k-j}) \\
&= 0 \qquad \text{if} \qquad k > \ell \\
&= G_{\ell-k} \sigma_a^2, \qquad \text{if} \qquad k \leq \ell
\end{aligned} \tag{3.3.5}$$

Also note the important relation

$$\gamma_k = E(X_t X_{t-k}) = E(X_{t-k} X_t) = E(X_t X_{t+k}) = \gamma_{-k} \tag{3.3.6}$$

3.3.3 The AR(1) Model

The autocovariance or autocorrelation function of an ARMA model can be derived in either of the two forms similar to the Green's function: the recursive form using the model or the explicit expression using the Green's function expansion. First, let us consider the AR(1) model

$$X_t = \phi_1 X_{t-1} + a_t$$

Implicit Expression

Multiplying both sides by X_{t-k} and taking expectation, we have

$$E(X_t X_{t-k}) = \phi_1 E(X_{t-1} X_{t-k}) + E(a_t X_{t-k})$$

and from Eqs. (3.3.4) to (3.3.6)

$$\gamma_0 = \phi_1 \gamma_1 + \sigma_a^2$$

$$\gamma_k = \phi_1 \gamma_{k-1}, \qquad k > 0$$

Substituting $\gamma_1 = \phi_1 \gamma_0$ from the second equation into the first one, we can write the relations for autocovariances as

$$\gamma_0 = \frac{\sigma_a^2}{1 - \phi_1^2}$$

$$\gamma_k = \phi_1 \gamma_{k-1}, \qquad k \geq 1 \tag{3.3.7}$$

This expression for the variance γ_0 agrees with Eq. (3.1.8). Dividing by γ_0, we get the relations for the autocorrelation function ρ_k as

$$\rho_0 = 1, \qquad \rho_k = \phi_1 \rho_{k-1} \tag{3.3.8}$$

The autocorrelation function always has the same form as the autocovariance function since it is only a scaled version of the latter with the scaling factor γ_0.

Explicit Expression

The explicit expression for γ_k can be obtained by the Green's function, which for the AR(1) model is

$$G_j = \phi_1^j$$

Therefore,

$$\gamma_k = E(X_t X_{t-k})$$

$$= E\left[\left(\sum_{i=0}^{\infty} G_i a_{t-i}\right)\left(\sum_{j=0}^{\infty} G_j a_{t-(j+k)}\right)\right]$$

$$= \left(\sum_{j=0}^{\infty} G_{j+k} G_j\right)\sigma_a^2 = \left(\phi_1^k \sum_{j=0}^{\infty} \phi_1^{2j}\right)\sigma_a^2$$

$$= \frac{\sigma_a^2}{(1-\phi_1^2)}\phi_1^k \qquad\qquad (3.3.9)$$

since the expectations in the second step are nonzero and each equal to $G_{j+k}G_j\sigma_a^2$ only when $i = j + k$. This expression can also be obtained from Eq. (3.3.7) either by direct substitution or as a solution of the homogeneous difference equation

$$(1-\phi_1 B)\gamma_k = 0, \qquad k > 0$$

with initial condition

$$\gamma_0 = \frac{\sigma_a^2}{1 - \phi_1^2}$$

Note that γ_k satisfies the homogeneous part of the AR(1) model for $k \geq 1$, as in the case of the Green's function G_j. In fact, its autocorrelation function ρ_k coincides with the Green's function and is illustrated in Fig. 3.14 for values of $\phi_1 = 0.5$ and $\phi_1 = -0.5$.

It is seen that when ϕ_1 is positive, the γ_k or ρ_k function for an AR(1) model exponentially decays to zero. We have seen in Chapter 2 that the AR(1) model with $\phi_1 = 0.8983$ is adequate for the papermaking data. If we assume this is the true value, its theoretical function ρ_k should be of the form

$$\rho_k = (0.8983)^k$$

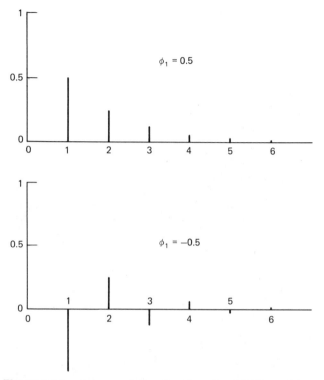

Figure 3.14 Autocorrelation functions for AR(1) models.

exponentially decaying at the rate 0.8983. The sample autocorrelation $\hat{\rho}_k$ computed from the data by Eq. (3.3.3) is shown in Fig. 3.15. Although it shows exponential decay initially, the $\hat{\rho}_k$ at larger lags is considerably distorted, exhibiting persistant wave patterns on the positive side of the axis, illustrating the undesirable properties of sample autocorrelations $\hat{\rho}_k$.

For the random walk model

$$(1 - B)X_t = a_t$$

since $\phi_1 = 1$ the ρ_k function is given by

$$\rho_k \equiv 1$$

Thus, the autocorrelation at any lag is the highest and the variance is infinite. These results are, of course, intuitively obvious. Since we treat this model as a limiting case of AR(1) with ϕ_1 tending to one, we consider it as a system with high variance and a very slowly decaying autocorrelation function. The $\hat{\rho}_k$ for the IBM data for which $\phi_1 = 0.999$ is also shown in Fig. 3.15.

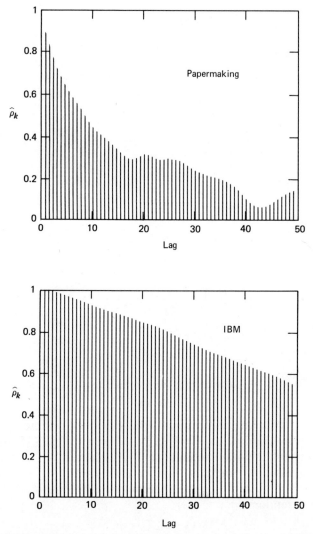

Figure 3.15 Sample autocorrelation $\hat{\rho}_k$ for papermaking and IBM data.

3.3.4 The MA(1) Model

For the MA(1) model

$$X_t = a_t - \theta_1 a_{t-1}$$

$$E(X_t X_{t-k}) = E(a_t X_{t-k}) - \theta_1 E(a_{t-1} X_{t-k})$$

and since

$$G_0 = 1, \quad G_1 = -\theta_1, \quad G_j = 0, \qquad j \geq 2$$

we have by Eqs. (3.3.5) and (3.3.6)

$$\left.\begin{array}{l} \gamma_0 = \sigma_a^2 + \theta_1^2 \sigma_a^2 = (1 + \theta_1^2) \sigma_a^2 \\ \gamma_1 = -\theta_1 \sigma_a^2 \\ \gamma_k = 0, \qquad k \geq 2 \end{array}\right\} \tag{3.3.10}$$

and the ρ_k function takes the form

$$\rho_0 = 1$$

$$\rho_1 = -\frac{\theta_1}{1 + \theta_1^2}$$

$$\rho_k = 0, \qquad k \geq 2 \tag{3.3.11}$$

Note that the ρ_k or γ_k function for the MA(1) model cuts off after lag one and becomes zero.

3.3.5 The ARMA(2,1) Model

Now consider the ARMA(2,1) model

$$X_t = \phi_1 X_{t-1} + \phi_2 X_{t-2} - \theta_1 a_{t-1} + a_t$$

Implicit Expression

$$E(X_t X_{t-k}) = \phi_1 E(X_{t-1} X_{t-k}) + \phi_2 E(X_{t-2} X_{t-k})$$
$$- \theta_1 E(a_{t-1} X_{t-k}) + E(a_t X_{t-k})$$

and since

$$G_1 = \phi_1 - \theta_1$$

we have by Eqs. (3.3.5) and (3.3.6)

$$k = 0 : \gamma_0 = \phi_1 \gamma_1 + \phi_2 \gamma_2 - (\phi_1 - \theta_1)\theta_1 \sigma_a^2 + \sigma_a^2$$
$$k = 1 : \gamma_1 = \phi_1 \gamma_0 + \phi_2 \gamma_1 - \theta_1 \sigma_a^2 \tag{3.3.12}$$
$$k : \gamma_k = \phi_1 \gamma_{k-1} + \phi_2 \gamma_{k-2}, \qquad k \geq 2$$

In order to be able to compute the values of γ_k recursively, we have to solve for γ_0 and γ_1, which can be done by using the first two linear equations in (3.3.12) after substituting for γ_2 from the third one. Thus,

$$\gamma_0 = \phi_1 \gamma_1 + \phi_2(\phi_1 \gamma_1 + \phi_2 \gamma_0) + \sigma_a^2(1 - \phi_1 \theta_1 + \theta_1^2)$$

that is,

$$(1 - \phi_2^2)\gamma_0 - \phi_1(1 + \phi_2)\gamma_1 = \sigma_a^2(1 - \phi_1 \theta_1 + \theta_1^2)$$

and

$$-\phi_1\gamma_0 + (1-\phi_2)\gamma_1 = -\theta_1\sigma_a^2$$

Therefore, using Cramer's rule to solve for γ_0,γ_1, we have

$$\gamma_0 = \sigma_a^2 \frac{\begin{vmatrix} (1-\phi_1\theta_1+\theta_1^2) & -\phi_1(1+\phi_2) \\ -\theta_1 & (1-\phi_2) \end{vmatrix}}{\begin{vmatrix} (1-\phi_2^2) & -\phi_1(1+\phi_2) \\ -\phi_1 & (1-\phi_2) \end{vmatrix}}$$

$$\gamma_1 = \sigma_a^2 \frac{\begin{vmatrix} (1-\phi_2^2) & (1-\phi_1\theta_1+\theta_1^2) \\ -\phi_1 & -\theta_1 \end{vmatrix}}{\begin{vmatrix} (1-\phi_2^2) & -\phi_1(1+\phi_2) \\ -\phi_1 & (1-\phi_2) \end{vmatrix}}$$

$$\gamma_k = \phi_1\gamma_{k-1} + \phi_2\gamma_{k-2}, \qquad k \geq 2 \tag{3.3.13}$$

Also,

$$\rho_0 = 1$$

$$\rho_1 = \frac{\begin{vmatrix} (1-\phi_2^2) & (1-\phi_1\theta_1+\theta_1^2) \\ -\phi_1 & -\theta_1 \end{vmatrix}}{\begin{vmatrix} (1-\phi_1\theta_1+\theta_1^2) & -\phi_1(1+\phi_2) \\ -\theta_1 & (1-\phi_2) \end{vmatrix}}$$

$$\rho_k = \phi_1\rho_{k-1} + \phi_2\rho_{k-2}, \qquad k \geq 2 \tag{3.3.14}$$

Explicit Expression

To obtain the explicit expression for γ_k, we have from Eq. (3.1.16a)

$$G_j = g_1\lambda_1^j + g_2\lambda_2^j$$

Hence,

$$\gamma_k = E(X_t X_{t-k})$$

$$= E\left[\left(\sum_{i=0}^{\infty} G_i a_{t-i}\right)\left(\sum_{j=0}^{\infty} G_j a_{t-(k+j)}\right)\right]$$

$$= \left(\sum_{j=0}^{\infty} G_{k+j} G_j\right)\sigma_a^2$$

$$= \sigma_a^2 \sum_{j=0}^{\infty} (g_1\lambda_1^{k+j} + g_2\lambda_2^{k+j})(g_1\lambda_1^j + g_2\lambda_2^j)$$

$$= \sigma_a^2 \sum_{j=0}^{\infty} [g_1^2\lambda_1^k\lambda_1^{2j} + g_2^2\lambda_2^k\lambda_2^{2j}$$

$$+ g_1g_2\lambda_1^j\lambda_2^j(\lambda_1^k + \lambda_2^k)]$$

$$= \sigma_a^2\left[\frac{g_1^2}{(1 - \lambda_1^2)}\lambda_1^k + \frac{g_2^2}{(1 - \lambda_2^2)}\lambda_2^k\right.$$

$$\left. + \frac{g_1g_2}{(1 - \lambda_1\lambda_2)}(\lambda_1^k + \lambda_2^k)\right] \tag{3.3.15}$$

In particular,

$$\gamma_0 = \sigma_a^2\left[\frac{g_1^2}{1 - \lambda_1^2} + \frac{g_2^2}{1 - \lambda_2^2} + \frac{2g_1g_2}{1 - \lambda_1\lambda_2}\right]$$

The autocorrelation function $\rho_k = \gamma_k/\gamma_0$ for some typical ARMA(2,1) models considered earlier are given in Figs. 3.16 to 3.18. If we collect the coefficients of λ_1^k and λ_2^k together,

$$\gamma_k = \sigma_a^2\left[\frac{g_1^2}{1 - \lambda_1^2} + \frac{g_1g_2}{1 - \lambda_1\lambda_2}\right]\lambda_1^k$$

$$+ \sigma_a^2\left[\frac{g_2^2}{1 - \lambda_2^2} + \frac{g_1g_2}{1 - \lambda_1\lambda_2}\right]\lambda_2^k$$

where, by (3.1.17)

$$g_1 = \frac{\lambda_1 - \theta_1}{\lambda_1 - \lambda_2}, \qquad g_2 = \frac{\lambda_2 - \theta_1}{\lambda_2 - \lambda_1}$$

It is thus seen that as in the case of the Green's function, the auto-covariance function is a linear combination of terms such as λ_i^k, that is,

$$\gamma_k = d_1\lambda_1^k + d_2\lambda_2^k \tag{3.3.16}$$

where

$$d_1 = \sigma_a^2 g_1\left[\frac{g_1}{1-\lambda_1^2} + \frac{g_2}{1-\lambda_1\lambda_2}\right] \tag{3.3.16a}$$

$$= \frac{\sigma_a^2(\lambda_1 - \theta_1)}{(\lambda_1 - \lambda_2)^2}\left[\frac{(\lambda_1 - \theta_1)}{(1 - \lambda_1^2)} - \frac{(\lambda_2 - \theta_1)}{(1 - \lambda_1\lambda_2)}\right]$$

$$d_2 = \sigma_a^2 g_2\left[\frac{g_2}{1-\lambda_2^2} + \frac{g_1}{1-\lambda_1\lambda_2}\right] \tag{3.3.16b}$$

$$= \frac{\sigma_a^2(\lambda_2 - \theta_1)}{(\lambda_1 - \lambda_2)^2}\left[\frac{(\lambda_2 - \theta_1)}{(1 - \lambda_2^2)} - \frac{(\lambda_1 - \theta_1)}{(1 - \lambda_1\lambda_2)}\right]$$

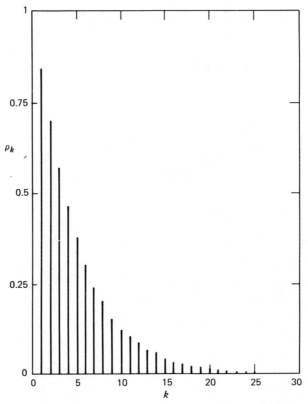

Figure 3.16 Autocorrelation functions for ARMA(2,1) model: $\phi_1 = 1.3$, $\phi_2 = -0.4$, $\theta_1 = 0.4$.

and, in particular,

$$\gamma_0 = d_1 + d_2 \qquad (3.3.16c)$$

which can also be obtained as a solution of the difference equation (3.3.13). When $\phi_1^2 + 4\phi_2 \geq 0$ and the roots λ_1 and λ_2 are real, γ_k is a sum of exponentials. When $\phi_1^2 + 4\phi_2 < 0$ and the roots are complex, γ_k is a damped sine wave for which the expression can be obtained in the same way as for G_j in Section 3.1.6. An important difference between the behaviors of G_j and γ_k is that G_1 can be larger than $G_0 = 1$ (whenever $\phi_1 - \theta_1 > 1$), but γ_1 can never be larger than γ_0. In fact, it can be verified that

$$\gamma_0 \geq \gamma_k$$

for all k and in particular for $k = 1$ [see Appendix A 2.1.2].

Special Case: AR(2)

The autocovariance function of the AR(2) model is obtained by setting $\theta_1 = 0$ in Eqs. (3.3.16a) and (3.3.16b). Note that the variance of the AR(2) process is

$$\gamma_0 = d_1 + d_2$$

$$= \frac{\sigma_a^2}{(\lambda_1 - \lambda_2)^2} \left[\frac{\lambda_1^2}{1 - \lambda_1^2} - \frac{2\lambda_1\lambda_2}{1 - \lambda_1\lambda_2} + \frac{\lambda_2^2}{1 - \lambda_2^2} \right]$$

$$= \frac{\sigma_a^2(1 + \lambda_1\lambda_2)}{(1 - \lambda_1\lambda_2)(1 - \lambda_1^2)(1 - \lambda_2^2)} \tag{3.3.17}$$

$$= \frac{\sigma_a^2(1 - \phi_2)}{(1 + \phi_2)[(1 - \phi_2)^2 - \phi_1^2]} \tag{3.3.17a}$$

by Eq. (3.1.15a), after some algebraic manipulations.

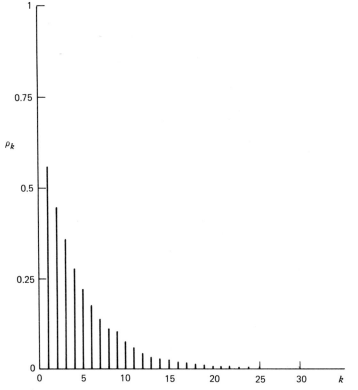

Figure 3.17 Autocorrelation function for ARMA(2,1) model: $\phi_1 = 0.85$, $\phi_2 = -0.04$, $\theta_1 = 0.4$.

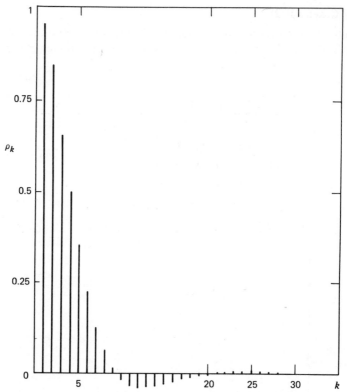

Figure 3.18   Autocorrelation function for ARMA(2,1) model: $\phi_1 = 1.5$, $\phi_2 = -0.6$, $\theta = -0.5$.

Special Case: ARMA(1,1)

The γ_k functions for the ARMA(1,1) model is obtained by putting $\lambda_2 = 0$ in Eqs. (3.3.16a) and (3.3.16b) as

$$\gamma_k = \frac{\sigma_a^2(\lambda_1 - \theta_1)}{\lambda_1^2}\left[\frac{\lambda_1 - \theta_1}{1 - \lambda_1^2} + \theta_1\right]\lambda_1^k$$

$$= \frac{\sigma_a^2(\lambda_1 - \theta_1)(1 - \lambda_1\theta_1)}{(1 - \lambda_1^2)}\lambda_1^{k-1}$$

$$= \frac{\sigma_a^2(\phi_1 - \theta_1)(1 - \phi_1\theta_1)}{(1 - \phi_1^2)}\phi_1^{k-1}, \qquad k \geq 1 \qquad (3.3.18a)$$

However, just like G_j for the ARMA(1,1) given by (3.1.19), this expression (3.3.18a) is true only for $k > 0$; γ_0 may be separately obtained by using

(3.1.19) and $G_0 = 1$ as

$$\gamma_0 = \sigma_a^2 \sum_{j=0}^{\infty} G_j^2$$

$$= \sigma_a^2 \left[1 + \sum_{j=1}^{\infty} (\phi_1 - \theta_1)^2 \phi_1^{2(j-1)} \right]$$

$$= \sigma_a^2 \left[1 + \frac{(\phi_1 - \theta_1)^2}{1 - \phi_1^2} \right]$$

$$= \sigma_a^2 \frac{(1 - 2\phi_1\theta_1 + \theta_1^2)}{(1 - \phi_1^2)} \tag{3.3.18b}$$

Substituting $\theta_1 = 0$, we can verify that we get the corresponding expressions for the AR(1) model. [See Problems 3.13 and 3.20 for alternative proofs of Eq. (3.3.18b).]

3.3.6 Representation of Dynamics

While discussing the Green's function, we pointed out that in representing the dynamics by an ARMA(2,1) model rather than by an AR(1) model, we are representing it by a linear combination of two roots rather than by only one root (see Section 3.1.8). The same is true of the γ_k function that has the form

$$\text{AR(1):} \quad \gamma_k = \gamma_0 \lambda_1^k$$

$$\text{ARMA(2,1):} \quad \gamma_k = d_1 \lambda_1^k + d_2 \lambda_2^k$$

The transition from AR(1) to ARMA(2,1) is needed when the sequential dependence or correlation in the data is not so simple as to be represented by a single exponential, but needs a weighted sum of two exponentials (real or complex).

The two coefficients d_1 and d_2 are always restricted by

$$\gamma_0 = d_1 + d_2$$

Only one of the two coefficients is free, say d_2, and the value d_1 must be

$$d_1 = \gamma_0 - d_2$$

If the free coefficient d_2 happens to take the value that satisfies [see Eq. (3.3.12)]

$$\gamma_1 = \phi_1 \gamma_0 + \phi_2 \gamma_1$$

$$\gamma_1 - \phi_2 \gamma_1 = \phi_1 \gamma_0$$

$$(1 - \phi_2)\gamma_1 = \phi_1 \gamma_0$$

substituting for γ_1 from Eq. (3.3.16)

$$(1-\phi_2)(d_1\lambda_1+d_2\lambda_2) = \phi_1\gamma_0$$

but $d_1 = \gamma_0 - d_2$, so

$$(1-\phi_2)[(\gamma_0-d_2)\lambda_1 + d_2\lambda_2] = \phi_1\gamma_0$$

substituting for ϕ_1 and ϕ_2 from Eq. (3.1.15a), we have

$$(1+\lambda_1\lambda_2)[(\gamma_0-d_2)\lambda_1 + d_2\lambda_2] = (\lambda_1+\lambda_2)\gamma_0$$

then the data can be represented by an AR(2) model. In general, however, when this equality is not satisfied, an ARMA(2,1) model would be required with $\theta_1\sigma_a^2$ given by Eq. (3.3.12)

$$\gamma_1 = \phi_1\gamma_0 + \phi_2\gamma_1 - \theta_1\sigma_a^2$$

that is,

$$\theta_1\sigma_a^2 = \phi_1\gamma_0 - \gamma_1 + \phi_2\gamma_1$$

$$= \phi_1\gamma_0 - (1-\phi_2)\gamma_1 \tag{3.3.19}$$

$$= (\lambda_1+\lambda_2)\gamma_0 - (1+\lambda_1\lambda_2)[(\gamma_0-d_2)\lambda_1 + d_2\lambda_2] \tag{3.3.19a}$$

3.3.7 Variance Decomposition

From the systems point of view, Eqs. (3.3.16) and (3.3.16c) have an interesting interpretation. They show that the total variance γ_0 in the system is decomposed into two components d_1 and d_2 respectively associated with the characteristic roots λ_1 and λ_2 in the autocovariance function. In other words, d_1 and d_2 provide an analysis of variance corresponding to the two roots and the parameter θ_1. For example, suppose we have a system with fixed $\sigma_a^2 = 1$. If the dynamics of this system is of a single exponential type so that it can be described by an AR(1) model, for example,

$$X_t - 0.8X_{t-1} = a_t$$

that is, by the Green's function with $\lambda_1=\phi_1=0.8$

$$G_j = (0.8)^j$$

then the variance of the system response will be

$$\gamma_0 = \frac{1}{1 - 0.8^2} = 2.78$$

and the covariance function is

$$\gamma_k = 2.78(0.8)^k$$

Next, suppose the dynamics can be made more complex by adding another root $\lambda_2=0.5$ and $\theta_1=0.4$, so that the Green's function is, as given by Eq. (3.1.16) [see Fig. 3.5]:

$$G_j = 1.33(0.8)^j - 0.33(0.5)^j$$

Then, the model becomes

$$X_t - 1.3X_{t-1} + 0.4X_{t-2} = a_t - 0.4a_{t-1}$$

and the variance components associated with $\lambda_1 = 0.8$ and $\lambda_2 = 0.5$ are, by Eqs. (3.3.16a) and (3.3.16b):

$$d_1 = \frac{0.4}{0.3^2} \left[\frac{0.4}{0.36} - \frac{0.1}{0.6} \right] = 4.2$$

$$d_2 = \frac{0.1}{0.3^2} \left[\frac{0.1}{0.75} - \frac{0.4}{0.6} \right] = -0.6$$

The variance of the system response increases to

$$\gamma_0 = 4.2 - 0.6 = 3.6$$

and the autocovariance function is

$$\gamma_k = 4.2(0.8)^k - 0.6(0.5)^k$$

which is dominated by the root $\lambda_1 = 0.8$, as is also the Green's function given above. This is not surprising; if, in fact, we use $\theta_1 = 0.5$ instead of 0.4, then we have

$$(1-0.8B)(1-0.5B)X_t = (1-0.5B)a_t$$

that is,

$$(1-0.8B)X_t = a_t$$

and we are back to the AR(1) model.

If the root $\lambda_2 = 0.5$ is added but $\theta_1 = 0$, that is, the model is now AR(2)

$$X_t - 1.3X_{t-1} + 0.4X_{t-2} = a_t$$

$$d_1 = \frac{0.8}{0.3^2} \left[\frac{0.8}{0.36} - \frac{0.5}{0.6} \right] = 12.3$$

$$d_2 = \frac{0.5}{0.75} \left[\frac{0.5}{0.75} - \frac{0.8}{0.6} \right] = -3.7$$

The variance now jumps to

$$\gamma_0 = 12.3 - 3.7 = 8.6$$

and the autocovariance function takes the form

$$\gamma_k = 12.3(0.8)^k - 3.7(0.5)^k$$

whereas the Green's function in this case is

$$G_j = 2.67(0.8)^j - 1.67(0.5)^j$$

3.3.8 Relation Between the Green's Function and the Autocovariance Function

The similarities between the Green's function G_j and the autocovariance function γ_k found in this section are not accidental. Once the basic covariance structure of the input a_t is defined by Eq. (3.3.1), it is the Green's function that completely determines the autocovariance structure of the output X_t. For the ARMA models the dynamics and the autocovariance represent the same phenomenon, namely the memory or the dependence in successive observations in a time series. Recall that given the input a_t, the dynamic response or output is given by its "convolution" with the Green's function:

$$X_t = \sum_{j=0}^{\infty} G_j a_{t-j}$$

Similarly, given the autocovariance of the input

$$E(a_t a_{t-k}) = \delta_k \sigma_a^2$$

the autocovariance of the output is also given by its double convolution with the Green's function:

$$\gamma_k = E(X_t X_{t-k})$$

$$= E\left(\sum_{i=0}^{\infty} G_i a_{t-i} \right)\left(\sum_{j=0}^{\infty} G_j a_{t-(k+j)} \right)$$

$$= \sum_{i=0}^{\infty} \sum_{j=0}^{\infty} G_i G_j \delta_{i-(k+j)} \sigma_a^2$$

$$= \sigma_a^2 \sum_{j=0}^{\infty} G_j G_{j+k} \tag{3.3.20}$$

Thus, the Green's function is really the essence of the autocovariance function.

3.3.9 The MA(2) Model

Now consider the MA(2) model

$$X_t = a_t - \theta_1 a_{t-1} - \theta_2 a_{t-2}$$

For this model

$$G_0 = 1, \quad G_1 = -\theta_1, \quad G_2 = -\theta_2, \quad G_j = 0, \quad j > 2$$

Therefore, either multiplying by X_{t-k} and taking expectation, or directly using Eq. (3.3.20), we can easily see that

$$\gamma_0 = (1 + \theta_1^2 + \theta_2^2)\sigma_a^2$$
$$\gamma_1 = (-\theta_1 + \theta_1\theta_2)\sigma_a^2$$
$$\gamma_2 = -\theta_2\sigma_a^2$$
$$\gamma_k = 0, \qquad k \geq 3 \tag{3.3.21}$$

and

$$\rho_0 = 1$$
$$\rho_1 = \frac{-\theta_1(1 - \theta_2)}{1 + \theta_1^2 + \theta_2^2}$$
$$\rho_2 = \frac{-\theta_2}{1 + \theta_1^2 + \theta_2^2}$$
$$\rho_k = 0, \qquad k \geq 3 \tag{3.3.22}$$

*3.3.10 General Results

By following similar methods, we can see that for a general ARMA(n,m) model

$$X_t = \phi_1 X_{t-1} + \phi_2 X_{t-2} + \ldots + \phi_n X_{t-n}$$
$$\quad + a_t - \theta_1 a_{t-1} - \theta_2 a_{t-2} - \ldots - \theta_m a_{t-m}$$
$$\gamma_0 = \phi_1\gamma_1 + \phi_2\gamma_2 + \ldots + \phi_n\gamma_n$$
$$\quad + (1 - \theta_1 G_1 - \theta_2 G_2 - \ldots - \theta_m G_m)\sigma_a^2$$
$$\gamma_1 = \phi_1\gamma_0 + \phi_2\gamma_1 + \ldots + \phi_n\gamma_{n-1}$$
$$\quad + (-\theta_1 - \theta_2 G_1 - \theta_3 G_2 - \ldots - \theta_m G_{m-1})\sigma_a^2 \tag{3.3.23}$$

$$\ldots$$

$$\gamma_m = \phi_1\gamma_{m-1} + \phi_2\gamma_{m-2} + \ldots + \phi_n\gamma_{|n-m|} - \theta_m\sigma_a^2$$
$$\gamma_k = \phi_1\gamma_{k-1} + \phi_2\gamma_{k-2} + \ldots + \phi_n\gamma_{k-n} \qquad k \geq m+1 \tag{3.3.24}$$

or

$$\Phi(B)\gamma_k = 0, \qquad k \geq m+1$$

For the ARMA$(n, n-1)$ model, Eq. (3.3.16) generalizes to

$$\gamma_k = d_1\lambda_1^k + d_2\lambda_2^k + \ldots + d_n\lambda_n^k \tag{3.3.25}$$

where

$$d_i = \left(\frac{g_i g_1}{1 - \lambda_i \lambda_1} + \frac{g_i g_2}{1 - \lambda_i \lambda_2} + \ldots + \frac{g_i g_n}{1 - \lambda_i \lambda_n} \right) \quad (3.3.26)$$

$$i = 1, 2, \ldots, n$$

and g_i is defined by Eq. (3.1.26). In particular, the variance (decomposition) is given by

$$\gamma_0 = d_1 + d_2 + \ldots + d_n \quad (3.3.27)$$

in which d_i is the contribution of the dynamic mode λ_i. When λ_i, λ_{i+1} are a pair of complex conjugate roots, $(d_i + d_{i+1})$ is the contribution due to the corresponding decaying sinusoidal mode, whose frequency ω and decay rate r can be calculated as illustrated in Section 3.1.6. Note that the explicit results for ARMA(n, m) models with $m < n - 1$ can be obtained from Eqs. (3.3.25) to (3.3.27) by setting the last $n - 1 - m$ θ_i's zero.

*3.4 PARTIAL AUTOCORRELATION AND AUTOSPECTRUM

In this section we briefly indicate two other characteristics of the ARMA systems derived from the autocovariance/autocorrelation function. In the existing literature they are obtained from the estimated sample autocorrelation. The first of these is the partial autocorrelation function which is mainly used to guess the order of the model when it is of pure autoregressive form. Since we do not use autocorrelation in modeling, we do not need the estimated sample partial autocorrelation in our study. The second is the autospectrum that gives a frequency decomposition of the output X_t. The autospectrum is particularly useful in engineering studies of various systems.

3.4.1 Partial Autocorrelation

It was shown in the last section that the autocovariance or autocorrelation of an MA(m) process cuts off (or becomes zero) after the mth lag. When the sample autocorrelation is close to the theoretical one, it will also be close to zero after the mth lag from which the order of a pure MA model can be guessed. No such indication is available for pure AR processes, for which the autocorrelation has no cut-off point. Therefore, partial autocorrelation ρ'_k is manufactured out of ρ_k so as to provide a cut-off after the nth lag for an AR(n) process. Suppose ϕ_{ki} denotes the ith autoregressive parameter of an AR(k) process; then,

$$\rho_i = \phi_{k1} \rho_{i-1} + \phi_{k2} \rho_{i-2} + \ldots$$

$$+ \phi_{kk} \rho_{i-k}, \quad i = 1, 2, \ldots, k \quad (3.4.1)$$

which gives the Yule-Walker equations

$$\rho_1 = \phi_{k1}\rho_0 + \phi_{k2}\rho_1 + \ldots + \phi_{kk}\rho_{k-1}$$
$$\rho_2 = \phi_{k1}\rho_1 + \phi_{k2}\rho_0 + \ldots + \phi_{kk}\rho_{k-2}$$

$$\ldots$$

$$\rho_k = \phi_{k1}\rho_{k-1} + \phi_{k2}\rho_{k-2} + \ldots + \phi_{kk}\rho_0$$

Using Cramer's rule successively for $k = 1, 2, \ldots$, since $\rho_0 = 1$, we have

$$\phi_{11} = \rho_1$$

$$\phi_{22} = \frac{\begin{vmatrix} 1 & \rho_1 \\ \rho_1 & \rho_2 \end{vmatrix}}{\begin{vmatrix} 1 & \rho_1 \\ \rho_1 & 1 \end{vmatrix}}$$

$$\phi_{33} = \frac{\begin{vmatrix} 1 & \rho_1 & \rho_1 \\ \rho_1 & 1 & \rho_2 \\ \rho_2 & \rho_1 & \rho_3 \end{vmatrix}}{\begin{vmatrix} 1 & \rho_1 & \rho_2 \\ \rho_1 & 1 & \rho_1 \\ \rho_2 & \rho_1 & 1 \end{vmatrix}}$$

$$\ldots$$

$$\phi_{kk} = \frac{\begin{vmatrix} 1 & \rho_1 & \rho_2 & \cdots & \rho_{k-2} & \rho_1 \\ \rho_1 & 1 & \rho_1 & \cdots & \rho_{k-3} & \rho_2 \\ \vdots & \vdots & \vdots & & \vdots & \vdots \\ \rho_{k-1} & \rho_{k-2} & \rho_{k-3} & \cdots & \rho_1 & \rho_k \end{vmatrix}}{\begin{vmatrix} 1 & \rho_1 & \rho_2 & \cdots & \rho_{k-1} \\ \rho_1 & 1 & \rho_1 & \cdots & \rho_{k-2} \\ \vdots & \vdots & \vdots & & \vdots \\ \rho_{k-1} & \rho_{k-2} & \rho_{k-3} & \cdots & 1 \end{vmatrix}} \qquad (3.4.2)$$

Now it is clear that for an AR(n) process for $k > n$ the determinant in ϕ_{kk} will vanish, since one of its rows can be written as a linear combination of others. The expression ϕ_{kk} is therefore called partial autocorrelation and denoted by ρ'_k. Simple recursive formulae for computing these are given in Durbin (1960).

3.4.2 Autospectrum

[handwritten: line dist of the power of the Data by frequency]

A deterministic signal or output from a system can be expressed as a mixture of sine cosine waves at different frequencies. When the output is in the form of a periodic function or is specified over a finite interval, one can use a Fourier series technique; otherwise, a Fourier transform can be used.

A time series or a stochastic system response treated as functions of time do not belong to the class of functions dealt with in the usual Fourier transform theory. Khintchin (1934) showed that the frequency decomposition of such a random function can be obtained by taking the Fourier transform of the autocovariance function that is a well-behaved deterministic function for which the usual Fourier transform can be used. This Fourier transform of the autocovariance function is known as autospectrum or power spectrum since it shows how the *variance* of the process X_t is distributed over the frequency bands. By taking the Fourier transform of the autocovariance function or by using the simpler methods of spectral representation given in Yaglom (1952), we can show that the autospectrum of an ARMA(n,m) process is given by

$$f(\omega) = \frac{\Delta\sigma_a^2}{2\pi} \frac{\left|e^{mi\omega\Delta} - \theta_1 e^{(m-1)i\omega\Delta} - \ldots - \theta_m\right|^2}{\left|e^{ni\omega\Delta} - \phi_1 e^{(n-1)i\omega\Delta} - \ldots - \phi_n\right|^2}, \qquad -\frac{\pi}{\Delta} \leq \omega \leq \frac{\pi}{\Delta}$$

(3.4.3)

where Δ is the sampling interval and ω is the angular frequency in radians per unit time.

As in the case of ρ_k, the autospectrum can be estimated from the sample by taking the Fourier transform of $\hat{\rho}_k$. However, for the reasons stated earlier in connection with $\hat{\rho}_k$, we prefer using Eq. (3.4.3) after fitting the model, and using the estimated values of the parameters. A better estimate of the spectrum over the entire bandwidth can be obtained from the continuous models that will be considered in Chapters 6 to 8.

APPENDIX A 3.1

PARTIAL FRACTIONS

Consider a fraction in which the denominator has been factored with *distinct* roots

$$\frac{f(x)}{(x-\mu_1)(x-\mu_2)\ldots(x-\mu_n)}$$

Then, the coefficient of any partial fraction with the denominator $(x-\mu_i)$ is obtained by substituting $x = \mu_i$ in the numerator and the denominator except the factor $(x - \mu_i)$. That is,

$$\frac{f(x)}{(x - \mu_1)(x - \mu_2) \ldots (x - \mu_n)} = \frac{f(\mu_1)}{(\mu_1 - \mu_2)(\mu_1 - \mu_3) \ldots (\mu_1 - \mu_n)}$$

$$\cdot \frac{1}{(x - \mu_1)} + \frac{f(\mu_2)}{(\mu_2 - \mu_1)(\mu_2 - \mu_3) \ldots (\mu_2 - \mu_n)}$$

$$\cdot \frac{1}{(x - \mu_2)}$$

$$\ldots + \frac{f(\mu_n)}{(\mu_n - \mu_1)(\mu_n - \mu_2) \ldots (\mu_n - \mu_{n-1})}$$

$$\cdot \frac{1}{(x - \mu_n)}$$

For example,

$$\frac{(x - 2)}{(x - 3)(x - 4)} = \frac{(3 - 2)}{(3 - 4)} \cdot \frac{1}{(x - 3)} + \frac{(4 - 2)}{(4 - 3)} \cdot \frac{1}{(x - 4)}$$

$$= -\frac{1}{(x - 3)} + \frac{2}{(x - 4)}$$

APPENDIX A 3.2

AN ALTERNATIVE POLAR FORM REPRESENTATION OF THE GREEN'S FUNCTION FOR THE ARMA(2,1) MODEL WITH COMPLEX ROOTS

A 3.2.1 Trigonometric Form of a Complex Number

Let us denote a complex number Z as

$$Z = a + ib \qquad (A\ 3.2.1)$$

where i is the imaginary unit. If we represent the complex number Z, corresponding to a point P on x-y complex plane as shown in Fig. A 3.1, the distance of the point P from the origin O is referred to as the absolute value or the modulus of Z and is denoted by

$$r = |Z| = \sqrt{a^2 + b^2} \qquad (A\ 3.2.2)$$

The directed angle measured from the positive x-axis to OP is called the argument of Z and is expressed as

$$\omega = \arg Z = \tan^{-1}\left(\frac{b}{a}\right) \qquad (A\ 3.2.3)$$

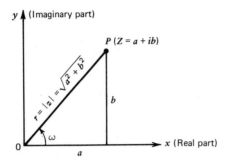

Figure A 3.1

Thus, the complex number Z in trigonometric form can also be written as

$$Z = a + ib$$

$$= \sqrt{a^2+b^2}\left[\frac{a}{\sqrt{a^2+b^2}} + i\frac{b}{\sqrt{a^2+b^2}}\right] \qquad \text{(A 3.2.4)}$$

$$= r[\cos \omega + i \sin \omega]$$

$$= re^{i\omega}$$

Similarly, it can be shown that the conjugate number $Z^* = a - ib$ in the trigonometric form can be denoted by

$$Z^* = a - ib = re^{-i\omega} \qquad \text{(A 3.2.5)}$$

Or, equivalently, the complex-conjugate numbers $a \pm ib$ can be written as $re^{\pm i\omega}$.

A 3.2.2 Polar Form Representation of the Green's Function (Eq. 3.1.16)

If the roots λ_1, λ_2 of the ARMA(2,1) model appear as a complex-conjugate pair, then from Eqs. (A 3.2.3) and (A 3.2.4) the roots can be represented as

$$\lambda_1, \lambda_2 = a \pm ib = \frac{1}{2}(\phi_1 \pm \sqrt{\phi_1^2 + 4\phi_2})$$

$$= r[\cos \omega \pm i \sin \omega]$$

$$= re^{\pm i\omega}$$

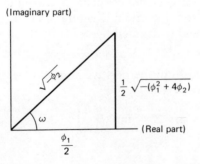

Figure A 3.2

Thus, by substituting the polar form of the roots in Eq. (3.1.16), we get

$$G_j = \left(\frac{\lambda_1 - \theta_1}{\lambda_1 - \lambda_2}\right)(re^{i\omega})^j + \left(\frac{\lambda_2 - \theta_1}{\lambda_2 - \lambda_1}\right)(re^{-i\omega})^j$$

$$= \frac{r^j}{(\lambda_1 - \lambda_2)}[(\lambda_1 - \theta_1)e^{+i(j\omega)} - (\lambda_2 - \theta_1)e^{-i(j\omega)}]$$

$$= \frac{r^j}{(\lambda_1 - \lambda_2)}[(\lambda_1 - \theta_1)(\cos j\omega + i \sin j\omega)$$

$$- (\lambda_2 - \theta_1)(\cos j\omega - i \sin j\omega)]$$

$$= \frac{r^j}{(\lambda_1 - \lambda_2)}[(\lambda_1 - \lambda_2) \cos j\omega$$

$$+ (\lambda_1 + \lambda_2 - 2\theta_1) i \sin j\omega]$$

$$= r^j\left[\cos j\omega + i \frac{\phi_1 - 2\theta_1}{(\lambda_1 - \lambda_2)} \sin j\omega\right]$$

$$= r^j\left[\cos j\omega + i \frac{(\phi_1 - 2\theta_1)}{i\sqrt{-(\phi_1^2 + 4\phi_2)}} \sin j\omega\right]$$

$$= r^j\left[\cos j\omega + \frac{\phi_1 - 2\theta_1}{\sqrt{-(\phi_1^2 + 4\phi_2)}} \sin j\omega\right] \qquad \text{(A 3.2.6)}$$

Let

$$-B = \frac{\phi_1 - 2\theta_1}{\sqrt{-(\phi_1^2 + 4\phi_2)}}$$

Thus, Eq. (A 3.2.6) can be rewritten as

$$G_j = r^j[\cos j\omega - B \sin j\omega] \qquad \text{(A 3.2.6a)}$$

If we define

$$A = \sqrt{1 + B^2}$$

Eq. (A 3.2.6a) could be rearranged as

$$= r^j A\left[\frac{1}{A} \cos j\omega - \frac{B}{A} \sin j\omega\right]$$

From Fig. A 3.3,

$$G_j = r^j A[\cos \beta \cos j\omega - \sin \beta \sin j\omega] \qquad \text{(A 3.2.7)}$$

$$= r^j A \cos (\beta + j\omega) = r^j 2g \cos (j\omega + \beta)$$

where

$$\beta = \tan^{-1}\left[\frac{-\phi_1 + 2\theta_1}{\sqrt{-(\phi_1^2 + 4\phi_2)}}\right]$$

and $g = \dfrac{A}{2}$ is defined by (3.1.17b)

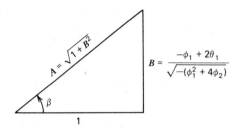

Figure A 3.3

APPENDIX A 3.3

STABILITY AND INVERTIBILITY CHECKS FROM PARAMETERS

Although the stability and invertibility can be checked by finding the roots and ensuring $|\lambda_i|<1$, $|\nu_i|<1$, a quicker check based on ϕ_i and θ_i can be made using a modification of the Routh-Hurwitz criterion of classical control theory [see, for example, Jury (1964)]. For stability of ARMA(n,m) models this criterion takes the following form. The parameters ϕ_i are used to construct the array given below, with $\phi_0 = -1$:

Row	Parameters				
1	ϕ_0	ϕ_1	ϕ_2	$\ldots\ldots$	ϕ_n
2	ϕ_n	ϕ_{n-1}	ϕ_{n-2}	$\ldots\ldots$	ϕ_0
3	a_0	a_1	a_2	$\ldots\ldots$	a_{n-1}
4	a_{n-1}	a_{n-2}	a_{n-3}	$\ldots\ldots$	a_0
5	b_0	b_1	b_2	$\ldots\ldots$	b_{n-2}
6	b_{n-2}	b_{n-3}	b_{n-4}	$\ldots\ldots$	b_0
$\ldots$	$\ldots$	$\ldots$	$\ldots$	$\ldots\ldots$	$\ldots$
$2n-3$	ℓ_0	ℓ_1	ℓ_2		

The elements of the third and fourth rows are found by

$$a_i = \begin{vmatrix} \phi_0 & \phi_{n-i} \\ \phi_n & \phi_i \end{vmatrix} = \phi_0\phi_i - \phi_n\phi_{n-i},$$

$$i = 0, 1, 2, \ldots, n-1, \ \phi_0 = -1$$

and those of the fifth and sixth rows by

$$b_i = \begin{vmatrix} a_0 & a_{n-1-i} \\ a_{n-1} & a_i \end{vmatrix} = a_0 a_i - a_{n-1}a_{n-1-i},$$

$$i = 0, 1, 2, \ldots, n-2$$

and so on until $2n-3$ rows are obtained. The last row will contain only three elements, say, ℓ_0, ℓ_1, and ℓ_2. The system is stable if and only if the following conditions are satisfied

(i) $\quad \phi_1 + \phi_2 + \phi_3 + \ldots + \phi_n < 1$

(ii) $\quad -\phi_1 + \phi_2 - \phi_3 + \ldots + (-1)^n \phi_n < 1$

(iii) $\quad |\phi_n| < 1$

$$|a_{n-1}| < |a_0|$$

$$|b_{n-2}| < |b_0|$$

$$\cdots \quad \cdots \quad \cdots$$

$$|\ell_2| < |\ell_0|$$

Invertibility can be checked in the same way if ϕ_i's are replaced by θ_i.

As an illustration, it is seen that for $n=2$ we need not construct any arrays. Conditions (i) and (ii) and the first of (iii) give

$$\phi_1 + \phi_2 < 1$$

$$\phi_2 - \phi_1 < 1$$

$$|\phi_2| < 1$$

as in Eqs. (3.1.21).

To illustrate the array computation for the Problem 3.2 (iii), the ARMA(3,2) model has $\phi_1 = 2$, $\phi_2 = -0.6$ and $\phi_3 = 0.2$ and the array takes the form

1	-1	2	-0.6	0.2
2	0.2	-0.6	2	-1
3	0.96	-1.78	0.2	

since

$$\ell_0 = a_0 = (-1)(-1) - (0.2)(0.2) = 0.96$$

$$\ell_1 = a_1 = (-1)(2) - (0.2)(-0.6) = -1.88$$

$$\ell_2 = a_2 = (-1)(-0.6) - (0.2)(2) = 0.2$$

Note that $\ell_0 = a_0$ is the determinant of the first and the last parameter columns, $\ell_1 = a_1$ that of the first and last but one, and $\ell_2 = a_2$ that of the first and last but two.

To check conditions (i) to (iii),

$$\phi_1 + \phi_2 + \phi_3 = 2 - 0.6 + 0.2 = 1.6 > 1$$

Since the condition (i) itself is not satisfied, no further checking needs to be done and we conclude that the system is unstable.

As these examples illustrate, conditions (i), (ii) and the first $|\phi_n| < 1$ of (iii) can be checked without constructing the array. If any one of them fails, the system is unstable and no further checking is needed. Only after these are satisfied, the array should be constructed and the remaining part of (iii) should be checked. The above array for Problem 3.2 (iii) was constructed only for illustration.

PROBLEMS

3.1 Plot G_j versus j for AR(1) model with ϕ_1 values
    (i) 0.2   (ii) 0.8   (iii) -0.8 (iv) $+1$ (v) -1

3.2 Find the first 5 Green's function values G_1, G_2, G_3, G_4, G_5 for the following models, using the implicit method of Section 3.1.5
    (i) $X_t - 1.5X_{t-1} + 0.6X_{t-2} = a_t - 0.5a_{t-1}$
    (ii) $X_t - 0.5X_{t-1} = a_t - 0.3a_{t-1}$
    (iii) $X_t - 2X_{t-1} + 0.6X_{t-2} - 0.2X_{t-3} = a_t - 0.6a_{t-1} - 0.5a_{t-2}$

3.3 Verify Eq. (3.1.12).

3.4 Give the explicit form of the Green's function for each of the following models and plot it similar to Fig. 3.8.
    (i) $X_t - 0.5X_{t-1} = a_t$
    (ii) $X_t - 0.9X_{t-1} + 0.2X_{t-2} = a_t - 0.3a_{t-1}$
    (iii) $X_t - 0.54X_{t-1} + 0.02X_{t-2} = a_t - 0.3a_{t-1}$
    (iv) $X_t - 0.54X_{t-1} + 0.02X_{t-2} = a_t$
    (v) $X_t - 1.5X_{t-1} + 0.6X_{t-2} = a_t - 0.5a_{t-1}$
    Comment on the similarities and differences in these Green's functions.

3.5 Using the Green's function of the ARMA(2,1) system, show that as one of the two roots λ_1, λ_2 approaches θ_1, the system tends to AR(1). Can you show this without using the Green's function?

3.6 Determine the asymptotic stability and stability of the following ARMA systems. Also determine whether each of the models is invertible or not.
    (i) $X_t - 1.3X_{t-1} + 0.4X_{t-2} = a_t - 0.9a_{t-1}$
    (ii) $X_t - 0.7X_{t-1} + 0.1X_{t-2} = a_t - 1.7a_{t-1} + 0.6a_{t-2}$
    (iii) $X_t - 1.6X_{t-1} + 0.6X_{t-2} = a_t - a_{t-1} + 0.24a_{t-2}$
    (iv) $\nabla^3 X_t = a_t - 1.9a_{t-1} + 0.7a_{t-2}$
    (v) $X_t - 1.1X_{t-1} = a_t$
    (vi) $(1 - B^2)X_t = a_t$
    (vii) $(1 - 1.1B)(1 - 0.5B)X_t = (1 - 1.1B)a_t$

3.7 Give the explicit form of the autocovariance function and compute $\rho_1, \rho_2, \rho_3, \rho_4$ by recursive relations for the ARMA(2,1) model,

$$X_t - 1.1X_{t-1} + 0.3X_{t-2} = a_t - 0.6a_{t-1}$$

3.8 Show that $\gamma_0 \geqslant |\gamma_k|$ or $|\rho_k| \leqslant 1$.

3.9 An engineer feels that the temperature variations in an insulated tank can be closely represented by a response characteristic that is the sum of two exponentials λ_1^t and λ_2^t, respectively, for the fluid and the insulation. He conjectures that the Green's function for the response to random temperature variations in the atmosphere may be described by

$$G_j = g_1\lambda_1^j + g_2\lambda_2^j$$

where $\lambda_1 = 0.6$, $\lambda_2 = 0.5$.

(a) If the "coupling" ratio g_1/g_2 is -2, what model will the fluid temperature variations follow?

(b) What should the ratio g_1/g_2 be so that the model will be AR(2)?

(c) In both cases above, compute the variance components d_1, d_2 and the variance γ_0, assuming $\sigma_a^2 = 1$.

(d) Suppose that the coupling ratio g_1/g_2 can be adjusted by a suitable design; what should its value be so that the variance γ_0 of the fluid temperature variation is minimum? What is the minimum variance?

3.10 Determine the value of A and the ARMA model with its parameters for which the following are Green's functions

(i) $A(0.4)^j \cos (0.6+0.8j)$

(ii) $A(0.8)^j \sin (0.6+0.8j)$

(iii) $A(0.4)^j \cos (0.6+0.8j) + 0.4(0.6)^j$

3.11 The Green's function G_j of an ARMA(n,m) model is given by:

$$G_j = 0.4(0.9)^{j-1}, \qquad j \geqslant 1$$

Assume $X_0 = 0$ and a_t's are:

$t =$	0	1	2	3	4	5
$a_t =$	0	0.5	-1	1	-2	2

(a) Compute X_5 by using the Green's function form Eq. (3.1.4).

(b) Find the corresponding ARMA model and its parameters.

(c) Check (a) by using the difference equation form of the model.

3.12 Give one example each of the ARMA models for the following systems:

(i)   Asymptotically stable but not invertible;

(ii) Unstable but invertible;

(iii)   Stable and invertible;

(iv)   Characteristic roots $|\lambda_i| \leq 1$ but unstable and not invertible;

(v)   Stable but not asymptotically stable.

3.13   Obtain explicit expressions for the autocovariance functions of an ARMA(1,1) model using implicit relations and verify that they agree with Eq. (3.3.18a) and (3.3.18b).

3.14   Six distinct regions are marked on the stability region of the AR(2) model in Fig. 3.10. Draw sketches of the autocovariance functions of AR(2) models with parameters in each of these regions and comment on their similarities and differences. Note that the situation of the parameters in each of these regions is determined by the roots λ_1, λ_2, which in turns determine the form of γ_k.

3.15   Substitute suitable values of m and n in Eqs. (3.1.24) and (3.1.26) to obtain explicit expressions for the Green's functions of ARMA(2,1), ARMA(3,2), and AR(3) models.

3.16   Prove Eqs. (3.1.17b) and (3.1.17c).

3.17   Using implicit expressions for γ_k, show that for an AR(2) model

$$\gamma_0 = \frac{\sigma_a^2(1-\phi_2)}{(1+\phi_2)[(1-\phi_2)^2 - \phi_1^2]}$$

$$\gamma_1 = \frac{\phi_1}{(1-\phi_2)}\gamma_0$$

$$\gamma_2 = \frac{(\phi_1^2+\phi_2-\phi_2^2)}{(1-\phi_2)}\gamma_0$$

3.18   Show that the stability conditions for the ARMA(3,m) model are given by [for invertibility of ARMA(n,3) replace ϕ by θ]

$$\phi_1 + \phi_2 + \phi_3 < 1$$

$$-\phi_1 + \phi_2 - \phi_3 < 1$$

$$|\phi_3| < 1$$

$$\phi_3(\phi_3 - \phi_1) - \phi_2 < 1$$

3.19   Find the stability conditions for ARMA(4,m) and ARMA(5,m) models in terms of ϕ_i's.

3.20   Prove Eq. (3.3.18b) directly from (3.3.16a) and (3.3.16b) [*Hint:* Refer to (3.1.19a).]

3.21   Suppose U_t and V_t are given by AR(1) models

$$(1 - \phi_{1u}B)U_t = a_{ut}, \quad (1 - \phi_{1v}B)V_t = a_{vt}$$

$$Ea_{ut}a_{vt-k} = \delta_k \sigma_{auv}$$

(a) Using the B operator, show that

$$X_t = U_t + V_t$$

will have an ARMA(2,1) model and determine its parameters ϕ_1 and ϕ_2.

(b) Assume $\sigma_{auv} = 0$ and use the covariance function of an MA(1) model for the moving average side to find the parameter θ_1. Is it unique? If not, which one would you choose and why?

(c) Under what conditions would the ARMA(2,1) model in (b) reduce to (i) AR(2), (ii) ARMA(1,1), (iii) AR(1)?

(d) Generalize these results for $\sigma_{auv} \neq 0$.

3.22 If U_t and V_t are given by ARMA(2,1) models and a_{ut} and a_{vt} are uncorrelated for different t, show that the model for $X_t = U_t + V_t$ would be in general ARMA(4,3). State special cases in which the autoregressive and or moving average orders would be less than 4 and/or 3.

3.23 Generalize the results of Problems 3.21 and 3.22. Show that if U_t and V_t are given by ARMA(n_u,m_u) and ARMA(n_v,m_v), and a_{ut} and a_{vt} are uncorrelated for different t, then the model for $X_t = U_t + V_t$ would be in general ARMA(n,m), with $n = n_u + n_v$ and $m = \max(n_u + m_v, n_v + m_u)$. State the special cases where the orders would be less than m and/or n.

4

MODELING

A modeling strategy that approximates the dependence in the data successively by a sequence of ARMA$(n, n-1)$ models was developed in Chapter 2. This strategy evolved out of the ordinary linear regression models by concentrating on the conditional regression aspect of the ARMA models. This static aspect also naturally leads to the conditional least squares method of estimating the parameters in the model.

However, as discussed in Chapter 3, the dynamic aspect of the ARMA models reveals the rather deep structure of the underlying systems. Is the above modeling strategy satisfactory in light of this dynamic aspect of the ARMA systems? The two principal characterizations of this dynamics or memory are the Green's functions G_j from the systems' point of view and the autocovariance function γ_k from the statistical point of view. Is the above strategy confirmed or contradicted by these two characterizations? Does the strategy need modification after taking into consideration G_j and γ_k? In this chapter we will try to answer these questions, aiming for a practical modeling procedure that can be executed on a computer.

In Section 4.1 we will first discuss a rational modeling strategy that is available for ARMA(n, m) models, with both n and m nonzero, when the ARMA$(n, n-1)$ strategy is *not* used. This will help us to see the rationale and simplicity of the ARMA$(n, n-1)$ strategy without other considerations. Further analysis with reference to G_j and γ_k will then reveal that not only do they not contradict the strategy, but they also strongly support it. In fact, the real justification for the ARMA$(n, n-1)$ strategy comes from treating the time series as the response of a dynamic system to random and independent or uncorrelated input, rather than trying to fit ARMA(n, m) models with unknown n and m to a series of dependent observations.

Therefore, our approach to modeling is not merely of regression or curve fitting, which is easily achieved by the strategy outlined in Chapter 2. We treat modeling as finding a representation of a stochastic dynamic system, in the form of difference equations, derived from and dependent on the time series data. Such an approach may be called a system approach to modeling as compared to the regression or correlation approach. If the data gathering procedure is such that the data truly represents the behavior of the underlying physical system, then the system approach provides its true representation and can be used for analyzing the physical system. This is especially true of the models in the form of differential equations considered in Chapters 7 and 8.

Section 4.1 of this chapter presents the implications of the new approach to modeling ARMA systems. The question of increments in the values of n in successively fitting the ARMA$(n,n-1)$ models is discussed in Section 4.2. A brief discussion of the estimation procedure and especially the methods of obtaining the initial values of the parameters required for the nonlinear regression routine are given in Section 4.3. A criterion for checking the models based on the residual sum of squares, together with some ad-hoc checks, is presented in Section 4.4. A *step-by-step* modeling procedure is given in Section 4.5 and illustrated by the detailed results for the five sets of data introduced in Chapter 2.

4.1 THE SYSTEM APPROACH TO MODELING

Before discussing the implications of the ARMA$(n,n-1)$ strategy with reference to the system dynamics represented by G_j and γ_k, leading to the system approach, let us consider what we can do if this approach is *not* available to us.

4.1.1 Comparison with the ARMA(*n,m*) Approach

Let us assume that the data comes from an ARMA(n,m) model and try to find out the values of n and m, which may be arbitrary positive integers. If either one of m and n is zero and the other is small, and the estimated sample correlations or partial autocorrelations are not distorted, we might be able to guess the nonzero order from the cut-off point in the $\hat{\rho}_k$ or $\hat{\rho}'_k$ functions.

In general, however, we may not be able to guess the orders m and n at all, or the guessed order may turn out to be wrong after fitting. Then the only method that remains is trial and error. A rational way in which to carry out trial and error is to successively fit all the models of orders less than one, two, and so on. Suppose that an ARMA(5,3) model is required for the data. Then we could have fitted as many as 27 models before reaching the adequate model.

On the other hand, if we successively fitted ARMA$(n,n-1)$ models, we would be fitting only five models before reaching the adequate model. As the autoregressive order n increases, the number of ARMA$(n,n-1)$ models to be fitted in the system approach increases only in steps of one, whereas the ARMA(n,m) models increase in steps of n. In addition, the 27 models in the above ARMA(n,m) models would have to be individually checked, whereas the five models in the system approach would automatically be checked by the procedure outlined in Section 4.4, and no special effort in checking would be required on our part.

4.1.2 Consideration of Dynamics

Now let us consider the dynamic aspect of the ARMA models characterized by G_j or γ_k and try to find out what kind of modeling strategy it leads to. Since we are considering a stable system, we know that the dynamic response of the system to a single a_t will eventually die out to zero. The trivial case of such a dynamic system is the one with no dynamics or memory, that is, a_t at time t has no influence at all after time t. Such a system should have the Green's function

$$G_j \equiv 0, j \geq 1$$

and we get the "AR(0)" model

$$X_t = a_t$$

First Order Dynamics

When the dynamics is nonzero so that a_t does have influence at subsequent values, it is not very likely that the influence of a_t at time, say, $t+1$ would be very large, and at the next moment $t+2$ it would drop to zero. What is more likely is that the influence dies down to zero more or less gradually. Such a gradual decay of the response is best represented by the exponential Green's function

$$G_j = \lambda_1^j$$

and we get the AR(1) model

$$(1 - \phi_1 B)X_t = a_t, \qquad \phi_1 = \lambda_1$$

Note that this model also provides representation for the dynamics which cuts off after a few lags. If $\phi_1 = \lambda_1 = 0.1$, then

$$G_0 = 1$$
$$G_1 = 0.1$$
$$G_2 = 0.01$$
$$G_j \approx 0 \qquad j \geq 2$$

and we get an almost equivalent MA(1) model

$$X_t = a_t - \theta_1 a_{t-1}, \qquad \theta_1 = -0.1$$

Similarly, if $\phi_1 = 0.3$ then

$$X_t = \frac{a_t}{(1 - \phi_1 B)}$$
$$= (1 + 0.3B + 0.09B^2 + 0.027B^3 + \ldots)a_t$$
$$\approx a_t + 0.3a_{t-1} + 0.09a_{t-2} + 0.027a_{t-3}$$

which is almost an MA(3) model for which the G_j practically cuts off after the third lag.

Second Order Dynamics

Next, suppose the dynamics is not simple enough to be represented by a single exponential. Then, we can try to represent it by a linear combination of two exponentials

$$G_j = g_1\lambda_1^j + g_2\lambda_2^j$$

As seen in Chapter 3, allowing λ_1 and λ_2 to be a real or complex conjugate can yield a function G_j of many different shapes, capable of representing a wide variety of dynamics patterns. It was also seen in Section 3.1.8 that it is only by coincidence, when the g_1, g_2, λ_1 and λ_2 satisfy the condition

$$G_1 = \phi_1 \qquad \text{or} \qquad g_1\lambda_1 + g_2\lambda_2 = \lambda_1 + \lambda_2$$

that we get an AR(2) model. In general, however, such a dynamics of two exponentials is represented by an ARMA(2,1) model

$$(1 - \phi_1 B - \phi_2 B^2)X_t = (1 - \theta_1 B)a_t$$

Because it is capable of representing diverse dynamics patterns, the ARMA(2,1) model is adequate for a large variety of data sets met in practice, especially when the number of observations is around 200. With real λ_1 and λ_2, it can represent a continuously decreasing memory similar to the AR(1) model or the "lagged memory" that increases for a while and then exponentially decays to zero. With complex λ_1 and λ_2, it can represent a damped sinusoidal dynamics with different degrees of damping. These diverse patterns together with the flexibility afforded by the three parameters ϕ_1, ϕ_2 (or equivalently, λ_1, λ_2), and θ_1 make it a key model in our modeling strategy. It will be seen in Chapter 7 that this model also arises as a result of uniformly sampling a one-degree-of-freedom vibration system excited by random inputs. This well-known mechanical system, consisting of mass, spring, and dashpot and represented by a second order differential equations, is also useful in representing many practical systems that are not mechanical. This is yet another reason why the ARMA(2,1) model plays such an important role in the strategy.

Just as an AR(1) model may be approximately equivalent to an MA(1), MA(2), or MA(3) model, it is also possible to get either pure MA(m) or ARMA(1,m) models as disguised forms of an ARMA(2,1) model. For example, consider an ARMA(2,1) model with $\lambda_1 = 0.2$, $\lambda_2 = 0.4$, and $\theta_1 = -0.1$

$$(1 - 0.6B + 0.08B^2)X_t = (1 + 0.1B)a_t$$

then this model may also be represented as

$$X_t = \frac{(1+0.1B)}{(1-0.2B)(1-0.4B)} a_t$$

$$= (1+0.1B)(1-0.2B)^{-1}(1-0.4B)^{-1}a_t$$

$$= (1+0.1B)(1+0.2B+0.04B^2+0.008B^3+0.0016B^4$$
$$+ \ldots)(1-0.4B)^{-1}a_t$$

$$\approx (1+0.7B+0.34B^2+0.148B^3+0.0616B^4+0.025B^5)a_t$$

$$= a_t+0.7a_{t-1}+0.34a_{t-2}+0.148a_{t-3}+0.0616a_{t-4}+0.025a_{t-5}$$

which is almost an MA(5) model. The readers can verify that the given ARMA(2,1) model may also be represented by an ARMA(1,4) model. Note that pure MA models are easily obtained from the Green's function, whereas pure AR are obtained from the inverse function.

Third Order Dynamics

Next suppose that the dynamics is so complex that the sum of two exponentials is also not adequate to represent its Green's function. Then, we can use a linear combination of three exponentials to represent it by

$$G_j = g_1\lambda_1^j + g_2\lambda_2^j + g_3\lambda_3^j$$

With the same argument as in the case of the ARMA(2,1) model, it is easy to see that we will get an ARMA(3,1) and an AR(3) model only under special circumstances. For example, we get an AR(3) model when the λ_i and g_i, $i=1,2,3$, happen to satisfy the condition obtained from Eq. (3.1.13) by setting $\theta_1 = \theta_2 = 0$ as

$$G_1 - \phi_1 = g_1\lambda_1 + g_2\lambda_2 + g_3\lambda_3 - (\lambda_1+\lambda_2+\lambda_3) = 0$$

$$G_2-\phi_1 G_1-\phi_2 G_0 = (g_1\lambda_1^2+g_2\lambda_2^2+g_3\lambda_3^2) - (\lambda_1+\lambda_2+\lambda_3)$$
$$\cdot (g_1\lambda_1+g_2\lambda_2+g_3\lambda_3) + (\lambda_1\lambda_2+\lambda_2\lambda_3+\lambda_3\lambda_1) = 0$$

In general, however, we get an ARMA(3,2) model

$$(1-\phi_1 B-\phi_2 B^2-\phi_3 B^3)X_t = (1-\theta_1 B-\theta_2 B^2)a_t$$

and an equivalent MA(m) model follows as before.

nth Order Dynamics

Proceeding in this way, more and more complex forms of dynamics can be represented by the Green's function

$$G_j = g_1\lambda_1^j + g_2\lambda_2^j + \ldots + g_n\lambda_n^j \qquad (4.1.1)$$

for values of $n = 1, 2, 3, \ldots$ Under special circumstances, when one or more of the conditions obtained from Eq. (3.1.13) by setting $\theta_i = 0$

$$G_1 - \phi_1 = 0$$

$$G_2 - \phi_1 G_1 - \phi_2 = 0$$

$$G_3 - \phi_1 G_2 - \phi_2 G_1 - \phi_3 = 0$$

$$\cdots$$

$$G_{n-1} - \phi_1 G_{n-2} - \cdots - \phi_{n-1} = 0 \qquad (4.1.2)$$

are satisfied, with $\phi_1, \phi_2, \ldots, \phi_n$ given by

$$(1 - \lambda_1 B)(1 - \lambda_2 B) \ldots (1 - \lambda_n B)$$

$$= (1 - \phi_1 B - \phi_2 B^2 - \cdots - \phi_n B^n) \qquad (4.1.3)$$

that is,

$$\phi_\ell = (-1)^{\ell+1} \sum_{\substack{i_1, i_2, \ldots, i_\ell = 1 \\ i_1 < i_2 < \ldots < i_\ell}}^{n} \lambda_{i_1} \lambda_{i_2} \ldots \lambda_{i_\ell}$$

$$\ell = 1, 2, \ldots, n \qquad (4.1.3a)$$

then the dynamics is represented by an ARMA(n,m) model with $m < n - 1$. In general, however, we get an ARMA($n, n-1$) model.

It is thus seen that the strategy of fitting ARMA($n, n-1$) models, purely based on an intuitive regression argument in Chapter 2, is also confirmed by the Green's function from the dynamics' point of view. The preceding analysis also shows that although the new approach only concentrates on ARMA($n, n-1$) models, it actually takes into account many other ARMA(n,m) models with $m < n - 1$ and $m > n - 1$, either by setting some parameters to zero or inverting some of the autoregressive or moving average factors.

*4.1.3 Adequacy of the System Approach

The preceding analysis still leaves an important question unanswered: Is it always possible to represent the dynamics of a stable stationary stochastic system, which we are considering, by a Green's function in the form of a linear combination of exponentials such as Eq. (4.1.1)? Is there a stable stationary stochastic system that cannot be represented by an ARMA($n, n-1$) model no matter how large an n we take, and therefore the modeling strategy fails?

Before answering this question, let us clarify what is meant by representing a dynamic system by a mathematical model. All that such a representation means is that the actual measured output of the system is very closely approximated by the model. A mathematical model is only a good approximation to the actual system behavior. Therefore, the above question may be more precisely formulated as: Can every stable stationary stochastic system be approximated as closely as we want by an ARMA($n,n-1$) model for sufficiently large n? As we will now show, the answer to this question is yes.

Suppose X_t' is the actual system response and we are trying to represent it by an ARMA($n,n-1$) response X_t. To see how closely X_t approximates X_t', we have to consider the difference $(X_t' - X_t)$. If this difference was a deterministic function, then X_t' and X_t would be close if $(X_t' - X_t)$ was small or close to zero in absolute value for every t. Since $(X_t' - X_t)$ is a random variable, it can be considered small only when its mean is zero and variance is close to zero. We can assume the mean μ to be zero, since if $\mu \neq 0$, we can subtract μ from each X_t and X_t', and consider the new difference that has zero mean. Therefore, to check whether X_t approximates X_t' as closely as we want, we have to make sure that given an arbitrary small number ε, we can choose an ARMA($n,n-1$) model with response X_t such that

$$E(X_t' - X_t)^2 < \varepsilon \tag{4.1.4}$$

Since X_t follows an ARMA($n,n-1$) model, we know that its dynamic response characteristic is G_j given by Eq. (4.1.1), but that of X_t' is unknown to us; in fact, it may be of any form. However, since X_t' is a stationary stochastic process, it can be expressed by Wold's decomposition (see Section 3.1.4) as a linear combination of past a_t's:

$$X_t' = a_t + W_1 a_{t-1} + W_2 a_{t-2} + \ldots$$

$$= \sum_{j=0}^{\infty} W_j a_{t-j}, \qquad W_0 = 1 \tag{4.1.5}$$

The unknown function W_j, which is the Green's function of X_t', may be of any form whatsoever, with the only restriction that, since the system is stable, its variance must be finite and therefore

$$E(X_t')^2 = \sigma_a^2 \sum_{j=0}^{\infty} W_j^2 < \infty$$

that is,

$$\sum_{j=0}^{\infty} W_j^2 < \infty \tag{4.1.6}$$

The functions or sequences W_j, $j = 0, 1, 2, \ldots$, subject to the condition (4.1.6) form a Hilbert space, and using the theory of linear operators it can be shown that any function W_j can be approximated as closely as we want by a function

$$G_j = g_1\lambda_1^j + g_2\lambda_2^j + \ldots + g_n\lambda_n^j$$

with a proper choice of λ_i, g_i, $i = 1, 2, \ldots, n$, and large enough n. This means that given $\varepsilon' > 0$ however small, we can find G_j of the above form with large enough n such that

$$\sum_{j=0}^{\infty} (W_j - G_j)^2 < \varepsilon'$$

We will now show that the process X_t, with the G_j as the Green's function, satisfies Eq. (4.1.4) and provides the desired approximation

$$E(X_t' - X_t)^2 = E\left[\sum_{j=0}^{\infty} W_j a_{t-j} - \sum_{j=0}^{\infty} G_j a_{t-j}\right]^2$$

$$= E\sum_{j=0}^{\infty} (W_j - G_j)^2 a_{t-j}^2$$

$$= \sigma_a^2 \sum_{j=0}^{\infty} (W_j - G_j)^2$$

$$\leqslant \sigma_a^2 \, \varepsilon'$$

$$= \varepsilon, \qquad \text{where} \quad \varepsilon = \sigma_a^2 \varepsilon'$$

Note that G_j given above is exactly the Green's function of an ARMA($n, n-1$) model, and X_t represented by this model approximates the actual response X_t' as closely as we want. These results form a special case of the Fundamental Theorem in Pandit (1973), which may be referred to for further details of the above derivation as well as the argument via the autocovariance function given below. It should be noted that we treat the models with the characteristic root $\lambda_i = 1$ (for example, random walk) as the limiting cases of stable models with $\lambda_i < 1$, and therefore the above development is applicable to them.

*4.1.4 Consideration of the Autocovariance Function

From the similarities in the form of the Green's function G_j and the autocovariance function γ_k discussed in Chapter 3, it is now easy to see that the modeling strategy can also be arrived at independently by considering γ_k. In the trivial case of independent data, when γ_k is zero for all $k \neq 0$, we have the "AR(0)" model. Since for dependent data the autocovariance or statistical dependence must decay for a large enough lag, the simplest form is exponential

$$\gamma_k = d_1\lambda_1^k$$

which gives the AR(1) model. More complex forms of autocovariances can then be represented by a linear combination of exponentials for increasing values of n

$$\gamma_k = d_1\lambda_1^k + d_2\lambda_2^k + \ldots + d_n\lambda_n^k \qquad (4.1.7)$$

If the coefficients $d_1, d_2, \ldots, d_n$ and the roots $\lambda_1, \lambda_2, \ldots, \lambda_n$ satisfy one or more of the conditions

$$\gamma_k - \phi_1\gamma_{k-1} - \ldots - \phi_n\gamma_{k-n} = 0, \qquad k = 1, 2, \ldots, n-1 \quad (4.1.8)$$

with $\gamma_{-k} = \gamma_k$ and ϕ_i's given by Eq. (4.1.3a), then we have ARMA(n,m) models with $m < n-1$. In general, however, we get ARMA($n,n-1$) models.

4.1.5 Reasons for the Moving Average Order $n-1$

The reason why, in general, a moving average order $n-1$ is associated with the autoregressive order n is most easily seen by considering G_j or γ_k given by Eqs. (4.1.1) and (4.1.7) as the solutions of difference equations. The question of finding a model for which Eqs. (4.1.1) and (4.1.7) are the Green's function and the autocovariance function is the same as finding a homogeneous difference equation with suitable initial conditions for which they are the solutions. Since there are n linearly independent roots $\lambda_1, \lambda_2, \ldots, \lambda_n$, the homogeneous equation is clearly nth order. Hence, the autoregressive order is n, and the autoregressive part is specified by

$$(1 - \phi_1 B - \phi_2 B^2 - \ldots - \phi_n B^n)G_j = 0, \qquad j \geq n$$

$$(1 - \phi_1 B - \phi_2 B^2 - \ldots - \phi_n B^n)\gamma_k = 0, \qquad k \geq n$$

with the relation between λ_i's and ϕ_i's given by Eq. (4.1.3). Now, the n constants $g_1, g_2, \ldots, g_n$ or $d_1, d_2, \ldots, d_n$ have to be specified by n initial conditions. However, one condition must always hold, namely

$$1 = G_0 = g_1 + g_2 + \ldots + g_n$$

or

$$\gamma_0 = d_1 + d_2 + \ldots + d_n$$

Hence, only $n-1$ conditions remain to be specified, which is done by Eq. (3.1.13) or (3.3.23), requiring $n-1$ θ_i's in general. Therefore, the moving average order is, in general, $n-1$. This gives an ARMA($n,n-1$) model, and only in exceptional cases when the constants g_i's or d_i's are such that these equations are satisfied with some or all θ_i's equal to zero, we have ARMA(n,m) models with $m < n-1$.

4.2 INCREMENT IN THE AUTOREGRESSIVE ORDER

The consideration of the static aspect of the ARMA models characterized by conditional regression in Chapter 2, as well as the dynamic aspect characterized by the Green's function and the autocovariance considered in the last section, leads to the system approach of successively fitting ARMA$(n,n-1)$ models. It therefore appears that in developing a specific modeling procedure, we should first fit an AR(1) model, then ARMA(2,1), ARMA(3,2) and so on, increasing n by steps of one.

Such a procedure is feasible in theory and may be used when it is known that the number of observations in each data set will be small, say around 200, and models of order higher than about $n=4$ are not likely. In general, however, increasing n by one is not economical in practice. In fact, empirical experience indicates that it is better to increase n in steps of two and fit ARMA$(2n,2n-1)$ models for $n = 1, 2, 3, \ldots$ In addition to empirical experience, there are two reasons for this choice of the sequence of models.

The first is the analogy of n degrees-of-freedom vibration system applicable to ARMA models. It was mentioned in the last section that the ARMA(2,1) model is obtained for a uniformly sampled one-degree-of-freedom spring mass dashpot system governed by second order differential equations and excited by white noise. Similarly, a two-degrees-of-freedom system is governed by a fourth order differential equation and when uniformly sampled, has a representation of an ARMA(4,3) model. Thus, increasing the degree-of-freedom by one amounts to advancing the autoregressive order by two. Since the n degrees-of-freedom vibration system is very useful in interpreting the ARMA models, it seems reasonable to use the sequence ARMA$(2n,2n-1)$ with $n = 1, 2, \ldots$, where n may be said to be the degrees-of-freedom. This can be appreciated more when we consider the continuous models in Chapters 6 and 7.

The second reason is the configuration of the characteristic roots λ_i. These roots, in general, may be real or complex. However, since the parameters ϕ_i's are always real, the complex roots can occur only in *conjugate pairs*. For example, for an ARMA(2,1) model since

$$(1 - \phi_1 B - \phi_2 B^2) = (1 - \lambda_1 B)(1 - \lambda_2 B)$$

$$\lambda_1, \lambda_2 = \frac{\phi_1}{2} \pm \frac{\sqrt{\phi_1^2 + 4\phi_2}}{2} \tag{4.2.1}$$

the roots are complex conjugate whenever $\phi_1^2 + 4\phi_2 < 0$. This is also clear from the fact that

$$\phi_1 = \lambda_1 + \lambda_2, \qquad \phi_2 = -\lambda_1 \lambda_2 \tag{4.2.2}$$

and if one of the to roots is complex, the other must be its conjugate if ϕ_1 and ϕ_2 are to be real. Similarly, for an ARMA(4,3) model, since by Eq. (4.1.3a)

$$\phi_1 = \lambda_1 + \lambda_2 + \lambda_3 + \lambda_4$$

$$\phi_2 = -(\lambda_1\lambda_2 + \lambda_2\lambda_3 + \lambda_3\lambda_4 + \lambda_1\lambda_4 + \lambda_1\lambda_3 + \lambda_2\lambda_4)$$

$$\phi_3 = \lambda_1\lambda_2\lambda_3 + \lambda_2\lambda_3\lambda_4 + \lambda_1\lambda_3\lambda_4 + \lambda_1\lambda_2\lambda_4$$

$$\phi_4 = -\lambda_1\lambda_2\lambda_3\lambda_4 \tag{4.2.3}$$

it is clear that if one of the roots is complex, its conjugate must be present in the remaining roots if ϕ_1, ϕ_2, ϕ_3, and ϕ_4 are to be real. This fact is true for all ARMA models.

This fact implies that whenever the autoregressive order is odd, one of the roots has to be real. Therefore, if we increase the order by one, allowing odd autoregressive orders, we are forcing one of the roots to be real. Suppose we have fitted an ARMA(2,1) model and want to increase the order. Suppose actually an ARMA(4,3) with two complex conjugate pairs is an adequate model for the data. If we try to fit an ARMA(3,2) model to these data, the third root is forced to be real, and hence this model might give a poorer approximation than the ARMA(2,1) model fitted earlier. This can be avoided by fitting an ARMA(4,3) model after the ARMA(2,1). If the actual adequate model is ARMA(3,2) with one of the roots being real, this will be indicated in the ARMA(4,3) model by giving two real roots, one of them very small so that ϕ_4 is small and its confidence interval includes zero. We can then drop the parameter ϕ_4 and any other moving average parameters close to zero and refit the model.

Finally, increasing the autoregressive order in steps of two is economical, particularly when higher order models are required to provide adequate representation of the data. One fits only half the number of models compared to the increase by one. For example, in the hypothetical case considered in Section 4.1.1, only three models need be fitted before reaching the adequate model compared to five models for the ARMA($n, n-1$) sequence. These models are: ARMA(2,1), ARMA(4,3), and ARMA(6,5). For the last model one would find $\phi_6, \theta_4, \theta_5$ close to zero (that is, their confidence interval includes zero), indicating that a model without these parameters may also be adequate. This would finally lead us to the adequate model ARMA(5,3).

A useful consequence of employing the ARMA($2n, 2n-1$) sequence is that we start with an ARMA(2,1) model. The importance of this model cannot be overemphasized. As explained in the last section, this model is suitable for a large number of practical systems and therefore fitting it first often provides an adequate model at once; fitting the next ARMA(4,3) model then merely becomes a check on this model.

Even when a higher order model such as ARMA(4,3) seems to provide a better model, the ARMA(2,1) model may still be useful as a good approximation to the system, due to its simplicity and ease of physical interpretation based on the one-degree-of-freedom vibration system. When an AR(1) model is actually adequate, it will be indicated by small values of ϕ_2 and θ_1 with confidence intervals including zero. For the same reason, whenever an ARMA(2,1) model is found to be the final adequate model, it is advisable to fit an AR(1) model and compare it with the ARMA(2,1) model. Although not strictly adequate, a simple AR(1) model may provide a representation good enough for practical purposes.

4.3 ESTIMATION

It was pointed out in Chapter 2 that since ARMA(n,m) models are conditional regression models, their parameters can be estimated by the least squares method that minimizes the sum of squares of the residuals a_t's. In particular, when there are no moving average parameters and we have a pure AR(n) model, the linear least squares method of estimation can be used.

4.3.1 AR Models

If we use the data after subtracting the average $\overline{X}$ then, using Eqs. (2.1.7) to (2.1.9) for N observations $X_1, X_2, \ldots, X_N$, we have

$$X_t = \phi_1 X_{t-1} + \phi_2 X_{t-2} + \ldots + \phi_n X_{t-n} + a_t \qquad (4.3.1)$$

therefore

$$\mathbf{Y} = \begin{bmatrix} X_{n+1} \\ X_{n+2} \\ \vdots \\ X_N \end{bmatrix},$$

$$\mathbf{X} = \begin{bmatrix} X_n & X_{n-1} & \cdots & X_1 \\ X_{n+1} & X_n & \cdots & X_2 \\ \vdots & \vdots & & \vdots \\ X_{N-1} & X_{N-2} & & X_{N-n} \end{bmatrix},$$

$$\phi = \begin{bmatrix} \phi_1 \\ \phi_2 \\ \vdots \\ \phi_n \end{bmatrix} \qquad (4.3.2)$$

This gives the least squares estimates as

$$\hat{\boldsymbol{\phi}} = (\mathbf{X}'\mathbf{X})^{-1}\mathbf{X}'\mathbf{Y} \qquad (4.3.3)$$

and

$$\hat{\sigma}_a^2 = \frac{1}{(N-n)} \sum_{t=n+1}^{N} (X_t - \hat{\phi}_1 X_{t-1} - \hat{\phi}_2 X_{t-2} - \ldots - \hat{\phi}_n X_{t-n})^2 \qquad (4.3.4)$$

4.3.2 ARMA Models

When a moving average parameter is present, the unconditional regression is nonlinear, and therefore the nonlinear least squares method needs to be used, as explained in Section 2.3.4. Standard library programs for nonlinear least squares methods are now available and hence we omit the details. We briefly discuss how it can be used to fit ARMA(n,m) models. Once such a program is available, it is better to use it even for AR(n) models, for which the routine converges very fast.

The nonlinear least squares routine generally starts with some initial or "guess" values of the parameters to be estimated. When these values are known, the a_t's can be recursively computed by

$$a_t = X_t - \phi_1 X_{t-1} - \phi_2 X_{t-2} - \ldots - \phi_n X_{t-n}$$
$$+ \theta_1 a_{t-1} + \theta_2 a_{t-2} + \ldots + \theta_m a_{t-m} \qquad (4.3.5)$$

Since X_t is not available for $t<0$, the first a_t to be computed is for $t=n+1$. The initial a_t's for $t<n$ are set equal to zero, their expected value. Once these starting values of a_t's are taken, the a_t's, $t = n+1, n+2, \ldots, N$, can be recursively generated by Eq. (4.3.5). This involves a slight loss of information, which is negligible when the number of observations N is large, say about 200. If desired, this loss can be remedied by employing maximum likelihood estimates in place of the conditional least squares estimates. It can be shown [see Pandit (1973)] that the conditional least squares estimates are also approximately maximum likelihood estimates.

When the nonlinear least squares routine is provided with Eq. (4.3.5) for computing a_t's or the "residual errors" and supplied with initial guess values of the parameters to be estimated, it monitors these values in the direction of the smaller sum of squares of a_t's. Once a point in the parameter space giving a smaller sum of squares is reached, it starts a new iteration with this point as the initial values. The iterations continue until some specified tolerances are reached, such as the relative reduction in the sum of squares, or the maximum change in the parameter values is below, say, 10^{-5}, or the number of iterations is greater than, say, 25. One of the algorithms that can be used for minimization is due to Marquardt (1963); it is a compromise between the Gauss method and the method of steepest descent. After we reach the minimum, we can obtain the approximate confidence intervals on the estimated parameters by the linear least squares theory, using local linear hypothesis (see Appendix A 2.2).

So far, we have taken the average of the data $\overline{X}$ as the estimate of the mean, subtracted it from the data, and assumed that the resulting data has zero mean. In the final estimation, however, it is more advisable to estimate the mean μ as an additional parameter by writing the ARMA(n,m) model as

$$(\dot{X}_t - \mu) - \phi_1(\dot{X}_{t-1} - \mu) - \phi_2(\dot{X}_{t-2} - \mu) - \cdots$$
$$- \phi_n(\dot{X}_{t-n} - \mu) = a_t - \theta_1 a_{t-1} - \theta_2 a_{t-2} - \cdots - \theta_m a_{t-m} \quad (4.3.6)$$

where $\dot{X}_t$ represents the original data. This change is easily incorporated in the nonlinear least squares routine by specifying μ as the $(n+m+1)$st parameter and replacing X_t in Eq. (4.3.5) by $\dot{X}_t - \mu$ for computing the a_t's. The initial guess value for μ is naturally the estimated average $\overline{X}$.

The main question that remains in estimation is how to obtain the initial guess values of the parameters ϕ_i's and θ_i's when m is nonzero. In many cases, when the number of observations and/or the number of parameters is small, the estimation situation is fairly simple and the routine will converge to the final values, even starting from far-off guess values. In such cases any reasonable set of values within the stability and invertibility region will suffice as initial guesses. In general, however, the sum of squares "surface" (treated as a function of the unknown parameters, with the data as known constants) may be of complex shape. Then, fairly close guess values are needed; otherwise, the routine may not converge or may converge to wrong estimates.

4.3.3 Initial Guess Values Based on the Inverse Function

Initial guess values can be obtained from sample autocorrelations $\hat{\rho}_k$ for ARMA(n,m) models with $m \leqslant 1$, as illustrated in Problems 4.11 to 4.13. However, for $m \geqslant 1$ this method becomes nonlinear. Therefore, it is advisable to use a method based on inverse function. The main advantage of this method is that it involves the solution of linear equations for an arbitrary ARMA(n,m) model. It is based on the fact that the relations in ϕ_i's and θ_i's expressed by the inverse function I_j (see Section 3.2.2) are linear for an arbitrary ARMA(n,m) model. The I_j coefficients themselves are autoregressive parameters of the infinite expansion of an ARMA model and therefore can be estimated well by the linear least squares method.

Recall from Section 3.2 that the ARMA(n,m) model can be written as

$$(1 - \phi_1 B - \phi_2 B^2 - \ldots - \phi_n B^n)X_t$$
$$= (1 - \theta_1 B - \theta_2 B^2 - \ldots - \theta_m B^m)a_t \quad (4.3.7)$$

and the inverse function coefficients are defined as

$$a_t = (1 - I_1 B - I_2 B^2 - \ldots)X_t \quad (4.3.8)$$

Substituting for a_t from Eq. (4.3.8) in Eq. (4.3.7), we get the operator identity

$$(1 - \phi_1 B - \phi_2 B^2 - \ldots - \phi_n B^n)$$
$$\equiv (1 - \theta_1 B - \theta_2 B^2 - \ldots - \theta_m B^m) \cdot (1 - I_1 B - I_2 B^2 - \ldots) \quad (4.3.9)$$

Equating the coefficients of equal powers of B, we get

$$\phi_1 = \theta_1 + I_1$$
$$\phi_2 = \theta_2 - \theta_1 I_1 + I_2$$
$$\phi_3 = \theta_3 - \theta_1 I_2 - \theta_2 I_1 + I_3 \quad (4.3.10)$$
$$\ldots$$
$$\phi_j = \theta_j - \theta_1 I_{j-1} - \theta_2 I_{j-2} - \ldots - \theta_{j-1} I_1 + I_j$$

for all j with the assumption that $\theta_j = 0$ for $j > m$ and $\phi_j = 0$ for $j > n$ for the ARMA(n,m) model. In particular, for $j > \max(n,m)$

$$(1 - \theta_1 B - \theta_2 B^2 - \theta_3 B^3 - \ldots - \theta_m B^m)I_j = 0 \quad (4.3.11)$$

It is clear that the initial values for the autoregressive as well as the moving average parameters can be found from the Eqs. (4.3.10) and (4.3.11) if the estimates of the inverse function are known. To get I_j's, let us consider the pure autoregressive model. The pure AR(p) model can be written as

$$X_t = \phi_1 X_{t-1} + \phi_2 X_{t-2} + \ldots + \phi_p X_{t-p} + a_t$$

or

$$a_t = X_t - \phi_1 X_{t-1} - \phi_2 X_{t-2} - \ldots - \phi_p X_{t-p}$$

or

$$a_t = (1 - \phi_1 B - \phi_2 B^2 - \ldots - \phi_p B^p)X_t \quad (4.3.12)$$

Comparing Eqs. (4.3.12) and (4.3.8) or by putting $\theta_i = 0$ in Eq. (4.3.10), we see that for a pure AR(p) model, $I_j = \phi_j$, $j = 1, 2, \ldots, p$, $I_j = 0$, $j > p$. Although this is not strictly true for an ARMA model, the invertibility condition ensures that I_j tends to zero for sufficiently large j. Therefore, fairly good estimates of I_j for an ARMA(n,m) model can be obtained from ϕ_j's of an AR(p) model for sufficiently large p; these ϕ_j's, in turn, can be estimated from the data by the linear least squares formulae. The I_j's so obtained from an AR(p) model can be used to get θ_i's by Eq. (4.3.11), which are linear and can be explicitly solved. Substituting these θ_i's in Eq. (4.3.10), we get ϕ_i's of the ARMA model, again as an explicit solution.

For the initial values of the m θ_i's of an ARMA(n,m) model, m equations from (4.3.11) are needed with $j>\max(n,m)$. Therefore, it suffices to take the order p of the AR(p) model in obtaining I_j's as

$$p = \max(n,m) + m \qquad (4.3.13)$$

When $m>n$, this involves a slight redundancy since only the first n equations from Eq. (4.3.10) are used successively in obtaining ϕ_i's and the next $m-n$ equations are redundant.

If p is chosen much larger than the value specified by Eq. (4.3.13), then one can use all these p equations (4.3.10) by treating them as a linear least squares set-up. Then the estimates of ϕ_i and θ_i are obtained by a double application of linear least squares: first, to get p I_j's from the data by fitting an AR(p) model and, second, to get ϕ_i's and θ_i's from Eqs. (4.3.10) using these estimated I_j's. In many engineering applications, where a large amount of data is available, the initial values of ϕ_i's and θ_i's obtained from Eq. (4.3.10) are good enough for practical purposes, and the additional nonlinear least squares estimation of the final values may be skipped. This is particularly advantageous in microprocessor hardware implementation.

After the ϕ_i's from the AR(p) model are substituted as I_j's in Eq. (4.3.11), the initial values of θ_i's are obtained, but they do not necessarily satisfy the invertibility condition. To check it, the roots v_i's are obtained as

$$(1 - \theta_1 B - \theta_2 B^2 - \ldots - \theta_m B^m)$$
$$= (1-v_1 B)(1-v_2 B) \ldots (1-v_m B) \qquad (4.3.14)$$

If any of the v_i's is greater than one in absolute value, it is replaced by its reciprocal that then satisfies invertibility. (It can be shown that such a replacement does not alter the autocovariance function γ_k.) The new invertible initial values of θ_i's are given by the reverse of Eq. (4.3.14)

$$\theta_\ell = (-1)^{\ell+1} \sum_{\substack{i_1,i_2,\ldots,i_\ell=1 \\ i_1<i_2\ldots<i_\ell}}^{m} v_{i_1} v_{i_2} \ldots v_{i_\ell}$$
$$\ell = 1, 2, \ldots, m \qquad (4.3.15)$$

Substituting these successively in Eq. (4.3.10), we get the initial values of the ϕ_i's of the ARMA(n,m) model. The procedure is illustrated by a numerical example in Section 4.3.4.

As a by-product of this method for finding the guess values, one gets a rough indication of (in addition to the more rigorous criteria discussed in Section 4.4) what value of n one should stop fitting ARMA($2n,2n-1$) models. To find the initial values for an ARMA($2n,2n-1$) model, one would first fit an AR(p) model with $p = 2n+2n-1 = 4n-1$. The residual sum of squares of this AR(p) model would generally be higher than that after fitting the corresponding ARMA($2n,2n-1$) model. However, if the difference is very large, it indicates that the residual sum of squares can be reduced by increasing the value of n. When the difference is small, it indicates that increasing n may not reduce the sum of squares appreciably; one may therefore decide to stop at the ARMA($2n,2n-1$) model when this conclusion is confirmed by the criteria given in Section 4.4. In the illustrative examples in Section 4.5.1, the sum of squares of the AR(p) model is given at the beginning for such a comparison.

4.3.4 An Example of Initial Values by the Inverse Function

To explain the steps followed in this method, let us consider the ARMA(2,1) model for the sunspot activity data. For the ARMA(2,1) model,

$$n = 2, \qquad m = 1$$

and

$$p = \max(n,m)+m = 2+1 = 3$$

Hence, the AR(3) model is fitted. The AR(3) model has three parameters ϕ_1, ϕ_2, and ϕ_3 that can be estimated using the linear least squares method. These ϕ_i's are taken as the inverse function coefficients I_j's (as discussed earlier).

The AR(3) model can be written as

$$X_t = \phi_1 X_{t-1} + \phi_2 X_{t-2} + \phi_3 X_{t-3} + a_t$$

For the linear least squares method,

$$\mathbf{Y} = \begin{bmatrix} X_4 \\ X_5 \\ \vdots \\ X_N \end{bmatrix} \qquad \mathbf{X} = \begin{bmatrix} X_3 & X_2 & X_1 \\ X_4 & X_3 & X_2 \\ \vdots & \vdots & \vdots \\ X_{N-1} & X_{N-2} & X_{N-3} \end{bmatrix}$$

and

$$\begin{bmatrix} \phi_1 \\ \phi_2 \\ \phi_3 \end{bmatrix} = (\mathbf{X'X})^{-1}\, \mathbf{X'Y}$$

Using the linear least squares methods, we get

$$\phi_1 = I_1 = 1.27$$
$$\phi_2 = I_2 = -0.50$$
$$\phi_3 = I_3 = -0.11$$

Residual sum of squares = 40843.66

Using the Eq. (4.3.11), we have

$$(1 - \theta_1 B)I_j = 0, \qquad j > 2$$
$$(1 - \theta_1 B)I_3 = 0$$
$$I_3 - \theta_1 I_2 = 0$$

$$\theta_1 = \frac{I_3}{I_2} = \frac{-0.11}{-0.50} = 0.22$$

(Since this is invertible, we can use it directly. If, however we obtain a noninvertible value, say 1.1, then we would use its reciprocal, $1/1.1 = 0.91$. For $m > 1$, Eqs. (4.3.14) and (4.3.15) need to be used.)

Knowing I_1, I_2, I_3 and θ_1, we can calculate ϕ_1 and ϕ_2 of the ARMA(2,1) model from Eq. (4.3.10) as follows:

$$\phi_1 = I_1 + \theta_1$$
$$= 1.27 + 0.22$$
$$= 1.49$$

$$\phi_2 = \theta_2 - \theta_1 I_1 + I_2$$
$$= 0 - 0.22 \times 1.27 - 0.50$$
$$\text{(since } \theta_2 = 0 \text{ for ARMA(2,1) model)}$$
$$= -0.78$$

Hence, the initial values of the parameters for the ARMA(2,1) model are

$$\phi_1 = \quad 1.49$$
$$\phi_2 = -0.78$$
$$\theta_1 = \quad 0.22$$

4.4 CHECKS OF ADEQUACY

The final question in modeling relates to the stopping value of n: At what n should the sequence ARMA($n,n-1$) or ARMA($2n,2n-1$) be terminated? In other words, for what value of n should the ARMA($n,n-1$)

model be considered adequate? As we will see in this section, one of the most important advantages of the system approach to modeling is that it leads to extremely simple checks of adequacy compared to those given in the existing literature on time series analysis.

4.4.1 Formulation as Testing of Hypothesis

The basic checking strategy in our approach was developed in Section 2.3.3 in Chapter 2. While evolving the ARMA$(n, n-1)$ strategy by the conditional regression argument in that chapter, we saw that the crucial assumption in an ARMA$(n, n-1)$ model was that a_t's are independent. If a_t's are independent, then they are clearly independent also of X_{t-1}, X_{t-2}, X_{t-3}, . . . since

$$X_t = a_t + G_1 a_{t-1} + G_2 a_{t-2} + \ldots$$

Therefore, for any ARMA(n,m) model, the checks of adequacy are really the checks of independence for a_t's. These checks are best formulated as a statistical testing of hypothesis. In a test of hypothesis, one usually assumes a conservative "null" hypothesis, say, H_0 and formulates an "alternative" hypothesis H_1. If the statistical criterion favors H_1, one concludes that there is evidence against H_0 at a preassigned significance level. Otherwise, one says that there is no evidence against H_0. H_0 is usually taken in the simplest possible form. The checking of the independence of a_t's is usually formulated as

$$H_0: a_t\text{'s are independent.}$$

$$H_1: a_t\text{'s are dependent.}$$

This is an example of a simple versus composite hypothesis, since the alternative H_1 incorporates many possibilities, a_t's may be dependent in many possible ways. Even in elementary statistics dealing with independent data, a test of simple versus composite hypothesis is difficult to make. With the additional complication of correlated data, such a test becomes more complicated and approximate ad-hoc tests are needed.

Checks on the independence of a_t's were made in Chapter 2 by plotting a_t versus a_{t-1} and computing their correlation, without involving sophisticated checking criteria. However, it was seen there that a simple and more reliable way of checking the independence of a_t's in order to test the adequacy of the model was to fit a higher order model. If the higher order model gives a substantial reduction in the residual sum of squares, it automatically means that the a_t's of the present model are not independent and therefore the model is not adequate. On the other hand, if the present model is adequate and we try to fit a higher order model, then the reduction in the sum of squares will be smaller and, in fact, in some practical cases we might even get a slightly larger sum of squares. The fact that we are introducing unnecessary additional parameters in the model will also be indicated by their small values, so small that their confidence intervals include zero, which serves as an additional check over the adequacy of the lower order model in the ARMA($2n$,$2n-1$) sequence.

Thus, the checking strategy reduces the checking procedure to a simple choice between two models: ARMA($2n$,$2n-1$) and ARMA($2n+2$,$2n+1$). However, an ARMA($2n$,$2n-1$) model is the same as an ARMA($2n+2$,$2n+1$) model with

$$\phi_{2n+2} = \phi_{2n+1} = \theta_{2n} = \theta_{2n+1} = 0$$

Therefore, if one of the two models is adequate, we can test it rigorously by a simple test of hypothesis

$$H_0: \phi_{2n+2} = 0, \quad \phi_{2n+1} = 0, \quad \theta_{2n} = 0, \quad \theta_{2n+1} = 0$$
$$H_1: \phi_{2n+2} \neq 0, \quad \phi_{2n+1} \neq 0, \quad \theta_{2n} \neq 0, \quad \theta_{2n+1} \neq 0$$

4.4.2 Checking Criterion

This is a test of the hypothesis that some of the parameters in a model are restricted to zero. In the case of linear regression models this is a well-known test for which the criterion may be found in a standard statistics text such as Rao (1965). If the linear regression model has r parameters and we want to test whether s of these are restricted to zero based on N observations, then the criterion is (see Appendix A 2.1.6)

$$F = \frac{A_1 - A_0}{s} \div \frac{A_0}{N - r} \sim F(s, N-r) \tag{4.4.1}$$

where A_0 is the (smaller) sum of squares of the unrestricted model, A_1 is the (larger) sum of squares of the restricted model, and $F(s, N-r)$ denotes F-distribution with s and $N-r$ degrees of freedom.

Since in the conditional or static aspect, the ARMA(n,m) model is exactly a linear regression model, the above criterion can be used to test the hypothesis that s out of its $(m+n) = r$ parameters are zero. Then, A_0 becomes the residual sum of squares of the ARMA(n,m) model and A_1 that of the same model with s parameters dropped out. The justification of the criterion for the unconditional or dynamic aspect of the ARMA models, together with its interpretation as a convergence criterion, may be found in Pandit (1973).

For the test of ARMA($2n$,$2n-1$) versus ARMA($2n+2$,$2n+1$) considered above, the number of parameters including the mean (or average) is

$$r = (2n+2) + (2n+1) + 1 = 4n+4$$
$$s = (4n+4) - [4(n-1) + 4] = 4$$
$$A_0 = \frac{\text{sum of squares of } a_t\text{'s for}}{\text{the ARMA}(2n+2, 2n+1) \text{ model}}$$
$$A_1 = \frac{\text{sum of squares of } a_t\text{'s for}}{\text{the ARMA}(2n, 2n-1) \text{ model}}$$
$$N = \text{number of observations}$$

and the F criterion (4.4.1) takes the form

$$F = \frac{A_1 - A_0}{4} \div \frac{A_0}{N - 4n - 4} \sim F(4, N-4n-4) \qquad (4.4.1a)$$

If the value of F obtained in this way exceeds the value of $F(4, N-4n-4)$ for, say, a 5% significance level obtained from the F-distribution table, then the improvement in the residual sum of squares in going from the ARMA$(2n, 2n-1)$ model to the ARMA$(2n+2, 2n+1)$ model is significant, and therefore there is evidence against the hypothesis that the ARMA$(2n, 2n-1)$ model is adequate. If the F value is less than the value from the table, then we may conclude that the model is adequate at that level of significance.

4.4.3 Application of the *F*-Criterion

The F-criterion defined by (4.4.1) may be used in a variety of ways. First of all, it can be used in the special form (4.4.1a) as a *stopping criterion* in fitting a sequence of ARMA$(2n, 2n-1)$ for $n = 1, 2, 3, \ldots$. Every time n is increased by one, we compare the sum of squares of the two models by this criterion and continue increasing n as long as we get significant F values. We stop at that value of n for which the F value, comparing ARMA$(2n, 2n-1)$ with ARMA$(2n+2, 2n+1)$, is found to be insignificant at a predetermined significance level such as 5%, and conclude that the ARMA$(2n, 2n-1)$ model is adequate.

Note that the use of the F criterion given by (4.4.1a) is rigorously justified only when one of the two models ARMA$(2n, 2n-1)$ and ARMA$(2n+2, 2n+1)$ *is* adequate. For example, suppose ARMA$(6,5)$ is the adequate model for the data. Then, the use of the criterion is strictly justified for ARMA$(4,3)$ to ARMA$(6,5)$ for which F will be found to be significant, and ARMA$(6,5)$ to ARMA$(8,7)$ for which F will be found to be insignificant. The criterion is not theoretically justified for ARMA$(2,1)$ to ARMA$(4,3)$ since *both* of them are inadequate. In such a case, the reduction in the sum of squares is often so large that even without the criterion, one would decide not to stop, which will be confirmed by the criterion. However, as a precautionary measure, the decision to stop should be made only after the autocorrelations of the residual a_t's are small, say within the $\pm 2/\sqrt{N}$ band.

Once a decision is made to stop at an ARMA$(2n,2n-1)$ model, the F-criterion can be further used to see if a model of odd autoregressive order such as ARMA$(2n-1,2n-2)$ or smaller moving average order, say $m < 2n-1$, may also be adequate. To do this we can use confidence intervals for guidance. If ϕ_{2n} and θ_{2n-1} are small and their confidence intervals include zero, we fit the odd order ARMA$(2n-1,2n-2)$ model and again check by the F criterion. In the same way, we can check the adequacy of a model with the smaller number of moving average parameters as indicated by the confidence intervals.

Once the final model is determined, the F-criterion can also be used to see if some other desired form of the model is adequate. For example, suppose we want to know what pure AR(n) model is adequate. Then, we can use the I_j function to see at what j the I_j is close to zero. Suppose this happens at $j = n$. Then, we can fit AR(n), AR$(n+1)$ models and check by the F-criterion to see if AR(n) is adequate. In Section 4.5 we will present illustrative examples for these various applications of the F-criterion.

4.4.4 Checks on Residual (a_t) Autocorrelations (see Appendixes A 2.1.4 and A 2.1.5)

As explained in Chapter 2 (Section 2.2.4), one can roughly check the independence of a_t's by ensuring that the autocorrelations of a_t's are small. To determine how small they should be to be considered zero, one can use the following formula due to Bartlett (1946) for the standard error of estimated autocorrelations when $\rho_k = 0$ for $k > p$:

$$\sigma(\hat{\rho}_k) \simeq \frac{1}{\sqrt{N}} [1 + 2(\hat{\rho}_1^2 + \hat{\rho}_2^2 + \ldots + \hat{\rho}_p^2)]^{1/2}, \quad k > p \quad (4.4.2)$$

Since the $\hat{\rho}_k$'s are asymptotically normally distributed, the hypothesis that their expected value is zero can be checked at a 5% level by verifying that $|\hat{\rho}_k|$ is less than $1.96\sigma(\hat{\rho}_k)$. Since $\sigma(\hat{\rho}_k) \simeq 1/\sqrt{N}$, $1.96\sigma(\hat{\rho}_k) \simeq 2/\sqrt{N}$. Thus, $\pm 2/\sqrt{N}$ provides a quick approximate check. Alternatively, when all the "unified" autocorrelations, $\hat{\rho}_k/\sigma(\hat{\rho}_k)$, are within ± 1.96 band, the a_t's can be approximately taken as independent. This should be further confirmed by the F-criterion.

A "portmanteau" test suggested by Box and Jenkins (1970) is to check that the statistic [see Eq. (A 2.1.72)]

$$Q = N \sum_{k=1}^{K} \hat{\rho}_k^2 \quad (4.4.3)$$

is less than $\chi^2(K-n-m)$ at appropriate probability level for checking the independence of a_t's from an ARMA(n,m) model. K should be large enough so that G_j is practically zero for $j \geqslant K$.

4.5 MODELING PROCEDURE WITH ILLUSTRATIVE EXAMPLES

Let us now summarize the developments concerning the modeling strategy in Chapter 2 and in this chapter, aiming at a specific step-by-step modeling procedure that can be automatically executed on a computer. The conditional regression or the static aspect of the ARMA models in the form of the difference equations in Chapter 2 leads to the ARMA($n,n-1$) strategy in trying to reduce a dependent series of data into an independent one. The unconditional regression or the dynamic aspect of the ARMA models in this chapter showed that increasingly complex dynamics can be represented with increasing n by the Green's function

$$G_j = g_1\lambda_1^j + g_2\lambda_2^j + \ldots + g_n\lambda_n^j$$

or the autocovariance function

$$\gamma_k = d_1\lambda_1^k + d_2\lambda_2^k + \ldots + d_n\lambda_n^k$$

Such a representation is again equivalent to the ARMA($n,n-1$) model, and it can be shown that an arbitrary stationary set of data (including limiting cases such as random walk) can be represented by such a model with large enough n. It was shown in Section 4.2 that it is advantageous to increase n in steps of two and fit a sequence of ARMA($2n,2n-1$) models until the reduction in the sum of squares of a_t's is insignificant as judged by the F-criterion given in Section 4.4.

It was seen in Section 4.3 that with $\overline{X}$ as the estimate of the mean, the estimation can be very simple if the model is pure AR. For an AR(p) model the estimates of $\phi_1, \phi_2, \ldots, \phi_p$ can be obtained by the linear least squares formulae. Therefore, in practical applications, for which large-scale computer facilities required by the nonlinear least squares routine are not available, it often suffices to fit AR(n) models with increasing n till the reduction in residual sum of squares is insignificant. This is especially true when one is not interested in the structure of the system underlying the data. Even in fitting ARMA($n,n-1$) models, the estimates obtained from AR(p) models are useful in obtaining initial values, as explained in Section 4.3.4. In specific applications, the extent of improvement in the residual sum of squares of an ARMA ($n,n-1$) model over the corresponding AR($2n-1$) model can often indicate whether computational effort will be saved by using the pure AR models on a routine basis.

We will now describe a step-by-step modeling procedure mainly given as a guideline. One may modify it to suit one's own requirement. For example, it is known from empirical experience that when the number of observations is small, the order of the models is not likely to be high and one may use the $ARMA(n,n-1)$ sequence, rather than the $ARMA(2n,2n-1)$ sequence recommended below. Similarly, sequences other than $ARMA(n,n-1)$ such as $ARMA(n,n)$ may be used, especially when justified by a mechanism generating the series. (See, for example, Problem 3.23).

Step 1

Fit the $ARMA(2n,2n-1)$ models using the estimation procedure outlined in Section 4.3. For every increase of n by one, check the improvement in the residual sum of squares of a_t's by the F-criterion, Eq. (4.1.1a). Confirm that the autocorrelations of a_t's are within the permissible band ($\pm 2/\sqrt{N}$ or the more precise Bartlett band, so that the unified autocorrelations are less than 2 in magnitude), stop as soon as the F value from $ARMA(2n,2n-1)$ to $ARMA(2n+2,2n+1)$ is insignificant at a predetermined value such as 5% and choose the $ARMA(2n,2n-1)$ model.

Step 2

Check the values of ϕ_{2n}, θ_{2n-1} to see if they are small compared to their largest absolute value one and if their confidence intervals include zero. If "no," the adequate model is $ARMA(2n,2n-1)$.

Step 3

If ϕ_{2n}, θ_{2n-1} are small and their confidence intervals include zero, fit an $ARMA(2n-1,2n-2)$ model and check it with an $ARMA(2n,2n-1)$ by the F-criterion. If the F value is not significant, then dropping the small MA parameters, fit an $ARMA(2n-1,m)$ model with $m < 2n-2$ and check by the F-criterion until the adequate model with the smallest number of parameters is reached.

Step 4

If the F value is significant, drop the small moving average parameters and determine an $ARMA(2n,m)$ model with $m < 2n-1$, as in Step 3.

Step 5

If necessary, fit the desired forms of models such as pure $AR(n)$ or pure $MA(m)$, etc. by the F-criterion for increasing order till an insignificant value is reached.

The procedure is schematically shown in Fig. 4.1. The computer programs for modeling are listed at the end of the book. The results for the

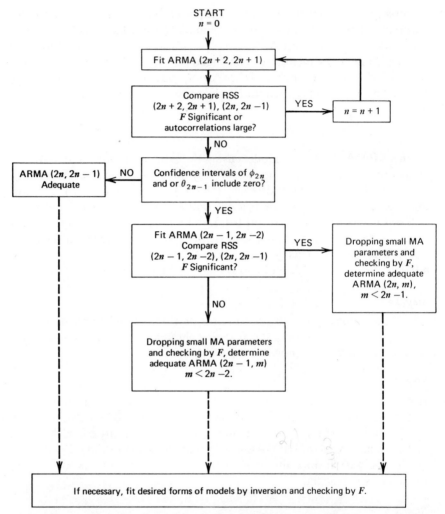

Figure 4.1 A flow chart of the modeling procedure.

five sets of data introduced in Chapter 2 are given in Tables 4.1 to 4.5. The mean is estimated as an additional parameter of the nonlinear least squares routine, using Eq. (4.3.6) to compute the residual a_t's. Since X_t are not available for $t \leq 0$, we can set $a_t = 0$, $t = 1$ to n and start the computation of a_t's from $t = n + 1$ involving slight loss of information, or set $X_t = a_t = 0$ for $t \leq 0$ and start with $a_1 = X_1$, $a_2 = X_2 - \phi_1 X_1 + \theta_1 a_1$, etc. The first method is preferable for stationary/stable series fluctuating around a fixed level and the second one for nonstationary/unstable series wandering away from the origin. Therefore the second method is used for the IBM and papermaking data and the first method for the remaining data sets. Both the options are available in the listed programs. Moreover, a further refinement of the former method to maximum likelihood estimates is possible by setting initial a_i's as additional parameters; this is the third option in the program. When convergence difficulties occur, it is sometimes necessary to alter the initial values, especially for high order models; therefore a separate option for user-supplied initial values has been provided in the program.

For illustration purposes, the example of sunspot activity data is discussed in detail and for other sets of data only the important points are discussed. For comparison, the corresponding AR models are also fitted. In general, the residual sum of squares of a pure AR model is larger than the corresponding ARMA model with the same number of parameters.

In many applications in science and engineering, Step 1 or, at most, Steps 1 to 3 suffice. This is particularly true when we are looking for major frequencies and/or eigen-values of the model, to see how they change with different conditions. Much of the work referred to in the Application Bibliography at the end of the book has been accomplished using ARMA($2n, 2n-1$) models. Low order models, even though strictly inadequate by the F-criterion, are often used in practice because of their simplicity and ease of interpretation. These are particularly attractive when justified by insignificantly small residual autocorrelations.

4.5.1 Sunspot Activity

To find an adequate model for the sunspot activity data, we first fit the ARMA(2,1) and ARMA(4,3) models. The initial and final values for the parameters are given below. The upper and lower 95% confidence bounds are shown in parenthesis after the estimated parameters.

(I) ARMA(2,1) model

(a) Initial Values

$$\phi_1 = 1.49, \ \phi_2 = -0.78, \ \theta_1 = 0.22$$

The residual sum of squares of the corresponding AR(3) model is = 42292.3.

(b) Final Values

	95% C.I.
$\phi_1 = 1.42$	$(\ 1.26 - 1.58)$
$\phi_2 = -0.72$	$(-0.86 - -0.58)$
$\theta_1 = 0.15$	$(-0.07 - 0.37)$

The residual sum of squares = 40879.0.

(II) ARMA(4,3) model

(a) Initial Values

$$\phi_1 = 0.79, \phi_2 = -0.22, \phi_3 = 0.28, \phi_4 = -0.38,$$
$$\theta_1 = -0.49, \theta_2 = -0.34, \theta_3 = -0.26$$

The residual sum of squares of the AR(7) model = 40828.7.

(b) Final Values

	95% C.I.
$\phi_1 = 0.42$	$(-0.58 - 1.42)$
$\phi_2 = 0.28$	$(-0.37 - 1.05)$
$\phi_3 = -0.08$	$(-0.57 - 0.41)$
$\phi_4 = -0.33$	$(-0.83 - 0.17)$
$\theta_1 = -0.89$	$(-1.89 - 0.11)$
$\theta_2 = -0.31$	$(-1.07 - 0.45)$
$\theta_3 = 0.14$	$(-0.09 - 0.37)$

The residual sum of squares = 39015.9.

The residual sum of squares of the ARMA(4,3) model is smaller than that of the ARMA(2,1) model and the F-criterion shows

$$A_1 = 40879.0$$
$$A_0 = 39015.9$$
$$s = 4$$
$$r = n+m+1 = 8 \quad \text{(including mean, } \mu \text{)}$$
$$N = 176$$
$$F = \frac{A_1 - A_0}{s} \div \frac{A_0}{N-r}$$
$$= \frac{40879.0 - 39015.9}{4} \div \frac{39015.9}{(176-8)}$$
$$= 2.01$$

From the F-distribution Table D at the end, $F_{0.95}(4,\infty) = 2.37$. Since the

F-test does not show significance, we need not go beyond the ARMA(2,1) model.

(III) AR(2) Model

Checking the parameters ϕ_2 and θ_1 of the ARMA(2,1) model, we found that the parameter θ_1 is small and its confidence limits include zero. If we drop the moving average parameter θ_1, the pure AR(2) model is fitted with the following results:

$$
\begin{array}{ll}
& 95\% \ C.I. \\
\phi_1 = \quad 1.34 & (\quad 1.23 \ — \quad 1.45) \\
\phi_2 = -0.65 & (-0.76 \ — \ -0.54)
\end{array}
$$

The residual sum of squares $= 41311.8$.

Also, the application of the F-criterion gives:

$$A_1 = 41311.8$$
$$A_0 = 40879.0$$
$$s = 1$$
$$r = 4$$
$$N = 176$$
$$F = \frac{41311.8 \ - \ 40879.0}{1} \ \div \ \frac{40879.0}{(176 \ - \ 4)}$$
$$= 1.82$$
$$F_{0.95}(1,\infty) = 3.84$$

As the computed F-value is less than the F-distribution value at a 5% significance level and the confidence limits of the parameter ϕ_2 of the AR(2) model do not include zero, an AR(2) model could be considered adequate for the sunspot activity data. The output from the computer program is given in Table 4.1. It should be noted for all such tables that the actual results obtained by the computer programs listed at the end may be somewhat different due to a variety of reasons such as subjective choices of convergence tolerances, initial values, method of computing a_t's etc.

4.5.2 IBM Stock Prices

The results obtained by following the modeling procedure outlined above are shown in Table 4.2. The adequate model using the F-criterion turns out to be ARMA(6,5). However, the confidence intervals are best for the AR(1) model and if we choose a 0.01 level of significance, using $F_{0.99}$

Table 4.1 Computer Output of Modeling Sunspot Series

Parameter	Order ARMA			
	(2,1)	(4,3)	(1,0)	(2,0)
ϕ_1	1.42 ± 0.16^a	0.42 ± 1.00	0.81 ± 0.09	1.34 ± 0.11
ϕ_2	-0.72 ± 0.14	0.28 ± 0.75		-0.65 ± 0.11
ϕ_3		-0.08 ± 0.49		
ϕ_4		-0.33 ± 0.50		
θ_1	0.15 ± 0.22	-0.89 ± 1.00		
θ_2		-0.31 ± 0.76		
θ_3		0.14 ± 0.23		
μ	44.52 ± 6.58	44.84 ± 6.58	42.99 ± 15.9	44.39 ± 7.31
Residual sum of squares	40879.0	39015.9	71688.6	41311.8

The adequate model is ARMA(2,0).
[a] The values in ($\pm$) indicate the $1.96 \times$ standard errors of the parameter estimates.

values from the Table D at the end of the book, then the AR(1) may be considered adequate.

As for the ARMA(2,1) model, ϕ_1 is close to one and ϕ_2 is small so the limiting ARMA(1,1) model with $\phi_1 = 1$

$$\nabla X_t = (1 - \theta_1 B) a_t$$

may also be adequate. For this model there is only one parameter since the mean μ gets cancelled

$$X_t - X_{t-1} = (X_t - \mu) - (X_{t-1} - \mu)$$

The estimate of the parameter θ_1 is

$$\theta_1 = -0.0848 \pm 0.1030$$

The residual sum of squares $= 19334.2$.

Thus, this limiting ARMA(1,1) model may also be considered adequate although the small value of θ_1 and its confidence interval again indicate the AR(1) model with $\phi_1 \to 1$ or the random walk. The ARMA(1,1) results will be useful in Chapter 5 in dealing with exponential smoothing.

4.5.3 Papermaking Data

It can be seen from Table 4.3 that the adequate model for the papermaking data is ARMA(2,1). However, an interesting result is found when the roots of the ARMA(2,1) model are examined. The ARMA(2,1) model can be

Table 4.2   Computer Output of Modeling IBM Stock Prices

Parameters	ARMA Order				
	(2,1)	(4,3)	(6,5)	(8,7)	(1,0)
ϕ_1	1.05 ± 1.16	2.40 ± 1.10	0.35 ± 0.24	-0.11 ± 0.55	0.998 ± 0.008
ϕ_2	-0.05 ± 1.16	-1.45 ± 2.10	-0.46 ± 0.25	-0.73 ± 0.65	
ϕ_3		-0.30 ± 1.17	0.17 ± 0.24	0.33 ± 0.36	
ϕ_4		0.36 ± 0.35	0.43 ± 0.22	0.36 ± 0.47	
ϕ_5			-0.32 ± 0.22	0.26 ± 0.62	
ϕ_6			0.83 ± 0.21	0.72 ± 0.50	
ϕ_7				-0.13 ± 0.52	
ϕ_8				0.28 ± 0.54	
θ_1	-0.03 ± 1.16	1.36 ± 1.07	-0.73 ± 0.27	-1.20 ± 0.55	
θ_2		0.04 ± 0.99	-1.21 ± 0.32	-1.97 ± 0.87	
θ_3		-0.43 ± 0.31	-0.97 ± 0.42	-1.63 ± 1.14	
θ_4			-0.53 ± 0.30	-1.22 ± 1.46	
θ_5			-0.76 ± 0.24	-0.77 ± 1.28	
θ_6				-0.05 ± 0.88	
θ_7				-0.17 ± 0.63	
μ	460.1 ± 14.2	459.7 ± 14.0	459.1 ± 13.5	458.9 ± 13.4	459.9 ± 14.2
RSS	19204.0	18964.2	17755.6	18263.3	19355.9

The adequate model is ARMA(6,5).

written as

$$X_t = 1.76X_{t-1} - 0.76X_{t-2} + a_t - 0.94a_{t-1}$$

or

$$(1 - 1.76B + 0.76B^2)X_t = (1 - 0.94B)a_t$$
$$(1 - B)(1 - 0.76B)X_t = (1 - 0.94B)a_t$$

that is, the roots are

$$\lambda_1 = 1, \qquad \lambda_2 = 0.76$$

We know from the Green function that if one of the roots of the ARMA(2,1) model is one, the system is not asymptotically stable but stable, which can be verified again in the present case as follows:

$$G_j = \frac{\lambda_1 - \theta_1}{\lambda_1 - \lambda_2} \lambda_1^j + \frac{\lambda_2 - \theta_1}{\lambda_2 - \lambda_1} \lambda_2^j$$

$$= \frac{1 - 0.94}{1 - 0.76} (1)^j + \frac{0.76 - 0.94}{0.76 - 1.0} (0.76)^j$$

$$\rightarrow \left(\frac{0.06}{0.24} \right) \qquad \text{as} \quad j \rightarrow \infty$$

Table 4.3 Computer Output of Modeling Papermaking Process Input Gate Opening

| Parameters | ARMA Order | | |
	(2,1)	(4,3)	(1,0)
ϕ_1	1.76 ± 0.16	−0.10 ± 0.28	0.98 ± 0.03
ϕ_2	−0.76 ± 0.16	1.63 ± 0.31	
ϕ_3		0.13 ± 0.27	
ϕ_4		−0.66 ± 0.29	
θ_1	0.94 ± 0.09	−0.95 ± 0.22	
θ_2		0.86 ± 0.18	
θ_3		0.86 ± 0.23	
μ	33.4 ± 0.54	33.4 ± 0.54	33.4 ± 0.61
Residual sum of squares	14.61	14.27	15.58

The adequate model is ARMA(2,1).

Therefore G_j remains bounded and the system is stable. It is also clear that the system is not asymptotically stable since G_j does not die out as $j \to \infty$. This point can also be verified if the AR(1) model is considered. The parameter $\phi_1 = 0.98$ is very close to its stability boundary value one, and therefore the system is very close to instability. (See Problems 10.10 and 10.11 for models with deterministic trends.)

4.5.4 Grinding Wheel Profile

Table 4.4 lists the results for this set of data. ARMA(6,5) is the adequate model since its unified residual autocorrelations are less than 2 in magnitude and the ARMA(8,7) model (not listed in the table) gives a larger residual sum of squares. However, the unified autocorrelations for the listed ARMA(2,1) model are also less than 2 in magnitude. Moreover, ϕ_2 and θ_1 of the ARMA(2,1) model include zero in their confidence intervals. Therefore, an AR(2) model was also fitted to the grinding wheel data and the results of the parameters and the residual sum of squares are given in the last column of Table 4.4. It is evident from the results that in this case, an AR(2) model may also be considered appropriate for the analysis purpose.

4.5.5 Mechanical Vibration Data

According to the modeling strategy, the sequence of ARMA(2,1), ARMA(4,3), ARMA(6,5), and ARMA(8,7) models is fitted since the reduction in the residual sum of squares is found to be significant for higher order models. With the tests of adequacy such as F-criterion, confidence

Table 4.4 Computer Output of Modeling Grinding Wheel Profile

Parameters	ARMA Order				
	(2,1)	(4,3)	(6,5)	(1,0)	(2,0)
ϕ_1	0.89 ± 0.39	0.54 ± 0.12	0.73 ± 0.13	0.63 ± 0.10	0.79 ± 0.12
ϕ_2	-0.29 ± 0.26	-0.62 ± 0.08	-0.55 ± 0.26		-0.22 ± 0.12
ϕ_3		0.85 ± 0.06	0.63 ± 0.13		
ϕ_4		-0.44 ± 0.08	-0.96 ± 0.16		
ϕ_5			0.38 ± 0.22		
ϕ_6			0.18 ± 0.14		
θ_1	0.11 ± 0.41	-0.25 ± 0.18	-0.09 ± 0.15		
θ_2		-0.70 ± 0.14	-0.43 ± 0.28		
θ_3		0.51 ± 0.18	0.45 ± 0.24		
θ_4			-0.61 ± 0.23		
θ_5			-0.17 ± 0.19		
μ	9.48 ± 0.68	9.41 ± 0.63	9.56 ± 0.68	9.41 ± 0.86	9.48 ± 0.69
Residual sum of squares	1472.81	1295.56	1217.31	1618.88	1473.52

The adequate model is ARMA(6,5).

Table 4.5 Computer Output of Modeling Mechanical Vibrations

Parameters	ARMA Order				
	(2,1)	(4,3)	(6,5)	(8,7)	(1,0)
ϕ_1	1.43 ± 0.16	0.65 ± 0.15	2.21 ± 0.38	1.46 ± 0.50	0.91 ± 0.007
ϕ_2	-0.61 ± 0.16	-0.41 ± 0.10	-2.13 ± 0.70	-0.67 ± 0.78	
ϕ_3		0.93 ± 0.05	1.94 ± 0.57	0.41 ± 0.84	
ϕ_4		-0.64 ± 0.12	-0.22 ± 0.57	-0.63 ± 0.90	
ϕ_5			1.46 ± 0.64	-0.04 ± 0.90	
ϕ_6			-0.33 ± 0.31	0.85 ± 0.85	
ϕ_7				-0.71 ± 0.78	
ϕ_8				0.27 ± 0.40	
θ_1	-0.54 ± 0.17	-1.36 ± 0.19	0.26 ± 0.36	-0.54 ± 0.46	
θ_2		-1.55 ± 0.19	-0.15 ± 0.22	-0.19 ± 0.73	
θ_3		-0.60 ± 0.19	0.80 ± 0.26	-0.34 ± 0.68	
θ_4			-0.20 ± 0.36	-0.16 ± 0.52	
θ_5			-0.56 ± 0.21	-0.47 ± 0.53	
θ_6				0.17 ± 0.64	
θ_7				0.26 ± 0.37	
μ	24.19 ± 3.01	24.89 ± 2.15	24.85 ± 4.08	24.40 ± 5.72	23.82 ± 6.46
RSS	535.65	395.77	355.47	349.64	1588.77

The adequate model is ARMA(6,5).

limits, etc., an ARMA(6,5) model is found to be adequate for the mechanical vibratory data. The detailed results are given in Table 4.5.

These data were collected from an actual mechanical vibration system governed by a second order differential equation, so the adequate model is expected to be ARMA(2,1). The higher order model ARMA(6,5) seems to be required because of additional modes (characteristic roots) possibly coming from corrupting noise, experimental error, etc. However, note that the parameters of the ARMA(2,1) model, although inadequate, give the natural frequency of 3.467 Hz (cycles per second), quite close to the "true" value of 3.316 Hz. These computations are further explained in Chapter 7, especially Section 7.7, where the proper continuous time model is fitted to these data.

PROBLEMS

4.1 A classical time series that is one of the first applications of moving average models is the Beveridge wheat price index series. Wold (1938) showed that an MA(2) model fits the series of 100 observations. With some rounding off the model is

$$X_t = a_t + 0.9a_{t-1} + 0.3a_{t-2}$$

If our modeling strategy is used for this series, we get an ARMA(2,1) model

$$X_t = 0.55X_{t-1} - 0.21X_{t-2} + a_t + 0.36a_{t-1}$$

Is one of the two models wrong; if so, which one? If both are correct, why? Explain your answer. If someone says that a pure AR model fits the data, can you say what that model would be?

4.2 Consider the ARMA(2,1) model fitted to the IBM data in Table 4.2. What ARMA(1,m) model would you suggest for these data? Suggest the value of m and obtain the approximate parameter values. Does your model agree with any other found adequate for these data?

4.3 The Green's function of an ARMA(n,m) model is given by

$$G_j = 0.4(0.6)^j + 0.8(0.5)^j - 0.2(0.7)^j, \qquad j \geq 0$$

Find the order n, m, and the values of the parameters ϕ and θ of the model. What is the order of an equivalent MA model?

4.4 From the following estimates $\hat{\phi}_1$, $\hat{\phi}_2$, $\hat{\phi}_3$ of an AR(3) model, find the *invertible* initial guess values for an ARMA(2,1) model

(i) 0.8, -0.4, -0.2

(ii) 0.5, -0.4, -0.5

4.5 From the following estimates $\hat{\phi}_1$, $\hat{\phi}_2$, $\hat{\phi}_3$, $\hat{\phi}_4$, and $\hat{\phi}_5$ of an AR(5) model, find the *invertible* initial guess values for an ARMA(3,2) model

(i)   0.1, -0.47, -0.15, -0.007, 0.051
(ii) 0.2, -0.48, -0.22, -0.123, -0.093

4.6 Compute the initial values of the parameters of ARMA(2,1) and ARMA(3,2) models by the inverse function method for the following sets of data and show your computations in detail:

(i)   Mechanical Vibration Data
(ii) Grinding Wheel Profile
      You may write your own program or use the ones listed at the end for getting the necessary I_j's. However, the computation of initial values should be done without using the computer.

4.7 Carry out the necessary *F*-tests on the models in Table 4.5 fitted to the mechanical vibration data and verify the adequate model. Which of these tests are theoretically justified?

4.8 An ARMA(2,1) model is fitted to 200 observations and gives a residual sum of squares of 202.3. What is the smallest sum of squares for an ARMA(4,3) model with which you can still conclude that the ARMA(2,1) model is adequate at 5% level, given $F_{0.95}(4,\infty) = 2.37$.

4.9 Compute the first few significantly large G_j and I_j weights for each of the models up to ARMA(6,5) fitted to the IBM stock data in Table 4.2. Comment on their similarities and differences for the different models in relation to the adequate model.

4.10 In a model fitting of an unknown set of data ($N=120$), the residual sum of squares of the AR(1) and ARMA(2,1) models are found to be 50.10 and 49.96, respectively. However, the parameter estimates of these models are $\hat{\phi}_1 = 0.90 \pm 0.06$ and $\hat{\phi}_1 = 1.32 \pm 0.15$, $\hat{\phi}_2 = -0.378 \pm 0.192$ and $\hat{\theta}_1 = 0.41 \pm 0.08$, respectively. From the *F*-criterion, it is found that the AR(1) model is the adequate model, whereas the parameters $\hat{\phi}_2$ and $\hat{\theta}_1$ of the ARMA(2,1) model suggest that the ARMA(2,1) model may be required since the estimated parameter values are not small and their 95% confidence intervals do not include zero. Could you explain this abnormal behavior of the ARMA(2,1) model or do you think that the ARMA(2,1) model is the adequate model for this set of data?

4.11 (a) Using Eq. (3.3.14) show that for an ARMA(2,1) model

$$\phi_1 = \frac{\rho_1\rho_2 - \rho_3}{\rho_1^2 - \rho_2}, \qquad \phi_2 = \frac{\rho_1\rho_3 - \rho_2^2}{\rho_1^2 - \rho_2}$$

(b) Generalize these results to an ARMA(n,m) model.

(c) For mechanical vibration data, $\hat{\rho}_1 = 0.90$, $\hat{\rho}_2 = 0.68$, $\hat{\rho}_3 = 0.42$. Find the initial guess values of ϕ_1 and ϕ_2.

4.12 (a) Using Eq. (3.3.14) show that the moving average parameter θ_1 satisfies the equation

$$\theta^2 - C\theta + 1 = 0$$

where

$$C = \phi_1 + \frac{1 - \phi_1\rho_1 - \phi_2\rho_2}{\phi_1 + (\phi_2 - 1)\rho_1}$$

(b) Use the results of Problem 4.11 to find the initial values of the parameter θ_1 for the vibration data.

(c) Of the two values in (b), which one would you use? Why? Can you always choose one of the two in this way? Why?

4.13 Given $\hat{\rho}_1 = 0.90$, $\hat{\rho}_2 = 0.70$, $\hat{\rho}_3 = 0.45$, obtain the initial values of the parameters of an ARMA(2,1) model.

4.14 Show that for an ARMA(n,m) model with $m > 1$, equations for θ_1, θ_2, etc., in terms of ρ_k, would be nonlinear and cannot be explicitly solved.

4.15 Tables 4.1, 4.4 and 4.5 are based on the first method computing a_t's, whereas Tables 4.2 and 4.3 are based on the second method. Switch these methods by choosing a different option in the computer programs listed at the end and construct your own tables deciding the adequate models. Comment on the similarities and differences between your results and those given in the text. Explain any anomalous results you may have. Is the dynamics of the models preserved in both methods? Do you now agree with the recommendation given in Section 4.5 about the data sets for which these methods should be used? Why? Explain, giving other fundamental reasons for or against this recommendation, in addition to those based on your results.

FORECASTING

One of the principal aims of system modeling or time series modeling is prediction or forecasting. It is often called extrapolation since it involves extrapolating the value $X_{t+\ell}$ ℓ steps ahead from the knowledge of the series X_t and its structure. In fact, the main reasons why one undertakes the analysis of a system from its observed data are to predict, to forecast, or to extrapolate its behavior at future times. The manipulation or control of the behavior is also based on prediction, so prediction may be viewed as the heart of control and regulation.

In Section 5.1 of this chapter we present a very brief historical review of the prediction theory and then offer a simple version of the basic Kolmogorov–Wiener prediction theory in Section 5.2 using Wold's decomposition of discrete stationary stochastic processes or time series. The forecasts thus obtained for the ARMA models are then related in Section 5.3 to those obtained by the even simpler device of conditional distribution and expectation introduced in Chapter 2. A procedure for obtaining the forecasts recursively, together with their probability limits, is also given in Section 5.3. The recursive method is further exploited in Section 5.4 for updating the forecasts when a new observation becomes known. Section 5.5 relates the forecasting by ARMA models to the well-known method of exponential smoothing, widely used in business and industry.

5.1 A BRIEF HISTORICAL REVIEW

The AR models were introduced by Yule (1927) to provide a more realistic description of the sunspot phenomena, based on the statistical analysis of Wolfer's sunspot numbers, so that one would be able to predict the activity in the sun. On the other hand, MA models arose as a result of the investigation by Slutsky (1927), aimed at revealing the dangers in the prevalent tendency to "smooth" the time series data by operations such as differencing or weighted averaging, and then fitting sine cosine waves or straight line trends that can be used for prediction or forecasting. (Note that we strictly avoided differencing in our modeling strategy in Chapter 4; it will be further discussed in later chapters—in particular, Chapter 9.)

Around 1933 the theoretical foundation for stochastic processes, basic to the time series models in particular, was presented by A. Kolmogorov. The papers Kolmogorov (1931, 1933) provide the theory of Markov processes, of which ARMA models are a very special case, and the measure theory on which the modern probability theory is founded. At the same time, the concept of stationary stochastic processes, which reduces the study of stochastic processes to their correlation, was advanced and developed by Khintchin (1934).

It was Wold (1938) who established a link between the statistical theory of time series analysis initiated by Yule and Slutsky and the probabilistic theory of the structure of time series initiated by Kolmogorov and Khintchin. Wold showed that an arbitrary time series can be decomposed into two parts, one (linearly) deterministic and the other stochastic, and that the stochastic part can be further decomposed as a finite or infinite moving average. This result is known as Wold's decomposition. We have seen in Chapter 3 that a MA model expresses X_t as a finite moving average (that is, a linear combination of a_t's), whereas an AR or an ARMA Model expresses X_t as an infinite moving average. This decomposition of a time series or stationary stochastic process X_t is an "orthogonal" decomposition since the concept of orthogonality of two random variables may be taken as equivalent to their independence or uncorrelatedness. Such an orthogonal decomposition completely reveals and therefore characterizes the structure of the stochastic system/process underlying the time series. For ARMA models it is the Green's function G_j, developed in Chapter 3, that summarizes this orthogonal decomposition in addition to the mean μ, which is the fixed deterministic component.

This concept of the orthogonal decomposition of a stationary stochastic process was fully generalized by Kolmogorov (1939, 1941a, 1941b), who used it to rigorously formulate the prediction problem in the geometry of Hilbert space and then solve it with the principle of orthogonal projection. The pioneer paper Kolmogorov (1941b) forms the starting point of the modern prediction theory; in this paper, he derived a general formula for the mean squared error of the best linear predictor of an arbitrary stationary sequence (that is, X_t with discrete t).

At about the same time, around 1941, Wiener independently developed the prediction theory using the spectral representation. His treatment was restricted to a much narrower class of processes, which have rational spectra, similar to those of the ARMA models. However, for this narrower class, Wiener succeeded in finding not only the mean squared error, but more generally an expression for the operator that performs the extrapolation. Since such an operator can easily be synthesized in hardware, Wiener's work is of immense practical value and was immediately applied to the prediction of aircraft flight paths, for the purpose of fire control during World War II. Wiener's results were not published until the appearance of his famous book: *The Extrapolation, Interpolation and Smooth-*

ing of Stationary Time Series with Engineering Applications (John Wiley, New York, 1949).

An account of the Kolmogorov-Wiener prediction theory, not involving the sophisticated mathematics of the former or the obstruse heuristic arguments of the latter, may be found in Yaglom (1952). The theory has been extended by Kalman (1960) to nonlinear processes by emphasizing conditional expectation and recursive algorithms. Whittle (1963) presents a simplified version of the theory using least squares methods and replacing the measure theoretic probabilistic arguments by the expectation operator. A combination of least squares methods and recursive algorithms for forecasting ARMA models may be found in Box and Jenkins (1970).

5.2 PREDICTION AS ORTHOGONAL PROJECTION

As we will see in the next section, prediction or forecasts for the ARMA models can be simply obtained by conditional expectation, as illustrated by the AR(1) model in Section 2.2.5. However, the geometric concept underlying the basic Kolmogorov-Wiener prediction theory is quite illuminating, and therefore we first present a simple version of its formulation and solution.

Suppose we are given two vectors $\mathbf{x}_1$ and $\mathbf{x}_2$, and we want to approximate a third vector $\mathbf{x}_3$ by a linear combination of the first two. In other words, the problem is to find the values of the constants g_1 and g_2 such that the vector $g_1\mathbf{x}_2 + g_2\mathbf{x}_1$ is closest to the vector $\mathbf{x}_3$ or the distance between them is minimum. As shown in Fig. 5.1a, the linear combinations of $\mathbf{x}_1$ and $\mathbf{x}_2$ form a plane and the solution to the problem is obtained by dropping a perpendicular from $\mathbf{x}_3$ onto this plane, that is, the orthogonal projection of $\mathbf{x}_3$ on this plane gives the vector $g_1\mathbf{x}_2 + g_2\mathbf{x}_1$ that is closest to $\mathbf{x}_3$.

5.2.1 Formulation

Now consider the problem of prediction of a stationary stochastic system or time series. Given observations $X_t, X_{t-1}, X_{t-2}, \ldots$, at time t, we want to predict the value of $X_{t+\ell}$ ℓ steps ahead. If we consider $X_{t+\ell}, X_t, X_{t-1}, X_{t-2}, \ldots$ as vectors and define the distance between the vectors as the expected (or mean) value of their squared difference, then the prediction problem can be formulated as discussed above. The best linear prediction of $X_{t+\ell}$ at time t, denoted by $\hat{X}_t(\ell)$ and given by a linear combination of $X_t, X_{t-1}, X_{t-2}, \ldots$, say

$$\hat{X}_t(\ell) = g_0^* X_t + g_1^* X_{t-1} + g_2^* X_{t-2} + \cdots \qquad (5.2.1)$$

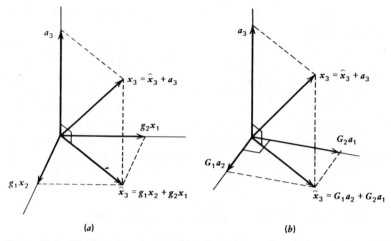

Figure 5.1 Geometric concept of prediction as orthogonal projection. (*a*) Decomposition by x_t. (*b*) Decomposition by a_t.

which minimizes the mean square of the prediction error

$$e_t(\ell) = X_{t+\ell} - \hat{X}_t(\ell) \tag{5.2.2}$$

that is,

$$E[e_t(\ell)]^2 = E[X_{t+\ell} - \hat{X}_t(\ell)]^2 \tag{5.2.3}$$

is given by the orthogonal projection of $X_{t+\ell}$ on the "plane" formed by the linear combination of X_t, X_{t-1}, X_{t-2}, ... Thus, we have to find the coefficients g_0^*, g_1^*, g_2^*, . . . such that $\hat{X}_t(\ell)$ given by Eq. (5.2.1) is the projection that minimizes the "distance" or the mean squared error given by Eq. (5.2.3).

This infinite dimensional problem can be visualized by means of a three-dimensional picture in Fig. 5.1(*a*). Vectors $\mathbf{x}_1$ and $\mathbf{x}_2$ are not necessarily orthogonal, just as X_t, X_{t-1}, . . . are dependent and therefore not orthogonal. Therefore, how to find $\hat{\mathbf{x}}_3$ (or g_1 and g_2) is not quite apparent. However, if we knew the coordinates of $\mathbf{x}_3$ in an orthogonal frame of reference, so that vectorially, as shown in Fig. 5.1(*b*), $\mathbf{x}_3 = \mathbf{a}_3 + G_1\mathbf{a}_2 + G_2\mathbf{a}_1$, then clearly the orthogonal projection $\hat{\mathbf{x}}_3 = G_1\mathbf{a}_2 + G_2\mathbf{a}_1$. The Green's function provides us with precisely such an orthogonal coordinate representation of time series X_t.

5.2.2 Solution

To solve this problem, that is, to obtain the orthogonal projection, we must know the structure of the stochastic system or time series X_t. Suppose this structure is specified by a fixed mean μ, which we take to be zero, and the Wold's decomposition.

$$X_t = a_t + G_1 a_{t-1} + G_2 a_{t-2} + \ldots \qquad (5.2.4)$$

The property of a_t's

$$E(a_t a_{t-k}) = 0, \qquad k \neq 0 \qquad (5.2.5)$$

specifies that the a_t's are uncorrelated or "orthogonal." Thus, the Wold's decomposition (5.2.4) is an orthogonal decomposition that expresses the vector X_t in terms of its "coordinates" G_j, along the "axes" a_{t-j}, $j = 0$, $1, 2, \ldots$. (See Fig. 3.4.)

Using (5.2.4), we can also express the vector $\hat{X}_t(\ell)$ in terms of the orthogonal vectors a_{t-j}, rather than in X_{t-j}'s as in (5.2.1)

$$\hat{X}_t(\ell) = G_0^* a_t + G_1^* a_{t-1} + G_2^* a_{t-2} + \ldots \qquad (5.2.6)$$

Knowing g_j^*, we can theoretically find G_j^* via (5.2.4) and vice versa. (See Problems 5.7 and 5.8.) Therefore, the problem of finding g_j^* is equivalent to that of finding G_j^*; as we will now see, G_j^* are much easier to find.

One way to find G_j^* is to explicitly use the concept of orthogonal projection. $\hat{X}_t(\ell)$ will be an orthogonal projection of $X_{t+\ell}$ on the plane formed by $X_t, X_{t-1}, \ldots$, that is, the plane formed by $a_t, a_{t-1}, a_{t-2}, \ldots$, if and only if the vector of prediction error

$$e_t(\ell) = X_{t+\ell} - \hat{X}_t(\ell)$$

is orthogonal to every a_{t-j} for $j = 0, 1, 2, \ldots$. This gives the condition

$$E[e_t(\ell) a_{t-j}] = E\{[X_{t+\ell} - \hat{X}_t(\ell)] a_{t-j} = 0, \qquad j = 0, 1, 2, \ldots$$

Substituting from (5.2.4) and (5.2.6)

$$E\{[(a_{t+\ell} + G_1 a_{t+\ell-1} + \ldots + G_\ell a_t + G_{\ell+1} a_{t-1} + \ldots)$$
$$- (G_0^* a_t + G_1^* a_{t-1} + \ldots)] a_{t-j}\} = 0$$
$$j = 0, 1, 2, \ldots \qquad (5.2.7)$$

for $j = 0$ we get, using Eq. (5.2.5)

$$E[(G_\ell - G_0^*) a_t^2] = \sigma_a^2 (G_\ell - G_0^*) = 0$$

that is, $G_0^* = G_\ell$.
Similarly for $j = 1$,

$$E[(G_{\ell+1} - G_1^*) a_{t-1}^2] = \sigma_a^2 (G_{\ell+1} - G_1^*) = 0$$

that is $G_1^* = G_{\ell+1}$,
and, in general,

$$G_j^* = G_{\ell+j} \qquad (5.2.8)$$

which solves the prediction problem.

5.2.3 An Alternative Solution

Another way to find G_j^* is to use the fact that $\hat{X}_t(\ell)$ is an orthogonal projection or the required minimum mean squared error prediction, when the mean squared prediction error (or the error variance)

$$V[e_t(\ell)] = E[e_t(\ell)]^2 = E[X_{t+\ell} - \hat{X}_t(\ell)]^2 \qquad (5.2.9)$$

is minimized. Using Eqs. (5.2.4) to (5.2.6), we have

$$X_{t+\ell} = a_{t+\ell} + G_1 a_{t+\ell-1}$$
$$+ \ldots + G_{\ell-1} a_{t+1} + G_\ell a_t + G_{\ell+1} a_{t-1} + \ldots$$
$$\hat{X}_t(\ell) = \qquad\qquad\qquad G_0^* a_t + G_1^* a_{t-1} + \ldots$$

and hence it is clear that (5.2.9) is minimized by setting

$$G_j^* = G_{\ell+j}$$

which again gives Eq. (5.2.8)

Thus, the minimum mean squared error prediction of $X_{t+\ell}$ is, by Eqs. (5.2.6) and (5.2.8),

$$\hat{X}_t(\ell) = G_\ell a_t + G_{\ell+1} a_{t-1} + G_{\ell+2} a_{t-2} + \ldots \qquad (5.2.10)$$

the prediction error is

$$e_t(\ell) = a_{t+\ell} + G_1 a_{t+\ell-1} + G_2 a_{t+\ell-2} + \ldots + G_{\ell-1} a_{t+1} \qquad (5.2.11)$$

and its variance [since a_t's are independent, see Eq. (A 2.1.15)]

$$V[e_t(\ell)] = \sigma_a^2 (1 + G_1^2 + G_2^2 + \ldots + G_{\ell-1}^2) \qquad (5.2.12)$$

which is minimum among all the possible linear predictions of $X_{t+\ell}$, at time t, based on $X_t, X_{t-1}, X_{t-2}, \ldots$.

Thus, we have solved the problem of finding $\hat{X}_t(\ell)$ in the orthogonal frame of reference similar to Fig. 5.1b. However, can we simplify it in terms of known data $X_t, X_{t-1}, \ldots$ similar to Fig. 5.1a? The answer is yes, because the geometric concept of projection is equivalent to the concept of "conditional expection" provided by the ARMA models.

5.3 FORECASTING BY CONDITIONAL EXPECTATION

We have seen in Chapter 2 that for the AR(1) model, simply using its conditional regression structure, we have

$$\hat{X}_t(1) = \phi_1 X_t$$

and the one step ahead prediction error

$$e_t(1) = a_{t+1}$$

This forecast error coincides with Eq. (5.2.11) for $\ell = 1$. To see whether the one step ahead forecast also coincides with Eq. (5.2.10), let us note that for the AR(1) model

$$G_j = \phi_1^j$$

Therefore, Eq. (5.2.10) gives

$$
\begin{aligned}
\hat{X}_t(\ell) &= \phi_1^\ell a_t + \phi_1^{\ell+1} a_{t-1} + \phi_1^{\ell+2} a_{t-2} + \dots \\
&= \phi_1^\ell (a_t + \phi_1 a_{t-1} + \phi_1^2 a_{t-2} + \dots) \\
&= \phi_1^\ell X_t
\end{aligned}
$$

and, in particular,

$$\hat{X}_t(1) = \phi_1 X_t$$

which is the same as above.

This suggests that there must be a simple method of computing the forecasts directly using the model without explicitly using Eq. (5.2.10). In fact, there is such a simple method; to obtain the forecasts by "conditional expectation." The forecast for the AR(1) model in Chapter 2 given above is simply a special case of such conditional expectation.

5.3.1 Conditional Expectation from Orthogonal Decomposition

Although the forecasts by conditional expectation can be derived in the framework of the orthogonal projection employed in the last section, we will derive them using the more fundamental concept of orthogonal decomposition and the analogy of 3-dimensional space. Suppose we decompose the vector $\mathbf{x}_3$ by a_t instead of $\mathbf{x}_t$ as shown in Fig. 5.1(b) and assume that the coordinates of the point $\mathbf{x}_3$ are (G_2, G_1, G_0), that is, the vector $\mathbf{x}_3$ has the orthogonal decomposition

$$\mathbf{x}_3 = \mathbf{a}_3 + G_1 \mathbf{a}_2 + G_2 \mathbf{a}_1$$

where $\mathbf{a}_1$, $\mathbf{a}_2$, and $\mathbf{a}_3$ denote the "unit" vectors along the orthogonal axes. Since the coordinates of $\hat{\mathbf{x}}_3$ are clearly (G_2, G_1), it is given by

$$\hat{\mathbf{x}}_3 = G_1 \mathbf{a}_2 + G_2 \mathbf{a}_1$$

and the error vector

$$\mathbf{x}_3 - \hat{\mathbf{x}}_3 = \mathbf{a}_3$$

Similarly, we know that for any ARMA model the orthogonal decomposition of $X_{t+\ell}$, which is to be predicted at time t, is given by

$$X_{t+\ell} = a_{t+\ell} + G_1 a_{t+\ell-1} + G_2 a_{t+\ell-2} + \dots$$

Therefore, the best linear forecast of $X_{t+\ell}$ based on $X_t, X_{t-1}, X_{t-2}, \dots$ or, equivalently, $a_t, a_{t-1}, a_{t-2}, \dots$ is the part of the decomposition containing a_{t-j}, $j \geq 0$ and the remaining part is naturally the prediction error. Thus,

$$X_{t+\ell} = \underbrace{a_{t+\ell} + G_1 a_{t+\ell-1} + \ldots + G_{\ell-1} a_{t+1}}_{} \underbrace{+ G_\ell a_t + G_{\ell+1} a_{t-1} + \ldots}_{}$$

$$= \qquad\qquad e_t(\ell) \qquad\qquad + \qquad\qquad \hat{X}_t(\ell)$$

$$= \qquad\qquad \text{Error} \qquad + \qquad \text{Forecast} \quad (5.3.1)$$

Note that the orthogonality of a_t's is to be interpreted as independence or uncorrelatedness. The a_{t-j}, $j \geq 0$ are known at time t, but since as random variables, they are independent or uncorrelated with the unknown a_{t+j}, $j > 0$, we cannot "predict" any of the a_{t+j}, $j > 0$ from a_{t-j}, $j \geq 0$. Therefore, $\hat{X}_t(\ell)$ is the best prediction and $e_t(\ell)$ is the prediction error with minimum variance. The conditional expectation is the same as orthogonal projection.

5.3.2 Rules for Conditional Expectation

At time t, $a_{t+\ell}, a_{t+\ell-1}, \ldots, a_{t+1}$ in Eq. (5.3.1) are random variables each with expectation zero, whereas $a_t, a_{t-1}, \ldots$ are known numbers. Therefore, if we take the conditional expectation at time t, denoted by $\underset{t}{E}$, on both sides of Eq. (5.3.1), we have

$$\underset{t}{E}(X_{t+\ell}) = G_\ell a_t + G_{\ell+1} a_{t-1} + \ldots$$

$$= \hat{X}_t(\ell)$$

Thus, we can frame the following rules. The conditional expectation of the:

1   Present or past observation is that (known) observation.
2   Future observation is its forecast.
3   Present or past "shock" a_t is that shock, which can be computed.
4   Future shock is zero.

That is,

$$\underset{t}{E}(X_{t-j}) = X_{t-j}, \; j = 0, 1, 2, \ldots$$

$$\underset{t}{E}(X_{t+j}) = \hat{X}_t(j) \; j = 1, 2, 3, \ldots$$

$$\underset{t}{E}(a_{t-j}) = a_{t-j}, \; j = 0, 1, 2, \ldots \qquad (5.3.2)$$

$$\underset{t}{E}(a_{t+j}) = 0, \quad j = 1, 2, 3, \ldots$$

5.3.3 The AR(1) Model—Forecasts and Probability Limits

For the AR(1) model considered above

$$\hat{X}_t(1) = \underset{t}{E}(X_{t+1}) = \underset{t}{E}(\phi_1 X_t + a_{t+1})$$

$$= \phi_1 X_t$$

$$\hat{X}_t(2) = E_t(X_{t+2}) = E_t(\phi_1 X_{t+1} + a_{t+2})$$
$$= \phi_1 \hat{X}_t(1) = \phi_1^2 X_t$$

and, in general,

$$\hat{X}_t(\ell) = E_t(X_{t+\ell}) = \phi_1 \hat{X}_t(\ell - 1) = \phi_1^\ell X_t$$

as before.

Prediction Error

The prediction error is

$$e_t(\ell) = a_{t+\ell} + G_1 a_{t+\ell-1} + \ldots + G_{\ell-1} a_{t+1}$$
$$= a_{t+\ell} + \phi_1 a_{t+\ell-1} + \ldots + \phi_1^{\ell-1} a_{t+1}$$

and, since a_t's are independent, by Eq. (A.2.1.15)

$$V[e_t(\ell)] = \sigma_a^2(1 + \phi_1^2 + \phi_1^4 + \ldots + \phi_1^{2(\ell-1)})$$

Conditional Distribution and Probability Limits

We have seen in Chapter 2 that since

$$X_{t+1} = \phi_1 X_t + a_{t+1}$$
$$= \hat{X}_t(1) + e_t(1)$$

the conditional distribution of X_{t+1}, given $X_t, X_{t-1}, \ldots$, is

$$(X_{t+1}|X_t, X_{t-1}, \ldots) \sim \text{NID}(\phi_1 X_t, \sigma_a^2)$$
$$\sim \text{NID}(\hat{X}_t(1), V[e_t(1)])$$

so that $\hat{X}_t(1)$ is the conditional expectation of X_{t+1} (at time t) given X_t, $X_{t-1}, \ldots$. Therefore, from Fig. 2.12, we see that we can use this conditional distribution to get the probability limits on the forecasts, which specify how close the true value will be to the forecasts by a probability statement. Thus, 95% probability limits on the forecast of X_{t+1} made at time t are

$$\hat{X}_t(1) \pm 1.96(V[e_t(1)])^{1/2}$$

that is,

$$\phi_1 X_t \pm 1.96\sigma_a$$

whereas 90% limits are

$$\phi_1 X_t \pm 1.65\,\sigma_a$$

Similarly, for the two step ahead forecast of X_{t+2} made at time t

$$X_{t+2} = \hat{X}_t(2) + e_t(2)$$
$$= \hat{X}_t(2) + (a_{t+2} + \phi_1 a_{t+1})$$

The conditional distribution of X_{t+2} given $X_t, X_{t-1}, \ldots$ is
$$(X_{t+2}|X_t, X_{t-1}, \ldots) \sim ND(\hat{X}_t(2), V[e_t(2)])$$
$$\sim ND(\hat{X}_t(2), \sigma_a^2(1 + \phi_1^2))$$

The forecast $\hat{X}_t(2)$ is again the mean or expectation of this conditional distribution of X_{t+2} given $X_t, X_{t-1}, \ldots$ The 95% probability limits are now given by
$$\hat{X}_t(2) \pm 1.96(V[e_t(2)])^{1/2}$$

that is,
$$\phi_1^2 X_t \pm 1.96\sigma_a(1 + \phi_1^2)^{1/2}$$

In general, for an AR(1) model
$$(X_{t+\ell}|X_t, X_{t-1}, \ldots) \sim ND[\hat{X}_t(\ell), \sigma_a^2(1 + \phi_1^2 + \phi_1^4 + \ldots + \phi_1^{2(\ell-1)})]$$
and hence the 95% probability limits are
$$\hat{X}_t(\ell) \pm 1.96\sigma_a (1 + \phi_1^2 + \phi_1^4 + \ldots + \phi_1^{2(\ell-1)})^{1/2}$$

5.3.4 The AR(1) Model—Correlation of Forecast Errors

The conditional distributions given above and the forecast errors for the same origin at different lead times (or different origin for the same lead time greater than one) are not independent.
For example,
$$e_t(2) = a_{t+2} + \phi_1 a_{t+1}$$
and
$$e_{t+1}(2) = a_{t+3} + \phi_1 a_{t+2}$$
are *not* independent; in fact, their covariance is
$$\text{Cov}[e_t(2), e_{t+1}(2)] = E(a_{t+2} + \phi_1 a_{t+1})(a_{t+3} + \phi_1 a_{t+2})$$
$$= \phi_1 \sigma_a^2$$
and the correlation
$$\text{Cor}[e_t(2), e_{t+1}(2)] = \frac{\text{Cov}[e_t(2), e_{t+1}(2)]}{\{\text{Var}[e_t(2)]\text{Var}[e_{t+1}(2)]\}^{1/2}}$$
$$= \frac{\phi_1 \sigma_a^2}{(1 + \phi_1^2)\sigma_a^2}$$
$$= \frac{\phi_1}{1 + \phi_1^2}$$

This correlation gives an indication about the error of prediction made from the origin $t + 1$, when that from the origin t is known. If the correlation is large and positive, then a large two step ahead error at t will tend to be followed by a large one of the same sign at $t + 1$. If the correlation is large and negative, then the signs of the errors at t and $t + 1$ will tend to be opposite, whereas the magnitude will tend to remain the same.

5.3.5 The AR(1) Model—Numerical Example

Consider the AR(1) model for the papermaking data, with $\phi_1 = 0.9$, $\sigma_a^2 = 0.094$. Subtracting the mean $\overline{X} = 32.02$, we have $X_4 = 1.78$, $X_5 = 1.48$, $X_6 = 0.98$, and $X_7 = 0.98$. The one step ahead forecast at time $t = 4$ is

$$\hat{X}_4(1) = \phi_1 X_4$$
$$= 0.9 \times 1.78$$
$$= 1.60$$

The actual one step ahead forecast error

$$e_4(1) = a_5 = X_5 - \hat{X}_4(1)$$
$$= 1.48 - 1.60$$
$$= -0.12$$

The conditional distribution of X_5 given X_4, X_3, . . .

$$(X_t | X_{t-1}, X_{t-2}, \ldots) \sim \text{NID}(\hat{X}_{t-1}(1), \sigma_a^2)$$
$$(X_5 | X_4, X_3, \ldots) \sim \text{NID}(\hat{X}_4(1), \sigma_a^2)$$
$$\sim \text{NID}(1.60, 0.094)$$

and the 95% probability limits on the forecast of X_5 made at time $t = 4$

$$\hat{X}_4(1) \pm 1.96\sigma_a = 1.60 \pm 1.96 \sqrt{0.094}$$
$$= 1.60 \pm 0.601$$

The two step ahead forecast of X_6 made at time $t = 4$ is

$$\hat{X}_4(2) = \phi_1^2 X_4$$
$$= 0.9^2 \times 1.78$$
$$= 1.44$$

The actual two step ahead forecast error

$$e_4(2) = X_6 - \hat{X}_4(2)$$
$$= 0.98 - 1.44$$
$$= -0.46$$

The conditional distribution of X_6, given X_4, X_3, . . .,

$$(X_t | X_{t-2}, X_{t-3}, \ldots) \sim \text{ND}[\hat{X}_{t-2}(2), (1 + \phi_1^2)\sigma_a^2]$$
$$(X_6 | X_4, X_3, \ldots) \sim \text{ND}[\hat{X}_4(2), (1 + \phi_1^2)\sigma_a^2]$$
$$\sim \text{ND}(1.44, 0.17)$$

and the 95% probability limits

$$\hat{X}_4(2) \pm 1.96\sigma_a\sqrt{1 + \phi_1^2} = 1.44 \pm 1.96\sqrt{0.17}$$

$$= 1.44 \pm 0.81$$

Similarly, the two step ahead forecast of X_7 made at time $t = 5$

$$\hat{X}_5(2) = \phi_1^2 X_5$$

$$= 0.9^2 \times 1.48$$

$$= 1.20$$

and the corresponding actual forecast error

$$e_5(2) = X_7 - \hat{X}_5(2)$$

$$= 0.98 - 1.20$$

$$= -0.22$$

The covariance between $e_4(2)$ and $e_5(2)$

$$\text{Cov}[e_4(2), e_5(2)] = \phi_1\sigma_a^2$$

$$= 0.9 \times 0.094$$

$$= 0.0846$$

and the correlation

$$\text{Cor}[e_4(2), e_5(2)] = \frac{\phi_1}{1 + \phi_1^2}$$

$$= \frac{0.9}{1 + 0.9^2}$$

$$= 0.49$$

which are also the values for the covariance and correlation between $(X_6|X_4,X_3, \ldots)$ and $(X_7|X_5,X_4, \ldots)$.

The forecasts and the 95% probability limits are plotted in Fig. 5.2 with a relevant portion of the papermaking data. Note that for large lead times the forecast approaches the mean value because the given AR(1) model is stable. The rate of approach is very slow since the parameter ϕ_1 = 0.9, close to the stability boundary value one. Although we have used $\overline{X}$ as the mean for an illustrative purpose, the value of μ estimated from the nonlinear estimation should be used in actual practice and should be added at the end to get actual forecasts.

5.3.6 General Results—ARMA(n,m) Model

These results can be generalized for an arbitrary ARMA(n,m) model. By Eq. (5.3.1)

$$X_{t+\ell} = \hat{X}_t(\ell) + e_t(\ell)$$

$$= \hat{X}_t(\ell) + (a_{t+\ell} + G_1 a_{t+\ell-1} + \ldots + G_{\ell-1} a_{t+1})$$

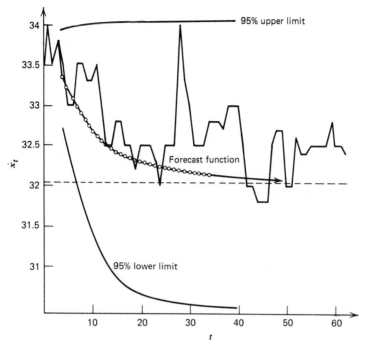

Figure 5.2   Forecasts and probability limits for the papermaking data.

and therefore the conditional distribution of $X_{t+\ell}$ given X_t, X_{t-1}, . . . is specified by

$$(X_{t+\ell}|X_t, X_{t-1}, \ldots) \sim \text{ND}(\hat{X}_t(\ell), V[e_t(\ell)])$$

$$\sim \text{ND}[\hat{X}_t(\ell), \sigma_a^2 \qquad (5.3.3)$$

$$(1 + G_1^2 + G_2^2 + \ldots + G_{\ell-1}^2)]$$

The forecast $\hat{X}_t(\ell)$ is the conditional mean or expectation of this distribution as shown in Fig. 5.3 (refer to Fig. 2.12 for comparison). The 95% probability limits on the forecasts are given by

$$\hat{X}_t(\ell) \pm 1.96 \, \sigma_a(1 + G_1^2 + G_2^2 + \ldots + G_{\ell-1}^2)^{1/2} \qquad (5.3.4)$$

Here, $\hat{X}_t(\ell)$ can be computed by using the rules for conditional expectation given by Eq. (5.3.2), and G_j should be computed by the *implicit* method of Section 3.1.5.

A few remarks about the probability limits for the forecasts can be made on the basis of (5.3.4). First of all, for the same model, the limits widen as we increase ℓ since increasing numbers of G_j^2's inflate the variance of the forecast error. This says that the further into the future we forecast, the more uncertain we are about the forecasts, which intuitively makes

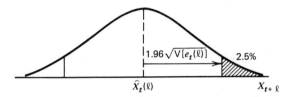

Figure 5.3 Conditional distribution of $X_{t+\ell}$ given X_t, X_{t-1}, ...

sense. Second, for a given lead time ℓ, the probability limits depend on G_j, that characterizes the memory of the dynamics of the systems. For systems with strong memory or slowly decaying G_j, the limits will be wider or the forecasts more uncertain than those with rapidly decaying G_j. For example, among the AR(1) systems, the random walk has the widest probability limits or the most uncertain forecasts. Third, for the same system or model and lead time ℓ, the limits are proportional to the variance of a_t's, σ_a^2.

The variance $V[e_t(\ell)]$ strictly represents the variance of the forecast error $e_t(\ell)$, which is a random variable when $X_{t+\ell}$ is not known. However, in spite of the fact that the $\hat{X}_t(\ell)$ is *not* a random variable and hence has no variance, $V[e_t(\ell)]$ is often loosely referred to as the variance of the forecast rather than that of the forecast error.

For an arbitrary ARMA model the ℓ step ahead forecast error is

$$e_t(\ell) = a_{t+\ell} + G_1 a_{t+\ell-1} + \ldots + G_{\ell-1} a_{t+1}$$

Since in the case of the ARMA model the forecast errors for a fixed value of lead time ℓ and different time origin (or fixed value of time origin t and different lead times) may be correlated, the covariance and the correlation between the forecast errors $e_t(\ell)$ and $e_{t+j}(\ell)$ for different values of j are

$$\text{Cov}[e_t(\ell), e_{t+j}(\ell)]$$

$$= E(a_{t+\ell} + G_1 a_{t+\ell-1} + \ldots + G_{\ell-1} a_{t+1})$$

$$\cdot (a_{t+\ell+j} + G_1 a_{t+\ell+j-1} + \ldots + G_{\ell-1} a_{t+j+1})$$

$$= \sigma_a^2 (G_j + G_1 G_{j+1} + G_2 G_{j+2} + \ldots$$

$$+ G_{\ell-j-1} G_{\ell-1}), \quad \text{for} \quad j < \ell$$

$$= 0, \quad \text{for} \quad j \geq \ell$$

$$\text{Cor}[e_t(\ell), e_{t+j}(\ell)]$$

$$= \frac{(G_j + G_1 G_{j+1} + G_2 G_{j+2} + \ldots + G_{\ell-j-1} G_{\ell-1})}{(1 + G_1^2 + G_2^2 + \ldots + G_{\ell-1}^2)}, \quad \text{for} \quad j < \ell$$

$$= 0, \quad \text{for} \quad j \geq \ell \tag{5.3.5}$$

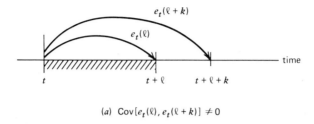

(a) $\text{Cov}[e_t(\ell), e_t(\ell + k)] \neq 0$

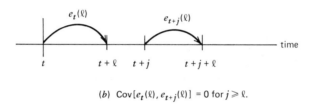

(b) $\text{Cov}[e_t(\ell), e_{t+j}(\ell)] = 0$ for $j \geq \ell$.

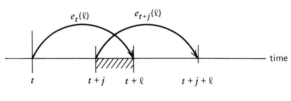

(c) $\text{Cov}[e_t(\ell), e_{t+j}(\ell)] \neq 0$ for $j < \ell$.

Figure 5.4   Graphical representation of the dependence of forecast errors. (*a*) $\text{Cov}[e_t(\ell), e_t(\ell+k)] \neq 0$. (*b*) $\text{Cov}[e_t(\ell), e_{t+j}(\ell)] = 0$ for $j \geq \ell$. (*c*) $\text{Cov}[e_t(\ell), e_{t+j}(\ell)] \neq 0$ for $j < \ell$.

Since a_t's are normally distributed, $e_t(\ell)$ will always have normal distribution.

The dependence between different forecast errors can be graphically depicted as shown in Fig. 5.4. The functions $e_t(\ell)$ and $e_t(\ell + k)$ denote the errors for the ℓ and $\ell + k$ steps ahead forecasts made at time t, that is, the forecasts made for time $t + \ell$ and $t + \ell + k$, respectively. Therefore, in Fig.

5.4(a), $e_t(\ell)$ is shown by an arrow starting at time t and ending at time $t + \ell$, whereas $e_t(\ell + k)$ is shown by the arrow starting at t and ending at time $t + \ell + k$. It is clear from Fig. 5.4(a) that the time interval t to $t + \ell$ is included in both the forecast errors, and hence $e_t(\ell)$ and $e_t(\ell + k)$ are not independent. In other words, their covariance is nonzero.

Figures 5.4b and 5.4c show the dependence between two forecast errors when the forecasts are made at different starting times, say t and $t + j$, but for the same time period ℓ, that is, the ℓ step ahead forecasts made at times t and $t + j$, respectively. It is seen from Figs. 5.4b and 5.4c that the covariance between $e_t(\ell)$ and $e_{t+j}(\ell)$ is zero for $j \geq \ell$ but nonzero for $j < \ell$. Hence, the forecast errors $e_t(\ell)$ and $e_{t+j}(\ell)$ are independent for $j \geq \ell$ and dependent for $j < \ell$.

5.3.7 Illustrative Examples—ARMA Models

Let us now consider some examples of ARMA models to show that the forecasts can be computed by conditional expectation, using Eq. (5.3.2). The AR(1) model was already considered above; hence, we will consider the ARMA(1,1) and ARMA(2,1) models.

(A) ARMA(1,1):

Forecasting

$$\ell = 1: \hat{X}_t(1) = \mathop{E}_t X_{t+1}$$

$$= \mathop{E}_t (\phi_1 X_t + a_{t+1} - \theta_1 a_t)$$

$$= \phi_1 X_t - \theta_1 a_t$$

where

$$a_t = X_t - \hat{X}_{t-1}(1)$$

can be computed recursively.

$$\ell = 2: \hat{X}_t(2) = \mathop{E}_t X_{t+2} = \mathop{E}_t(\phi_1 X_{t+1} + a_{t+2} - \theta_1 a_{t+1})$$

$$= \phi_1 \hat{X}_t(1)$$

and

$$\hat{X}_t(\ell) = \phi_1 \hat{X}_t(\ell - 1), \qquad \ell \geq 2$$

Probability Limits

For this model since [see Eq. (3.1.13)]

$$G_1 = \phi_1 - \theta_1, \; G_2 = \phi_1 G_1 = (\phi_1 - \theta_1)\phi_1,$$

$$G_j = \phi_1 G_{j-1}, \, j \geq 1$$

the 95% probability limits of the above forecast are, by Eq. (5.3.4),

$$\hat{X}_t(1) \pm 1.96\, \sigma_a$$

$$\hat{X}_t(2) \pm 1.96\, \sigma_a[1 + (\phi_1 - \theta_1)^2]^{1/2}$$

$$\ldots$$

$$\hat{X}_t(\ell) \pm 1.96\, \sigma_a\{1 + (\phi_1 - \theta_1)^2$$

$$(1 + \phi_1^2 + \phi_1^4 + \ldots + \phi_1^{2(\ell-2)})\}^{1/2}, \qquad \ell \geqslant 2$$

Numerical Example—IBM Stock Prices

We have seen in Section 4.5.2 that the ARMA(1,1) model with

$$\phi_1 = 1$$

$$\theta_1 = -0.0848$$

$$\sigma_a^2 = 52.4$$

provides an adequate representation for the IBM data. We will use this model with rounded-off numbers to numerically illustrate the computation of the forecasts and the forecast errors.

$$\phi_1 = 1,\ \theta_1 = -0.09,\ \sigma_a^2 = 52$$

$$X_t - X_{t-1} = a_t + 0.09a_{t-1}$$

Let us consider the forecasts at the origin $t = 22$. Since $\phi_1 = 1$, it is seen from the form of the model that the mean μ gets cancelled and need not be taken into account. The actual data values around this origin $t = 22$ are shown in the fourth column of Table 5.1. Assume

$$a_{t-3} = 0$$

Then,

$$a_{t-2} = X_{t-2} - X_{t-3} - 0.09a_{t-3}$$

$$= 491 - 487$$

$$= 4$$

$$a_{t-1} = X_{t-1} - X_{t-2} - 0.09a_{t-2}$$

$$= 487 - 491 - 0.09(4)$$

$$= -4.36$$

$$a_t = X_t - X_{t-1} - 0.09a_{t-1}$$

$$= 482 - 487 - 0.09 \times (-4.36)$$

$$= -5.39$$

$$X_{t+1} = X_t + a_{t+1} + 0.09a_t$$

$$\hat{X}_t(1) = 482 + 0.09 \times (-5.39)$$
$$= 482 - 0.48$$
$$= 481.52$$
$$\hat{X}_t(\ell) = \hat{X}_t(1)$$
$$= 481.52, \quad \ell = 2, 3, \ldots$$
$$\text{Var}\,[e_t(1)] = \sigma_a^2 = 52$$
$$G_1 = \phi_1 - \theta_1 = 1 - (-0.09) = 1.09$$
$$\text{Var}\,[e_t(2)] = [1 + G_1^2]\,\sigma_a^2$$
$$= (1 + 1.09^2) \times 52 = 113.78$$
$$\hat{X}_t(1) \pm 1.96\sqrt{\text{Var}\,[e_t(1)]} = 481.52 \pm 1.96\sqrt{52}$$
$$= 481.52 \pm 14.13$$
$$= 495.65 \quad \text{or} \quad 467.39$$
$$\hat{X}_t(2) \pm 1.96\sqrt{\text{Var}\,[e_t(2)]} = 481.52 \pm 1.96\sqrt{113.78}$$
$$= 481.52 \pm 20.91$$
$$= 502.43 \quad \text{or} \quad 460.61$$

The results are shown in Table 5.1. Fig. 5.5 shows a section of the IBM data, together with several steps ahead forecasts (indicated by circles) from the origin $t = 22$. Also shown are the 95% probability limits for $X_{22+\ell}$. Since the autoregressive root $\lambda_1 = \phi_1 = 1$, the forecast remains constant for large lags, $\ell \geq 2$. Due to the small value of θ_1, even $\hat{X}_t(1)$ is

Table 5.1

ℓ	G_ℓ	$t=22$	X_t	$\hat{X}_t(\ell)$	± 1.96 $\sqrt{\text{Var}\,e_t(\ell)}$	$\hat{X}_{t+1}(\ell-1)$
		$t-3$	487			
		$t-2$	491			
		$t-1$	487			
0	1	$t=22$	482			
1	1.09	$t+1$	479	481.52	± 14.13	
2	1.09	$t+2$	478	481.52	± 20.91	478.77
3	1.09	$t+3$	479	481.52	± 25.97	478.77
4	1.09	$t+4$	477	481.52	± 30.19	478.77
5	1.09	$t+5$	479	481.52	± 33.90	478.77
6	1.09	$t+6$	475	481.52	± 37.23	478.77
7	1.09	$t+7$	479	481.52	± 40.29	478.77
8	1.09	$t+8$	476	481.52	± 43.14	478.77

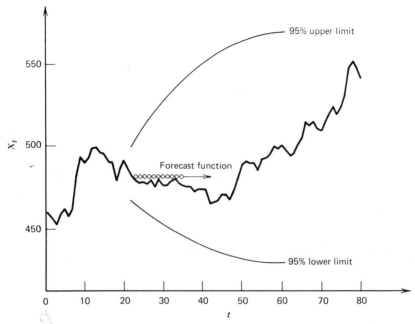

95% upper limit

Forecast function

95% lower limit

Figure 5.5   IBM stock prices—ARMA(1,1) model: forecasts and probability
limits at origin, $t = 22$.

almost the same as X_t; for a random walk with $\theta_1 = 0$, $\hat{X}_t(\ell) = X_t$, $\ell =$
$1, 2, \ldots$.

(B) ARMA(2,1)

   Forecasting

$$\ell = 1:\ \hat{X}_t(1) = \underset{t}{E}(\phi_1 X_t + \phi_2 X_{t-1} + a_{t+1} - \theta_1 a_t)$$

$$= \phi_1 X_t + \phi_2 X_{t-1} - \theta_1 a_t$$

where

$$a_t = X_t - \hat{X}_{t-1}(1)$$

can be computed recursively. Note that in the special case of an AR(2)
model with $\theta_1 = 0$, such recursive computation is not necessary and the
forecasts can be obtained by just knowing X_t, X_{t-1}.

$$\ell = 2:\ \hat{X}_t(2) = \underset{t}{E}(\phi_1 X_{t+1} + \phi_2 X_t + a_{t+2} - \theta_1 a_{t+1})$$

$$= \phi_1 \hat{X}_t(1) + \phi_2 X_t$$

and

$$\hat{X}_t(\ell) = \phi_1 \hat{X}_t(\ell - 1) + \phi_2 \hat{X}_t(\ell - 2), \qquad \ell \geq 3$$

Probability Limits

The 95% probability limits on the forecast are

$$\hat{X}_t(1) \pm 1.96 \, \sigma_a$$

$$\hat{X}_t(2) \pm 1.96 \, \sigma_a[1 + (\phi_1 - \theta_1)^2]^{1/2}$$

$$\cdots$$

$$\hat{X}_t(\ell) \pm 1.96 \, \sigma_a[1 + G_1^2 + G_2^2 + \ldots + G_{\ell-1}^2]^{1/2} \qquad \ell = 1, 2, 3, \ldots, S$$

where

$$G_0 = 1, \quad G_1 = \phi_1 - \theta_1$$

$$G_j = \phi_1 G_{j-1} + \phi_2 G_{j-2}, \qquad j \geq 2$$

by Eq. (3.1.11).

Numerical Example

For example, for the AR(2) model with $\phi_1 = 0.764$, $\phi_2 = -0.210$, $\sigma_a^2 = 6.254$, and $\hat{\mu} = 9.51$

$$G_0 = 1$$

$$G_1 = \phi_1 = 0.764$$

$$G_2 = \phi_1 G_1 + \phi_2$$

$$\quad = 0.764 \times 0.764 - 0.210 = 0.374$$

Also let $\dot{X}_3 = 4.00$, $\dot{X}_4 = 4.50$. After subtracting the mean μ, we obtain $X_3 = -5.51$ and $X_4 = -5.01$. Hence, the one step, two step, and three step ahead forecasts made at time $t = 4$ are given by

$$\hat{X}_4(1) = \phi_1 X_4 + \phi_2 X_3$$

$$\quad = 0.764 \times (-5.01) + (-0.210) \times (-5.51)$$

$$\quad = -2.671$$

$$\hat{X}_4(2) = \phi_1 \hat{X}_4(1) + \phi_2 X_4$$

$$\quad = 0.764 \times (-2.671) + (-0.210) \times (-5.01)$$

$$\quad = -0.989$$

$$\hat{X}_4(3) = \phi_1 \hat{X}_4(2) + \phi_2 \hat{X}_4(1)$$

$$\quad = 0.764 \times (-0.989) + (-0.210) \times (-2.671)$$

$$\quad = -0.195$$

The 95% probability limits on these forecasts are:

$$\hat{X}_4(1) \pm 1.96 \, \sigma_a = -2.671 \pm 1.96\sqrt{6.254}$$

$$= -2.671 \pm 4.902$$

$$\hat{X}_4(2) \pm 1.96 \, \sigma_a \sqrt{1 + G_1^2} = -0.989$$
$$\pm 1.96\sqrt{6.254(1 + 0.764^2)}$$
$$= -0.989 \pm 6.168$$
$$\hat{X}_4(3) \pm 1.96 \, \sigma_a \sqrt{1 + G_1^2 + G_2^2} = -0.195$$
$$\pm 1.96\sqrt{6.254(1 + 0.764^2 + 0.374^2)}$$
$$= -0.195 \pm 6.435$$

Note that $\hat{\mu}$ should be added back to the values of the forecasts in order to obtain the actual forecasts.

5.3.8 Eventual Forecasts and Stability

These examples show that the forecasts of any lead time for any ARMA model can be most simply computed by conditional expectation, and the corresponding error variances and the probability limits obtained by the Green's function G_j. It is not difficult to see from these examples that the eventual forecasts for large enough ℓ satisfy the autoregressive part of the difference equations in the same manner as γ_k or G_j. In fact, for an ARMA(n,m) model

$$\hat{X}_t(\ell) = \phi_1 \hat{X}_t(\ell - 1) + \phi_2 \hat{X}_t(\ell - 2) + \ldots$$
$$+ \phi_n \hat{X}_t(\ell - n), \quad \ell \geq \max(n + 1, m + 1) \quad (5.3.6)$$

The moving average parameters affect the forecast equation only for small lead times $\ell \leq m + 1$; for larger lead times, it is the autoregressive parameters ϕ_j that control the forecasts.

Recall from the discussion of stability in Section 3.1 that the Green's function G_j, which also satisfies Equation (5.3.6), tends to zero eventually for an asymptotically stable system, that is, for an ARMA models with all roots λ_i less than one in absolute value. Thus, for such an asymptotically stable system the eventual forecast $\hat{X}_t(\ell)$ as ℓ tends to ∞ will be zero. Actually, it will be the mean of X_t, μ, which we have taken to be zero for convenience. In fact, all the actual forecasts are $\hat{X}_t(\ell) + \mu$.

For example, for the ARMA(1,1) model with $\phi_1 = 0.5$, we have
$$\hat{X}_t(1) = 0.5X_t - \theta_1 a_t$$

and

$$\hat{X}_t(\ell) = 0.5\hat{X}_t(\ell - 1), \qquad \ell \geq 2$$
$$= 0.5^{\ell-1}\hat{X}_t(1)$$
$$= 0.5^{\ell-1}(0.5X_t - \theta_1 a_t)$$

As ℓ approaches infinity, $0.5^{\ell-1}$ approaches zero, and hence $\hat{X}_t(\ell)$ approaches zero.

For a system which is stable but not asymptotically stable, that is, when at least one of the roots λ_i's is one in absolute value, the eventual forecast will be constant. For the ARMA(1,1) model considered above, if $\phi_1 = 1$, then

$$\hat{X}_t(1) = X_t - \theta_1 a_t$$

and

$$\hat{X}_t(\ell) = \hat{X}_t(1), \qquad \ell \geq 2$$

which is constant.

Finally, for an unstable system the eventual forecast will increase in absolute value as ℓ increases and will tend to infinity as $\ell \to \infty$. As an example, consider the ARMA(1,1) model; if $\phi_1 = 1.5 > 1$, then

$$\hat{X}_t(1) = 1.5X_t - \theta_1 a_t$$

and

$$\begin{aligned}
X_t(\ell) &= 1.5\hat{X}_t(\ell - 1) \\
&= 1.5^{\ell-1}\hat{X}_t(1) \\
&= 1.5^{\ell-1}(1.5X_t - \theta_1 a_t)
\end{aligned}$$

which approaches infinity as $\ell \to \infty$.

5.4 UPDATING THE FORECASTS

Once the model for a set of data is known, the forecasts for an arbitrary lead time can be computed by the method of conditional expectation outlined in the last section. Suppose we have the forecasts $\hat{X}_t(\ell)$ at time t for, say, $\ell = 1, 2, 3$, which forecast the observations X_{t+1}, X_{t+2}, and X_{t+3}, respectively. At time $t + 1$ the observation X_{t+1} and therefore

$$a_{t+1} = X_{t+1} - \hat{X}_t(1)$$

become known. Based on this new piece of information, the forecasts of X_{t+2} and X_{t+3} can now be changed and, in fact, improved to $\hat{X}_{t+1}(1)$ and $\hat{X}_{t+1}(2)$. Although these new improved or updated forecasts $\hat{X}_{t+1}(1)$, $\hat{X}_{t+1}(2)$ can again be obtained by conditional expectation as before, there are simpler ways in which they can be computed from the old forecasts of the same observations $\hat{X}_t(2)$, $\hat{X}_t(3)$. This is possible because of the recursive method of computing forecasts for the ARMA models.

To see how such an updating can be done, note that for an arbitrary ARMA model, by Eq. (5.3.1),

$$\begin{aligned}
\hat{X}_{t+1}(\ell) &= G_\ell a_{t+1} + \quad G_{\ell+1} a_t + G_{\ell+2} a_{t-1} + \cdots \\
\hat{X}_t(\ell + 1) &= \qquad\qquad\quad G_{\ell+1} a_t + G_{\ell+2} a_{t-1} + \cdots \\
\therefore \hat{X}_{t+1}(\ell) &= \hat{X}_t(\ell + 1) + G_\ell a_{t+1}
\end{aligned} \qquad (5.4.1)$$

where

$$a_{t+1} = X_{t+1} - \hat{X}_t(1)$$

The updating equation (5.4.1) says that at time $t + 1$, the new updated forecasts $\hat{X}_{t+1}(\ell)$ of the observation $X_{t+\ell+1}$ is obtained from its old forecast $\hat{X}_t(\ell + 1)$ by simply adding G_ℓ times the new shock a_{t+1}, which becomes known as soon as X_{t+1} is known.

For example, consider the IBM stock prices with an ARMA(1,1) model for, say, $\ell = 1$, one step ahead forecast. Since $G_1 = \phi_1 - \theta_1$, the updating equation (5.4.1) becomes

$$\hat{X}_{t+1}(1) = \hat{X}_t(2) + (\phi_1 - \theta_1)a_{t+1}$$
$$= \hat{X}_t(2) + (1 - \theta_1)[X_{t+1} - \hat{X}_t(1)] \qquad \text{for} \qquad \phi_1 = 1$$

or

$$\hat{X}_t(1) = \hat{X}_{t-1}(2) + (1 - \theta_1)[X_t - \hat{X}_{t-1}(1)]$$

Since $\hat{X}_t(\ell) = \hat{X}_t(1)$ for an ARMA(1,1) model with $\phi_1 = 1$, then

$$\hat{X}_t(1) = \hat{X}_{t-1}(1) + (1 - \theta_1)[X_t - \hat{X}_{t-1}(1)] \qquad (5.4.2)$$

Equation (5.4.2) will be useful in explaining exponential smoothing in the next section.

From the data used in the last section,

$$a_{t+1} = X_{t+1} - \hat{X}_t(1)$$
$$= 479 - 481.52 = -2.52$$

and the updated forecasts of X_{t+2} is

$$\hat{X}_{t+1}(1) = \hat{X}_t(2) + (1 - \theta_1)a_{t+1}$$
$$= 481.52 + 1.09(-2.52)$$
$$= 478.77$$

The updated forecasts $\hat{X}_{t+1}(\ell - 1)$ are also shown in the last column of Table 5.1.

5.5 EXPONENTIAL SMOOTHING

Side by side with the prediction theory based on mathematical analysis described in the beginning of this chapter, empirical methods of forecasting for the business and industrial data were being developed. The most important and widely used among these is the technique of exponential smoothing, which seems to have been first suggested in unpublished reports by C. Holt. The details of this technique may be found in Brown (1962). In this section, we will briefly explain the technique and show that forecasting by exponential smoothing is a very restricted special case of forecasting by ARMA models. The time series model for which exponential smoothing yields optimal prediction was first pointed out by Muth (1960).

5.5.1 The Concept of Exponentially Weighted Moving Average

The idea behind exponential smoothing is quite simple but intuitively appealing. For the time being, let us suppose that we have no knowledge of the ARMA models and their forecasts. In the absence of such knowledge, what can we do if we have a record of past data on, say, sales of a product or its price fluctuations, and our purpose is only to forecast its next sales volume or price? Let us assume that there are no obvious trends in the data, and if there are, they have been taken out.

The simplest and safest method appears to be to take the average of the past data as the forecast. This would smooth out any random fluctuations in the data, and we would then hope to get a good smoothed estimate of the forecast.

On closer thought, taking the average of the entire past data seems unreasonable. The data in the distant past may be outmoded and should be discarded in obtaining the forecasts. For example, the sales data ten years ago may not be relevant in the present day forecasts. This leads to the idea of discounting the past data by taking "moving averages." That is, we choose a number N depending upon how far we want to go into the past and take the average of the most recent N observations as the forecast of the next observation.

$$\hat{X}_t(1) = \frac{1}{N} \sum_{j=0}^{N-1} X_{t-j} = \sum_{j=0}^{N-1} \frac{1}{N} X_{t-j} \qquad (5.5.1)$$

As t increases and more and more new data come in, this moving average automatically drops or discounts the distant past data before N time units.

The moving average Eq. (5.5.1) is a weighted average of the past data with equal weights $1/N$ given to all observation from X_t to X_{t-N+1} and zero to observations before X_{t-N+1}, that is, to $X_{t-j}, j \geq N$. This procedure of giving equal weight $1/N$ to N recent observations and then suddenly cutting it off to zero seems unreasonable. In fact, if the recent data is more relevant to the forecast than the past one, why should this principle not be applied within the N observations also, rather than giving them all equal weight $1/N$?

This suggests that we should use continually decreasing weights to average the data and obtain the forecasts. The simplest way this can be done is to choose a number θ such that

$$|\theta| < 1$$

depending upon how fast we want the weights to decrease and then use exponential weights

$$\theta^j$$

However, if we multiplied X_{t-j} by θ^j and added them to get the forecast, it is no longer an "average" since the weights do not add to one. In fact, they add to

$$\sum_{j=0}^{\infty} \theta^j = \frac{1}{1 - \theta}$$

So if we want a weighted *average*, we should take the weights

$$(1 - \theta)\theta^j$$

that now add to one and therefore we can write the forecast as

$$\hat{X}_t(1) = \sum_{j=0}^{\infty} (1 - \theta)\theta^j X_{t-j} \qquad (5.5.2)$$

This is known as an Exponentially Weighted Moving Average (EWMA) and is often written in the form

$$\hat{X}_t(1) = \sum_{j=0}^{\infty} \lambda(1 - \lambda)^j X_{t-j} \qquad (5.5.2a)$$

where

$$\lambda = 1 - \theta \qquad (5.5.3)$$

The operation in equation (5.5.2a) yields a smoothed average; hence, it is called exponential smoothing. The constant λ is called the exponential smoothing constant or parameter since it determines how the past observations are smoothed or averaged to obtain the forecast.

5.5.2 Interpretation and Advantages

When the value of λ is small, the weights $\lambda(1 - \lambda)^j$ are small and decrease very slowly. The forecast then depends almost entirely on the weighted average of a large amount of past data. For very small values of λ the EWMA is therefore almost equivalent to a simple moving average. In the extreme case of $\lambda \to 0$, the EWMA is the average of the entire past data. Thus, these two cases discussed earlier are included in the EWMA. The weights for $\lambda = 0.1$ are shown in Fig. 5.6.

On the other hand, if λ is large, the present observation is given more weight and the past observations have less influence on the forecast. For example, if $\lambda = 0.9$, the weights are $0.9, 0.09, 0.009, \ldots$, which are shown in Fig. 5.6. In the extreme case, when $\lambda = 1$, the past data has no influence at all, and

$$\hat{X}_t(1) = X_t$$

For intermediate values the influence of the past data is somewhere between these extreme cases. The weights for $\lambda = 0.5$ are also shown in Fig. 5.6.

In addition to its intuitive appeal, another reason why exponential

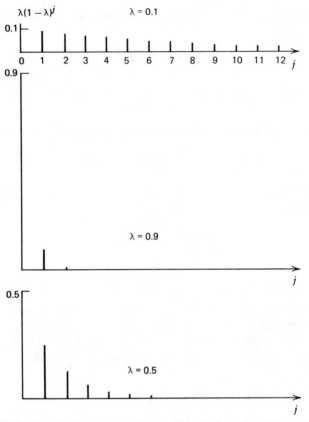

Figure 5.6   Exponential smoothing weights for $\lambda = 0.1, 0.9,$ and 0.5.

smoothing is so popular is that one need not store the entire past data to compute the forecasts. One needs to store only the latest forecast, from which the new forecast can be obtained by a simple recursive formula that involves very little computations. To derive this formula from Eq. (5.5.2a), since

$$\hat{X}_{t-1}(1) = \sum_{j=0}^{\infty} \lambda(1 - \lambda)^j X_{t-j-1}$$

$$\hat{X}_t(1) = \sum_{j=0}^{\infty} \lambda(1 - \lambda)^j X_{t-j}$$

$$= \lambda X_t + \sum_{j=1}^{\infty} \lambda(1 - \lambda)^j X_{t-j}$$

$$= \lambda X_t + (1 - \lambda) \sum_{j=0}^{\infty} \lambda(1 - \lambda)^j X_{t-j-1}$$

$$= \lambda X_t + (1 - \lambda) \hat{X}_{t-1}(1) \qquad (5.5.4a)$$

$$= \hat{X}_{t-1}(1) + \lambda[X_t - \hat{X}_{t-1}(1)] \qquad (5.5.4)$$

which shows that $\hat{X}_t(1)$ can be computed simply from $\hat{X}_{t-1}(1)$ and the new observation X_t without knowing the past observations at time t, which therefore need not be stored.

Formula (5.5.4) has an interesting interpretation. Knowing that we have made a forecast error $[X_t - \hat{X}_{t-1}(1)]$ at time t, we adjust the old forecast $\hat{X}_{t-1}(1)$ by an amount λ times the forecast error, to get the new forecast. Eq. (5.5.4a), on the other hand, says that the new forecast is an interpolation between the old forecast and the new observation.

If the exponential smoothing technique is applied to the IBM example of Section 5.3, then we have

$$\lambda = 1 - \theta = 1 + 0.09 = 1.09$$
$$1 - \lambda = \theta = -0.09$$

Hence, by Eq. (5.5.2a) we have

$$\hat{X}_{t-1}(1) = \sum_{j=0}^{\infty} \lambda(1 - \lambda)^j X_{t-j-1}$$

$$= \sum_{j=0}^{\infty} 1.09(-0.09)^j X_{t-j-1}$$

$$= 1.09(1)487 + 1.09(-0.09)491$$
$$\quad + 1.09(-0.09)^2 487 + 1.09(-0.09)^3 478$$
$$\quad + 1.09(-0.09)^4 489 + \ldots$$

$$= 530.83 - 48.167 + 4.3 - 0.38 + 0.035 - \ldots$$

$$\approx 486.618$$

and by Eq. (5.4.4)

$$\hat{X}_t(1) = 486.618 + 1.09[482 - 486.618]$$

$$= 486.618 - 5.034$$

$$= 481.58$$

The recursive exponential smoothing formula (5.5.4a) or (5.5.4) shows that once a smoothing constant λ is chosen, the new forecasts can be generated by updating the old forecasts, and one does not need to use the entire past data every time. Thus, only the forecasts need to be stored in the computation; this is one of the most important advantages of the technique of exponential smoothing.

Note that this advantage of storing only the forecasts rather than the entire past data is shared by all the ARMA model as seen from the updating equation (5.4.1). This equation shows that once the ARMA model underlying the data is determined so that G_ℓ is known, one needs to store only the forecasts $\hat{X}_t(\ell + 1)$ and $\hat{X}_t(1)$ to compute the new forecasts as the new observations keep coming in.

5.5.3 Relation with ARMA Models

Comparing the special case of the updating equation (5.4.1), given by Eq. (5.4.2), with Eq. (5.5.4), we can clearly see that forecasting by exponential smoothing is the same as that by the ARMA(1,1) model, with $\phi_1 = 1$ and the relation between the moving average parameter θ and the exponential smoothing constant λ specfied by Eq. (5.5.3) as $\lambda = 1 - \theta$. Another way of deriving this result is to answer the question: For what model would the forecast given by Eq. (5.5.2a) be optimal in the sense of a mean squared error? If Eq. (5.5.2) is to be a one step ahead forecast for an ARMA model, then the observation X_{t+1} is this forecast plus a one step ahead forecast error, which is a_{t+1}. Hence,

$$X_{t+1} = \sum_{j=0}^{\infty} (1 - \theta)\theta^j X_{t-j} + a_{t+1}$$

that is

$$X_t = \sum_{j=1}^{\infty} (1 - \theta)\theta^{j-1} X_{t-j} + a_t$$

which is an ARMA model with

$$I_j = (1 - \theta)\theta^{j-1}$$

Now Eq. (3.2.15) shows that this is an ARMA(1,1) model with $\lambda = 1 - \theta$. Similarly, if we take the ARMA(1,1) forecast equation

$$
\begin{aligned}
\hat{X}_t(1) &= \phi_1 X_t - \theta a_t \\
&= \phi_1 X_t - \theta[X_t - \hat{X}_{t-1}(1)] \\
&= (\phi_1 - \theta)X_t + \theta \hat{X}_{t-1}(1) \\
&= (\phi_1 - \theta)X_t + \theta[(\phi_1 - \theta)X_{t-1} + \theta \hat{X}_{t-2}(1)] \\
&= (\phi_1 - \theta)X_t + (\phi_1 - \theta)\theta X_{t-1} \\
&\quad\vdots \quad + \theta^2[(\phi_1 - \theta)X_{t-2} + \theta \hat{X}_{t-3}(1)] \\
&= \sum_{j=0}^{\infty} (\phi_1 - \theta)\theta^j X_{t-j}
\end{aligned}
$$

which again reduces to exponential smoothing when $\phi_1 = 1$.

It is thus seen that forecasting by exponential smoothing is a very restricted special case of that by ARMA models. It corresponds to the limiting ARMA(1,1) model

$$(1 - B)X_t = (1 - \theta B)a_t \tag{5.5.5}$$

that is,

$$\nabla X_t = (1 - \theta B)a_t \tag{5.5.5a}$$

As explained in Section 3.1.9, this model is also called an Integrated Moving Average (IMA) model; we, however, treat it as a limiting case of the ARMA(1,1) model as ϕ_1 tends to one.

* 5.5.4 Genesis of ARMA Models from Exponential Smoothing

We have introduced the ARMA models, starting from the concept of conditional regression, emphasizing the system aspect, and converging on the Green's function G_j as the central characterization. The discussion in this section shows that one can also introduce the ARMA models starting from the concept of an exponentially weighted moving average, emphasizing the forecasting aspect, and converging on the inverse function I_j as the central characterization. Just as the function G_j determines how the system response or output X_t is obtained as a weighted sum of inputs a_{t-j}, the function I_j determines how the one step ahead forecast $\hat{X}_t(1)$ is obtained as a weighted sum of the observations X_{t-j}. The stability condition on G_j assures that the weights given to a_{t-j} do not increase with j; the invertibility condition on I_j affirms that the weights given to X_{t-j} do not increase as we go further into the past. Note that for an arbitrary ARMA model, if we take a conditional expectation on the I_j weight form of the model (3.2.1), that is,

$$X_{t+1} = \sum_{j=1}^{\infty} I_j X_{t+1-j} + a_{t+1}$$

we have

$$\hat{X}_t(1) = \sum_{j=1}^{\infty} I_j X_{t+1-j} \tag{5.5.6}$$

We can start with Eq. (5.5.6) as expressing the forecast $\hat{X}_t(1)$ as a weighted sum of the past observations and develop ARMA models. If the weights are assumed to be exponentially decaying and are restricted (1) to be given by a single parameter, say $\theta = v_1$, and (2) to add to one so that the forecast is not only a weighted sum but also a weighted average, then we get

$$I_j = (1 - \theta)\theta^{j-1} = \frac{(1 - v_1)}{v_1} v_1^j$$

as shown earlier. This I_j, which is the same as EWMA weights, gives the ARMA(1,1) model (5.5.5).

If we still assume that the weights are exponentially decaying but give up the restrictions (1) and (2) of the EWMA, then we can make the I_j function a little more flexible by introducing a new parameter, say ϕ_1,

$$I_j = (\phi_1 - \theta)\theta^{j-1}, j = 1, 2, \ldots$$

This I_j function yields the unrestricted ARMA(1,1) model

$$X_t - \phi_1 X_{t-1} = a_t - \theta a_{t-1}$$

If we now require that the weights follow the exponential function only for $j \geq 2$ and let I_1 be "free," by introducing an additional parameter ϕ_2, the I_j function can be written as

$$I_1 = \phi_1 - \theta$$

$$I_j = (\phi_2 + \phi_1 \theta - \theta^2)\theta^{j-1}, \qquad j \geq 2$$

which gives an ARMA(2,1) model

$$X_t - \phi_1 X_{t-1} - \phi_2 X_{t-2} = a_t - \theta a_{t-1}$$

Note that, as in the case of the general ARMA(1,1) model, the forecasts are no longer "averages" since the weights do not add to one. The forecasts are simply weighted sums, which is also true for the general ARMA models.

* 5.5.5 Exponential Smoothing, ARMA Models and Wiener-Kolmogorov Prediction Theory

It is interesting to note that the forecasting by ARMA models connects the profound prediction theory initiated by Kolmogorov and Wiener in a very abstract framework with empirical methods of forecasting, starting with the simple concept of averaging or smoothing. The discussion on the systems approach of modeling given in Section 4.1 explains why these two methods of forecasting are basically the same. The most natural orthogonal decomposition of a stationary stochastic process or time series, implicitly or explicitly used for prediction by orthogonal projection in the Kolmogorov-Wiener theory, is provided by an ARMA model, which can approximate the "true" stochastic process to any degree of accuracy. The modeling procedure of Chapter 4 ensures such an approximation by an ARMA(n,n-1) model. The results of this chapter show that forecasting by exponential smoothing is equivalent to using a special case of the ARMA(2,1) model with $\phi_2 = 0$, $\phi_1 = 1$ for the purpose of forecasting.

A question now arises: If we are given a set of data with the sole purpose of forecasting, should we use exponential smoothing or an ARMA model? A complete answer to this question can be given after we have studied the continuous systems in later chapters. However, just considering the discrete systems studied so far, it is easy to see that the best procedure is to first fit an ARMA model by the method outlined in Chapter 4. If the adequate model turns out to be ARMA(2,1) with ϕ_1 close to one and ϕ_2 close to zero, there is a possibility that the ARMA(1,1) model (5.5.5) may be adequate, which may be checked by actually fitting it. If it is found to be adequate, we can use EWMA for forecasting with the optimal value of the smoothing constant $\lambda = 1 - \theta$. This value is optimal in the sense that it gives the minimum mean squared error among all the admissible values of λ. If the ARMA(1,1) model (5.5.5) is not adequate, then one should use the fitted ARMA model for forecasting.

Since exponential smoothing is a very restricted special case of ARMA models, it might appear that if the above procedure is used, one would be rarely using exponential smoothing for forecasting. This is not the case. In fact, empirical experience with business and economic series indicates that exponential smoothing often gives good forecasts. The theoretical explanation of this empirical fact will be revealed when we study the continuous systems in later chapters, in particular, Chapter 8. As an example, the IBM data was adequately fitted by an ARMA(1,1) model with $\phi_1 = 1$ in Chapter 4, and therefore it can be forecast well by exponential smoothing with $\lambda = 1 - (-0.09) = 1.09$. (Also see Problem 5.11)

PROBLEMS

5.1 For the ARMA(2,1) model
$$X_t - X_{t-1} + 0.3X_{t-2} = a_t + 0.4a_{t-1}, \quad a_t \sim \text{NID}(0,256)$$
given $X_{t-5} = 34$, $X_{t-4} = 36$, $X_{t-3} = 1$, $X_{t-2} = 1$, $X_{t-1} = -16$, $X_t = -35$, and assuming $a_{t-4} = 0$,
(i) compute $\hat{X}_t(\ell)$, $\ell = 1, 2, 3, 4$, and their 90% probability limits;
(ii) update $\hat{X}_t(\ell)$, $\ell = 2, 3, 4$, given $X_{t+1} = -37$.

5.2 For the ARMA(1,1) model
$$X_t - \phi_1 X_{t-1} = a_t - \theta_1 a_{t-1}$$
derive $\hat{X}_t(3)$ by taking conditional expectation on the
(i) difference equation form of the model;
(ii) G_j weight form of the model;
and show that these two expressions for $\hat{X}_t(3)$ are equal.

5.3 It was found that a sequence of 100 observations, X_t, followed an AR(2) model:
$$X_t = 0.5X_{t-1} - 0.5X_{t-2} + a_t$$

Suppose $X_{100} = 0.8$, $X_{99} = 1.8$ and a 95% confidence interval of one step ahead forecast made at origin $t = 100$ is $(-1.5, 0.5)$.
(i)   Find $\hat{X}_t(2)$, $\hat{X}_t(3)$ and their 95% confidence limits.
(ii)   If $X_{101} = 0$, update the above forecasts.

5.4   (i)   From the ARMA(1,1) model

$$\nabla X_t = (1 - \theta_1 B)a_t$$

derive the EWMA prediction formula

$$\hat{X}_t(1) = \hat{X}_{t-1}(1) + \lambda[X_t - \hat{X}_{t-1}(1)]$$

where $\lambda = 1 - \theta_1$.

(ii)   For this model, given $X_{t-4} = 460$, $X_{t-3} = 457$, $X_{t-2} = 452$, $X_{t-1} = 459$, $X_t = 462$, and $\lambda = 0.9$, obtain $\hat{X}_t(\ell)$, $\ell = 1, 2, 3$.

5.5   A data analyst has been using exponential smoothing with $\lambda = 0.3$ for forecasting. From his past experience, however, he feels that he can get better forecasts for two different sets of data by using the weighting functions
(i) $I_j = 0.2(0.7)^{j-1}$, $j = 1, 2, \ldots$
(ii) $I_1 = 0.7$, $I_j = 0.21(0.7)^{j-2}$, $j = 2, 3, \ldots$

$$\hat{X}_t(1) = \sum_{j=1}^{\infty} I_j X_{t-j+1}$$

(a)   If these weighting functions give optimal forecasts in the sense of a minimum mean squared error, what are the corresponding ARMA models in each case?
(b)   Assuming the models are correct, give a recursive formula for one step ahead forecasts that does not require the past data and therefore can be used as simply as exponential smoothing.
(c)   If the past data were available, would you still advise him to use the models in (b) or suggest some other procedure? Explain.

5.6   State whether it is possible to have ARMA models, with characteristic roots $|\lambda_i| \leqslant 1$ and X_t having zero mean, for which the eventual forecast $\hat{X}_t(\ell)$ as $\ell \to \infty$
(i)   tends to zero;
(ii)   tends to a constant;
(iii)   tends to infinity.
In each case, if the answer is "yes," give an example; if the answer is "no," explain why not.

5.7   For $\ell > 1$, the recursive formulae for the forecasts use the preceding forecasts. The forecast $\hat{X}_t(\ell)$ may also be expressed purely as a linear combination for the present and past values of the series using the G_j and I_j functions. Show that

$$\hat{X}_t(\ell) = \sum_{j=1}^{\infty} I_j^{(\ell)} X_{t-j+1}, \qquad \ell \geqslant 1$$

where

$$I_j^{(\ell)} = I_{j+\ell-1} + \sum_{k=1}^{\ell-1} I_k I_j^{(\ell-k)}$$

5.8 Alternatively, show that the relation between G_j and I_j is given by

$$G_j = \sum_{h=1}^{j} G_{j-h} I_h, \qquad j \geqslant 1$$

or

$$I_j = G_j - \sum_{h=1}^{j-1} G_{j-h} I_h, \qquad j \geqslant 1$$

hence, in the notation of Problem 5.7

$$I_j^{(\ell)} = \sum_{h=1}^{\ell} G_{\ell-h} I_{j+h-1},$$

or

$$I_j^{(\ell)} = I_{j+1}^{(\ell-1)} + G_{\ell-1} I_j$$

5.9 Show that the variance of the forecast error $V[e_t(\ell)]$ for an ARMA(n,m) model increases with the increasing values of lead time ℓ, and reaches the variance of the observations X_t as $\ell \to \infty$. Verify your answer by considering an AR(1) model with $\phi_1 = 0.9$ and $\sigma_a^2 = 5.0$.

5.10 Give a graphical representation of Eq. (5.4.1) to show the updating of the forecasts for an arbitrary ARMA model.

5.11 Prove the following results from Muth (1960).

(a) The EWMA model can also be written as

$$X_t = a_t + \lambda \sum_{j=1}^{\infty} a_{t-j}$$

where λ is the smoothing constant. Therefore a part of the "shock" a_t in any period has a permanent effect while the rest affects the system only in the current period. Thus the exponentially weighted average can be interpreted as the expected value of a time series made up of two random components: one lasting a single time period (transitory) and the other lasting through all subsequent periods (permanent). Many business and economic systems might generate time series in this way.

(b) Such a time series underlying exponential smoothing may also be regarded as a random walk with "noise" superimposed. Show that if y_t is a random walk, for example,

$$(1 - B)y_t = \varepsilon_t$$

and

$$X_t = y_t + \eta_t$$

with

$$E\eta_t\eta_{t-k} = \delta_k\sigma_\eta^2, \qquad E\varepsilon_t\eta_{t-k} = \sigma_{\varepsilon\eta}\delta_k$$

then

$$(1 - B)X_t = (1 - \theta B)a_t$$

where

$$\theta = 1 + \frac{\sigma_\varepsilon^2}{2(\sigma_\eta^2 + \sigma_{\varepsilon\eta})} - \frac{\sigma_\varepsilon}{\sqrt{\sigma_\varepsilon^2 + \sigma_{\varepsilon\eta}}}\sqrt{1 + \frac{\sigma_\varepsilon^2}{4(\sigma_\varepsilon^2 + \sigma_{\varepsilon\eta})}}$$

and

$$\sigma_a^2 = \frac{\sigma_\eta^2 + \sigma_{\varepsilon\eta}}{\theta}$$

(*Hint*: Use autocovariance function of the MA(1) model.)

6

UNIFORM SAMPLING OF CONTINUOUS SYSTEMS—FIRST ORDER

In the preceding chapters, we have considered only discrete models of stochastic systems. This is primarily because such models can be readily obtained from a discrete set of data by extending the linear regression methods. Moreover, such models are adequate when one is interested only in predicting a system at discrete points.

However, many of the systems met in practice are continuous. For such systems continuous models in the form of differential equations are more meaningful than the discrete models in the form of difference equations. The differential equations provide a "live" description of the system in the sense that the coefficients in the equation can be related with the known characteristics of the system. The continuous models are best-suited for characterization, and they are therefore extremely useful in system analysis and system design, in addition to system prediction and control.

We will now show how such continuous models can be obtained from the discrete set of observed data taken at uniform intervals. The discrete systems will be treated as uniformly sampled continuous systems by developing the relations between the differential equation representing a continuous system and the difference equation representing the uniformly sampled system.

This approach enables us to obtain the continuous and discrete models simultaneously from the discrete data. The discrete models for the uniformly sampled system turn out to be special cases of the ARMA$(n, n-1)$ models. The modeling procedure developed therein can therefore be used also for continuous systems, with appropriate modifications in the estimation procedure to incorporate the parametric relationships between the continuous and the discrete models.

The discrete models so obtained can be utilized for prediction and forecasting, whereas the continuous models are useful for characterization and systems analysis including stability analysis, system design, etc. We will restrict ourselves to stochastic systems governed by linear differential equations with constant coefficients. To distinguish the continuous and discrete systems, the response of a continuous system at time t will be denoted by $X(t)$, whereas that of a discrete system will be denoted by X_t as before.

The simplest system governed by differential equations with constant coefficients is the first order system. From the point of view of continuous discrete relationships, the first order system also turns out to be the simplest one, because its uniformly sampled process has the discrete representation of the AR(1) model, as might be expected. It will be shown in the next chapter that this is no longer true for higher order systems, for example, a uniformly sampled second order system does *not*, in general, have an AR(2) representation.

In spite of its simplicity, the first order differential equation is capable of representing a large number of systems in practice. The basic assumption underlying the equation is that the rate of change of the variable is proportional to its amount at any time; this assumption is very nearly true in many physical systems such as chemical relaxation, rotational mechanics, etc. The assumption is true for any system in equilibrium subjected to small disturbance, and therefore the first order model is useful for many stochastic systems.

A preliminary review of the differential equation basic to the first order system, together with its stability analysis and the physical interpretation of the parameters, is given in Section 6.1. Just like the discrete white noise a_t in discrete systems, the continuous white noise $Z(t)$ plays a crucial part in continuous systems. Whereas the discrete white noise a_t is defined by a covariance function with the Kronecker delta δ_k [see Eq. (3.4.1)], the continuous white noise $Z(t)$ is defined by a covariance function with the Dirac delta $\delta(u)$. Therefore, the properties of the Dirac delta and related functions are briefly discussed in Section 6.2. The introduction of the Dirac delta function not only makes the transition from the discrete to the continuous systems smooth, but also considerably simplifies the analysis of the continuous systems.

The homogeneous first order differential equation and $Z(t)$ as the forcing function are employed in Section 6.3 to formulate the first order continuous autoregressive system. The analogy with the approximation of a continuous curve by rectangles and the resultant limiting representation as an integral with delta function is exploited to introduce and explain the orthogonal decomposition of a continuous time stochastic system. This orthogonal decomposition leads to the Green's function and the autocovariance function.

Using the autocovariance function, we show in Section 6.4 that the uniformly sampled representation of the first order continuous system is AR(1). The relations between the parameters of the continuous and discrete models are also derived; these can be used to obtain the continuous model from the discrete data. The limiting cases of the discrete model for extreme values of the parameters and the sampling interval, including the important case of random walk, are derived in Section 6.5.

6.1 FIRST ORDER DIFFERENTIAL EQUATION

Consider the system response $X(t)$ for which the rate of change at time t is proportional to its value at the time and opposite in sign. If the constant of proportionality is denoted by α_0, then the equation of the system is

$$\frac{dX(t)}{dt} = -\alpha_0 X(t) \qquad (6.1.1)$$

or

$$\frac{dX(t)}{dt} + \alpha_0 X(t) = 0 \qquad (6.1.1a)$$

6.1.1 Solution

Recall that the solution of an nth order linear differential equation has the form of a sum of n exponentials. Thus, the solution of Eq. (6.1.1) is of the form (differential equations and their solutions needed here are briefly reviewed in Appendix A 6.1)

$$Ce^{\mu t}$$

To obtain the value of μ we substitute this function in (6.1.1a) to get

$$(\mu + \alpha_0)Ce^{\mu t} = 0$$

This gives the characteristic equation

$$\mu + \alpha_0 = 0$$

for which the characteristic root is

$$\mu = -\alpha_0$$

Hence, the solution of Eq. (6.1.1) takes the form

$$X(t) = Ce^{-\alpha_0 t} \qquad (6.1.2)$$

The constant C can be evaluated from the known "initial conditions" of the system. For example, if

$$X(0) = C_0$$

then it follows from Eq. (6.1.2) that

$$C = C_0$$

and the solution may be written as

$$X(t) = C_0 e^{-\alpha_0 t} \qquad (6.1.2a)$$

The form of the response curve (6.1.2a) is shown in Fig. 6.1 for non-negative values of α_0. Note that $-C_0\alpha_0$ represents the slope of the curve at the origin. It reflects how rapidly the response dies down to zero from the initial condition C_0. The larger the value of α_0, the more rapidly the response dies down to zero. For very large values of α_0, the response is

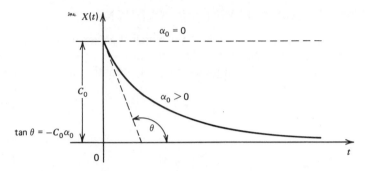

Figure 6.1   Response of a first order system, $\alpha_0 \geq 0$.

practically zero everywhere except at the origin. On the other hand, the smaller the value of α_0, the more slowly the response approaches zero. For very small values of α_0 close to zero, the response approaches a horizontal straight line near the origin, as shown by the dotted line in Fig. 6.1.

6.1.2 Time Constant τ

Equation (6.1.2a), which represents the solution of the first order differential equation, can be written in the form

$$X(t) = C_0 e^{-t/\tau} \tag{6.1.2b}$$

where $\tau = \dfrac{1}{\alpha_0}$ is called the time constant because it has the dimension of time, which can be seen as follows: The terms $\dfrac{dX}{dt}$ and $\alpha_0 X$ in Eq. (6.1.1) have the same dimension; hence, α_0 has the same dimension as $\dfrac{1}{X(t)} \cdot$ $\dfrac{dX(t)}{dt}$, that is $\dfrac{1}{time}$. Therefore $\tau = \dfrac{1}{\alpha_0}$ has the dimension of time. Equation (6.1.2b) can also be expressed as

$$\frac{X(t)}{C_0} = e^{-t/\tau} \tag{6.1.2c}$$

The plot of t/τ versus $X(t)/C_0$ is shown in Fig. 6.2. Note that both t/τ and $X(t)/C_0$ are dimensionless. It can be seen from Eqs. (6.1.2b) and (6.1.2c) or Fig. 6.2 that when $t = \tau$, the response reaches

$$X(\tau) = C_0 e^{-1} = 0.368 \, C_0$$

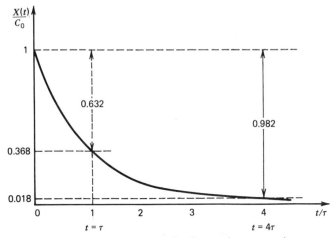

Figure 6.2   Response characteristic of a first order system in terms of time constant $\tau = 1/\alpha_0$.

or

$$\frac{X(\tau)}{C_0} = 0.368$$

of its initial value; when $t = 4\tau$, the response reaches

$$X(4\tau) = C_0 e^{-4} = 0.018\, C_0$$

or

$$\frac{X(4\tau)}{C_0} = 0.018$$

of its initial value. Hence, when the value of time reaches one time constant, the response drops by 63.2% (100 − 36.8), while it takes four time constants for the response to drop by 98.2% of its initial value and almost return to the equilibrium position. Therefore, a reasonable estimate of the time required for a physical system to respond completely is about four time constants. These properties of the time constant are quite useful in estimating the response time of a first order system.

6.1.3 Stability

It can be seen that for positive values of α_0, the system eventually or asymptotically returns to its equilibrium position, more or less slowly depending on the magnitude of α_0. When $\alpha_0 > 0$, the system is said to be

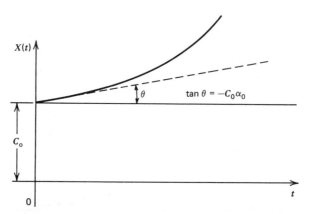

Figure 6.3   Response of a first order system, $\alpha_0 < 0$.

"asymptotically stable." When $\alpha_0 = 0$, the system does not return to its equilibrium position at all, and it is not asymptotically stable. However, it remains at a finite distance from the equilibrium position and does not "explode." Therefore, when $\alpha_0 = 0$, the system is said to be "stable." Note that an asymptotically stable system is always stable but the reverse is not always true.

On the other hand, when $\alpha_0 < 0$, the system "explodes," or the response tends to infinity with increasing time as shown in Fig. 6.3. To put it more precisely, for large enough t, the system response will exceed any given bound however large. Therefore, under the condition $\alpha_0 < 0$, the system is said to be "unstable." Thus, the stability condition for a first order system may be summarized as follows:

$$\alpha_0 > 0 \quad \text{Asymptotically stable} \quad (6.1.3a)$$

$$\alpha_0 \geq 0 \quad \text{Stable} \quad (6.1.3b)$$

$$\alpha_0 < 0 \quad \text{Unstable} \quad (6.1.3c)$$

Note that the asymptotic stability is the strongest form of stability.

The simple stability analysis given above for the first order system readily extends to higher order systems. For the first order system the characteristic root also happens to be the coefficient in the differential equation that is always real. This is no longer true for higher order equations in which the coefficients are real but the characteristic roots may be complex. Then, the above conditions are applicable to the real parts of the roots. This is treated in more detail in Chapter 7.

6.2 DIRAC DELTA FUNCTION AND ITS PROPERTIES

We have seen earlier that the entire analysis of discrete stochastic systems was based on the most important property

$$E(a_t a_{t-k}) = \sigma_a^2 \delta_k \qquad (6.2.1)$$

of the discrete white noise or uncorrelated sequence a_t. It is this property that provided an orthogonal decomposition of the response X_t, which reveals the dynamic structure of the system. This property was also central to the derivation of the auto-covariance/autocorrelation function in Chapter 3 and the forecasting theory in Chapter 5. The discrete delta function appearing in this property is the Kronecker delta defined by

$$\delta_k = 1, \qquad k = 0 \qquad (6.2.2)$$
$$= 0, \qquad k \neq 0$$

6.2.1 Definition

The continuous time analog of a_t is the "continuous white noise," denoted by $Z(t)$, and defined by the property

$$E[Z(t)Z(t - u)] = \sigma_z^2 \delta(u) \qquad (6.2.3)$$

This property plays the same crucial role in continuous time systems as Eq. (6.2.1) does in discrete time systems. The delta function $\delta(u)$ appearing in Eq. (6.2.3) is called the "Dirac delta" and is much more complicated than its discrete counterpart δ_k, the Kronecker delta. The Dirac delta function is defined by

$$\delta(u) = \infty, \qquad u = 0 \qquad (6.2.4a)$$
$$= 0, \qquad u \neq 0$$

and

$$\int_{-\tau}^{\tau} \delta(u)\, du = 1 \qquad (6.2.4b)$$

for arbitrary τ. [In fact (6.2.4b) implies (6.2.4a).]

This function, called unit impulse in control theory, has played an important role in the analysis of physical systems since its use by Dirac in his work on quantum mechanics. The function is generally approximated in physical phenomena by a pulse for which a quantity changes by a very large magnitude over a very short time in response to a certain excitation, yet the total change resulting from the excitation has a fixed finite value. As shown in Fig. 6.4b, the unit pulse is zero everywhere except in the finite region $-\tau < t < \tau$, where it has the value $\dfrac{1}{2\tau}$ to make the area under it unity $(2\tau \cdot \dfrac{1}{2\tau} = 1)$. As $\tau \to 0$, the maximum value of the function approaches infinity (see Fig. 6.4a), but its "strength" indicated by the area under the curve remains unity and justifies the defining property (6.2.4b). Note that any function that satisfies Eqs. (6.2.4a) and (6.2.4b) is a delta

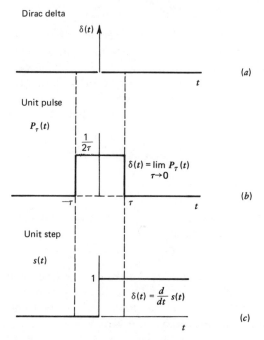

Figure 6.4   Dirac delta or unit impulse as limiting form of unit pulse and derivative of unit step.

function. For example, it can be shown that $e^{3t}\delta(t)$ is also a delta function (see Problem 6.1).

6.2.2 Relation With Unit Step Function

As shown in Fig. 6.4c, the unit step function is defined as

$$s(t) = \begin{cases} 0, & t < 0 \\ 1, & t \geq 0 \end{cases} \qquad (6.2.5)$$

The derivative of the unit step function is zero everywhere except at $t = 0$, where it approaches infinity because of a discontinuity. The Dirac delta function is also zero everywhere except at $t = 0$, where it approaches infinity. Hence, the delta function can be considered as the derivative of the unit step function, that is,

$$\delta(u) = \frac{d}{dt} s(u) \qquad (6.2.4c)$$

Note that both the unit step and the delta function have a discontinuity at $t = 0$. The discontinuity for the step function is a jump of height one, whereas for the delta function it is a more complicated jump (of height infinity). The integral of the step function is also defined as

$$r(t) = \int_{-\infty}^{t} s(u)\, du = \begin{cases} 0, & t < 0 \\ t, & t \geqslant 0 \end{cases} \qquad (6.2.6)$$

and is referred to as a "unit ramp." This is a continuous function, and it is easy to see that further integrals of the unit ramp will also be continuous.

6.2.3 Properties

The two important properties of the delta function which will be used later in this chapter are described as follows:

For any continuous function $f(t)$

$$(1) \int_{-\infty}^{\infty} f(t - u)\delta(u)\, du = f(t) = \int_{-\infty}^{\infty} f(u)\delta(t - u)\, du \quad (6.2.7)$$

and, if its kth derivative is continuous, then

$$(2) \int_{-\infty}^{\infty} f(t - u)\delta^{(k)}(u)\, du = (-1)^k f^{(k)}(t) \qquad (6.2.8)$$

where a superscripted k in parentheses denotes the kth derivative, that is,

$$f^{(k)}(t) = \frac{d^{(k)}}{dt^{(k)}}[f(t)]$$

The proofs of the above two properties are given in Appendix A 6.2. Note that the first property follows from the second by putting $k = 0$.

6.3 THE FIRST ORDER AUTOREGRESSIVE SYSTEM A(1)

We are now in a position to formulate the equation for a continuous time stochastic system governed by a first order linear differential equation with constant coefficients. Recall that the discrete system governed by a first order linear difference equation may be obtained by starting with a homogeneous difference equation

$$X_t - \phi_1 X_{t-1} = 0 \qquad (6.3.1)$$

and converting it into a nonhomogeneous stochastic difference equation by introducing a forcing function a_t, that is,

$$X_t - \phi_1 X_{t-1} = a_t \qquad (6.3.2)$$

which is the AR(1) model. The homogeneous Eq. (6.3.1) characterizes the discrete system as an operator that is deterministic. The nonhomogeneous Eq. (6.3.2) provides a complete description of the system, by specifying that the input and output to the system are random functions or stochastic processes, which makes the system stochastic.

The relation between the output and the input can be expressed more explicitly, but in a way equivalent to Eq. (6.3.2), by its solution in terms of the Green's function G_j as [see Eq. (3.1.4)]

$$X_t = \sum_{j=0}^{\infty} G_j a_{t-j} = \sum_{j=-\infty}^{t} G_{t-j} a_j \qquad (6.3.3)$$

that is,

$$X_t = (1 - \phi_1 B)^{-1} a_t$$

$$= \sum_{j=0}^{\infty} \phi_1^j a_{t-j} \qquad (6.3.4)$$

which gives

$$G_j = \phi_1^j$$

for the AR(1) model. Equation (6.3.3) or (6.3.4) is an orthogonal decomposition (or Wold's decomposition) of the random function X_t. It can also be looked upon as the decomposition of a complicated random function X_t into a simple or elementary random function a_t, by expressing X_t as a convolution of a_t with G_j. This decomposition was extremely useful in the analysis of discrete systems since the function a_t with the simple defining property (6.2.1) is much easier to deal with than X_t. That is why the orthogonal decomposition (6.3.3) was used in almost every aspect of discrete system analysis: in obtaining its statistical characterization γ_k, in modeling, in forecasting, etc.

6.3.1 The Stochastic Differential Equation and Its Solution

To formulate a continuous time first order autoregressive stochastic system, we start with a homogeneous differential equation

$$\frac{dX(t)}{dt} + \alpha_0 X(t) = 0$$

that is,

$$(D + \alpha_0)X(t) = 0 \qquad (6.3.5)$$

where

$$D^n \equiv \frac{d^n}{dt^n}$$

and convert it into a nonhomogeneous stochastic differential equation by introducing a forcing function $Z(t)$ with the property (6.2.3)

$$(D + \alpha_0)X(t) = Z(t)$$

$$E[Z(t)] = 0 \qquad (6.3.6)$$

$$E[Z(t)Z(t - u)] = \sigma_z^2 \sigma(u)$$

Equation (6.3.6) represents a continuous time first order autoregressive system, and to avoid using the word continuous every time to distinguish it from the discrete AR systems, we will denote it by A(1).

Here again, the input $Z(t)$ is the simplest or elementary continuous time stochastic process, which has zero mean and is uncorrelated at different times. The ouput $X(t)$ is also a stationary stochastic process with zero mean since the operator in Eq. (6.3.5) is linear; its correlation or covariance function is now a function of continuous time and can be derived knowing the orthogonal decomposition of $X(t)$.

To obtain the "solution" of Eq. (6.3.6) and to express the relation between the output $X(t)$ and the input $Z(t)$ more explicitly, we have to invert the differential operator in Eq. (6.3.6), that is, symbolically,

$$X(t) = (D + \alpha_0)^{-1} Z(t)$$

Since the inversion of a difference operator is given by a summation, expressed as a convolution of the input or the forcing function with the Green's function in discrete time, it follows that the inversion of a differential operator will be given by integration, expressed as a convolution of the input or the forcing function $Z(t)$, with the Green's function in continuous time denoted by $G(v)$. Therefore, corresponding to Eq. (6.3.3) in discrete time we have

$$X(t) = \int_0^\infty G(v)\, Z(t - v)\, dv = \int_{-\infty}^t G(t - v) Z(v)\, dv \quad (6.3.7)$$

in continuous time. Fig. 6.5 graphically illustrates the convolution of the input $Z(t)$ with the Green's function $G(v)$ in continuous time. Equation (6.3.7) is the orthogonal decomposition of the stochastic process or time series $X(t)$ since $Z(t)$'s are uncorrelated or independent at different times. As in the discrete case, this decomposition will yield all the necessary characteristics of the A(1) system, including its autocovariance function. It should be emphasized that $Z(t)$ in Fig. 6.5 has been somewhat smoothed for graphical clarity and therefore appears correlated. Actual Z(t) would be so erratic that it cannot be graphed. Note that white noise $Z(t)$ is physically unrealizable.

6.3.2 Orthogonal Decomposition

This concept of orthogonal decomposition, by expressing a complicated function as a linear combination of simple or elementary functions that can be easily dealt with, is the key to linear systems analysis. In the existing literature dealing with deterministic systems, the "orthogonal" elementary function used for such decomposition are the Kronecker delta δ_k for discrete time and the Dirac delta $\delta(u)$ for continuous time. In fact, Eq. (6.2.7),

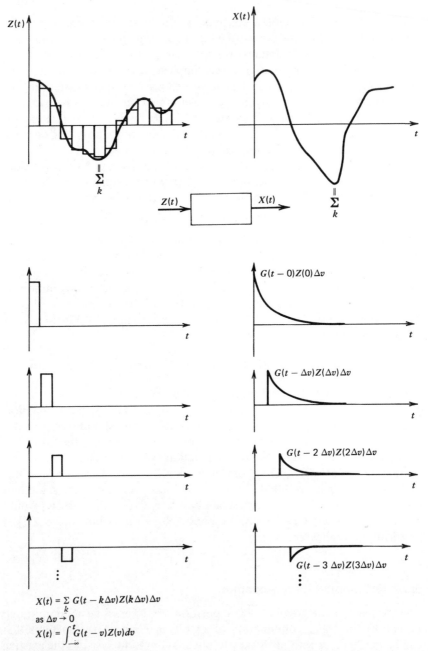

Figure 6.5   Convolution of the input $Z(t)$ with the Green's function.

rewritten as

$$f(t) = \int_{-\infty}^{\infty} f(t - u)\delta(u) \, du = \int_{-\infty}^{\infty} f(u)\delta(t - u) \, du$$

is itself an orthogonal decomposition, where t being continuous, we have an integral representation rather than a summation. The "axes" $\delta(u)$, $-\infty < u < \infty$ are orthogonal since they are independent in the sense that no $\delta(u)$ can be expressed as a linear combination of $\delta(v)$'s, $v \neq u$, either in summation or integration form.

This idea is graphically represented in Fig. 6.6. It was pointed out in the last section that the unit impulse or $\delta(u)$ can be approximated by a unit pulse. The function $f(t)$ can be approximated in any finite interval $-T \leq t < T$ by a finite number of unit pulses of width Δu, $p_{\Delta u}(t)$, occurring at $t = k\Delta u$ ($k = 0, \pm 1, \pm 2, \ldots \pm N = T/\Delta u$). This approximation may be represented by

$$f(t) \simeq \sum_{k=-N}^{N} f(k\Delta u)p_{\Delta u}(t - k\Delta u)\Delta u$$

As $N \to \infty$ or $\Delta u \to 0$, the pulses become impulses, the above summation becomes integration, and finally with $T \to \infty$, we get Eq. (6.2.7). Similarly, a discrete function has orthogonal decomposition in terms of δ_k:

$$f_t = \sum_{j=0}^{\infty} f_j\delta_{t-j} = \sum_{j=-\infty}^{t} f_{t-j}\delta_j \qquad (6.3.8)$$

If we compare Eq. (6.3.8) with Eq. (6.3.3) and the property (6.2.1) of a_t's, it should be easy to appreciate the central role played by the

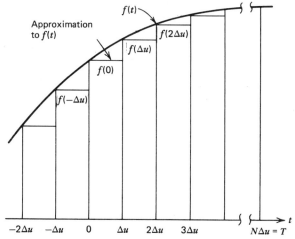

Figure 6.6   Approximation of a function by unit pulses.

"coordinate" function G_j. Since X_t is the response or output of a stochastic system in discrete time, it is natural to decompose it in terms of the elementary random function a_t, which has a discrete impulse δ_k of constant strength σ_a^2 as its correlation function. Therefore, it is also natural to decompose $X(t)$, the response of a stochastic system in continuous time, in terms of the elementary random function $Z(t)$, which has a "continuous" impulse $\delta(u)$ of constant strength σ_z^2 as its correlation function.

In spite of these similarities, an important difference between σ_a^2 and σ_z^2 should be carefully noted. σ_a^2 is the variance of a_t since

$$E(a_t a_{t-k}) = \delta_k \sigma_z^2 \Rightarrow E(a_t^2) = \sigma_a^2$$

However, σ_z^2 is *not* the variance of $Z(t)$ since

$$E[Z(t)Z(t - u)] = \delta(u)\sigma_z^2 \Rightarrow E[Z(t)]^2 = \delta(0)\sigma_z^2 = \infty$$

Thus, the variance of $Z(t)$ is "infinite," which is a mathematical translation of the physical fact that white noise does not exist. However, the integral of white noise over a finite interval, say Δ, does exist and, in fact, has a variance $\Delta\sigma_z^2$ because it can be shown that (see Section 6.5.3)

$$E\left[\int_{t-\Delta}^{t} Z(t)dt\right]^2 = \Delta\sigma_z^2$$

6.3.3 The Green's Function   *read optional*

Our main task therefore is to obtain the Green's function of the differential operator in Eq. (6.3.6). Note that as in the discrete case, the Green's function $G(v)$ is independent of the input or output. In fact, $G(v)$ may be symbolically defined for a first order system by the equivalent relations

$$(D + \alpha_0)X(t) = h(t) \tag{6.3.9a}$$

$$X(t) = (D + \alpha_0)^{-1}h(t) = \int_0^{\infty} G(v)h(t - v)dv \tag{6.3.9b}$$

for any continuous time forcing function $h(v)$. Although the Green's function $G(v)$ can be obtained by the classical theory of linear differential equations [see, for example, Coddington and Levinson (1955)], we will derive it by a simple method exploiting the properties of the Dirac delta function given in Section 6.2.

What input or forcing function $h(t)$ in Eq. (6.3.9a) will give us the Green's function $G(t)$ as the output? That is, we want the function $h(t)$ that gives

$$G(t) = \int_0^{\infty} G(v)h(t - v)\, dv$$

Since we are interested in the integral from zero to infinity, we can take

$$G(v) = 0, \qquad v < 0$$

and write the above integral as

$$G(t) = \int_{-\infty}^{\infty} G(v)h(t - v) \, dv$$

Now comparing with the important property

$$f(t) = \int_{-\infty}^{\infty} f(u)\delta(t - u) \, du$$

given by Eq. (6.2.7), we see that the answer is precisely the delta function

$$h(t) \equiv \delta(t)$$

and we have the equivalent relations corresponding to Eqs. (6.3.9a) and (6.3.9b)

$$(D + \alpha_0)G(t) = \delta(t) \tag{6.3.10a}$$

$$G(t) = \int_{0}^{\infty} G(v)\delta(t - v) \, dv \tag{6.3.10b}$$

Thus, the required $G(v)$ is the solution of the nonhomogeneous differential equation (6.3.10a).

This equation (6.3.10a) can be solved by replacing it with a homogeneous differential equation with initial conditions, which can be obtained from the properties of the delta function. The homogeneous equation for (6.3.10a) is obviously (6.3.5). To find the initial conditions, let us observe what Eq. (6.3.10a) says about the continuity properties of the function $G(t)$. First of all, since the input $\delta(t)$ is zero up to $t = 0$,

$$G(t) = 0, \qquad t < 0$$

At $t = 0$, the derivative of $G(t)$ contains the same discontinuity as that of a delta function. Therefore, the function $G(t)$, which is the integral of its derivative, contains the same discontinuity at $t = 0$, as that of the integral of $\delta(t)$ that is the unit step function $s(t)$ in Section 6.2. This is a jump discontinuity of height one, and therefore

$$G(0) = 1$$

Thus, Eq. (6.3.10a) is equivalent to the homogeneous equation

$$(D + \alpha_0)G(t) = 0$$

with initial condition

$$G(0) = 1$$

for which we get the solution from Section 6.1, with $C_0 = 1$, as

$$G(t) = e^{-\alpha_0 t}, \qquad t \geqslant 0 \tag{6.3.11}$$
$$= 0, \qquad t < 0$$

that is,

$$G(t) = s(t)e^{-\alpha_0 t} \tag{6.3.11a}$$

where $s(t)$ is the unit step function. Checking that this function satisfies Eq. (6.3.10a) is left as an exercise.

Using the Green's function of the A(1) system, we may write the orthogonal decomposition from Eq. (6.3.7) as

$$X(t) = \int_0^\infty e^{-\alpha_0 v} Z(t-v)\, dv = \int_{-\infty}^t e^{-\alpha_0(t-v)} Z(v)\, dv \tag{6.3.12}$$

This is also the Wold's decomposition of the stochastic process or time series $X(t)$ and represents an explicit solution or equivalent form of the nonhomogeneous differential equation (6.3.6). We thus have Eq. (6.3.6), (6.3.7), and (6.3.12) for the continuous time A(1) system, which respectively correspond to Eqs. (6.3.2), (6.3.3), and (6.3.4) for the discrete time AR(1) system.

6.3.4 The Autocovariance Function

We can now use the decomposition (6.3.12) together with the property

$$E[Z(t)Z(t-s)] = \delta(s)\,\sigma_z^2$$

to derive the covariance function of the A(1) system in exactly the same way as γ_k for the AR(1) model was derived in Section 3.3.3 [see Eq. (3.3.9)], using the decomposition (6.3.4) and the property

$$E[a_t a_{t-k}] = \delta_k \sigma_a^2$$

However, it is equally simple to first derive a general relation [corresponding to Eq. (3.3.20)] and then derive the A(1) special case. If the covariance of $X(t)$ at lag s (a continuous variable) is denoted by $\gamma(s)$, then

$$\gamma(s) = E[X(t)X(t-s)]$$

$$= E\left[\int_0^\infty G(v')Z(t-v')\, dv' \int_0^\infty G(v)Z(t-s-v)\, dv \right]$$

$$= \int_0^\infty \int_0^\infty G(v')G(v)E[Z(t-v')Z(t-s-v)]\, dv\, dv' \tag{6.3.13}$$

$$= \sigma_z^2 \int_0^\infty \left[\int_0^\infty G(v')G(v)\delta(v+s-v')\, dv' \right] dv \qquad \text{by (6.2.3)}$$

$$= \sigma_z^2 \int_0^\infty G(v)G(v+s)\, dv \tag{6.3.13}$$

In the last step we have used the property (6.2.7) for the integral in the brackets. Using the Green's function for the A(1) system given by Eq. (6.3.11), we have

$$\gamma(s) = \sigma_z^2 \int_0^\infty e^{-\alpha_0 v} e^{-\alpha_0(v+s)} \, dv$$

$$= \sigma_z^2 e^{-\alpha_0 s} \int_0^\infty e^{-2\alpha_0 v} \, dv$$

$$= \sigma_z^2 e^{-\alpha_0 s} \left[\frac{-e^{-2\alpha_0 v}}{2\alpha_0} \right]_0^\infty$$

$$= \frac{\sigma_z^2}{2\alpha_0} e^{-\alpha_0 s}, \qquad s \geq 0 \tag{6.3.14}$$

assuming that the system is stable, that is, $\alpha_0 > 0$. In particular, the variance of $X(t)$ that is the covariance at zero lag is given by

$$\gamma(0) = \frac{\sigma_z^2}{2\alpha_0} \tag{6.3.15}$$

and therefore the correlation function

$$\rho(s) = \frac{\gamma(s)}{\gamma(0)} = e^{-\alpha_0 s}, \qquad s > 0$$

with

$$\rho(-s) = \rho(s) \tag{6.3.16}$$

Note that the relation $\rho(-s) = \rho(s)$ follows from

$$\gamma(-s) = \gamma(s)$$

which can be derived for the A(1) model, as above, by starting with $X(t)$ and $X(t + s)$. It is also true, in general, since $\gamma(s)$ is the covariance between $X(t)$ and $X(t - s)$, and

$$\gamma(-s) = E[X(t)X(t + s)] = E[X(t + s)X(t)] = \gamma(s)$$

The autocovariance and autocorrelation functions can also be written in the form

$$\gamma(s) = \frac{\sigma_z^2}{2\alpha_0} e^{-\alpha_0 |s|} \tag{6.3.14a}$$

$$\rho(s) = e^{-\alpha_0 |s|} \tag{6.3.16a}$$

where now s may vary from $-\infty$ to $+\infty$.

It follows from Eqs. (6.3.14) to (6.3.16) that the stability condition $\alpha_0 > 0$ may be interpreted as the condition of finite variance or decreasing autocovariance/autocorrelation, as in the discrete case. A detailed consideration of the case $\alpha_0 \to 0$ will be given in Section 6.5.

* 6.3.5 The Spectrum

By taking the Fourier transform of the autocovariance function (6.3.14a), we can show that the spectrum or auto-spectrum $f(\omega)$ of the A(1) system is given by

$$f(\omega) = \frac{\sigma_z^2}{2\pi} \frac{1}{|i\omega + \alpha_0|^2} = \frac{\sigma_z^2}{2\pi} \frac{1}{(\omega^2 + \alpha_0^2)}$$

$$-\infty < \omega < \infty$$

(6.3.17)

Note that the frequency decomposition is now available over the entire band width as opposed to over the limited band width $(-\pi/\Delta, \pi/\Delta)$ as in the discrete case [see Eq. (3.4.3)]. Therefore, the spectrum of a continuous system provides more complete information than the discrete system. This is one of the reasons for having continuous models. The method used to obtain continuous models from the discrete data will be discussed in Chapters 6, 7, and 8; the necessary results for the case of A(1) system will be given in the next section.

6.4 UNIFORMLY SAMPLED FIRST ORDER AUTOREGRESSIVE SYSTEM

In practice, of course, one would not be given a model such as A(1), as is commonly assumed in classical systems analysis for deterministic systems. At the most, what one can expect is a conjecture, based on physical reasoning, that the stochastic system may be A(1). We are, in fact, considering the problem at the other end; we assume no prior knowledge of the system except for the observed data, at least in the beginning, and first try to obtain a model. Then, we may use the available physical knowledge to interpret, modify, or choose among the models.

In the physical sciences and engineering, it is sometimes possible to get a continuous plot of the data. However, both from the point of view of the accuracy of observation and the ease of data processing on a digital computer, it is advantageous to use discrete data obtained at uniform sampling intervals. Most of the data in the physical sciences and engineering, as in business, economics, and other sciences, is already available in discrete form at uniform intervals. To obtain the continuous system models from such a discrete data, we have to know the representation of such a system sampled at uniform intervals. In this section, we will show that for the A(1) model, the uniformly sampled system has a discrete representation that is of the same form as the AR(1) model and then establish the parametric relations between the discrete and the continuous representations.

6.4.1 The Sampled System Model

A discrete representation can be found using the principle of covariance equivalence, that is, requiring that the covariance of the discrete model coincide with that of the continuous model at sampling points. Therefore, to obtain the discrete model for the sampled system, let us find its auto-covariance function. If the sampling interval is Δ, then the autocovariance k intervals apart, γ_k, is the same as $\gamma(k\Delta)$, that is,

$$\gamma(s) = E[X(t)X(t - s)]$$
$$\gamma(k\Delta) = \gamma_k = E[X(t)X(t - k\Delta)] \qquad (6.4.1)$$
$$= E[X_t X_{t-k}]$$

and hence the autocovariance function of the uniformly sampled discrete system is obtained by substituting $s = k\Delta$. Since, by Eq. (6.3.14),

$$\gamma(s) = \frac{\sigma_z^2}{2\alpha_0} e^{-\alpha_0 s}$$

therefore

$$\gamma_k = \gamma(k\Delta) = \frac{\sigma_z^2}{2\alpha_0} e^{-\alpha_0 k\Delta} \qquad (6.4.2a)$$

$$= \frac{\sigma_z^2}{2\alpha_0} \phi^k \qquad (6.4.2b)$$

where

$$\phi = e^{-\alpha_0 \Delta} \qquad (6.4.3)$$

Equation (6.4.2b) shows that the sampled system has the form of an AR(1) model

$$X_t - \phi X_{t-1} = a_t \qquad (6.4.4)$$

where ϕ is given by Eq. (6.4.3). Note that although every uniformly sampled A(1) model has the form of an AR(1) model (6.4.4), every AR(1) is not necessarily a uniformly sampled A(1) model. *This is because the parameter ϕ of the sampled system cannot be negative by Eq. (6.4.3).* Thus

$$X_t - 0.5X_{t-1} = a_t$$

can be a sampled first order system model but

$$X_t + 0.5X_{t-1} = a_t$$

cannot be, since ϕ is negative.

6.4.2 Expression for σ_a^2

To complete the description of the discrete model (6.4.4), we have to find the variance of a_t's, σ_a^2. For this purpose, let us write down its autocovariance function, that is,

$$E[X_t X_{t-k}] = \phi E[X_{t-1} X_{t-k}] + E[a_t X_{t-k}]$$

and from Eqs. (3.3.4) to (3.3.6)

$$\gamma_0 = \phi \gamma_1 + \sigma_a^2 \tag{6.4.5}$$

$$\gamma_1 = \phi \gamma_0, \qquad k > 0$$

Substituting $\gamma_1 = \phi \gamma_0$ from the second equation into the first, one can write

$$\gamma_0 = \phi^2 \gamma_0 + \sigma_a^2$$

$$\sigma_a^2 = (1 - \phi^2)\gamma_0$$

$$\gamma_0 = \frac{\sigma_z^2}{2 d_0}$$

Using Eq. (6.3.15) for γ_0, we obtain

$$\sigma_a^2 = \frac{\sigma_z^2}{2\alpha_0} (1 - \phi^2) \tag{6.4.6}$$

It is now clear that if an AR(1) model, found to be adequate for the data, has a positive autoregressive parameter, then it is automatically a sampled first order system model and the corresponding parameters α_0 and σ_z^2 of an A(1) model are simply obtained by

$$\alpha_0 = -\frac{\ln \phi}{\Delta} \tag{6.4.7}$$

and

$$\sigma_z^2 = \frac{2\alpha_0 \sigma_a^2}{(1 - \phi^2)} \tag{6.4.8}$$

which follow immediately from Eqs. (6.4.3) and (6.4.6).

6.4.3 Illustrative Example

As an example, consider the AR(1) model for the papermaking data in Chapter 2, with $\phi = 0.8983$ and $\sigma_a^2 = 0.094$. Since ϕ is positive, this AR(1) model is also a sampled model, for which the parameters can be obtained by Eqs. (6.4.7) and (6.4.8) with $\Delta = \frac{1}{3}$ hour as

$$\alpha_0 = -\frac{\ln \phi}{\Delta} = 0.322$$

and

$$\sigma_z^2 = \frac{2\alpha_0\sigma_a^2}{(1 - \phi^2)} = 0.3133$$

Thus, the A(1) model for the papermaking data (after subtracting $\overline{X}$) is

$$\frac{dX(t)}{dt} + 0.322X(t) = Z(t)$$

Or the adjustment of the gate opening has a time constant of $\tau = 1/\alpha_0 = 3.1$ hours.

6.5 LIMITING CASES—EFFECT OF SAMPLING INTERVAL Δ AND THE PARAMETER α_0

For very large and very small values of Δ and α_0, the sampled discrete model takes some simple forms. In particular, the random walk discussed in Chapter 2 arises in this way.

6.5.1 Sampling Interval Δ

Let us first consider the limiting cases of the sampled discrete model for extreme values of the sampling interval, Δ.

(a) Large Δ

When the sampling interval is large, we see from Eq. (6.4.3) that

$$\phi = e^{-\alpha_0\Delta} \to 0, \quad \text{as} \quad \Delta \to \infty \tag{6.5.1}$$

and therefore the model becomes

$$X_t = a_t \tag{6.5.2}$$

or an AR(0) model or a sequence of uncorrelated random variables with variance, by Eq. (6.4.6),

$$\sigma_a^2 = \frac{\sigma_z^2}{2\alpha_0}(1 - \phi^2) \to \frac{\sigma_z^2}{2\alpha_0} = \gamma_0, \quad \text{as} \quad \Delta \to \infty \tag{6.5.3}$$

This result is intuitively obvious and can also be guessed considering the $\gamma(s)$ function, given by Eq. (6.3.14), as

$$\gamma(s) = E[X(t)X(t - s)] = \frac{\sigma_z^2}{2\alpha_0}e^{-\alpha_0 s}$$

When the observations are taken so far apart that the autocovariance or autocorrelation function decays to zero in the sampling interval time, then there is practically no correlation between the successive sampled observations; this is what the AR(0) model (6.5.2) says.

From the above reasoning, it should be clear that for a large sampling interval *any* continuous time stable stationary stochastic system would be represented by the AR(0) model, since for large enough Δ the $\gamma(s)$ function would decay to zero over the interval Δ. In Chapters 7 and 8 we will show that this result is indeed true for higher order models.

The above result also explains the conditions under which observations made on a continuous system may be considered independent or uncorrelated, an assumption most commonly made in statistical analyses. This assumption may be justified when the sampling interval is so large that the autocorrelation function or the Green's function decays practically to zero over that interval. For the first order system, A(1), under consideration, the sampling interval that yields uncorrelated observations depends on the single parameter α_0, the inverse of the "time constant" of the system. The larger the parameter α_0, the smaller the sampling interval to produce uncorrelated observations will be and vice versa, since the exponent in Eq. (6.5.1) is $-\alpha_0\Delta$.

The discussion of the time constant $\tau = 1/\alpha_0$ given in Section 6.1.2 throws some light on a reasonable estimate of the sampling interval Δ, in terms of τ, which can practically justify the independence or AR(0) assumption. Since $e^{-4} = 0.018$ and $e^{-5} = 0.0067$, it would seem that a sampling interval larger than four times the time constant would practically suffice for this purpose.

(b) Small Δ

When the sampling interval is small, we have from Eq. (6.4.3)

$$\phi = e^{-\alpha_0\Delta} \to 1, \quad \text{as} \quad \Delta \to 0 \tag{6.5.4}$$

and therefore the model becomes

$$X_t - X_{t-1} = a_t$$

that is,

$$\nabla X_t = a_t \tag{6.5.5}$$

which is a random walk in discrete time. Also from Eq. (6.4.6)

$$\sigma_a^2 = \frac{\sigma_z^2}{2\alpha_0} (1 - \phi^2) \to 0 \quad \text{as} \quad \Delta \to 0 \tag{6.5.6}$$

This result illustrates one of the ways in which the random walk in discrete time may arise in practice. It also explains the reason for our treatment of the differenced models such as (6.5.5) as limiting cases with $\phi \to 1$ rather than ϕ being exactly one. Note that Eq. (6.5.4) says that ϕ is very close to one but not exactly one.

6.5.2 The Parameter α_0

The limiting cases (6.5.2) and (6.5.5) may also arise in a more fundamental way for extreme values of the time constant, $1/\alpha_0$.

(a) Large α_0: As $\alpha_0 \to \infty$

$$\phi = e^{-\alpha_0 \Delta} \to 0 \tag{6.5.7}$$

with

$$\sigma_a^2 = \frac{\sigma_z^2}{2\alpha_0}(1 - \phi^2) \to 0 \tag{6.5.8}$$

(b) Small α_0: As $\alpha_0 \to 0$

$$\phi = e^{-\alpha_0 \Delta} \to 1 \tag{6.5.9}$$

with

$$
\begin{aligned}
\sigma_a^2 &= \frac{\sigma_z^2}{2\alpha_0}(1 - e^{-2\alpha_0 \Delta}) \\
&= \frac{\sigma_z^2}{2\alpha_0}\left\{ 1 - \left[1 - (2\alpha_0 \Delta) + \frac{(2\alpha_0 \Delta)^2}{2!} - \cdots \right] \right\} \\
&\to \sigma_z^2 \Delta
\end{aligned}
\tag{6.5.10}
$$

Before giving a physical interpretation of these results, let us see what form the basic A(1) model takes under these extreme cases. For the case (a), $\alpha_0 \to \infty$, we can rewrite the A(1) model in the form

$$
\begin{aligned}
\frac{1}{\alpha_0}\frac{dX(t)}{dt} + X(t) &= \frac{1}{\alpha_0} Z(t) \\
&= Z'(t)
\end{aligned}
\tag{6.5.11}
$$

if we denote $\dfrac{1}{\alpha_0} Z(t)$ by $Z'(t)$. It follows from Eq. (6.2.3) that

$$
\begin{aligned}
E[Z'(t)Z'(t - u)] &= \frac{1}{\alpha_0^2} E[Z(t)Z(t - u)] \\
&= \frac{\sigma_z^2}{\alpha_0^2} \delta(u)
\end{aligned}
\tag{6.5.12}
$$

Thus, as $\alpha_0 \to \infty$, Eq. (6.5.11) becomes

$$X(t) = Z'(t) \tag{6.5.13}$$

where $Z'(t)$ is such that $\sigma_z^2 \to 0$ by Eq. (6.5.12), which really means that there is no response or motion and the system stays in the equilibrium position. To avoid this trivial case, we assume that although α_0 is large, σ_z^2 is so large that $\dfrac{\sigma_z^2}{\alpha_0^2}$ is a nonzero constant, there is some response or motion, and

$$\sigma_a^2 = \frac{\sigma_z^2}{2\alpha_0}(1 - \phi^2) \to \frac{\sigma_z^2}{2\alpha_0} \tag{6.5.14}$$

which is also a nonzero finite constant. Actually, because of the exponential relationship, the limit $\phi \to 0$ is attained much faster than the limit $\sigma_a^2 \to 0$, so that for reasonably large values of α_0, we get $X_t = a_t$, with nonzero finite σ_a^2. This will be further elaborated upon in Sections 6.5.4 and 6.5.5 and numerically illustrated in Problem 6.5.

For the case (b), $\alpha_0 \to 0$, it is easy to see that the A(1) model

$$\frac{dX(t)}{dt} + \alpha_0 X(t) = Z(t)$$

takes the form

$$\frac{dX(t)}{dt} = Z(t) \tag{6.5.15}$$

which is the continuous time random walk. As we have seen in the discrete case, random walk is simply a summation of uncorrelated variables a_t. Naturally, in the continuous case it is an integration of the uncorrelated white noise $Z(t)$. Either directly from Eq. (6.5.15) or taking a limit as α_0 tends to zero in (6.3.12), we have the equivalent representation of (6.5.15) as

$$X(t) = \int_{-\infty}^{t} Z(v)\, dv \tag{6.5.16a}$$

$$= \int_{t-\Delta}^{t} Z(v)\, dv + \int_{t-2\Delta}^{t-\Delta} Z(v)\, dv$$

$$+ \int_{t-3\Delta}^{t-2\Delta} Z(v)\, dv + \ldots \tag{6.5.16b}$$

that is,

$$X_t = a_t + a_{t-1} + a_{t-2} + \ldots$$

which is precisely the discrete time random walk, Eq. (6.5.5), the limiting case of the discrete model as $\alpha_0 \to 0$.

The limiting cases of the discrete model for extreme values of the sampling interval Δ and α_0 are summarized in Table 6.1. The limiting cases are also illustrated by Fig. 6.7, which shows how ϕ and σ_a^2 change with Δ and α_0. It can be seen from Fig. 6.7a that when α_0 or Δ approaches infinity, ϕ approaches zero; when α_0 or Δ approaches zero, ϕ approaches one. Figure 6.7b shows that as α_0 approaches infinity, σ_a^2 approaches zero; as α_0 approaches zero, σ_a^2 approaches $\sigma_z^2 \Delta$. Figure 6.7c shows that as Δ approaches infinity, σ_a^2 approaches γ_0; as Δ approaches zero, σ_a^2 approaches zero.

Table 6.1   Limiting Cases of the Sampled First Order System Model

$\Delta \to \infty$	$\phi \to 0$ $\sigma_a^2 \to \gamma_0$ $X_t = a_t$	White noise
$\Delta \to 0$	$\phi \to 1$ $\sigma_a^2 \to 0$ $\nabla X_t = a_t$	Random walk
$\alpha_0 \to \infty$	$\phi \to 0$ $\sigma_a^2 \to 0$ $X_t = a_t$	White noise
$\alpha_0 \to 0$	$\phi \to 1$ $\sigma_a^2 \to \sigma_z^2 \Delta$ $\nabla X_t = a_t$	Random walk

* 6.5.3 Independent Increment Process

The continuous time random walk in the form of Eq. (6.5.15) is also called an independent increment process since it possesses the property that the increment in the process over disjoint intervals are independent. If the two intervals (t_1, t_2), (t_3, t_4) are disjoint, that is, $t_1 < t_2 < t_3 < t_4$, then the two increments

$$X(t_2) - X(t_1) = \int_{t_1}^{t_2} Z(v) \, dv$$

and

$$X(t_4) - X(t_3) = \int_{t_3}^{t_4} Z(v) \, dv$$

are independent, as easily verified by taking expectation.

On the other hand, if $t_1 < t_2$, $t_3 < t_4$, but $t_3 < t_2$, that is, the intervals (t_1, t_2), (t_3, t_4) overlap, then the increments are correlated. Thus,

$$E\{[X(t_2) - X(t_1)][X(t_4) - X(t_3)]\}$$

$$= E\left\{\left[\int_{t_1}^{t_2} Z(v) \, dv\right]\left[\int_{t_3}^{t_4} Z(v) \, dv\right]\right\}$$

$$= 0 \quad \text{if} \quad t_2 \leq t_3 \qquad (6.5.17a)$$

$$= E\left\{\left[\int_{t_3}^{t_2} Z(v) \, dv\right]^2\right\}$$

$$= \sigma_z^2 (t_2 - t_3), \quad \text{if} \quad t_2 > t_3 \qquad (6.5.17b)$$

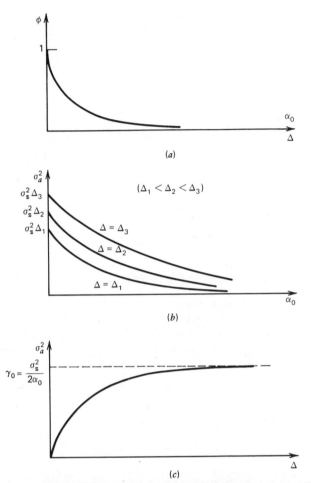

Figure 6.7 Limiting behavior of the sampled first order system. (a) ϕ versus A(1) model parameter α_0 and sampling interval Δ. (b) σ_a^2 versus A(1) model parameter α_0. (c) σ_a^2 versus sampling interval Δ.

As mentioned in Section 2.2.6, the random walk model was conjectured by Bachelier, in 1900, in connection with stock prices. The continuous time random walk (6.5.15) was used by Wiener and others in describing the Brownian motion of a particle under the random bombardment of the molecules of the fluid in which it is suspended. Hence, the continuous time random walk is also called a Bachelier-Wiener process. In this section we have presented a rather simple derivation of this process, which is generally derived as an independent increment process starting with Eqs. (6.5.17a) and (6.5.17b) as the definition [see, for example, Doob (1953)].

6.5.4 Physical Interpretation   *Read*

These limiting cases for the extreme values of the parameter α_0 of the system also have an intuitive physical interpretation, similar to that for the extreme values of Δ given earlier. The parameter α_0 represents the damping, inertia, or resistance to change in the equilibrium poisition of the system. The parameter σ_z^2 represents, in a probabilistic sense, the strength or intensity of the random forcing function, noise, or disturbance $Z(t)$ that causes the change and generates the response $X(t)$.

When α_0 increases, that is, the damping or the resistance to change increases, the change caused by a single disturbance is quickly damped and affects the response for a shorter time. Therefore, for large values of α_0, the system will have a very short memory or dynamics; in other words, observations at even short intervals will be uncorrelated, as the limit $\phi \rightarrow 0$ and the limiting AR(0) form (6.5.2) indicate.

However, if the damping or the resistance of the system measured by α_0 increases, whereas σ_z^2 measuring the strength or intensity of $Z(t)$ causing the change remains fixed, then the forcing function $Z(t)$ fails to cause appreciable change in the system. Thus, for very large values of α_0, there is practically no response, that is, the variance $\gamma_0 = \gamma(0)$ tends to zero as seen from Eq. (6.3.15), and so does σ_a^2 by (6.5.8). On the other hand, even with large α_0 if $Z(t)$ is so strong that it overcomes this large resistance and causes motion, that is, σ_z^2 is also large, then $X(t)$ tends to the white noise $Z'(t)$ as given by Eq. (6.5.13). The corresponding model is again an AR(0) model (6.5.2), but with a nonzero σ_a^2 by (6.5.14). As both the limiting models (6.5.2) and (6.5.13) show, there is no dynamics, memory, or correlation in such a stochastic system.

When α_0 decreases, that is, the damping or resistance to change is small, the change caused by a single disturbance affects the response for a long time. Hence, for small values of α_0, the system will have a very strong memory or dynamics, and observations at even long intervals will still be highly correlated, as the limit $\phi \rightarrow 1$ and the discrete time random walk (6.5.5) will result.

Recall that the homogeneous equation

$$\frac{dX(t)}{dt} = -\alpha_0 X(t)$$

can be interpreted as saying that the rate of change of $X(t)$ is proportional to $X(t)$ and opposite in sign. When the forcing function $Z(t)$ is introduced, the equation

$$\frac{dX(t)}{dt} = -\alpha_0 X(t) + Z(t)$$

says that the rate of change is also affected by the random component $Z(t)$. The term $-\alpha_0 X(t)$ acts like a restoring or damping force pulling the system back to its equilibrium or mean position, which we have assumed to be zero. As long as this term dominates, the response stays more or less close to the mean position fluctuating around it.

When α_0 becomes small, the random term $Z(t)$ starts dominating the rate of change. The noise $Z(t)$ may have either a positive or negative sign and so may produce a rate of change either toward or away from the equilibrium position in a random manner. When $-\alpha_0$ is close to zero, the rate of change is totally dominated by the random term $Z(t)$ as shown by Eq. (6.5.15) and therefore produces a random walk in which the response may drift away from the equilibrium position for long periods of time. Equation (6.5.15) says that the derivative of $X(t)$ is random, whereas its sampled representation Eq. (6.5.5) says that the difference between two observations at equal intervals is random. A comparison of Eqs. (6.5.16a) and (6.5.16b) and the one for the discrete time random walk following it clearly brings out the relation between the continuous and discrete processes.

6.5.5   How Large Is "Infinity" and How Small Is "Zero"?

It should be noted that the limits with respect to Δ or α_0 have been considered in this section for the values tending to infinity and zero for the sake of mathematical analysis and precision. Actually, due to the exponential relation $\phi = e^{-\alpha_0 \Delta}$, its convergence to the respective limiting values is very fast and occurs for moderately large and small values of Δ or α_0. Thus, for example, $e^{-5} = 0.006738$, $e^{-8} = 0.0003355$, and $e^{-10} = 0.0000454$ which are already close to the limiting value zero.

Second, from a finite amount of data, one cannot say whether the parameter is either zero or one with certainty. The best one can do is to see whether the confidence interval on the parameter ϕ includes the respective values. In the case of a zero value, one can check whether the parameter is zero by testing the adequacy of the model with the parameter dropped out. Therefore, the parameter values below 0.01 or, at the most, 0.001, and above 0.99 or, at the most, 0.999 may be taken as the limiting values in practice. Note that $e^{-4.6} = 0.010052$ and $e^{-0.01} = 0.990050$. Thus, for the limiting cases to occur in practice, neither is the infinity very "large" nor the zero very "small." This is especially important for the cases $\alpha_0 \to \infty$ and $\Delta \to \infty$ discussed above. For $\sigma_a^2 \neq 0$ or $\gamma_0 \neq 0$ in practice, it is not necessary that σ_z^2 increase with Δ or α_0 since ϕ approaches one much faster than σ_z^2 or γ_0 approaches zero with increasing α_0 or Δ. Thus, for moderately large values of α_0 or Δ the parameter ϕ is close to one, whereas σ_a^2 or γ_0 is not close to zero. (See Problems 6.6 and 2.18.)

6.5.6 Illustrative Example: IBM Stock Prices

To illustrate the limiting cases, let us consider the IBM stock price data, for which the AR(1) model with $\phi = 0.999$ and $\sigma_a^2 = 52.61$ was fitted in Chapter 2. Since the parameter ϕ is positive, this AR(1) model is also a sampled system model. Therefore, we can get the parameters of the A(1) model by Eqs. (6.4.7) and (6.4.8). If we take the sampling interval of one day as the unit of time so that $\Delta = 1$, then

$$\alpha_0 = -\frac{\ell n\ \phi}{\Delta} = -\ell n\ (0.999)$$
$$= 0.001$$

and

$$\sigma_z^2 = \frac{2\alpha_0\ \sigma_a^2}{(1 - \phi^2)} = \frac{2 \times 0.001 \times 52.61}{(1 - 0.999^2)}$$
$$= 52.66$$

Thus, the A(1) model for the IBM stock prices (after substracting $\overline{X}$) is

$$\frac{dX(t)}{dt} + 0.001\ X(t) = Z(t)$$

This example illustrates the limiting case of random walk as $\alpha_0 \rightarrow 0$, discussed in the last section. The parameter $\phi = 0.999$ verifies Eq. (6.5.9), whereas since $\Delta = 1$,

$$\sigma_a^2 = 52.61 \approx \sigma_z^2\ \Delta = \sigma_z^2 = 52.66$$

which corresponds to Eq. (6.5.10). The A(1) model given above can also be approximately represented by

$$\frac{dX(t)}{dt} = Z(t)$$

or

$$X(t) = \int_{-\infty}^{t} Z(v)\ dv$$

which is a continuous time random walk, and the continuous time process underlying the IBM stock prices is a Bachelier-Wiener process.

APPENDIX A 6.1

DIFFERENTIAL EQUATIONS AND THE EXPONENTIAL FUNCTION

Differential equations play the same role in continuous time as difference equations in discrete time. Recall that the derivative of a function is defined from its difference by

$$\frac{d}{dt} f(t) = \lim_{\Delta t \to 0} \frac{f(t + \Delta t) - f(t)}{\Delta t}$$

For the trivial constant function $f(t) = c$

$$\frac{d}{dt} f(t) = \lim_{\Delta t \to 0} \frac{c - c}{\Delta t} = 0 \qquad \text{(A 6.1.1)}$$

For the straight line $f(t) = ct$

$$\frac{d}{dt} f(t) = \lim_{\Delta t \to 0} \frac{c(t + \Delta t) - ct}{\Delta t} = \lim_{\Delta t \to 0} \frac{c\Delta t}{\Delta t} = c \qquad \text{(A 6.1.2)}$$

For the parabola $f(t) = ct^2$

$$\frac{d}{dt} f(t) = \lim_{\Delta t \to 0} \frac{c(t + \Delta t)^2 - ct^2}{\Delta t} = \lim_{\Delta t \to 0} \frac{c(t^2 + 2t\Delta t + \Delta t^2) - ct^2}{\Delta t}$$

$$= \lim_{\Delta t \to 0} (2ct + c\Delta t) = 2ct \qquad \text{(A 6.1.3)}$$

In general, for $f(t) = ct^n$ we can use the Binomial theorem to show that

$$\frac{d}{dt} f(t) = \lim_{\Delta t \to 0} \frac{c}{\Delta t} [(t + \Delta t)^n - t^n]$$

$$= \lim_{\Delta t \to 0} \frac{c}{\Delta t} \left[\left\{ t^n + nt^{n-1}\Delta t + \frac{n(n-1)}{2!} t^{n-2}\Delta t^2 + \dots \right\} - t^n \right] \qquad \text{(A 6.1.4)}$$

$$= \lim_{\Delta t \to 0} c \left[nt^{n-1} + \frac{n(n-1)}{2!} t^{n-2}\Delta t + \dots \right] = cnt^{n-1}$$

We can also take the derivative of a derivative to get higher order derivatives. Thus, for $f(t) = ct$

$$\frac{d^2}{dt^2} f(t) = \frac{d}{dt} \left[\frac{d}{dt} ct \right] = \frac{d}{dt} [c] = 0 \qquad \text{(A 6.1.5)}$$

Similarly, for $f(t) = ct^n$

$$\frac{d^{n+1}}{dt^{n+1}} f(t) = 0 \qquad \text{(A 6.1.6)}$$

Equations (A 6.1.1) to (A 6.1.6) are examples of simple linear differential equations with constant coefficients. The functions that satisfy them are solutions of these equations. Thus, $f(t) = c$ is a solution of

$$\frac{df(t)}{dt} = 0$$

Equations such as (A 6.1.1), (A 6.1.5), and (A 6.1.6) are examples of *homogeneous* equations whose left-hand side involves derivatives of arbitrary order or the function itself and whose right-hand side is zero. Equations such as (A 6.1.2) to (A 6.1.4), whose right-hand side is not zero, are *nonhomogeneous* equations.

A solution of a homogeneous equation always contains an arbitrary multiplicative constant. Thus,

$$f(t) = ct^2$$

is a solution of

$$\frac{d^3f}{dt^3} = 0$$

for any value of c. This arbitrariness can be eliminated by specifying an "initial condition." Thus, the only solution of

$$\frac{d^3f}{dt^3} = 0, \qquad f(0) = 1$$

is

$$f(t) = t^2$$

No such arbitrary constant is present in the solution of a nonhomogeneous equation. Thus, the only solution of

$$\frac{df}{dt} = 2t$$

is

$$f(t) = t^2$$

It is seen from this example that a nonhomogeneous equation can be replaced by a homogeneous equation with some initial condition so that both the equations have the same solution. Such a method is often followed in solving nonhomogeneous differential equations.

Now consider a slightly difficult homogeneous differential equation

$$\frac{df}{dt} = f \qquad \text{that is,} \qquad \frac{df}{dt} - f = 0$$

To solve this, we want a function that gives back the same function after differentiation. Other than the trivial solution $f(t) = 0$, the simplest function we can consider is

$$f(t) = 1$$

If it is differentiated, we do not get back 1. To get back 1, let us add t and consider

$$f(t) = 1 + t$$

Now, we do get back 1 but not t. We can get back t by adding $\dfrac{t^2}{2}$; thus, we have

$$f(t) = 1 + t + \frac{t^2}{2}$$

After differentiation, we are only missing $\dfrac{t^2}{2}$ that can be obtained by adding $\dfrac{t^3}{2 \times 3}$; thus, we arrive at

$$f(t) = 1 + t + \frac{t^2}{2} + \frac{t^3}{3 \times 2}$$

It is now clear that the desired function is the infinite sum

$$1 + t + \frac{t^2}{2!} + \frac{t^3}{3!} + \ldots + \frac{t^n}{n!} + \ldots$$

where

$$n! = n(n - 1)\,(n - 2) \ldots 3.2.1$$

This is the *exponential* function

$$e^t = \sum_{n=0}^{\infty} \frac{t^n}{n!} \tag{A 6.1.7}$$

which is a solution of

$$\frac{df}{dt} - f = 0$$

and this being a homogeneous equation, its general solution is

$$f(t) = ce^t$$

By similar reasoning, we can see that a solution of

$$\frac{df}{dt} = \mu f$$

is

$$e^{\mu t} = 1 + \mu t + \frac{(\mu t)^2}{2} + \ldots + \frac{(\mu t)^n}{n!} + \ldots \tag{A 6.1.8}$$

and hence the general solution is

$$f(t) = ce^{\mu t} \tag{A 6.1.9}$$

which was used in solving Eq. (6.1.1) of the text. This is the solution of the first order homogeneous equation

$$(D - \mu)f = 0 \tag{A 6.1.10}$$

where

$$D \equiv \frac{d}{dt}$$

From the definition (A 6.1.8) of the exponential function, it is easy to verify that

$$\frac{d^n}{dt^n}(e^{\mu t}) = \mu^n e^{\mu t} \qquad \text{(A 6.1.11)}$$

This property makes the exponential function most useful in solving differential equations.

Next, consider the second order homogeneous equation

$$\frac{d^2 f}{dt^2} + \alpha_1 \frac{df}{dt} + \alpha_0 f = 0$$

that is,

$$(D^2 + \alpha_1 D + \alpha_0)f = 0 \qquad \text{(A 6.1.12)}$$

that is,

$$(D - \mu_1)(D - \mu_2)f = 0$$

where μ_1 and μ_2 (assumed to be distinct) satisfy

$$(D^2 + \alpha_1 D + \alpha_0) = (D - \mu_1)(D - \mu_2)$$

and hence are the roots of the polynomial in D

$$D^2 + \alpha_1 D + \alpha_0 = 0 \qquad \text{(A 6.1.13)}$$

Hence, a function $f(t)$ will be a solution of Eq. (A 6.1.12) if it satisfies either

$$(D - \mu_1)f = 0 \qquad \text{or} \qquad (D - \mu_2)f = 0$$

However, we know from Eqs. (A 6.1.9) and (A 6.1.10) that the functions that satisfy these two equations are $C_1 e^{\mu_1 t}$ and $C_2 e^{\mu_2 t}$, and therefore the general solutions of the homogeneous Eq. (A 6.1.12) is

$$f(t) = C_1 e^{\mu_1 t} + C_2 e^{\mu_2 t} \qquad \text{(A 6.1.14)}$$

where C_1 and C_2 are arbitrary constants with at least one of them nonzero so that we have a nontrivial solution. As before, this arbitrariness can be eliminated by specifying two initial conditions. For example, if the conditions are

$$f(0) = 0, f'(0) = \frac{df}{dt}\bigg|_{t=0} = 1$$

then, since

$$f'(t) = C_1 \mu_1 e^{\mu_1 t} + C_2 \mu_2 e^{\mu_2 t}$$

we have

$$C_1 + C_2 = 0$$

$$C_1\mu_1 + C_2\mu_2 = 1$$

Therefore,

$$C_1 = \frac{1}{\mu_1 - \mu_2}, \qquad C_2 = -\frac{1}{\mu_1 - \mu_2}$$

It can be easily confirmed that (A 6.1.14) is indeed a solution of Eq. (A 6.1.12) by substituting (A 6.1.14) in (A 6.1.12). Equation (A 6.1.13) is the *characteristic equation* of the differential Equation (A 6.1.12) and μ_1, μ_2 are the *characteristic roots* or the *eigen-values*.

Since integration is the reverse of differentiation, the exponential function is equally easy to manipulate under integration. Thus, the property

$$\frac{d}{dt} e^{\mu t} = \mu e^{\mu t}$$

is expressed by the indefinite integral

$$\int e^{\mu t} = \frac{e^{\mu t}}{\mu} \qquad (A\ 6.1.14)$$

or more correctly by the definite integral

$$\int_a^b e^{\mu t} = \left[\frac{e^{\mu t}}{\mu}\right]_a^b = \frac{e^{\mu b} - e^{\mu a}}{\mu} \qquad (A\ 6.1.15)$$

Also, note that by the definition (A 6.1.7)

$$e^0 = 1$$

$$e^1 = e = 2.718281828$$

$$e^{-1} = \frac{1}{e} = 0.3678794412$$

$$1 - e^{-1} = 0.6321205588$$

$$e^{-4} = 0.0183156389$$

$$e^{-5} = 0.006737947$$

$$e^{-\infty} = \frac{1}{e^\infty} = 0$$

The values of e^0 and $e^{-\infty}$ are useful in obtaining the limiting cases, e^1 and e^{-1} provide a rough estimate of μ in $e^{\mu t}$ as the value of the function at $t = \pm\frac{1}{\mu}$, and e^{-4}, e^{-5} provide the values of t at which the function is "practically" zero.

Appendix A 6.2

PROPERTIES OF DIRAC DELTA FUNCTION

(1) For any continuous function $f(t)$,

$$\int_{-\infty}^{\infty} f(t-u)\delta(u)\ du = f(t) = \int_{-\infty}^{\infty} f(u)\delta(t-u)\ du \qquad (6.2.7)$$

Proof:

$$\int_{-\infty}^{\infty} f(t-u)\delta(u)\ du = \int_{-\infty}^{0^-} f(t-u)\delta(u)\ du$$

$$+ \int_{0^-}^{0^+} f(t-u)\delta(u)\ du + \int_{0^+}^{\infty} f(t-u)\delta(u)\ du$$

Since the delta function is zero everywhere except at $u = 0$, its product by any function $f(t-u)$ will also be zero except at $u = 0$, that is,

$$\int_{-\infty}^{\infty} f(t-u)\delta(u)\ du = 0 + \int_{0^-}^{0^+} f(t-0)\delta(u)\ du + 0$$

$$= f(t) \int_{0^-}^{0^+} \delta(u)\ du$$

$$= f(t)$$

from the property (6.2.4b) that

$$\int_{0^-}^{0^+} \delta(u)\ du = 1$$

By letting $t - u = v$, we obtain

$$\int_{-\infty}^{\infty} f(t-u)\delta(u)\ du = \int_{-\infty}^{\infty} f(v)\delta(t-v)\ dv$$

$$= f(t) \qquad (6.2.7)$$

(2) For any function $f(t)$ whose kth derivative is continuous,

$$\int_{-\infty}^{\infty} f(t-u)\delta^{(k)}(u)\ du = (-1)^k f^{(k)}(t) \qquad (6.2.8)$$

where the superscripted k in parentheses denotes the kth derivative, that is,

$$f^{(k)}(t) = \frac{d^{(k)}}{dt^{(k)}} [f(t)]$$

Proof: Using integration by parts, we get

$$\int_{-\infty}^{\infty} f(t-u)\delta^{(k)}(u)\ du = f(t-u)\delta^{(k-1)}(u)\Big|_{u=-\infty}^{u=+\infty}$$

$$-\int_{-\infty}^{\infty} f^{(1)}(t-u)\delta^{(k-1)}(u)\ du$$

$$= -\int_{-\infty}^{\infty} f^{(1)}(t-u)\delta^{(k-1)}(u)\ du$$

$$= -f^{(1)}(t-u)\delta^{(k-2)}\Big|_{u=-\infty}^{u=\infty}$$

$$+\int_{-\infty}^{\infty} f^{(2)}(t-u)\delta^{(k-2)}(u)\ du$$

$$= \int_{-\infty}^{\infty} f^{(2)}(t-u)\delta^{(k-2)}(u)\ du$$

$$\cdots$$

$$= (-1)^k \int_{-\infty}^{\infty} f^{(k)}(t-u)\delta(u)\ du$$

Using Eq. (6.2.7), we obtain

$$\int_{-\infty}^{\infty} f^{(k)}(t-u)\delta(u)\ du = (-1)^k f^{(k)}(t) = (-1)^k \frac{d^{(k)}}{dt^{(k)}}[f(t)]   \quad (6.2.8)$$

PROBLEMS

6.1   Show that the following functions satisfy (6.2.4a) and (6.2.4b), and therefore they are also delta functions.
(i) $e^{3t}\delta(t)$ (ii) $\cos(4t)\delta(t)$ (iii) $5^t\delta(t)$

6.2   Check that (6.3.11a) is a solution of the nonhomogeneous differential equation (6.3.10a).

6.3   Considering the AR(1) models in Tables 4.2 to 4.4 as fairly good approximations, find the time constant and the parameter σ_z^2 for each of the following systems, taking $\Delta = 1$.
(i)   IBM stock prices;
(ii)   Papermaking process;
(iii)   Grinding wheel profile.
Which of these may be taken as random walk with, say, a 95% confidence?

6.4   An A(1) system with the model

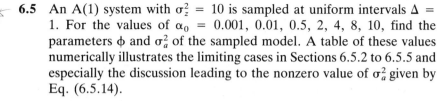

$$\frac{dX(t)}{dt} + 5.5X(t) = Z(t), \qquad \sigma_z^2 = 10$$

is sampled at uniform intervals Δ.
(a)   Find the sampled discrete model for each of the following sampling intervals: $\Delta = 1, 0.1, 0.01, 0.001$.
(b)   Suppose the estimated value of ϕ in the sampled first order model lies within ± 0.01 of the true value. For which sampling intervals is one likely to conclude that the system is a random walk or independent increment process, considering only the discrete model?

6.5   An A(1) system with $\sigma_z^2 = 10$ is sampled at uniform intervals $\Delta = 1$. For the values of $\alpha_0 = 0.001, 0.01, 0.5, 2, 4, 8, 10$, find the parameters ϕ and σ_a^2 of the sampled model. A table of these values numerically illustrates the limiting cases in Sections 6.5.2 to 6.5.5 and especially the discussion leading to the nonzero value of σ_a^2 given by Eq. (6.5.14).

6.6   An engineer suspects that her system follows a continuous time random walk and wants to confirm this from a uniformly sampled set of data. She decides to conclude that the system is a random walk if the confidence interval on ϕ includes 1. If her sampling interval is $\Delta = 0.2$, what is the largest value of α_0 for which she will still conclude that the system is a continuous time random walk, assuming that the estimated value of ϕ is, in fact, the true value and that the confidence limits are ± 0.01?

6.7   (a)   An A(1) system that is very close to continuous time random walk, with $\sigma_z^2 = 1$, is sampled at uniform intervals $\Delta = 0.2$. Give the form of the sampled process and its parameters, including σ_a^2.
(b)   For an AR(1) model $\sigma_a^2 = 7.5$ and $\gamma_0 = 10$. Give the parameters of the AR(1) model and those of the corresponding A(1) model, assuming the sampling interval $\Delta = 1$.

6.8   An A(1) system has $\alpha_0 = 5.1$. If the sampling interval is changed from 0.1 to 0.2, what percentage of change in the residual sum of squares can you expect in modeling uniformly sampled data with the same number of observations?

SECOND ORDER SYSTEM AND RANDOM VIBRATION

The most important and widely applicable stochastic system is the one governed by a second order linear differential equation with constant coefficients. In this chapter we will treat such a system formulated as random vibration, by considering the response of a spring-mass-dashpot system to a continuous time white noise input or forcing function. By themselves, random vibrations are important in many engineering systems, and the treatment in this chapter provides a time domain approach to their analysis, in contrast to the existing frequency domain approach.

What is more important, however, is that any other scalar second order system may also be hypothesized as random vibration and analyzed on the basis of the detailed consideration given here. The interpretation of the two coefficients of a second order differential equation as the restoring and damping forces is useful for many systems. On the basis of such an interpretation, it becomes easier to analyze the effect of various constituents of the system on the response, which, in turn, leads to the necessary information for design and control.

It should be emphasized that although for the purpose of mathematical derivation and analysis we start with the formulation of a second order random vibration equation, the problem we want to tackle will be the reverse. From a given set of discrete data, we want to find a model in the form of a differential equation. When this equation turns out to be of the second order, then we can interpret it as random vibration and use the analogy to analyze the parameters in terms of the system constituents.

The formulation of the mass-spring-dashpot system of vibration and the solution of the corresponding homogeneous equation together with its stability are discussed in Section 7.1. The nonhomogeneous equation for the second order autoregressive stochastic system $A(2)$ is formulated in Section 7.2 by introducing the white noise forcing function. The Green's function and the autocovariance function of the $A(2)$ system are also derived in Section 7.2, and the response for simpler inputs such as an impulse or step function are discussed for illustrative purposes. The discrete representation for the uniformly sampled $A(2)$ system and the expression for its parameters are derived in Section 7.3. The stability region for the discrete representation is considered in Section 7.4. The modeling of the differential equation from the discrete data is discussed in Section 7.5, with special reference to the uniqueness of the model. The effects of the sampling interval, the natural frequency, and the damping factor on the parameters of the discrete model are derived and illustrated in Section 7.6. Section 7.7 presents a simple experimental demonstration and an illustrative example.

7.1 DIFFERENTIAL EQUATION FOR A DAMPED SPRING MASS SYSTEM

Consider a spring-mass-dashpot system, as shown schematically in Fig. 7.1. A mass M is subjected to a forcing function $f(t)$, and the motion caused by $f(t)$ is opposed by the spring and the dashpot. Let $X(t)$ denote the distance of the mass from the equilibrium position and assume that $X(0) = 0$. If we assume that the distance from the equilibrium position is small, then the variation of gravitational force may be neglected, and the force due to the spring may be considered proportional to the displacement $X(t)$, and that due to the dashpot proportional to the velocity $\dfrac{dX(t)}{dt}$. The forces measured in the direction of $f(t)$ and their signs are

$\qquad f(t)$: forcing function
$\qquad -KX(t)$: spring force with K the spring constant
$\qquad -C\dfrac{dX(t)}{dt}$: damping force with C the dashpot constant

The positive direction of motion $X(t)$ of the mass M and the free body diagram of the system are shown in Fig. 7.1.

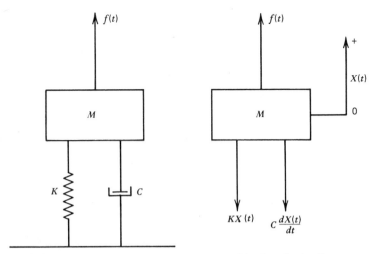

Figure 7.1 Spring-mass-dashpot system and its free body diagram.

7.1.1 Formulation of the Nonhomogeneous Equation

By Newton's law we get the equation of the motion of the mass M as

$$f(t) - KX(t) - C\frac{dX(t)}{dt} = M\frac{d^2X(t)}{dt^2}$$

or

$$\frac{d^2X(t)}{dt^2} + \frac{C}{M}\frac{dX(t)}{dt} + \frac{K}{M}X(t) = \frac{1}{M}f(t) \tag{7.1.1}$$

Let

$$\omega_n^2 = \frac{K}{M}, \quad \text{or} \quad \omega_n = \sqrt{\frac{K}{M}} \tag{7.1.2}$$

and

$$\zeta = \frac{C}{2\sqrt{KM}} = \frac{\text{actual damping}}{\text{critical damping}} \tag{7.1.3}$$

Then ω_n is called the natural frequency and ζ the damping ratio of the spring-mass-dashpot vibration system, and Eq. (7.1.1) now takes the familiar form

$$\frac{d^2X(t)}{dt^2} + 2\zeta\,\omega_n\frac{dX(t)}{dt} + \omega_n^2\,X(t) = \frac{1}{M}f(t)$$

or

$$(D^2 + 2\zeta\,\omega_n D + \omega_n^2)X(t) = \frac{1}{M}f(t) \tag{7.1.4}$$

The reason for the transformations (7.1.2) and (7.1.3) and the terminology introduced above will become clear in the next section. Equation (7.1.4) represents the forced vibration of a one-degree-of-freedom vibration system subjected to the forcing function $f(t)$.

7.1.2 Solution of the Homogeneous Equation

The homogeneous equation for the one-degree-of-freedom vibration system is

$$(D^2 + 2\zeta\,\omega_n D + \omega_n^2)\,X(t) = 0 \tag{7.1.5}$$

Since this is a second order linear differential equation with constant coefficients, its solution is a sum of two exponentials [see Eq. (A 6.1.14)]:

$$X(t) = C_1 e^{\mu_1 t} + C_2 e^{\mu_2 t} \tag{7.1.6}$$

where C_1 and C_2 are arbitrary constants, which can be determined by the initial conditions, and μ_1 and μ_2 are the distinct characteristic roots, which are the solutions of the characteristic equation

$$(D^2 + 2\zeta\,\omega_n D + \omega_n^2) = (D - \mu_1)(D - \mu_2) = 0 \qquad (7.1.7)$$

that is,

$$\mu_1 = -\zeta\omega_n + \omega_n\sqrt{\zeta^2 - 1} \qquad (7.1.7a)$$

$$\mu_2 = -\zeta\omega_n - \omega_n\sqrt{\zeta^2 - 1} \qquad (7.1.7b)$$

When $\zeta \neq 1$, that is, the actual damping is not equal to the critical damping and the roots μ_1 and μ_2 are distinct, then the solution (7.1.6) may be written as

$$X(t) = e^{-\zeta\omega_n t}\left(C_1 e^{(\omega_n\sqrt{\zeta^2-1})t} + C_2 e^{-(\omega_n\sqrt{\zeta^2-1})t}\right) \qquad (7.1.8)$$

For $\zeta > 1$, $X(t)$ is a sum of exponentials, whereas for $\zeta < 1$, it can represent a sine wave damped when $\zeta \neq 0$ and undamped when $\zeta = 0$. If, however, the characteristic roots are equal, that is, if $\zeta = 1$ or the actual damping is equal to the critical damping [see Eq. (7.1.3)], then the solution is given by

$$X(t) = C_1 e^{-\omega_n t} + C_2 t e^{-\omega_n t} \qquad (7.1.9)$$

7.1.3 Stability

Equations (7.1.8) and (7.1.9) give a complete solution of the homogeneous equation (7.1.5) and represent the response of the vibration system, respectively, for $\zeta \neq 1$ and $\zeta = 1$. This response immediately gives us the stability conditions. It is easy to see from Eq. (7.1.8) that the response tends to zero asymptotically with increasing t as long as $\zeta\omega_n > 0$, which is also true for Eq. (7.1.9) since $e^{-\omega_n t}$ tends to zero faster than t tends to infinity. Therefore, for $\zeta\omega_n > 0$ the system is clearly asymptotically stable. Similarly, when $\zeta\omega_n < 0$, it is clear that the system is unstable.

It is when $\zeta\omega_n = 0$ that the difference between the stability behavior of (7.1.8) and (7.1.9) becomes important. Recall a similar difference in the stability of a second order discrete model, discussed in Section 3.1.9, between equal and distinct roots, when both have absolute value one. Note that by Eqs. (7.1.2), (7.1.3), and (7.1.7),

$$\zeta\omega_n = \frac{C}{2M} = -Re(\mu_1) = -Re(\mu_2) \qquad (7.1.10)$$

each term representing half the total damping of the system per unit mass, per unit velocity.

Now $\zeta\omega_n$ may be zero because $\zeta = 0$, but $\omega_n \neq 0$. This is possible only when $C = 0$, $K \neq 0$ in Eq. (7.1.1) and we have an undamped vibration system. In that case Eq. (7.1.8) gives the solution, and if we put $\zeta = 0$, this equation takes the form

$$X(t) = C_1 e^{i\omega_n t} + C_2 e^{-i\omega_n t}$$
$$= (C_1 + C_2) \cos \omega_n t + i(C_1 - C_2) \sin \omega_n t$$

where

$$i = \sqrt{-1}$$

Since the response $X(t)$ is real, the constants C_1 and C_2 must be complex conjugate so that $(C_1 + C_2)$ is twice the real part, and $C_1 - C_2$ is twice the imaginary part times i. Let

$$C_1 = a + ib, \qquad C_2 = a - ib$$

then

$$X(t) = 2A \sin (\omega_n t + \beta) \tag{7.1.11}$$

where

$$A = \sqrt{a^2 + b^2}$$
$$\beta = \tan^{-1}\left(-\frac{a}{b}\right)$$

Equation (7.1.11) shows that the response is an undamped sine wave with angular frequency ω_n radians per unit time, the phase β and amplitude A being dependent upon the initial conditions that determine C_1 and C_2, that is, a and b. Thus, when $\zeta = 0$, $\omega_n \neq 0$, the system is stable but not asymptotically stable since the response remains finite but does not decay to the equilibrium position zero. This result is, of course, clear from the physical considerations of the spring-mass-dashpot system. When there is no damping and the system is disturbed to some initial conditions, it will perform a perpetual undamped vibration due to the spring.

In contrast, when $\zeta \neq 0$, $\zeta < 1$, starting from Eq. (7.1.8) and following the above steps, we have

$$X(t) = 2A e^{-\zeta \omega_n t} \sin (\omega_n \sqrt{1 - \zeta^2}\, t + \beta) \tag{7.1.12}$$

a damped sine wave which eventually decays to zero giving an asymptotically stable system. This is the reason why ω_n, which is the frequency of the undamped motion (7.1.11), is called the natural frequency and $\omega_n \sqrt{1 - \zeta^2}$ in Eq. (7.1.12) is called the damped (natural) frequency of the system. The critical damping for which the motion changes from a damped sine wave to damped exponentials is

$$\zeta = 1, \quad \text{or} \quad C = 2\sqrt{KM}$$

which explains the terminology of Eq. (7.1.3). A system with a response given by a sum of damped exponentials is, of course, asymptotically stable.

On the other hand, if $\zeta \omega_n$ is zero because both ζ and ω_n are zero, that is, both the spring constant C and the dashpot constant K are zero, then both the roots μ_1 and μ_2 are zero. The solution is then given by Eq. (7.1.9) which takes the form

$$X(t) = C_1 + C_2 t \tag{7.1.13}$$

for $\omega_n = 0$. Note the similarity of this expression with the Green's function G_j for the limiting AR(2) model (3.1.22), which has $\lambda_1 = \lambda_2 = 1$. Equation (7.1.13) clearly shows that the system is unstable.

The stability conditions of the continuous time second order system may now be summarized as follows:

$$Re(\mu_1) < 0, \; Re(\mu_2) < 0, \quad \text{Asymptotically stable} \tag{7.1.14a}$$

$$\left. \begin{array}{l} Re(\mu_1) \leq 0, \; Re(\mu_2) \leq 0, \\ \text{and, if } \mu_2 = \mu_1, \text{ then } Re(\mu_1) = Re(\mu_2) < 0 \end{array} \right\} \quad \text{Stable} \tag{7.1.14b}$$

Compare these conditions with Eq. (3.1.23). As in the discrete case, unless we want to specifically distinguish between asymptotic stability and stability, we will take Eq. (7.1.14a) as the condition of stability. Henceforth we will also assume that the roots are distinct, that is, $\zeta \neq 1$, since the case $\zeta = 1$ may be obtained by a limiting argument from the results of distinct roots. A graphical illustration of the various solutions discussed above under specific forcing functions will be given in Section 7.2.

7.1.4 An Experimental Example

As an example, consider an experimental spring-mass-dashpot system with

$$M = 0.0023 \text{ lb-sec}^2/\text{in.}$$
$$K = 1 \text{ lb/in.}$$
$$C = 0.0440 \text{ lb-sec/in.}$$

Then by Eqs. (7.1.2) and (7.1.3)

$$\omega_n^2 = \frac{K}{M} = \frac{1}{0.0023} = 434.78 \text{ rad/sec}$$

$$\therefore \omega_n = 20.85 \text{ rad/sec}$$

$$\zeta = \frac{C}{2\sqrt{KM}} = \frac{0.044}{2\sqrt{0.0023}} = 0.459 < 1$$

The homogeneous equation is

$$(D^2 + 19.13D + 434.78)X(t) = 0$$

and

$$\mu_1 = -\zeta\omega_n + \omega_n\sqrt{\zeta^2 - 1} = -9.570 + i\,18.524$$

$$\mu_2 = -\zeta\omega_n - \omega_n\sqrt{\zeta^2 - 1} = -9.570 - i\,18.524$$

The solution of the homogeneous equation by Eqs. (7.1.6), (7.1.8), and (7.1.12) is

$$X(t) = C_1 e^{\mu_1 t} + C_2 e^{\mu_2 t}$$

$$= e^{-9.570t} (C_1 e^{i18.524t} + C_2 e^{-i18.524t})$$

$$= 2A e^{-9.570t} \sin (18.524t + \beta)$$

Thus, $X(t)$ is a damped sine wave, which eventually decays to zero giving an asymptotically stable system.

Also,

$$Re(\mu_1) = Re(\mu_2) = -9.570 < 0$$

which again shows that the system is asymptotically stable.

7.2 SECOND ORDER AUTOREGRESSIVE SYSTEM A(2)

To formulate a continuous time second order autoregressive stochastic system, we start with the homogeneous equation (7.1.5), which for the sake of uniformity of notation of $A(n)$ systems we write as

$$(D^2 + \alpha_1 D + \alpha_0)X(t) = 0 \qquad (7.2.1)$$

where

$$\alpha_1 = 2\zeta\omega_n = \frac{C}{M} \qquad (7.2.2)$$

$$\alpha_0 = \omega_n^2 = \frac{K}{M} \qquad (7.2.3)$$

The A(2) system equation is obtained by introducing the forcing function $Z(t)$ as

$$(D^2 + \alpha_1 D + \alpha_0)X(t) = Z(t) \qquad (7.2.4)$$

$$E[Z(t)] = 0$$

$$E[Z(t)Z(t - u)] = \sigma_z^2 \delta(u)$$

It should be noted that when we fit the A(2) model (7.2.1) to the data, α_0 and α_1 are obtained as regression coefficients. The break-up of these coefficients in ζ and ω_n is for convenience of interpretation of the response as a random vibration. Equations (7.2.2) and (7.2.3) provide the break-up and also equate α_1 with the system damping force per unit mass per unit velocity, and α_0 with the restoring force per unit mass per unit displacement. These equations are useful in interpreting the visual records of the data and relating the model to the physical constituents of the system. However, this analogy should not be stretched too far. Unless the record comes from an actual mechanical vibration system, the physical interpretation based on it is only an abstraction and should be used with care in dealing with the properties and characteristics of the system on which the data has been taken.

7.2.1 The Green's Function of the A(2) System

The solution of the nonhomogeneous equation or the orthogonal (Wold's) decomposition of the stochastic-process/time-series $X(t)$ may be symbolically written as

$$X(t) = (D^2 + \alpha_1 D + \alpha_0)^{-1} Z(t)$$

$$= \int_0^\infty G(v)\, Z(t-v)\, dv$$

$$= \int_{-\infty}^t G(t-v)\, Z(v)\, dv \tag{7.2.5}$$

As in the case of the A(1) system, it follows from the property (6.2.7) of the delta function that the Green's function $G(t)$ is defined by the equivalent relations

$$(D^2 + \alpha_1 D + \alpha_0)G(t) = \delta(t) \tag{7.2.6a}$$

$$G(t) = \int_0^\infty G(v)\delta(t-v)\, dv \tag{7.2.6b}$$

The nonhomogeneous equation (7.2.6a) can be solved by reducing it to a homogeneous equation with initial conditions. The homogeneous equation is clearly (7.2.1). The initial conditions can be obtained by considering the continuity behavior of $G(t)$ and its derivatives from Eq. (7.2.6a). Since the delta function input is zero up to time $t = 0$,

$$G'(t) = G(t) = 0, \qquad t < 0$$

At $t = 0$, $G''(t)$, the second derivative of $G(t)$, contains the same discontinuity as that of a delta function. Therefore, $G'(t)$, which is the integral of $G''(t)$, contains the same discontinuity as that of the integral of the delta function, which is a unit step function. Hence,

$$G'(0) = 1$$

Similarly, $G(t)$, the integral of $G'(t)$, behaves at $t = 0$ like the integral of the step function, which is the ramp $r(t)$ (see Section 6.2). Therefore, like $r(t)$, $G(t)$ is continuous at $t = 0$ and hence

$$G(0) = 0$$

Thus, the nonhomogeneous equation (7.2.6a) is equivalent to the homogeneous equation

$$(D^2 + \alpha_1 D + \alpha_0)G(t) = 0$$

with initial conditions

$$G(0) = 0, \qquad G'(0) = 1$$

Since the solution of the homogeneous equation is

$$G(t) = C_1 e^{\mu_1 t} + C_2 e^{\mu_2 t}$$

where

$$(D^2 + \alpha_1 D + \alpha_0) = (D - \mu_1)(D - \mu_2) = 0 \qquad (7.2.7a)$$

that is,

$$\mu_1, \mu_2 = \frac{1}{2}(-\alpha_1 \pm \sqrt{\alpha_1^2 - 4\alpha_0}) = \omega_n(-\zeta \pm \sqrt{\zeta^2 - 1}) \qquad (7.2.7b)$$

substituting the initial conditions, we have

$$C_1 + C_2 = 0$$

$$C_1\mu_1 + C_2\mu_2 = 1$$

Solving these, we get

$$C_1 = \frac{1}{\mu_1 - \mu_2}$$

$$C_2 = -\frac{1}{\mu_1 - \mu_2}$$

and this gives

$$G(t) = \frac{e^{\mu_1 t} - e^{\mu_2 t}}{\mu_1 - \mu_2}, \qquad \text{for} \quad t \geqslant 0$$

$$= 0, \qquad \text{otherwise} \qquad (7.2.8)$$

It can be verified by substitution that Eq. (7.2.8) satisfies Eq. (7.2.6a) and is therefore the solution. (This may also be alternatively obtained by Laplace transform.)

Since the Green's function $G(t)$ is the response or output of the system when the input is the unit impulse function, it is often called the impulse response function. Physically it is the motion of the mass in the vibration system when a sudden shock is applied to it. When $\zeta \geqslant 1$, the roots μ_1 and μ_2 are real and the impulse response (7.2.8) is a linear combination of exponentials as shown in Fig. 7.2 and Fig. 7.3. Figure 7.2 shows how two exponentials are combined to give the impulse response when $\zeta > 1$. Note that the response stays on the same side as the impulse and decays

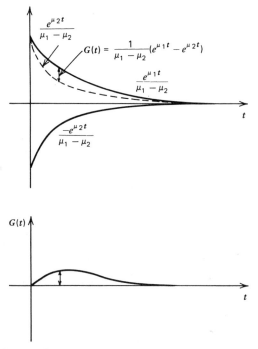

Figure 7.2 Green's function or impulse response of a second order system
$(\zeta > 1)$. (Compare with Fig. 3.7.)

to zero exponentially. When $\zeta < 1$, the roots μ_1 and μ_2 are complex and
$G(t)$ is a damped sine wave. Thus, we have

$$G(t) = \frac{e^{-\zeta\omega_n t}}{2\omega_n\sqrt{\zeta^2 - 1}}\left(e^{\omega_n\sqrt{\zeta^2 - 1}t} - e^{-\omega_n\sqrt{\zeta^2 - 1}t}\right)$$

$$= \frac{e^{-\zeta\omega_n t}}{\omega_n\sqrt{\zeta^2 - 1}}\sinh\left(\omega_n\sqrt{\zeta^2 - 1}\,t\right), \qquad \zeta > 1 \qquad (7.2.9a)$$

$$= \frac{e^{-\zeta\omega_n t}}{(\omega_n\sqrt{1 - \zeta^2}}\sin\left(\omega_n\sqrt{1 - \zeta^2}t\right), \qquad \zeta < 1 \qquad (7.2.9b)$$

In particular, for $\zeta = 0$, Eq. (7.2.9b) shows that $G(t)$ is an undamped sine
wave. These impulse reponses are also shown in Fig. 7.3 Note that Eqs.
(7.2.9a) and (7.2.9b) are special cases of Eqs. (7.1.8) and (7.1.12), respectively. These responses shown in Fig. 7.3 also illustrate graphically the
distinction between asymptotic stability and stability. In contrast to the
case of $\zeta \geq 1$, the response undershoots below zero and oscillates around
it when $\zeta < 1$.

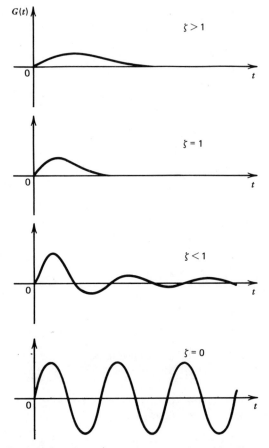

Figure 7.3 Green's function or impulse response of a second order system.

For the experimental spring-mass-dashpot example of Section 7.1.4, we have

$$\alpha_1 = 2\zeta\omega_n = \frac{C}{M} = 19.13$$

$$\alpha_0 = \omega_n^2 = \frac{K}{M} = 434.78$$

and the Green's function by Eq. (7.2.9b) is:

$$G(t) = \frac{e^{-9.570t}}{18.524} \sin (18.524t) \qquad (\zeta < 1)$$

which is a damped sine wave, decaying to zero exponentially.

Now if ω_n was 20.85 rad/sec, but ζ is changed to 1.2, then

$$\alpha_1 = 2\zeta\omega_n = 2 \times 1.2 \times 20.85 = 50.04$$

$$\alpha_0 = \omega_n^2 = 434.78$$

Since $\zeta > 1$, the Green's function is obtained from Eq. (7.2.9a) as

$$G(t) = \frac{e^{-\zeta\omega_n t}}{\omega_n\sqrt{\zeta^2 - 1}} \sinh (\omega_n\sqrt{\zeta^2 - 1}t)$$

$$= \frac{e^{-25.02t}}{13.83} \sinh (13.83t)$$

which is a damped hyperbolic sine wave or a linear combination of two exponentials, as shown in Fig. 7.2.

7.2.2 The Solution of the Nonhomogeneous Second Order Equation

As indicated in the last chapter, the Green's function allows us to solve the nonhomogeneous equation with an arbitrary forcing function by expressing the solution as a convolution. The solution of the general second order nonhomogeneous equation (7.1.4) can be expressed as

$$X(t) = \int_0^\infty \frac{(e^{\mu_1 v} - e^{\mu_2 v})}{(\mu_1 - \mu_2)} \cdot \frac{1}{M} f(t - v) \, dv \qquad (7.2.10)$$

In particular, when $\frac{1}{M}f(t) = \delta(t)$, we have $X(t) = G(t)$ by the property of the delta function, and this was used in deriving $G(t)$.

For illustrative purposes, let us find the response when $f(t)$ is a simple step forcing function, say,

$$f(t) = 0, \qquad t < 0$$
$$= F, \qquad t \geq 0$$

If we substitute it in Eq. (7.2.10), the integral reduces to

$$X(t) = \frac{F}{M} \int_0^t \frac{e^{\mu_1 v} - e^{\mu_2 v}}{\mu_1 - \mu_2} \, dv$$

$$= \frac{F}{M\mu_1\mu_2} \left[\frac{\mu_2 e^{\mu_1 v} - \mu_1 e^{\mu_2 v}}{\mu_1 - \mu_2} \right]_0^t$$

$$= \frac{F}{K} \left[\frac{\mu_2 e^{\mu_1 t} - \mu_1 e^{\mu_2 t}}{\mu_1 - \mu_2} + 1 \right] \qquad (7.2.11)$$

since $K = M\omega_n^2 = M\mu_1\mu_2$. When the real part of μ_1 or μ_2 is negative, that is, when $\zeta\omega_n > 0$ or $\zeta > 0$, the first part in the square brackets, which is called the transient response, dies out; the response eventually stabilizes to the steady state value F/K. When $\zeta = 0$, the system performs an undamped sinusoidal vibration around the steady state value. The response given by Eq. (7.2.10) for different values of ζ is shown in Fig. 7.4.

For example, consider the experimental spring-mass-dashpot system. The step response by Eq. (7.2.11) is

$$
\begin{aligned}
X(t) &= \frac{F}{K}\left[\frac{\mu_2 e^{\mu_1 t} - \mu_1 e^{\mu_2 t}}{\mu_1 - \mu_2} + 1\right] \\
&= \frac{F}{K}\left[\frac{(-9.570 - 18.524i)e^{(-9.570+18.524i)t}}{2 \times 18.524i}\right. \\
&\qquad \left. \frac{-(-9.570 - 18.524i)e^{(-9.570+18.524i)t}}{}\right] + \frac{F}{K} \\
&= \frac{F}{K}\left[e^{-9.570t}\right]\left[\cos 18.524t - \frac{9.570}{18.524}\sin 18.524t\right] + \frac{F}{K}
\end{aligned}
$$

which is a damped sine wave ($\zeta < 1$), decaying eventually to the steady state value F/K.

7.2.3 Orthogonal Decomposition and Autocovariance of the A(2) System

When the forcing function $(1/M)f(t)$ is equal to the white noise $Z(t)$, we get the A(2) system for which the solution or the orthogonal decomposition may now be written from Eq. (7.2.5) or (7.2.10)

$$
\begin{aligned}
X(t) &= \int_0^\infty \frac{e^{\mu_1 v} - e^{\mu_2 v}}{\mu_1 - \mu_2} Z(t-v)\, dv \\
&= \int_{-\infty}^t \frac{e^{\mu_1(t-v)} - e^{\mu_2(t-v)}}{\mu_1 - \mu_2} Z(v)\, dv \qquad (7.2.12)
\end{aligned}
$$

This integral may be viewed as a superposition or summation of impulse responses generated by impulses $Z(t)$, which have random strengths (or areas under the approximating pulses), each with mean zero and variance σ_z^2. This is graphically represented in Fig. 7.5, where the impulses are shown at finite distances for clarity in drawing; actually, they come in continuously over time. Each impulse of random strength $Z\delta(t)$ generates its own response $ZG(t)$ and the actual response $X(t)$ is a limiting summation or integration of such responses.

From the orthogonal decomposition (7.2.12) or directly by substituting $G(t)$ given by Eq. (7.2.8) in the general formula (6.3.13), we get the covariance function of the A(2) system. It can be verified that the autocovariance

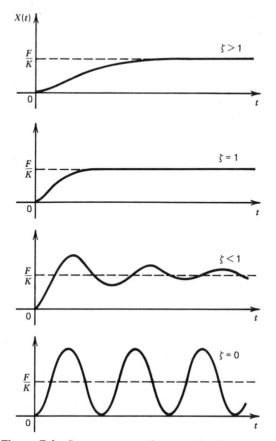

Figure 7.4   Step response of a second order system.

$$\gamma(s) = \sigma_z^2 \int_0^\infty G(v)G(v+s)\, dv$$

$$= \frac{\sigma_z^2}{2\mu_1\mu_2(\mu_1^2 - \mu_2^2)} (\mu_2 e^{\mu_1 s} - \mu_1 e^{\mu_2 s}) \qquad (7.2.13)$$

and the variance

$$\gamma(0) = -\frac{\sigma_z^2}{2\mu_1\mu_2(\mu_1 + \mu_2)} = \frac{\sigma_z^2}{4\zeta\omega_n^3} \qquad (7.2.14)$$

which gives the autocorrelation

$$\rho(s) = \frac{\mu_2 e^{\mu_1 s} - \mu_1 e^{\mu_2 s}}{\mu_2 - \mu_1} \qquad (7.2.15)$$

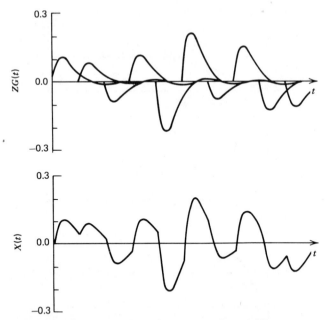

Figure 7.5   Generation of response of an A(2) system.

It follows from the expression for the variance (7.2.14) that as the real part of μ_1 or μ_2 approaches zero, that is, as ζ and/or ω_n tends to zero, the variance becomes infinite and the stochastic system tends to be unstable. This is in agreement with our earlier stability analysis based purely on the deterministic homogeneous equation. Intuitively, this makes sense; as the restoring and/or damping forces become smaller, the probability of the system moving far away from the equilibrium position increases.

Expressions (7.2.13) and (7.2.15) are applicable for nonnegative s; for negative s they are, as usual, defined by $\gamma(-s) = \gamma(s)$ and $\rho(-s) = \rho(s)$. Therefore, over the entire range of s from $-\infty$ to $+\infty$ they may be written as

$$\gamma(s) = \frac{\sigma_z^2}{2\mu_1\mu_2(\mu_1^2 - \mu_2^2)} \left(\mu_2 e^{\mu_1|s|} - \mu_1 e^{\mu_2|s|}\right) \quad -\infty < s < \infty \quad (7.2.13a)$$

and

$$\rho(s) = \frac{\mu_2 e^{\mu_1|s|} - \mu_1 e^{\mu_2|s|}}{\mu_2 - \mu_1} \quad (7.2.15a)$$

For the experimental spring-mass-dashpot example, we have

$$\rho(s) = e^{-9.570|s|} \left[\cos(18.524|s|) - \frac{9.570}{18.524} \sin(18.524|s|) \right]$$

which is a damped sine wave function of s approaching zero as s approaches infinity.

7.2.4 Physical Examples of A(2) Systems

Although in this chapter we will be concerned with obtaining a model such as A(2) from the data by conditional regression, there are examples of physical systems for which the A(2) model can be conjectured considering the physics of a system. One such example is the Brownian motion of a particle suspended in a fluid. This particle is constantly bombarded by the molecules in the liquid and traverses a path known as the Brownian motion, which can be represented by the A(2) system. The "purely random" force due to the fluctuations in the number of molecular collisions can be represented by $Z(t)$, which is the forcing function causing the Brownian displacement $X(t)$. The frictional forces opposing the motion are proportional to the velocity and can be represented as a damping force by the term $2\zeta\omega_n[dX(t)/dt]$. Similarly, the elastic forces acting on the particle and proportional to the displacement are represented by the term $\omega_n^2 X(t)$.

A phenomenon closely related to the Brownian motion occurs in electrical circuits. The thermal motion of the electrons produces fluctuations in the current and voltage that are called "thermal noise." This thermal noise can be very closely represented by $Z(t)$. When such a noise passes through a circuit containing a resistance, a capacitance, and an inductance, the output fluctuations can be represented by an A(2) system. The RLC circuit is then represented by the homogeneous part of the A(2) equation.

* 7.2.5 Spectrum of the A(2) System

Taking the Fourier transform of $(7.2.13a)$, we can show that the spectrum or autospectrum of the A(2) system is given by

$$f(\omega) = \frac{\sigma_z^2}{2\pi} \frac{1}{|(i\omega)^2 + \alpha_1(i\omega) + \alpha_0|^2}$$

$$= \frac{\sigma_z^2}{2\pi} \frac{1}{[(\omega^2 - \omega_n^2)^2 + 4\zeta^2\omega_n^2\omega^2]} \qquad (7.2.16)$$

$$-\infty < \omega < \infty$$

Since the frequency decomposition is available over the entire frequency band, rather than $-\dfrac{\pi}{\Delta}$ to $\dfrac{\pi}{\Delta}$ as in the discrete case, the continuous time spectrum provides much more complete information in the frequency domain. The A(2) model can be fitted to the data and the parameters α_0, α_1, and σ_z^2 estimated by the methods discussed in Section 7.5.5. Once these parameters are estimated, the estimates of the spectra at arbitrary frequencies can be obtained by substituting the estimated values in Eq. (7.2.16).

It can be seen from Eq. (7.2.16) that when α_1 or ζ is zero, the spectral density $f(\omega)$ has a sharp peak at $\omega = \omega_n$, which flattens as ζ increases. It can be shown that there is no peak when $\zeta > 1/\sqrt{2}$. Thus, the smooth plot of $f(\omega)$ by Eq. (7.2.16) provides a clear visual signature of the system, and its relative state of damping can be judged at a glance. In fact, the autospectrum $f(\omega)$ plays the same role for the random vibration system as the frequency response "resonance" curves for usual vibratory systems.

7.3 UNIFORMLY SAMPLED SECOND ORDER AUTOREGRESSIVE SYSTEM

Our aim is to obtain the A(2) model from a discrete set of data for the same reasons as explained in Section 6.4. To estimate the A(2) model parameters we need a discrete model for the sampled data. Such a model can be obtained either by reparameterizing the form (7.2.12) or by requiring that the discrete model be an ARMA model and that its covariance function coincide with (7.2.13) at all the sampled points $k\Delta$, $k = 0, 1, 2,$ ..., where Δ is the sampling interval. In this chapter and the next, we will follow the second, more conventional, route of finding a covariance equivalent ARMA model. Reference may be made to Pandit (1973) or Pandit and Wu (1975) for the first route.

7.3.1 Representation of the Sampled System

At the sampled points $s = k\Delta$, $k = 0, 1, 2, \ldots$, the autocovariance $\gamma(s)$ given by Eq. (7.2.13) takes the form

$$\gamma_k = \gamma(k\Delta)$$

$$= \frac{\sigma_z^2}{2\mu_1\mu_2(\mu_1^2 - \mu_2^2)} (\mu_2 e^{\mu_1 k\Delta} - \mu_1 e^{\mu_2 k\Delta}) \qquad (7.3.1a)$$

$$= d_1\lambda_1^k + d_2\lambda_2^k \qquad (7.3.1)$$

where

$$d_1 = \frac{\sigma_z^2}{2\mu_1(\mu_1^2 - \mu_2^2)} \qquad (7.3.2a)$$

$$d_2 = \frac{-\sigma_z^2}{2\mu_2(\mu_1^2 - \mu_2^2)} \qquad (7.3.2b)$$

and

$$\lambda_1 = e^{\mu_1\Delta} \qquad (7.3.3)$$

$$\lambda_2 = e^{\mu_2\Delta}$$

Note that

$$\gamma_0 = \gamma(0) = d_1 + d_2 \qquad (7.3.4)$$

$$= -\frac{\sigma_z^2}{2\mu_1\mu_2(\mu_1 + \mu_2)} = \frac{\sigma_z^2}{4\zeta\,\omega_n^3}$$

Since the γ_k function (7.3.1) is a linear combination of two exponentials, it now follows from the results of Section 3.3 that the sampled A(2) system can be represented by a difference equation of the same form as an ARMA(2,1) model with

$$X_t - \phi_1 X_{t-1} - \phi_2 X_{t-2} = a_t - \theta_1 a_{t-1} \qquad (7.3.5)$$

where

$$\phi_1 = \lambda_1 + \lambda_2 = e^{\mu_1\Delta} + e^{\mu_2\Delta} \qquad (7.3.6)$$

$$\phi_2 = -\lambda_1\lambda_2 = -e^{(\mu_1+\mu_2)\Delta} \qquad (7.3.7)$$

By Eqs. (7.3.6) and (7.3.7), the values of ϕ_1, ϕ_2 for the spring-mass-dashpot example of Section 7.1.4 can be obtained as

$$\phi_1 = e^{-9.570\Delta + i18.524\Delta} + e^{-9.570\Delta - i18.524\Delta}$$

$$= 2e^{-9.570\Delta} \cos 18.524\Delta$$

$$\phi_2 = -e^{-19.14\Delta}$$

If $\Delta = 0.02$, then we have

$$\phi_1 = 1.541$$

$$\phi_2 = -0.682$$

7.3.2 Expressions for θ_1 and σ_a^2

The expressions for the coefficients d_1 and d_2 of the ARMA(2,1) model (7.3.5) have been derived in Section 3.3 and are given by Eqs. (3.3.16a) and (3.3.16b). We can equate these with expressions (7.3.2a) and (7.3.2b) to find θ_1 and σ_a^2.

$$\frac{\sigma_z^2}{2\mu_1(\mu_1^2 - \mu_2^2)} = \frac{\sigma_a^2(\lambda_1 - \theta_1)}{(\lambda_1 - \lambda_2)^2}\left[\frac{\lambda_1 - \theta_1}{1 - \lambda_1^2} - \frac{\lambda_2 - \theta_1}{1 - \lambda_1\lambda_2}\right] \qquad (7.3.8a)$$

$$-\frac{\sigma_z^2}{2\mu_2(\mu_1^2 - \mu_2^2)} = \frac{\sigma_a^2(\lambda_2 - \theta_1)}{(\lambda_1 - \lambda_2)^2}\left[\frac{\lambda_2 - \theta_1}{1 - \lambda_2^2} - \frac{\lambda_1 - \theta_1}{1 - \lambda_1\lambda_2}\right] \qquad (7.3.8b)$$

Dividing Eq. (7.3.8a) by Eq. (7.3.8b), we have

$$-\frac{\mu_2}{\mu_1} = \frac{(\lambda_1 - \theta_1)\left[\dfrac{\lambda_1 - \theta_1}{1 - \lambda_1^2} - \dfrac{\lambda_2 - \theta_1}{1 - \lambda_1\lambda_2}\right]}{(\lambda_2 - \theta_1)\left[\dfrac{\lambda_2 - \theta_1}{1 - \lambda_2^2} - \dfrac{\lambda_1 - \theta_1}{1 - \lambda_1\lambda_2}\right]} \qquad (7.3.9)$$

After rearrangement, Eq. (7.3.9) is a quadratic equation in θ_1, (see Problem 4.12)

$$\theta_1^2 + 2P\theta_1 + 1 = 0 \qquad (7.3.10)$$

that is,

$$2P = -\left(\theta_1 + \frac{1}{\theta_1}\right) \qquad (7.3.10a)$$

where

$$2P = \frac{-\mu_1(1 + \lambda_1^2)(1 - \lambda_2^2) + \mu_2(1 + \lambda_2^2)(1 - \lambda_1^2)}{\mu_1\lambda_1(1 - \lambda_2^2) - \mu_2\lambda_2(1 - \lambda_1^2)} \qquad (7.3.11)$$

and θ_1 is given by

$$\theta_1 = -P \pm \sqrt{P^2 - 1} \qquad (7.3.12)$$

Note that, by Eq. (7.3.10a), if θ_1 is a solution, so is $\dfrac{1}{\theta_1}$. Since the value of P is always real because θ_1 is real, an expression for P consisting of only real values and hence easy-to-compute (even by a hand calculator) is derived from Eq. (7.3.11). In fact, it will be seen later that in addition to the simplicity, the following expression of P will be of help when dealing with uniqueness.

Let a and b represent the respective real and imaginary parts of the complex roots, that is,

$$\mu_{1,2} = -a \pm ib$$

and

$$a = \frac{\alpha_1}{2} = \zeta\omega_n \qquad (7.3.13)$$

$$b = \frac{1}{2}\sqrt{|\alpha_1^2 - 4\alpha_0|} = \omega_n\sqrt{|\zeta^2 - 1|} \qquad (7.3.14)$$

then the expression for P can be written in terms of a, b, and Δ as follows:

$$P = \begin{cases} \dfrac{b\sinh(2a\Delta) - a\sinh(2b\Delta)}{2a\sinh(b\Delta)\cosh(a\Delta) - 2b\sinh(a\Delta)\cosh(b\Delta)} & (7.3.15a) \\ \text{for } \alpha_1^2 \geqslant 4\alpha_0 \\[2ex] \dfrac{b\sinh(2a\Delta) - a\sin(2b\Delta)}{2a\sin(b\Delta)\cosh(a\Delta) - 2b\sinh(a\Delta)\cos(b\Delta)} & (7.3.15b) \\ \text{for } \alpha_1^2 < 4\alpha_0 \end{cases}$$

It is left as an exercise to derive Eq. (7.3.10), using the implicit expression of the autocovariance function of the ARMA(2,1) model given by

Eq. (3.3.12), and show that an alternative expression for P is

$$2P = \left[\frac{\gamma_0 - \phi_1\gamma_1 - \phi_2\gamma_2 - \phi_1[(1 - \phi_2)\gamma_1 - \phi_1\gamma_0]}{(1 - \phi_2)\gamma_1 - \phi_1\gamma_0} \right] \quad (7.3.16)$$

Furthermore, by substituting the values of γ_0, γ_1, γ_2 and ϕ_1, ϕ_2 from Eqs. (7.3.1a), (7.3.6) and (7.3.7), respectively, it can be shown that expressions (7.3.11) and (7.3.16) coincide.

Once θ_1 is known, σ_a^2 can be expressed from either Eq. (7.3.8a) or Eq. (7.3.8b) as

$$\sigma_a^2 = \frac{\mu_2(1 + \lambda_2^2)(1 - \lambda_1^2) - \mu_1(1 + \lambda_1^2)(1 - \lambda_2^2)}{2\mu_1\mu_2(\mu_1^2 - \mu_2^2)(1 + \theta_1^2)}\sigma_z^2 \quad (7.3.17)$$

Similar to the alternative forms of the expression for θ_1, the other two expressions for σ_a^2 can be derived from Eqs. (7.3.13), (7.3.14), (7.3.17), and (3.3.12) as

$$\sigma_a^2 = \begin{cases} \dfrac{b\sinh(2a\Delta) - a\sinh(2b\Delta)}{2ab(a^2 - b^2)(1 + \theta_1^2)e^{2a\Delta}} \cdot \sigma_z^2 & \text{for } \alpha_1^2 \geqslant 4\alpha_0 \quad (7.3.18a) \\[3mm] \dfrac{b\sinh(2a\Delta) - a\sin(2b\Delta)}{2ab(a^2 + b^2)(1 + \theta_1^2)e^{2a\Delta}} \cdot \sigma_z^2 & \text{for } \alpha_1^2 < 4\alpha_0 \quad (7.3.18b) \end{cases}$$

$$\sigma_a^2 = \frac{\gamma_0 - \phi_1\gamma_1 - \phi_2\gamma_2 - \phi_1[(1 - \phi_2)\gamma_1 - \phi_1\gamma_0]}{(1 + \theta_1^2)} \quad (7.3.19)$$

7.3.3 Illustrative Examples

The calculation of θ_1 and σ_a^2 from the parameters of the continuous model for the overdamped system and the underdamped system is illustrated in (a) and (b) below:

(a) Overdamped System (μ_1 and μ_2 Are Real)

Suppose that $\alpha_1 = 2.527$

$$\alpha_0 = 1.571$$
$$\Delta = 1$$

then the values of θ_1 and σ_a^2 can be obtained as follows:

$$a = \frac{\alpha_1}{2} = 1.2635$$

$$b = \sqrt{|\alpha_1^2 - 4\alpha_0|} = 0.1595$$

If we use Eq. (7.3.15a)

$$P = 2.352$$

and from Eq. (7.3.12) there are two values of θ_1, that is,

$$\theta_1 = -P \pm \sqrt{P^2 - 1}$$
$$= -2.352 \pm 2.129$$
$$= -0.223, \; -4.481$$

and we choose the invertible value -0.223. (See Problem 4.12.) If we use Eq. (7.3.18a)

$$\sigma_a^2 \simeq 0.07\sigma_z^2$$

(b) Underdamped System (μ_1 and μ_2 Are Complex Conjugate)

Using the spring-mass-dashpot example in which μ_1 and μ_2 are

$$\mu_1 = -9.570 + i18.524$$
$$\mu_2 = -9.570 - i18.524$$

We can easily see that

$$a = 9.570$$
$$b = 18.524$$

If $\Delta = 0.02$, then the value of P is found by using Eq. (7.3.15b) to be

$$P = 1.980$$

The corresponding invertible θ_1 is

$$\theta_1 = -0.271$$

and if we use Eq. (7.3.18b)

$$\sigma_a^2 = (3.324 \times 10^{-6})\sigma_z^2$$

7.4 STABILITY REGION

We have seen in Section 7.2 that it is the real part of the roots μ_1, μ_2, or $\zeta\omega_n$ that controls the stability of the A(2) system. It follows from the basic relation (7.3.3) that the equivalent stability conditions may be written as

$$Re(\mu_i) < 0, \qquad i = 1,2 \qquad (7.4.1a)$$
$$|\lambda_i| < 1, \qquad i = 1,2 \qquad (7.4.1b)$$

where Re denotes the real part. It is known from the corresponding conditions in discrete models [see Eqs. (3.1.20) and (3.1.21)] that (7.4.1b) is equivalent to

$$\phi_1 + \phi_2 < 1$$
$$\phi_2 - \phi_1 < 1 \qquad (7.4.2)$$
$$|\phi_2| < 1$$

7.4.1 Additional Restrictions

The relations (7.3.6) and (7.3.7) impose some additional restrictions on the parameters of the covariance equivalent ARMA(2,1) model. First of all, since μ_1 and μ_2 are complex conjugate, $(\mu_1 + \mu_2)$ is real; in fact,

$$\mu_1 + \mu_2 = -2\zeta\omega_n = -\alpha_1$$

and therefore by Eq. (7.3.7), $\phi_2 = -exp\,(-\alpha_1\Delta)$ immediately implies that

$$\phi_2 < 0 \qquad (7.4.3)$$

because the exponential of a real number cannot be negative. With the same reasoning it follows from Eq. (7.3.6) that since $\phi_1 = \lambda_1 + \lambda_2 = e^{\mu_1\Delta} + e^{\mu_2\Delta}$, if μ_1 and μ_2 are real, λ_1 and λ_2 are real positives and therefore ϕ_1 cannot be negative. Thus,

$$\phi_1 \geqslant 0, \qquad \text{for real} \qquad \mu_1, \mu_2 \qquad \text{or} \qquad \lambda_1, \lambda_2 \qquad (7.4.4)$$

However, if μ_1 and μ_2 are complex conjugate, ϕ_1 can be positive or negative. The conditions (7.4.2) to (7.4.4) define the stability region of the covariance equivalent ARMA(2,1) model shown in Fig. 7.6. The corresponding region for the usual ARMA(2,1) model is shown by a dotted line for comparison.

7.4.2 Reduced Stability Region

The reduction of the stability region of the covariance equivalent ARMA(2,1) model in Fig. 7.6 from the dotted line to the solid line should be carefully noted. The triangular region above the line $\phi_2 = 0$ drops out because of the restriction (7.4.3). Similarly, the region below the line $\phi_2 = 0$, to the left and above the curve $\phi_1^2 + 4\phi_2 = 0$ is excluded because of the restriction (7.4.4). It is easy to see that the stability region is strongly dominated by complex roots giving pseudo-periodic motion. The exponential motion is possible only in the small region to the right between $\phi_2 = 0$ and $\phi_1^2 + 4\phi_2 = 0$. It should also be noted that the upper boundary of the region, given by the line $\phi_2 = 0$ joining the solid curve $\phi_1^2 + 4\phi_2 = 0$, is not really a stability boundary but arises because of the physical limitations imposed by the process of sampling.

For the experimental spring-mass-dashpot example, the values of ϕ_1 and ϕ_2 by Eqs. (7.3.6) and (7.3.7) are

$$\phi_1 = 2e^{-9.570\Delta} \cos 18.524\Delta$$

$$\phi_2 = -e^{-2\times9.570\Delta} = -e^{-19.14\Delta}$$

Hence,

$$\phi_1 = 2\sqrt{-\phi_2} \cos 18.524\Delta$$

$$\therefore \phi_1^2 = -4\phi_2 \cos^2 18.524\Delta$$

$$\therefore \phi_1^2 \leqslant -4\phi_2$$

or

$$\phi_1^2 + 4\phi_2 \leqslant 0$$

also

$$-1 < \phi_2 < 0$$

7.4.3 Static and Dynamic Stability

The stability region in Fig. 7.6 has an interesting physical interpretation in light of the classification of static and dynamic stability given by Den-Hartog (1956) for vibration systems. Extending this classification to random vibrations, we can say that a random vibration system is statically stable if a displacement $X(t)$ from the equilibrium position sets up a force (or a torque) that tends to bring the system back to its mean or equilibrium position. In terms of Eq. (7.1.1) this requires K/M to be positive; therefore, static instability means negative α_0 in Eq. (7.2.4), and the condition of static stability may be written as

$$\alpha_0 > 0 \qquad (7.4.5)$$

Similarly, the system is dynamically stable if its velocity sets up a force (or a torque) that tends to bring the system back to its equilibrium position. Thus, dynamic instability means negative α_1 in Eq. (7.2.4) and the condition of dynamic stability may be written as

$$\alpha_1 > 0 \qquad (7.4.6)$$

Now consider Fig. 7.6. The parabolic line $\phi_1^2 + 4\phi_2 = 0$ in the figure corresponds to the critical damping $\zeta = 1$ of the A(2) system since for $\zeta = 1$

$$\phi_1^2 + 4\phi_2 = (e^{-\omega_n\Delta} + e^{-\omega_n\Delta})^2 - 4e^{-2\omega_n\Delta} = 0$$

Below this line the roots μ_1 and μ_2 and therefore λ_1 and λ_2 are complex, and the random vibration is a pseudo-periodic or damped periodic motion. The parameters ϕ_1 and ϕ_2 lie in this region only when the damping of the A(2) system is less than critical, that is, $\zeta < 1$. In this case, the instability occurs when damping becomes zero or negative, that is, $\alpha_1 < 0$, and the stability boundary is the bottom line $\phi_2 = -1$. Below this line the system is dynamically unstable. Thus, the condition of dynamic stability (7.4.6)

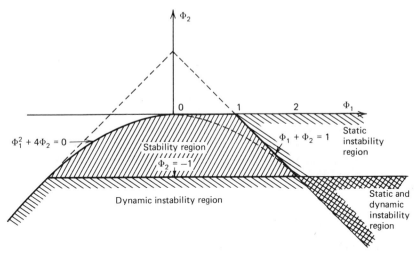

Figure 7.6 Stability region of the covariance equivalent ARMA(2,1) model.

may be equivalently written as

$$\phi_2 > -1 \qquad (7.4.6a)$$

However, the random vibration system A(2) may also become unstable when the damping is above the critical level or the damping factor ζ is greater than one. For such a damping the parameters ϕ_1 and ϕ_2 lie above the parabolic curve, and instability can occur when the parameters lie beyond the line $\phi_1 + \phi_2 = 1$. Since the damping is positive so that $\alpha_1 > 0$, the only way this can happen, that is, $\phi_1 + \phi_2 > 1$ or $Re(\mu_i) > 0$ for $i = 1$ or 2, is when the constant α_0 is negative, as by Eq. (7.2.7b): $\mu_1, \mu_2 = \frac{1}{2}(-\alpha_1 \pm \sqrt{\alpha_1^2 - 4\alpha_0})$. Beyond the line $\phi_1 + \phi_2 = 1$, therefore, $\alpha_0 < 0$, and the system is statically unstable as shown in Fig. 7.6. The equivalent condition of static stability of a random vibration system in terms of the discrete model parameter is

$$\phi_1 + \phi_2 < 1 \qquad (7.4.5a)$$

When both the conditions (7.4.5a) and (7.4.6a) are violated, the system becomes both statically and dynamically unstable. In a sense this is the "worst" instability of the A(2) system. The stability boundary of such a case is the right-hand corner of the stability region in Fig. 7.6, with $\phi_1 = 2$, $\phi_2 = -1$.

Negative constants α_0 and/or α_1 leading static and/or dynamic instabilities are impossible in the case of an actual spring-mass-dashpot system. However, there are many mechanical systems in which the vibratory motion itself induces "self-excited" vibrations. These may either aid or oppose the original vibratory motion. When they aid the original motion, then there is a possibility that the constants α_0 and α_1 of the combined vibratory motion may become negative, making the system statically and/or dynamically unstable. Since the nature of self-excited vibrations is generally unknown, it is advisable to formulate such a motion as random vibration and ascertain its stability from the model obtained from the operating data, using the results of this chapter. Such analysis of the stability of self-excited random vibrations with application to the problem of machine-tool chatter is given in Pandit, Subramanian, and Wu (1975), from which the above discussion and Fig. 7.6 have been taken.

7.5 THE A(2) MODEL FROM DISCRETE DATA

So far we have discussed the sampling of a given A(2) system and shown that the resultant discrete representation is unique. In other words, given ζ, ω_n, and σ_z^2 that characterize a second order vibration system, the characteristic parameters ϕ_1, ϕ_2, θ_1, and σ_a^2 are uniquely determined.

7.5.1 Nonuniqueness of the A(2) Model Parameters

However, the problem in our case is the reverse one. From the available data a discrete model of the form Eq. (7.3.5) is fitted, and a continuous model of the form Eq. (7.2.4) is obtained from it concurrently. The parameters of the continuous model corresponding to the given values of ϕ_1 and ϕ_2 are not necessarily unique. This is most easily seen from the relationship Eq. (7.3.3), which connects the characteristic roots of the differential equation (7.2.4) with those of the difference equation (7.3.5). The reverse relations are

$$\mu_1 = \frac{1}{\Delta}\, \ell n(\lambda_1), \qquad \mu_2 = \frac{1}{\Delta}\, \ell n(\lambda_2) \qquad (7.5.1)$$

If λ_1 and λ_2 are real, that is, if $\phi_1^2 + 4\phi_2 \geq 0$, then μ_1 and μ_2 are also real and uniquely determined. However, if λ_1 and λ_2 are complex (in fact, complex conjugate since ϕ_1 and ϕ_2 are real), then μ_1 and μ_2 are also complex (conjugate) and since the complex logarithmic function is multiple valued, there are infinite values of μ_1 and μ_2 corresponding to the same values of λ_1 and λ_2.

These conclusions can also be verified from the following explicit relationships between the parameters of the continuous and discrete models, obtained from Eqs. (7.3.6), (7.3.7) and (7.2.7b).

For $\phi_1^2 + 4\phi_2 \geq 0$,

$$\zeta = \sqrt{\frac{[\ell n(-\phi_2)]^2}{[\ell n(-\phi_2)]^2 - 4\left[\cosh^{-1}\left(\frac{\phi_1}{2\sqrt{-\phi_2}}\right)\right]^2}} \geq 1 \qquad (7.5.2)$$

$$\omega_n = \frac{1}{\Delta}\sqrt{\frac{[\ell n(-\phi_2)]^2}{4} - \left[\cosh^{-1}\left(\frac{\phi_1}{2\sqrt{-\phi_2}}\right)\right]^2} \qquad (7.5.3)$$

and for $\phi_1^2 + 4\phi_2 < 0$,

$$\zeta = \sqrt{\frac{[\ell n(-\phi_2)]^2}{[\ell n(-\phi_2)]^2 + 4\left[\cos^{-1}\left(\frac{\phi_1}{2\sqrt{-\phi_2}}\right)\right]^2}} < 1 \qquad (7.5.4)$$

$$\omega_n = \frac{1}{\Delta}\sqrt{\frac{[\ell n(-\phi_2)]^2}{4} + \left[\cos^{-1}\left(\frac{\phi_1}{2\sqrt{-\phi_2}}\right)\right]^2} \qquad (7.5.5)$$

The multiplicity of the values of $\cos^{-1}$ function leads to the multiplicity of values of ζ and ω_n when $\phi_1^2 + 4\phi_2 < 0$.

7.5.2 Multiplicity in Parameters ζ and ω_n

It can be discerned from Eqs. (7.5.2) and (7.5.3) that due to the real values of ϕ_1 and ϕ_2, the transformation from the discrete model parameters to continuous model parameters ζ and ω_n is unique if $\phi_1^2 + 4\phi_2 \geq 0$. However, if $\phi_1^2 + 4\phi_2 < 0$, the transformation of the parameters given by Eqs. (7.5.4) and (7.5.5) is nonunique because they involve the $\cos^{-1}$ function that has multiple values. Therefore, the multiplicity of A(2) model parameters arises when $\phi_1^2 + 4\phi_2 < 0$.

Let us rewrite Eqs. (7.3.6) and (7.3.7)

$$\phi_1 = \lambda_1 + \lambda_2 = e^{\mu_1\Delta} + e^{\mu_2\Delta}$$

$$= 2e^{-a\Delta}\cos(b\Delta) \qquad (7.5.6)$$

$$\phi_2 = -\lambda_1\lambda_2 = -e^{(\mu_1 + \mu_2)\Delta}$$

$$= -e^{-2a\Delta} \qquad (7.5.7)$$

where, by Eqs. (7.3.13) and (7.3.14)

$$\mu_{1,2} = -a \pm ib$$

$$= -\zeta\omega_n \pm i\omega_n\sqrt{1 - \zeta^2}$$

$$= -\frac{\alpha_1}{2} \pm i\sqrt{\alpha_0 - \left(\frac{\alpha_1}{2}\right)^2}$$

From Eqs. (7.5.6) and (7.5.7) and Eqs. (7.3.13) and (7.3.14),

$$a = \zeta\omega_n = \frac{\alpha_1}{2} = -\frac{\ell n\,(-\phi_2)}{2\Delta} \qquad (7.5.8)$$

and

$$\cos\,(b\Delta) = \cos\,[(\omega_n\sqrt{1 - \zeta^2})\Delta] = \frac{\phi_1}{2\sqrt{-\phi_2}} \qquad (7.5.9)$$

It can be seen from Eqs. (7.5.8) and (7.5.9) that the parameters ζ and ω_n are functionally related to the values of a and b. Therefore, the multiplicity in a and b will cause the multiplicity of the $A(2)$ model parameters ζ and ω_n.

From Eq. (7.5.8), the value of a ($= \alpha_1/2$, half the system damping) can be uniquely determined because ϕ_2 is real and the log of a real number is unique. On the other hand, the value of b can be determined from Eq. (7.5.9) if we write

$$\cos\,(b\Delta) = \cos\,(2n\pi + b\Delta) = \cos\,(2n\pi - b\Delta)$$

$$= \frac{\phi_1}{2\sqrt{-\phi_2}},\, n = 0,\, \pm 1,\, \pm 2,\, \ldots$$

or

$$b = \frac{\left|\pm\cos^{-1}\left(\dfrac{\phi_1}{2\sqrt{-\phi_2}}\right) + 2n\pi\right|}{\Delta} \qquad (7.5.10)$$

which shows the multiplicity in b, or essentially in $\alpha_0 = \omega_n^2$. It is also evident from Fig. 7.7 that for a given value of $b\Delta$, the value of $\dfrac{\phi_1}{2\sqrt{-\phi_2}}$ is unique, but for an assigned value of $\dfrac{\phi_1}{2\sqrt{-\phi_2}}$, there are (infinitely) many corresponding values of $b\Delta$. In other words, the value of b cannot be determined uniquely unless n is known or unless we know *a priori* that the sampling interval is sufficiently small, so that the damped (natural) frequency is smaller than the highest frequency (which in the usual spectral analysis is known as Nyquist frequency) of $\dfrac{1}{2\Delta}$, that is,

$$\frac{\omega_n\sqrt{1 - \zeta^2}}{2\pi} < \frac{1}{2\Delta} \qquad (7.5.11a)$$

or

$$\omega_n\sqrt{1 - \zeta^2}\Delta = b\Delta < \pi \qquad (7.5.11)$$

Note that in this special case, the value of n in Eq. (7.5.10) is zero.

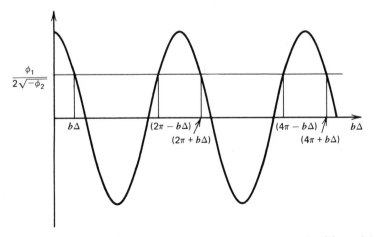

Figure 7.7   A graphical illustration of the multiplicity of A(2) model parameters.

7.5.3 Resolution of Multiplicity in b (or in $\alpha_0 = \omega_n^2$)

Before we suggest a procedure to resolve the multiplicity of values of b, let us reconsider the computation of b given by Eq. (7.5.9), that is,

$$\cos b\Delta = \frac{\phi_1}{2\sqrt{-\phi_2}}$$

It is clear from Eq. (7.5.9) that the computation of the value of b involves the discrete model parameters ϕ_1 and ϕ_2 but does not make use of the moving average parameters θ_1. If we consider the moving average parameter θ_1 along with the autoregressive parameters ϕ_1 and ϕ_2 the value of b can be uniquely determined.

From Eq. (7.3.15b),

$$b = \frac{a[2P \sin (b\Delta) \cosh (a\Delta) + \sin (2b\Delta)]}{2P \sinh (a\Delta) \cos (b\Delta) + \sinh (2a\Delta)} \tag{7.5.12}$$

Using Eqs. (7.3.10a), (7.5.8), and (7.5.9) in (7.5.12), we can verify that

$$b = \frac{-\ell n(-\phi_2)}{2\Delta} \sqrt{-(\phi_1^2 + 4\phi_2)}$$

$$\cdot \left[\frac{2\phi_1 - (1 - \phi_2)\left(\theta_1 + \dfrac{1}{\theta_1}\right)}{2(1 - \phi_2^2) - \phi_1(1 + \phi_2)\left(\theta_1 + \dfrac{1}{\theta_1}\right)} \right] \tag{7.5.13}$$

Although Eq. (7.5.13) is more complicated than Eq. (7.5.10), it gives a unique value of b for given values of ϕ_1, ϕ_2, and θ_1. In fact, it is easy to see that Eq. (7.5.13) only involves the log of a real parameter $-\phi_2$ that gives a unique value of b. Therefore, by considering ϕ_1, ϕ_2 and θ_1, the multiplicity of the value of b is resolved.

As mentioned earlier, once the values of a and b are uniquely determined, the continuous model's [A(2)'s] parameters ζ and ω_n can be obtained as

$$\omega_n = \sqrt{a^2 + b^2} \qquad (7.5.14a)$$

$$\zeta = \frac{a}{\omega_n} \qquad (7.5.14b)$$

or

$$\alpha_1 = 2a \qquad (7.5.15a)$$

$$\alpha_0 = a^2 + b^2 = \omega_n^2 \qquad (7.5.15b)$$

7.5.4 Illustrations for Multiplicity

As discussed in Section 7.5.1, our objective is to determine the continuous process via its discrete data. Two examples are given in this section to illustrate that one could obtain erroneous results if the multiplicity problem is not properly recognized.

Example A

We wish to determine the A(2) parameters from the known covariance equivalent ARMA(2,1) and sampling interval

$$\phi_1 = \ \ 1.54$$
$$\phi_2 = -0.68$$
$$\theta_1 = -0.27$$
$$\Delta = \ \ 0.02$$

Since $\phi_1^2 + 4\phi_2 = -0.41 < 0$, one could use Eqs. (7.5.4) and (7.5.5) to obtain

$$\zeta = \ \ 0.466$$
$$\omega_n = 20.68$$

However, we can verify these results by further computing the value of parameters ζ and ω_n from Eqs. (7.5.8), (7.5.13), and (7.5.14) as

$$a = \ \ 9.64$$
$$b = 18.30$$
$$\omega_n = 20.68$$
$$\zeta = \ \ 0.466$$

Note that the parameter values are exactly the same as those obtained previously from Eqs. (7.5.4) and (7.5.5). The reason why the two results came out to be the same is because the sampling interval is small, thereby ensuring that the damped natural frequency is smaller than the Nyquist frequency $1/2\Delta$ equal to 25. In essence, the value of n in Eq. (7.5.10) is zero.

Example B

Suppose that we have known the discrete parameters and the sampling interval to be

$$\phi_1 = \quad 0.15589$$
$$\phi_2 = -0.00674$$
$$\theta_1 = \quad 0.083149$$
$$\Delta = \quad 2.5$$

and we wish to determine the A(2) parameters. From Eqs. (7.5.4) and (7.5.5) the values of the parameters are calculated as

$$\zeta = 0.992$$
$$\omega_n = 1.008$$

Now let us compute the A(2) model parameters using the method discussed in Section 7.5.3. From Eqs. (7.5.8), (7.5.13), and (7.5.14), the parameter values are obtained as

$$a = 1.0$$
$$b = 4.899$$
$$\omega_n = 5.0$$
$$\zeta = 0.2$$

Note that the results of the parameter values obtained from the two methods do not match. In fact, the results obtained using Eqs. (7.5.4) and (7.5.5) are incorrect because n in Eq. (7.5.10) is -2 instead of zero. This claim could be further justified by carrying out the following calculation (where $n = -2$).

$$b = \frac{1}{\Delta} \left| \cos^{-1}\left(\frac{\phi_1}{2\sqrt{-\phi_2}} \right) + 2n\pi \right|$$
$$= |0.12758 - 5.02655|$$
$$= 4.899$$

7.5.5 Estimation

The estimation of the discrete covariance equivalent ARMA(2,1) cannot be separated from that of the A(2) model since θ_1 cannot be determined

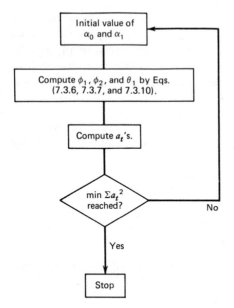

Figure 7.8 Flowchart for estimation of the covariance equivalent ARMA (2,1) model.

unless μ_1 and μ_2 or implicitly a and b are known. The estimation of the A(2) model can be done by a nonlinear least squares search algorithm to minimize the sum of squares of a_t's in the model (7.3.5). The parameters to be estimated are α_1 and α_0. The values of ϕ_1, ϕ_2, and θ_1 can be computed using Eq. (7.3.6), (7.3.7), and (7.3.11), (7.3.12) or (7.3.13) to (7.3.15). The procedure starts with initial values for α_1 and α_0; then, the values of ϕ_1, ϕ_2, and θ_1 are computed. The sum of squares of a_t's is computed using Eq. (7.3.5). The values of α_1 and α_0 are changed until the minimum sum of squares of a_t's is reached. The estimation procedure is given in the flowchart of Fig. 7.8.

* 7.5.6 Implications in Spectral Estimation

An important fact that emerges from the above analysis relates to the frequencies that can be detected from a set of discrete sampled data. The usual spectral estimation procedure (even theoretically) cannot yield any information about frequencies higher than $1/2\Delta$. The uniqueness of the A(2) model discussed above shows that it does yield information about frequencies higher than $1/2\Delta$, at least in theory. This is a significant advantage to using the modeling approach for spectral estimation.

Note that this advantage of providing spectra beyond the frequencies $\pm 1/2\Delta$ is also clear from the expression for the spectra (7.2.16) for which the parameters are estimated via the discrete model. Since the expression

(7.2.16) is true for frequencies $-\infty$ to ∞, the spectra at arbitrary frequencies can be obtained from it.

This advantage results from the assumption of a continuous time system underlying the discrete data at uniform intervals. The time domain conditional least square fitting provides the best fitting continuous time "stochastic curve" *interpolating* the discrete observations by means of the A(2) model. In the frequency domain, this is equivalent to *extrapolating* the spectrum $f(\omega)$ given by (7.2.16) beyond the frequency range $\pm\ 1/2\Delta$, within which the information provided by the data lies.

7.6 EFFECT OF SAMPLING INTERVAL, NATURAL FREQUENCY, AND DAMPING RATIO

For the extreme values of the parameters ζ and ω_n and the sampling interval Δ, the discrete model takes some simple forms. Some of these forms are the so-called accumulated process in Whittle (1963) or the integrated processes treated as nonstationary processes in Box and Jenkins (1970). These are the particular cases of the processes with stationary increments in Yaglom (1955). We have treated them as limits of the stationary processes (see Section 3.1.9). The additional reasons for this choice will become clear from the results of this section.

In practice, the choice of a sampling interval for digitizing a given continuous system, say A(2), may affect the final form of the discrete model. In fact, in extreme cases when the sampling interval is too large or too small (a small sampling interval means that there are too many samples per natural period of oscillation, $2\pi/\omega_n$), the information extracted from the sampled data can be misleading. To illustrate this point, a set of continuous data ($\omega_n = 1$ Hz, $\zeta = 0.01$) is generated and shown in Fig. 7.9a. The digitized data using three different sampling intervals ($\Delta = 0.15$ sec, $\Delta = 0.75$ sec, and $\Delta = 0.0025$ sec) are also plotted in Figs. 7.9b to 7.9d. Note the Fig. 7.9d represents only 0.6 sec of sampled data. It is apparent from the plots that the sampled data with $\Delta = 0.15$ sec Fig. 7.9b, approximately 7 points/natural period of oscillation $= 1$ sec, looks similar to the original continuous signal, whereas the sampled data with large $\Delta = 0.75$ sec and small $\Delta = 0.0025$ sec look quite different from the original signal.

It is important to note that although the data sampled at the small sampling interval Δ differ from the original signal, in cases of continuous signals with high frequency, the smaller sampling interval may be approprite to correctly represent the original signal. Similarly, the large sampling interval may be needed to sample a continuous signal with very low frequency. In this section we will investigate the limiting cases of $(\omega_n\Delta)$ and

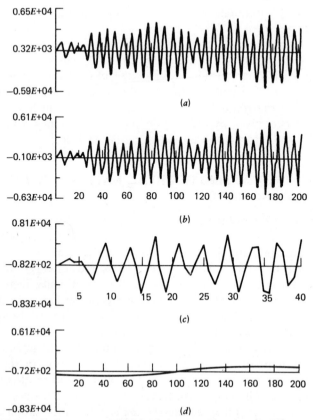

Figure 7.9 Continuous and digitized data profiles (ω_n = 1Hz, ζ = 0.01). (a) Continuous profile (30 seconds). (b) Digitized data (Δ = 0.15 seconds). (c) Digitized data (Δ = 0.75 seconds). (d) Digitized data (Δ = 0.0025 seconds).

damping factor ζ, and their effect on the form of the discrete sampled system model. The effect of the limiting cases of the parameters ω_n or Δ alone, while the other is fixed, could be derived as a special case of the limiting effect of the joint parameters.

7.6.1 Small $\omega_n\Delta$

As $\omega_n\Delta \to 0$, from Eqs. (7.2.7b) and (7.3.3),

$$\lambda_1 = e^{\mu_1\Delta} = e^{-\omega_n\Delta(\zeta+\sqrt{\zeta^2-1})} \to 1$$

$$\lambda_2 = e^{\mu_2\Delta} = e^{-\omega_n\Delta(\zeta-\sqrt{\zeta^2-1})} \to 1$$

$$\lambda_1 + \lambda_2 = \phi_1 \to 2 \tag{7.6.1}$$

$$-\lambda_1\lambda_2 = \phi_2 \to -1 \tag{7.6.2}$$

It can be verified using Eqs. (7.3.11), (7.3.12), and (7.3.17) that as $\omega_n \Delta \rightarrow 0$,

$$P \rightarrow 2$$
$$\theta_1 \rightarrow -2 + \sqrt{3} \tag{7.6.3}$$

and

$$\sigma_a^2 \rightarrow \frac{\sigma_z^2 \Delta^3}{6(2 - \sqrt{3})} \tag{7.6.4}$$

Thus, for $\omega_n \Delta \rightarrow 0$, the discrete model takes the form

$$\nabla^2 X_t = X_t - 2X_{t-1} + X_{t-2} = a_t + (2 - \sqrt{3})a_{t-1} \tag{7.6.5}$$

Note that $\omega_n \Delta$ tends to zero means either or both of ω_n and Δ tend to zero. Suppose $\Delta \rightarrow 0$, from Eq. (7.6.4)

$$\sigma_a^2 \rightarrow 0$$

Hence, when Δ is actually zero, Eq. (7.6.5) merely equates two random variables degenerate at zero. This is logical since a zero sampling interval is meaningless. Therefore, Eq. (7.6.5) should only be treated as an approximation to the discrete model for very small sampling intervals.

7.6.2 Large $\omega_n \Delta$

As $\omega_n \Delta \rightarrow \infty$

$$\lambda_1 \rightarrow 0$$
$$\lambda_2 \rightarrow 0$$
$$\phi_1 \rightarrow 0 \tag{7.6.6}$$
$$\phi_2 \rightarrow 0 \tag{7.6.7}$$

since

$$\gamma_k = d_1 \lambda_1^k + d_2 \lambda_2^k$$
$$\gamma_1 \rightarrow 0$$
$$\gamma_2 \rightarrow 0$$

Using Eq. (7.3.16), we can verify that for large $\omega_n \Delta$

$$P \rightarrow \infty$$

and therefore

$$\theta_1 \rightarrow 0 \tag{7.6.8}$$

Finally, it immediately follows from Eq. (7.3.17) that

$$\sigma_a^2 \rightarrow \gamma_0 = \frac{-\sigma_z^2}{2\mu_1 \mu_2 (\mu_1 + \mu_2)} = \frac{\sigma_z^2}{4\zeta \omega_n^3} \tag{7.6.9}$$

Therefore, as $\omega_n \Delta \to \infty$, the discrete model takes the form

$$X_t = a_t \qquad (7.6.10)$$

which is a discrete white noise. As discussed in Section 6.5.1, this result is intuitively obvious. If the sampling interval Δ is large, the $\gamma_k = \gamma(k\Delta)$ is nearly zero, and X_t and X_{t-k} are uncorrelated. As a result, the discrete process becomes a sequence of uncorrelated random variables with variance γ_0. Furthermore, if ω_n is large, the variance of X_t is $\gamma_0 = \sigma_z^2/4\zeta\omega_n^3 \to 0$. This result is obvious since for a mechanical system, large ω_n forces it to remain at equilibrium, and hence the sampled process tends to zero in spite of the white noise disturbance.

7.6.3 Intermediate Values of $\omega_n \Delta$

For the intermediate values of $\omega_n \Delta$ the behavior of the parameters of the discrete model depends largely upon the value of ζ. For an overdamped system ($\zeta > 1$), the parameters generally exhibit an increasing or decreasing trend, whereas for an underdamped system ($\zeta < 1$), the parameters (except ϕ_2) show an oscillatory nature.

(a) Overdamped Systems ($\zeta > 1$)

It can be shown from (7.3.6) that

$$\phi_1 = 2e^{-a\Delta} \cosh (b\Delta)$$

$$= 2e^{-\zeta\omega_n\Delta} \cosh (\sqrt{\zeta^2 - 1} \cdot \omega_n\Delta) \qquad (7.6.11)$$

Thus, ϕ_1 is a strictly decreasing function of $\omega_n \Delta$, and it is easy to see that

$$0 < \phi_1 < 2$$

Similarly, from (7.3.7)

$$\phi_2 = -2e^{-2\zeta\omega_n\Delta} \qquad (7.6.12)$$

and it is a strictly increasing function of $\omega_n \Delta$. From Eqs. (7.3.10), (7.3.16), and (7.3.19), we can show that both θ_1 and σ_a^2 are increasing functions of $\omega_n \Delta$. In fact,

$$(-2 + \sqrt{3}) < \theta_1 < 0 \qquad (7.6.13)$$

Figure 7.10 shows the behavior of ϕ_1, ϕ_2, θ_1 and σ_a^2 of an overdamped system with $\zeta = 1.1$.

(b) Underdamped System ($\zeta < 1$)

For an underdamped system, the expressions for parameters ϕ_1, θ_1, and σ_a^2 [(7.5.6), (7.3.10), (7.3.15b), and (7.3.18b)] involve sine and cosine functions. Since the trigonometric functions are oscillatory in nature, the parameters ϕ_1, θ_1, and σ_a^2 generally show the oscillatory behavior with the changing value of $\omega_n \Delta$. The parameter ϕ_2, however, is a strictly increasing function of $\omega_n \Delta$.

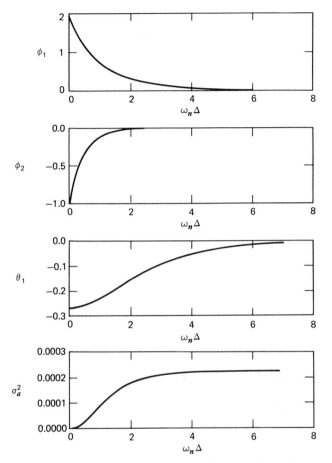

Figure 7.10 Variation of parameters with $\omega_n\Delta$—overdamped system with $\zeta = 1.1$.

For illustration purposes, the behavior of ϕ_1, ϕ_2, θ_1 and σ_a^2 for the A(2) system with $\zeta = 0.6$ and $\zeta = 0.1$ are shown in Figs. 7.11 and 7.12, respectively. It can be seen from the plots that the oscillatory behavior of ϕ_1, θ_1, and σ_a^2 is more pronounced for $\zeta = 0.1$ than for $\zeta = 0.6$.

The ranges of the parameters ϕ_1, ϕ_2, θ_1, and σ_a^2 depend upon the magnitude of ζ. It can be shown that, in general,

$$-2 < \phi_1 < 2 \tag{7.6.14}$$

$$-1 < \phi_2 < 0 \tag{7.6.15}$$

$$-1 < \theta_1 < 1 \tag{7.6.16}$$

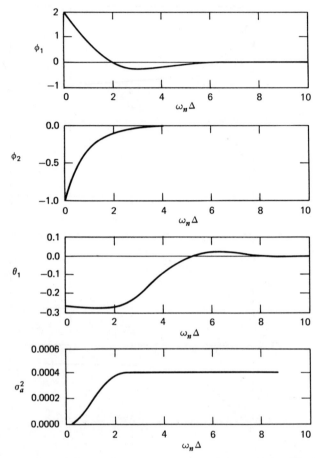

Figure 7.11 Variation of parameters with $\omega_n\Delta$—underdamped system with $\zeta = 0.6$.

and

$$0 < \sigma_a^2 < \gamma_0 \qquad (7.6.17)$$

A graphical representation of the combined variation of ϕ_1 and ϕ_2 with sampling interval Δ for two values of ζ, that is, $\zeta = 0.1$ and $\zeta = 1.4$, and $\omega_n = 10$ Hz is given in Fig. 7.13. The wiggly curve shows the variation of ϕ_1 and ϕ_2 with the sampling interval for $\zeta = 0.1$, whereas the exponentially rising curve represents the variation of ϕ_1 and ϕ_2 for $\zeta = 1.4$. The difference in the behavior of these two curves can be clearly observed from Figs. 7.10 and 7.11. As seen from Figs. 7.10 and 7.11, ϕ_2 is always an exponential function of $\omega_n\Delta$, whereas ϕ_1 is a sum of exponentials (strictly

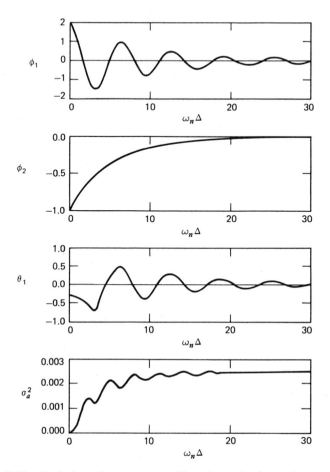

Figure 7.12   Variation of parameters with $\omega_n\Delta$—underdamped system with $\zeta = 0.1$.

decreasing function) when $\zeta > 1$, and an oscillatory function (sine wave) when $\zeta < 1$.

In addition to the depiction of the parameters' behavior, the stability region for ϕ_1 and ϕ_2 obtained in Eqs. (7.6.14) and (7.6.15) can be observed from Fig. 7.13. The figure also provides the ranges of ϕ_1 and ϕ_2 for $\zeta = 1$. In fact, it is seen that for $\zeta = 1$, parameters lie *on* the parabola

$$\phi_1^2 + 4\phi_2 = 0$$

Note that for the underdamped system with $\zeta = 0.1$, the wiggly curve lies under the parabola, $\phi_1^2 + 4\phi_2 = 0$, whereas the exponentially rising curve for $\zeta = 1.4$ lies above the parabola.

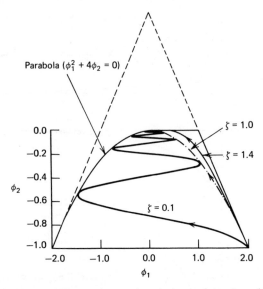

Figure 7.13 Stability region and variations of ϕ_1, ϕ_2, with $\omega_n\Delta$.

7.6.4 Limiting Values of ζ

We will now discuss the behavior of discrete model parameters for the extreme values of ζ (that is, $\zeta \to 0$ and $\zeta \to \infty$) and a fixed $\omega_n\Delta$.

(i) As $\zeta \to 0$

$$\mu_1 \to i\omega_n$$
$$\mu_2 \to -i\omega_n$$

Thus,

$$\phi_1 = e^{\mu_1\Delta} + e^{\mu_2\Delta} \to 2\cos\omega_n\Delta \qquad (7.6.18)$$
$$\phi_2 \to -1$$

Using Eq. (7.3.15b) and L'Hôspital's rule,

$$\lim_{\zeta \to 0} P = \frac{2\omega_n\Delta - \sin 2\omega_n\Delta}{2(\sin \omega_n\Delta - \omega_n\Delta \cos \omega_n\Delta)} \qquad (7.6.19)$$

Also,

$$1 \le |P| \le \infty$$

Thus,

$$-1 < \theta_1 \le 1$$

The limit of σ_a^2 can be obtained from Eq. (7.3.18b) by applying L'Hôspital's rule. The limit is

$$\sigma_a^2 = \frac{2\omega_n\Delta - \sin 2\omega_n\Delta}{2\omega_n^3(1 + \theta_1^2)} \tag{7.6.20}$$

It should be noted that γ_0 increases without bound as $\zeta \to 0$, yet σ_a^2 is finite unless $\omega_n\Delta \to \infty$. This can be explained by the fact that, for an ARMA process,

$$X_t = \sum_{j=0}^{\infty} G_j a_{t-j}$$

$$\gamma_0 = \sigma_a^2 \sum_{j=0}^{\infty} G_j^2$$

If $\zeta \to 0$, then using Eq. (3.1.17), we can verify that
$$G_j = A \sin (\beta + j\omega_n\Delta)$$

where

$$A = \sqrt{1 - \frac{(\phi_1 - 2\theta_1)^2}{\phi_1^2 + 4\phi_2}}$$

$$\beta = \tan^{-1}\left[\frac{\sqrt{-(\phi_1^2 + 4\phi_2)}}{\phi_1 - 2\theta_1}\right]$$

Therefore, it is obvious that, as $\zeta \to 0$, $\gamma_0 \to \infty$ in spite of finite σ_a^2.

It has been pointed out previously that as $\omega_n\Delta \to \infty$, $\sigma_a^2 \to \gamma_0$. This observation leads to the conclusion that $\sigma_a^2 \to \infty$ as $\omega_n\Delta \to \infty$. As $\omega_n\Delta$ increases, the behaviors of ϕ_1, ϕ_2, θ_1 and σ_a^2 of a system with no damping are illustrated in Figure 7.14.

(ii) As $\zeta \to \infty$

For a fixed $\omega_n\Delta$,

$$\mu_1 \to 0$$

$$\mu_2 \to -\infty$$

$$\lambda_1 \to 1$$

$$\lambda_2 \to 0$$

and from Eqs. (7.3.6) and (7.3.7)

$$\phi_1 = \lambda_1 + \lambda_2 \to 1 \tag{7.6.21}$$

$$\phi_2 = -\lambda_1\lambda_2 \to 0 \tag{7.6.22}$$

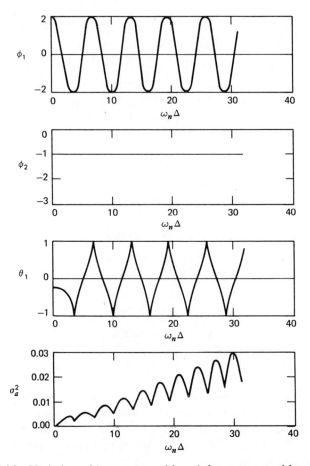

Figure 7.14   Variation of parameters with $\omega_n\Delta$ for a system with no damping.

The limiting value of θ_1 can be found by approximating P from Eq. (7.3.11) for small μ_1 and large negative μ_2. Then, as $\mu_1 \to 0$ and $\mu_2 \to -\infty$, $P \to -\infty$; therefore by Eq. (7.3.12) θ_1 is negative and θ_1 tends to zero, that is

$$\theta_1 \to 0,\ \theta_1 < 0 \qquad (7.6.23)$$

The limit of σ_a^2 can be shown from Eq. (7.3.18a) as

$$\sigma_a^2 \to 0 \qquad (7.6.24)$$

Thus, in the limiting case, the discrete model takes the form

$$\nabla X_t = a_t \qquad (7.6.25)$$

which is a random walk.

The foregoing results offer an explanation of the conditions leading to random walk. These results also suggest that it is better to fit an A(2) model when the random walk model (7.6.25) is found to fit well. Both models will yield practically the same results for forecasting due to the fact that θ_1 is close to zero, but the parameters ϕ_1 and ϕ_2 of the A(2) model, although close to one and zero respectively, may offer valuable information characterizing the system. This will be illustrated in Section 7.6.5.

7.6.5 An Illustrative Application of Limiting Cases—IBM Stock Prices

It was seen in Chapter 2 that when $\overline{X}$ is taken as the mean, the random walk model fits the IBM data well. This confirms the Bachelier jhypothesis that the stock prices behave like a random walk, and therefore the best forecast of tomorrow's price is today's price. The modeling procedure in Chapter 4 showed that AR(1) is also adequate when the mean is estimated. However, the ARMA(1,1) model may also provide a reasonably good fit, although the estimate $\theta_1 = -0.0848 \pm 0.1030$ in Section 4.5.2 shows that the 95% confidence interval includes zero and the model is tending toward a random walk. It is thus apparent that the IBM data should provide a good illustration of the results (7.6.21) to (7.6.25) if an A(2) model is fitted.

The A(2) model fitted to the IBM data supports the hypothesis that it is a random walk, as in the case of the discrete model fitting. Hence, as far as forecasting is concerned, both the A(2) model and the random walk model will practically give the same results. Being close to a random walk, the forecast of tomorrow's stock price is today's price. Therefore, once the model is known to be close to a random walk, neither the discrete nor the continuous model can provide significantly better forecasts than today's price.

However, the continuous A(2) model can provide additional insight into the stock price history by analogy with the physically meaningful vibration system. The parameters of the A(2) model can lead to important indices about the behavior of the company as reflected in the stock price history. Therefore, although stock prices cannot be "usefully" predicted, they can be characterized by means of an A(2) model. The A(2) model parameters tell us what the stock price history can say about the performance of the company, the stability of its stock prices, its potential for profitable investment, particularly related to other companies, and so on. If we simply fit the stock prices with a random walk model, all these questions remain unanswered. We will now show that by fitting an A(2) model, we can attempt to answer some of these questions quantitatively.

The parameters of the A(2) model were estimated by the method discussed in Section 7.5.5 to minimize the sum of squares of residual a_t's. θ_1 was computed as a dependent parameter in each iteration. The estimated parameters of the fitted discrete model are given below with their 95% confidence limits:

$$\phi_1 = 0.9985 \pm 0.0052$$

$$\phi_2 = -0.00078 \pm 0.00266$$

$$\theta_1 = -0.08742$$

$$\text{mean} = 460.0590 \pm 7.2117$$

$$\text{Residual Variance} : 51.999$$

Hence, the discrete model is (X_t denotes deviation from the mean)

$$X_t - 0.9985X_{t-1} + 0.00078X_{t-2} = a_t + 0.08742a_{t-1}$$

and considering the confidence interval, this model may be represented by

$$\nabla X_t = a_t$$

which is the same as the random walk model.

To see whether the conditions necessary for a discrete model to become a random walk model are satisfied or not, we note that the roots λ_i and μ_i are

$$\lambda_1 = 0.99776$$

$$\lambda_2 = 0.000782$$

$$\mu_1 = -0.00224$$

$$\mu_2 = -7.1528$$

It is thus clear that the conditions $\mu_1 \to 0$, $\mu_2 \to -\infty$ or $\lambda_1 \to 1$, $\lambda_2 \to 0$ are well-satisfied and we get the random walk as a limiting case. Note that $\theta_1 \to 0$, $\theta_1 < 0$ is also closely satisfied as predicted by the relation (7.6.23).

The underlying A(2) model is

$$\frac{d^2X}{dt^2} + 7.155\frac{dX}{dt} + 0.016X = Z$$

for which

$$\omega_n = 0.1265$$

$$\zeta = 2.826$$

Hence, the IBM stock prices are characterized by a high damping ratio 28.26 and comparatively low restoring force $\omega_n^2 = 0.016$. Both these facts are intuitively clear. IBM's large share in the computer market, its long standing prestige and reputation account for the high damping ratio, showing that the disturbances in its stock prices will not have a lasting effect and will be damped quickly. On the other hand, its huge size and consequent loss of flexibility may account for the small restoring force or weak "spring" action. These two numbers ω_n and ζ may be used to compare the stock price histories of two different companies, or those of the same company at two different periods, in terms of performance and stability.

7.7 EXPERIMENTAL DEMONSTRATION AND ILLUSTRATIVE APPLICATION

The experimental example in Section 7.1.4 comes from a mechanical dynamic system, consisting of a mass, a dashpot, and a spring, which was used to demonstrate the theory. A load cell is fixed on the frame structure and the other three components are allowed to move along the vertical direction. The load cell functions as a signal transducer while the moving components constitute the mechanical dynamic system. The moving components were chosen as follows:

mass	M	= 0.0023 lb-sec^2/in.
stiffness constant	K	= 1.0000 lb/in.
viscous coefficient	C	= 0.0440 lb-sec/in.

The natural frequency ω_n and the damping ratio ζ of the system, computed in Section 7.1.4, are

$$\omega_n = 20.85 \text{ rad/sec}$$

and

$$\zeta = 0.459$$

By Eqs. (7.3.6), (7.3.7), (7.3.13), (7.3.14), and (7.3.15) we have calculated the following values of parameters in Sections 7.3.1 and 7.3.3

$$\phi_1 = 1.541$$
$$\phi_2 = -0.682$$
$$\theta_1 = -0.271$$

Thus, the system is characterized by the following A(2) or discrete covariance equivalent ARMA(2,1) models

$$\frac{d^2X}{dt^2} + 19.13 \frac{dX}{dt} + 434.78X = Z$$

$$X_t - 1.541X_{t-1} + 0.682X_{t-2} = a_t + 0.271 a_{t-1}$$

The mass of the system was bombarded with random impulse shocks. These shocks were manually produced by drumming on the mass in a manner that was intended to produce an effect analogous to a random impulse function. With the drumming-excitation forces acting on the mass, the movements of the mass exerted forces on the load cell ring. Two strain gages were attached on the ring to measure the elastic deformation of the ring. The electrical signals of the gages were amplified and transmitted to a strip chart recorder. The record was then digitized with a uniform sampling interval of 0.02 seconds. A total of 130 sampled observations were collected and are listed in Appendix I at the end. The plot of the data is shown in Fig. 2.4. The usual ARMA models fitted to the data are given in Chapter 4 (Table 4.5).

7.7.1 Experimental Verification

The experimental data were then used to see whether the results based on the proposed method agree with the "true" values given above. The procedure given in Section 7.5.5 was used to estimate the parameters of the model from the collected data of 130 observations. The results of estimation, with a 95% confidence interval on the parameters, are as follows

$$\zeta = 0.3766 \pm 0.1983$$

$$\omega_n = 22.95 \pm 3.69$$

Residual Variance: 4.289
(Variance of original series: $\gamma_0 = 69.552$)
The corresponding discrete parameters are:

$$\phi_1 = 1.533$$

$$\phi_2 = -0.7076$$

$$\theta_1 = -0.2726$$

Close agreement between the true values and estimated values, and, especially, the fact that the true values lie within the 95% confidence interval of the parameters ζ and ω_n afford a clear experimental demonstration of the theory.

7.7.2 Application to Grinding Wheel Profile

To illustrate how the theory developed in preceding sections can be used to model and characterize systems other than mechanical vibrations, we will analyze the grinding wheel profile data used in Chapters 2 to 5. This data consists of 250 observations on the profile of a 32A80H8 grinding wheel. The data are taken from Stralkowski, Wu, and DeVor (1969) wherein a discrete AR(2) model has been fitted to it.

We now use the results of this chapter to fit an A(2) model to this data, employing the estimation procedure described in Section 7.5.5. The parameters for the regression routine are α_1 and α_0, whereas ϕ_1, ϕ_2, and

θ_1 are computed at each step by Eqs. (7.3.6), (7.3.7), and (7.3.15). Since the series has a nonzero mean, the mean was also estimated as an additional parameter. The estimated parameters and 95% confidence interval, based on a local linear hypothesis, are as follows:

$$\zeta = 1.008 \pm 0.36$$
$$\omega_n = 1.254 \pm 0.208$$
$$\text{mean} = 9.420 \pm 0.374$$

The covariance equivalent ARMA(2,1) model has

$$\phi_1 = 0.5725$$
$$\phi_2 = -0.0799$$
$$\theta_1 = -0.2232$$
$$\sigma_a^2 = 5.919$$

The roots μ_i and λ_i are (assuming $\Delta = 1$ for computational accuracy)

$$\mu_1 = -1.423 \qquad \lambda_1 = 0.2410$$
$$\mu_2 = -1.104 \qquad \lambda_2 = 0.3315$$

The underlying A(2) system is therefore

$$\frac{d^2X}{dt^2} + 2.527 \frac{dX}{dt} + 1.572X = Z$$

By Eq. (7.3.8)

$$\sigma_z^2 = 84.68$$

The main purpose of modeling grinding wheel profiles is to characterize the wheels by only a few parameters and relate these parameters to the physical constituents. Pandit and Wu (1973) have shown how the A(2) model parameters ζ and ω_n can be used for this purpose.

PROBLEMS

7.1 Find the homogeneous solution of Eq. (7.1.1) for the special case of a second order system with $K = 0$, $C > 0$ and discuss its stability.

7.2 Verify that Eq. (7.2.8) gives the solution of Eq. (7.2.6a).

7.3 Show that Eqs. (7.3.11), (7.3.15a), (7.3.15b), and (7.3.16) give the same P.

7.4 An A(2) system of the form

$$\frac{d^2X(t)}{dt^2} + 3\frac{dX(t)}{dt} + 1.9X(t) = Z(t)$$

is sampled at uniform intervals. Find the covariance equivalent ARMA(2,1) for each of the following sample intervals

(i) $\Delta = 1$

(ii) $\Delta = 0.1$

and discuss the effect of the sampling interval on the covariance equivalent ARMA(2,1) model.

7.5 An A(2) system of the form:

$$\frac{d^2X(t)}{dt^2} + 0.46 \frac{dX(t)}{dt} + 0.177X(t) = Z(t)$$

is sampled at a uniform interval, $\Delta = 0.1$ sec. Find the covariance equivalent ARMA(2,1) model and discuss its stability.

7.6 Verify Eqs. (7.5.2) to (7.5.5), (7.5.12), and (7.5.13).

7.7 Derive an expression for the sampling interval below which the parameters of a continuous second order system can be uniquely determined from only ϕ_1 and ϕ_2 of the sampled model.

7.8 The sampling intervals and ARMA models for five time series are given below.

(a) $X_t = 1.5188X_{t-1} - 0.5488X_{t-2} + a_t + 0.2637a_{t-1}$
     $\Delta = 0.02, \hat{\sigma}_a^2 = 2.814$

(b) $X_t = 1.3477X_{t-1} - 0.4066X_{t-2} + a_t - 0.2517a_{t-1}$
     $\Delta = 0.03, \hat{\sigma}_a^2 = 7.30$

(c) $X_t = -0.2954X_{t-1} - 0.4065X_{t-2} + a_t$
     $\Delta = 0.45, \hat{\sigma}_a^2 = 1995$

(d) $X_t = 0.8222X_{t-1} - 0.8026X_{t-2} + a_t$
     $\Delta = 0.11, \hat{\sigma}_a^2 = 500$

(e) $X_t = -.93394X_{t-1} - 0.36788X_{t-2} + a_t + 0.49664a_{t-1}$
     $\Delta = 0.5, \hat{\sigma}_a^2 = 98.78$

Discuss which of the above models are also covariance equivalent ARMA(2,1) and find the parameters of the A(2) models for the corresponding covariance equivalent ARMA(2,1). [*Hint:* A simple computer program may be useful in solving the problem.]

7.9 A student used the estimation procedure given in Section 7.5.5 to estimate an A(2) system, and he found that $\zeta = 0.707$ and $\omega_n = 78.6$ rad/sec. He later discovered that the true sampling interval was 0.0015 instead of 0.003 sec. What is the true ζ and ω_n? Would the value of σ_z^2 change?

8

AM(2,1) MODEL
AND ITS APPLICATION
TO EXPONENTIAL SMOOTHING

A second order model extended by introducing a derivative of the white noise $Z(t)$ on the right-hand side of the A(2) model is a continuous time Autoregressive Moving Average model denoted by AM(2,1). The corresponding sampled system leads to many important discrete models, in particular the EWMA model discussed in Chapter 5. The results in this chapter relating the AM(2,1) model to the technique of exponential smoothing are based on Pandit and Wu (1974) with some modifications.

This chapter analyzes the continuous stochastic system underlying exponential smoothing. The systems in economics, business, quality control, etc. for which exponential smoothing works well are generally so complex that it is not possible to formulate the differential equations for such systems *a priori*, as in the case of mechanical or electrical systems. However, many of these complex systems can be approximated by a "lumped" second order system analogous to the familiar spring-dashpot system of mechanical vibrations.

The spring-dashpot interpretation can be applied even in systems that are not mechanical. For example, in dealing with a model for stock market prices, the "spring" may be interpreted as the flexibility in the policies of the firm that enables it to react to external shocks. The "dashpot" may be interpreted as the collection of factors such as the financial stability of the firm, its reputation and prestige, monopoly of its products etc., which can damp the effect of external pressure on stock prices to some extent.

When such a continuous stochastic system is sampled at uniform intervals, it has a discrete representation of the same form as an ARMA(2,1) model. It is shown in this chapter that exponential smoothing is optimal for a limiting case of the discrete model, which therefore represents the basic underlying stochastic system. If the limiting conditions are satisfied, then exponential smoothing will yield optimal prediction; otherwise, the discrete model obtained from the data should be used.

This approach via the AM(2,1) model has the following advantages:

1 The circumstances and conditions under which exponential smoothing is appropriate become clear.

2 If exponential smoothing is appropriate, the optimal value of the smoothing parameter λ can be obtained from the data and the extent of increase in prediction error for nonoptimal values ascertained. If not, the discrete model that will give a minimum mean squared prediction error can be used.

3 The differential equation associated with the model is helpful in characterizing the system and interpreting it by analogy with a spring-dash-pot system. Moreover, the coefficients of the differential equations can be used for meaningful comparison between two systems.

4 A theoretical explanation is revealed for the empirical fact that many business, economic, and quality control systems can be predicted well by exponential smoothing.

The differential equation for an AM(2,1) model is formulated in Section 8.1 and the discrete model derived. It is shown in Section 8.2 that the time series model for which the exponential predictor is optimal can be derived as a limiting case of the discrete model. This derivation is used in Section 8.3 to obtain the optimal values of the exponential smoothing parameter and its sensitivity to mean squared prediction error. The interpretation of the AM(2,1) model as a result of a first order system with first order feedback is developed in Section 8.4. A discussion of why this model would be basic for many complex systems is also given. Other limiting cases of the discrete model are discussed in Section 8.5. A practical example dealing with actual quality control data on a machined part is given in Section 8.6.

8.1 AM(2,1) MODEL AND ITS SAMPLED REPRESENTATION

Consider the continuous time autoregressive moving average model AM(2,1) obtained by introducing a derivative of $Z(t)$ on the right-hand side of an A(2) model:

$$(D^2 + 2\zeta\,\omega_n D + \omega_n^2)X(t) = (b_1 D + 1)Z(t)$$

$$E[Z(t)] = 0 \qquad (8.1.1)$$

$$E[Z(t)Z(t-u)] = \sigma_z^2\,\delta(u)$$

The Green's function of the AM(2,1) system is given by

$$(D^2 + 2\zeta\omega_n D + \omega_n^2)G(t) = (b_1 D + 1)\delta(t) \qquad (8.1.2)$$

and is thus the sum of the responses of the A(2) system for the inputs $\delta(t)$ and $b_1\delta'(t)$, which can be obtained by the same discontinuity argument as in the A(2) case. The Green's function can also be obtained more directly from that of the A(2) system by

$$G(t) = (D^2 + 2\zeta\omega_n D + \omega_n^2)^{-1}[(b_1 D + 1)\delta(t)]$$

$$= \int_{-\infty}^{t} \left[\frac{e^{\mu_1(t-v)} - e^{\mu_2(t-v)}}{\mu_1 - \mu_2} \right] [b_1\delta'(v) + \delta(v)] \, dv \qquad (8.1.3)$$

$$= \frac{1}{(\mu_1 - \mu_2)} [(b_1\mu_1 + 1)e^{\mu_1 t} - (b_1\mu_2 + 1)e^{\mu_2 t}]$$

using the property (6.2.7) of the delta function, and this gives the orthogonal decomposition of $X(t)$ as

$$X(t) = \frac{1}{\mu_1 - \mu_2} \int_{-\infty}^{t} [(b_1\mu_1 + 1)e^{\mu_1(t-v)}$$

$$- (b_1\mu_2 + 1)e^{\mu_2(t-v)}]Z(v) \, dv \qquad (8.1.4)$$

where

$$\mu_1 = -\zeta\omega_n + \omega_n\sqrt{\zeta^2 - 1}, \qquad \mu_2 = -\zeta\omega_n - \omega_n\sqrt{\zeta^2 - 1} \qquad (8.1.5)$$

Assuming a stable system, the autocovariance function $\gamma(s)$ is given by

$$\gamma(s) = E[X(t)X(t-s)]$$

$$= \sigma_z^2 \int_0^{\infty} G(v)G(v+s) \, dv$$

$$= \frac{\sigma_z^2(1 - b_1^2\mu_1^2)}{2\mu_1(\mu_1^2 - \mu_2^2)} e^{\mu_1 s} - \frac{\sigma_z^2(1 - b_1^2\mu_2^2)}{2\mu_2(\mu_1^2 - \mu_2^2)} e^{\mu_2 s} \qquad (8.1.6)$$

$$\gamma(s) = \gamma(-s)$$

If the AM(2,1) process given by Eq. (8.1.1) is observed at uniform intervals Δ, then the autocovariance γ_k of the sampled discrete process can be obtained by replacing s by $k\Delta$ in Eq. (8.1.6).

$$\gamma_k = d_1\lambda_1^k + d_2\lambda_2^k$$

$$= \gamma(k\Delta)$$

$$= \frac{\sigma_z^2(1 - b_1^2\mu_1^2)}{2\mu_1(\mu_1^2 - \mu_2^2)} e^{\mu_1 k\Delta} - \frac{\sigma_z^2(1 - b_1^2\mu_2^2)}{2\mu_2(\mu_1^2 - \mu_2^2)} e^{\mu_2 k\Delta} \qquad (8.1.7)$$

where d_1 and d_2 from Chapter 3 are given by Eqs. (3.3.16a) and (3.3.16b). Therefore,

$$\lambda_1 = e^{\mu_1\Delta} \qquad (8.1.8)$$

$$\lambda_2 = e^{\mu_2\Delta} \qquad (8.1.9)$$

$$d_1 = \frac{\sigma_a^2(\lambda_1 - \theta_1)}{(\lambda_1 - \lambda_2)^2} \left[\frac{(\lambda_1 - \theta_1)}{(1 - \lambda_1^2)} - \frac{(\lambda_1 - \theta_1)}{(1 - \lambda_1\lambda_2)} \right]$$

$$= \frac{\sigma_z^2(1 - b_1^2\mu_1^2)}{2\mu_1(\mu_1^2 - \mu_2^2)} \qquad (8.1.10)$$

$$d_2 = \frac{\sigma_a^2(\lambda_2 - \theta_1)}{(\lambda_1 - \lambda_1)^2}\left[\frac{(\lambda_2 - \theta_1)}{(1 - \lambda_2^2)} - \frac{(\lambda_1 - \theta_1)}{(1 - \lambda_1\lambda_2)}\right] \tag{8.1.11}$$

$$= -\frac{\sigma_z^2(1 - b_1^2\mu_2^2)}{2\mu_2(\mu_1^2 - \mu_2^2)}$$

can be equated and the sampled process may be represented by the co-variance equivalent ARMA(2,1) model

$$X_t - \phi_1 X_{t-1} - \phi_2 X_{t-2} = a_t - \theta_1 a_{t-1} \tag{8.1.12}$$

where

$$\phi_1 = \lambda_1 + \lambda_2 = e^{\mu_1 \Delta} + e^{\mu_2 \Delta} \tag{8.1.13}$$

$$\phi_2 = -\lambda_1\lambda_2 = -e^{(\mu_1 + \mu_2)\Delta}$$

The expression for θ_1 can be obtained by imposing the condition that the autocovariance function of the discrete model (8.1.12) be the same as the sampled autocovariance function of the AM(2,1) model as given by Eqs. (8.1.7) to (8.1.11). Using the same procedure as that in Chapter 7 and solving the simultaneous equations (8.1.10) and (8.1.11), we can show the expression of θ_1 to be

$$\theta_1 = -P \pm \sqrt{P^2 - 1}, \qquad |\theta_1| < 1 \tag{8.1.14}$$

where

$$P = \frac{\mu_2(1 + \lambda_2^2)(1 - \lambda_1^2)(1 - b_1^2\mu_1^2) - \mu_1(1 + \lambda_1^2)(1 - \lambda_2^2)(1 - b_1^2\mu_2^2)}{2\{\mu_1\lambda_1(1 - \lambda_2^2)(1 - b_1^2\mu_2^2) - \mu_2\lambda_2(1 - \lambda_1^2)(1 - b_1^2\mu_1^2)\}} \tag{8.1.15a}$$

Note that μ_1 and μ_2 could be either real or complex, but P is always real. An alternative expression for P is derived as follows. Let

$$a = \zeta\omega_n$$

$$b = \omega_n\sqrt{|\zeta^2 - 1|}$$

Then for real μ_1 and μ_2, (that is, $\zeta > 1$)

$$P = \frac{b(1 + b_1^2\omega_n^2)\sinh(2a\Delta) - a(1 - b_1^2\omega_n^2)\sinh(2b\Delta)}{2[a\sinh(b\Delta)\cosh(a\Delta)(1 - b_1^2\omega_n^2) - b\sinh(a\Delta)\cosh(b\Delta)(1 + b_1^2\omega_n^2)]} \tag{8.1.15b}$$

For complex μ_1 and μ_2 (that is, $\zeta < 1$),

$$P = \frac{b(1 + b_1^2\omega_n^2)\sinh(2a\Delta) - a(1 - b_1^2\omega_n^2)\sin(2b\Delta)}{2[a\sin(b\Delta)\cosh(a\Delta)(1 - b_1^2\omega_n^2) - b\sinh(a\Delta)\cos(b\Delta)(1 + b_1^2\omega_n^2)]} \tag{8.1.15c}$$

The reader should note that b_1 represents the moving average parameter of the AM(2,1) model, whereas $b = \omega_n\sqrt{|\zeta^2 - 1|}$ denotes the (imaginary) part of the roots μ_1 and μ_2.

It can be shown from the implicit expressions of the discrete auto-covariance function that

$$\sigma_a^2 = \frac{\mu_2(1+\lambda_2^2)(1-\lambda_1^2)(1-b_1^2\mu_1^2) - \mu_1(1+\lambda_1^2)(1-\lambda_2^2)(1-b_1^2\mu_2^2)}{2\mu_1\mu_2(\mu_1^2-\mu_2^2)(1+\theta_1^2)}\sigma_z^2 \qquad (8.1.16a)$$

Similar to P, the expressions for σ_a^2 due to real or complex roots can be further written as

$$\sigma_a^2 = \frac{b(1+b_1^2\omega_n^2)\sinh(2a\Delta) - a(1-b_1^2\omega_n^2)\sinh(2b\Delta)}{2ab(a^2-b^2)(1+\theta_1^2)e^{2a\Delta}}\sigma_z^2, \qquad (8.1.16b)$$

$$\text{for } \zeta \geq 1$$

$$\sigma_a^2 = \frac{b(1+b_1^2\omega_n^2)\sinh(2a\Delta) - a(1-b_1^2\omega_n^2)\sin(2b\Delta)}{2ab(a^2+b^2)(1+\theta_1^2)e^{2a\Delta}}\sigma_z^2, \qquad (8.1.16c)$$

$$\text{for } \zeta < 1$$

Thus, we have a complete representation of the sampled process by Eq. (8.1.12), the parameters ϕ_1, ϕ_2, θ_1, and σ_a^2 being given by Eqs. (8.1.13), (8.1.14), and (8.1.16) respectively.

When the parameter b_1 in Eq. (8.1.1) is zero, it takes the familiar form of a vibration system with white noise excitation discussed in Chapter 7:

$$\frac{d^2X}{dt^2} + 2\zeta\omega_n\frac{dX}{dt} + \omega_n^2X = Z$$

This is the same as the autoregressive model of order 2, denoted by A(2). When sampled at a uniform interval Δ, it still has discrete representation (8.1.12) with ϕ_1, ϕ_2 given by Eq. (8.1.13), but the expressions for θ_1 and σ_a^2 are different as given in Chapter 7, which can be obtained from Eqs. (8.1.14) to (8.1.16) by substituting $b_1 = 0$.

8.2 DERIVATION OF EXPONENTIAL SMOOTHING

Let us consider the time series model for which exponential smoothing is optimal. Such a model is called the Exponentially Weighted Moving Average (EWMA) model. As shown in Section 5.5.3, the EWMA model is [see Eq. (5.5.5)]

$$\nabla X_t = X_t - X_{t-1} = a_t - \theta_1 a_{t-1} \qquad (8.2.1)$$

We now show how the EWMA model (8.2.1) arises when the damping factor ζ in the AM(2,1) model becomes large. For large ζ, the model (8.1.12) tends to EWMA.

It is seen from Eqs. (8.1.5), (8.1.8), and (8.1.9) that for a fixed $\omega_n\Delta$

when $\zeta \to \infty$

$$\mu_1 \to 0$$
$$\mu_2 \to -\infty \qquad (8.2.2)$$
$$\lambda_1 \to 1$$
$$\lambda_2 \to 0$$

Hence, by Eq. (8.1.13)

$$\phi_1 = \lambda_1 + \lambda_2 \to 1 \qquad (8.2.3)$$
$$\phi_2 = -\lambda_1 \lambda_2 \to 0$$

The limiting value of θ_1 can be found if we note from Eq. (8.1.15a) that when $b_1 \neq 0$

$$P \to -1$$

Therefore, by Eq. (8.1.14)

$$\theta_1 \to 1, \qquad \theta_1 > 0 \qquad (8.2.4)$$

[Compare and contrast this result with the one for $b_1 = 0$ in Section 7.6.4(ii).]

The results (8.2.3) and (8.2.4) show that for a large damping factor ζ, model (8.1.12) tends to the EWMA model with positive θ_1 very close to one, or equivalently, $\lambda (=1-\theta_1)$ very close to zero. Based on empirical experience, Brown (1962) has recommended values of λ between 0 and 0.3 as general guidance for most business and economic systems. The foregoing results offer a possible theoretical justification for these values and suggest that the underlying time series model for such systems is AM(2,1). In Section 8.4 we will interpret the AM(2,1) model as a first order system with feedback and substantiate this hypothesis.

8.3 OPTIMAL VALUE OF λ AND ITS SENSITIVITY

Many attempts have been made [see Brown (1962)] to find the value of λ that is optimal in the sense of a minimum mean squared error of prediction. The reason why such attempts have been somewhat unsuccessful is clear from the preceding analysis. First of all, it shows that the exponential predictor is appropriate only when the underlying second order system has a high damping factor. Second, even when it is appropriate, its optimal value would be close to zero if the underlying model is AM(2,1).

Hence, a method of obtaining the optimal value for a given set of data is to fit an AM(2,1) model via Eqs. (8.1.12) to (8.1.14) to the data. If the parameters ϕ_1 and ϕ_2 are close to 1 and 0, then the EWMA model (8.2.1) may be fitted to the data and the estimated value of θ_1 gives the optimal value of λ. Otherwise, the fitted model (8.1.12) rather than the exponential predictor should be used for prediction.

It is common in the existing literature on exponential smoothing to treat the EWMA predictor as a technique by itself and apply it to arbitrary time series. In such a case, it is necessary to assume some model or probabilistic structure for the time series in order to derive the optimal value of the smoothing constant λ. The preceding analysis shows that the model for the series should be AM(2,1).

Cox (1961) has derived the optimal smoothing parameter by assuming a discrete autoregressive model of order one. Using this model, he shows that for a given value of the autoregressive parameter, the mean squared prediction error of the EWMA predictor varies extremely slowly with λ so that the choice of λ is not critical.

This insensitivity to the choice of λ is in accord with practical experience in exponential smoothing and can be readily explained by the foregoing results. Suppose that an exponential smoothing constant λ is being used for prediction. Then, for $\theta_1 = 1 - \lambda$, the one step ahead prediction is given by

$$\hat{X}_t(1) = X_t - \theta_1 a_t$$

If the underlying model is AM(2,1), then the observation X_{t+1} is given by

$$X_{t+1} \simeq X_t - a_t + a_{t+1}$$

so that the expected prediction mean squared error is

$$E[X_{t+1} - \hat{X}_t(1)]^2 = [1 + (1 - \theta_1)^2]\sigma_a^2 = (1 + \lambda^2)\sigma_a^2 \qquad (8.3.1)$$

Since σ_a^2 is the optimal prediction mean squared error (m.s.e.), it is seen that the m.s.e. varies very slowly for appreciable deviations in λ from optimal values. For example, if a value of $\lambda = 0.5$ is chosen instead of $\lambda_{opt} = 0$, the m.s.e. increases only by 25%. In particular, the recommended values are $0 < \lambda < 0.3$, with $\lambda_{opt} = 0$, and the largest increase in m.s.e. is only 9%. The sensitivity of the mean squared prediction error for errors in the choice of λ is plotted in Fig. 8.1. This shows that as long as exponential smoothing is appropriate, the largest increase in the m.s.e. remains under 10% for arbitrary values of λ within the recommended range of 0 to 0.3. This makes the exponential smoothing technique somewhat robust and therefore popular in business and industry. The reason why the AM(2,1) model leading the EWMA would be appropriate is explained in the next section.

8.4 AM(2,1) MODEL AS A RESULT OF FEEDBACK

In this section, we will interpret the AM(2,1) model (8.1.1) as a first order system with first order feedback. The limiting case of the resultant discrete model then explains why many business and economic systems follow an EWMA model.

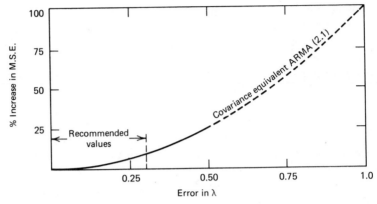

Figure 8.1 Sensitivity of mean squared error of prediction to error in choice of EWMA Constant λ.

Consider a first order system governed by an A(1) model

$$\frac{dX}{dt} + \alpha_0'X = Z \tag{8.4.1}$$

that may also be written as

$$\frac{dX}{dt} = Z - \alpha_0'X \tag{8.4.2}$$

This form shows that the coefficient α_0' represents the resistance to change caused by the white noise disturbance Z, and the larger the parameter α_0', the larger the resistance. This resistance, which counteracts the disturbance Z, is "internal" to the system. The influence of Z on X can also be counteracted by a feedback Y that depends on X. If we assume this feedback is also first order, the system equations may be written as

$$\frac{dX}{dt} + \alpha_0'X = Z - Y \tag{8.4.3}$$

$$\frac{dY}{dt} + \alpha_f Y = cX \tag{8.4.4}$$

where α_f is the parameter of the feedback system. A block diagram of this system is shown in Figure 8.2a.

Eliminating Y from the two equations, we get the equation for the complete system

$$\frac{d^2X}{dt^2} + (\alpha_0' + \alpha_f)\frac{dX}{dt} + (c + \alpha_0'\alpha_f)X = \frac{dZ}{dt} + \alpha_f Z \tag{8.4.5}$$

(a)

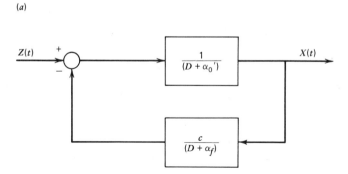

(b)

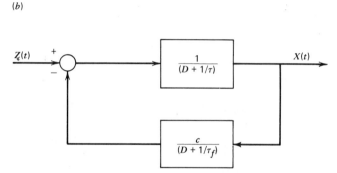

Figure 8.2   Block diagram of the feedback system for an AM(2,1) model.

This is an AM(2,1) model for which

$$\mu_1 = \frac{1}{2}[-(\alpha_0' + \alpha_f) + \sqrt{(\alpha_0' + \alpha_f)^2 - 4(c + \alpha_0'\alpha_f)}]$$

$$\mu_2 = \frac{1}{2}[-(\alpha_0' + \alpha_f) - \sqrt{(\alpha_0' + \alpha_f)^2 - 4(c + \alpha_0'\alpha_f)}]$$

(8.4.6)

It now follows from the reasoning of Section 8.2 that if α_f is very small compared to α_0' or vice versa so that $\alpha_0'\alpha_f$ is small, and moreover c is small, then $\mu_1 \to 0$, $\mu_2 \to -\infty$, and we get an EWMA model as the limiting case of (8.4.5) with $\theta_1 \to 1$, $\theta_1 > 0$. For example, if $\alpha_f = 0.01$, $\alpha_0' = 10$, $c = 0.1$, then $\mu_1 \simeq -0.02$ and $\mu_2 \simeq -9.99$. Note that the result is exactly the same for $\alpha_f = 10$ and $\alpha_0' = 0.01$, as is also clear from the symmetry of (8.4.6) in α_0' and α_f. These requirements can be relaxed when c is negative, as discussed at the end of Section 8.6.

Such a situation is easily obtained in practice in many systems of which precise knowledge is lacking, particularly in business and economic systems. In such systems, both the internal resistance to disturbances and the external corrective action in the form of feedback based on output deviations from the desired level are present. Hence, these systems are well-represented by the AM(2,1) model.

If the system response to the disturbance is very quick, or in other words, the resistance is strong leading to a large α'_0, then the output variable remains more or less in control. Therefore, there is very little need for corrective action by feedback, the feedback response is very slow, and the value of α_f is very small.

In contrast, if the system response to disturbance is very slow, that is, the resistance is low, or α'_0 is very small, then the output will be almost out of control unless corrective action is quickly taken. Heavy feedback will be used in such cases, the feedback response will be very quick, and the value of α_f will be very large.

In many business and economic systems the knowledge of the system response is so limited that they are controlled almost totally by considering output levels. If the variation in the output is not critical, very little feedback is used, thus giving large α'_0 and small α_f. If the variation is large, the situation is reversed. This may explain why many business and economic systems can be predicted well by using the EWMA predictor.

Another instance in which the AM(2,1) model (8.1.1) could be appropriate is quality control. Here, the equipment or machinery is the primary system and the operator serves as the feedback loop. When the equipment is in good condition, its resistance to disturbances is very large, the quality of the product is fairly under control, and the operator makes much less use of feedback leading to large α'_0 and small α_f. When the equipment is in bad condition, the opposite is the case. The success of Shewhart control charts and the fact that the EWMA model (8.2.2) is the right model underlying Shewhart charts indicate that most of the series used in quality control follow EWMA. The above discussion explains why the underlying model for quality control systems is EWMA.

Recall from Section 6.1.2 that for a first order system, the time constant τ is equal to $1/\alpha_0$ (see Fig. 6.2). Therefore, the feedback system of the AM(2,1) model may also be alternatively represented in terms of the respective time constants as shown in Figure 8.2b. The above discussion shows that for a first order stochastic system with first order feedback, if the time constant of either the system $\tau = 1/\alpha'_0$ or of the feedback loop $\tau_f = 1/\alpha_f$ is much larger compared to the other, then the sampled representation tends to an EWMA model. For such a system, exponential smoothing will provide good forecasts and can be used in preference to ARMA or AM models. The choice of the smoothing constant λ can be made on the basis of the AM model occasionally fitted to the data as a check.

8.5 OTHER LIMITING CASES AND INTERMEDIATE VALUES OF THE DISCRETE PARAMETERS

As in Chapter 7, the extreme values of ω_n and Δ will be considered together to study the behavior of the discrete model parameters.

(a) Large $\omega_n\Delta$

As $\omega_n\Delta \to \infty$

$$\phi_1 \to 0$$
$$\phi_2 \to 0$$
$$\theta_1 \to 0$$
$$\sigma_a^2 \to \gamma(0)$$

and

$$\gamma(0) \to 0 \qquad \text{if} \qquad \omega_n \to \infty$$

(b) Small $\omega_n\Delta$

As $\omega_n\Delta \to 0$

$$\phi_1 \to 2$$
$$\phi_2 \to -1$$

and after using L'Hospital's rule for taking the limit of θ_1, we can verify that, if $b_1 \neq 0$,

$$\theta_1 \to 1$$

Hence, for $\omega_n\Delta \to 0$, the discrete model (8.1.12) takes the form:

$$X_t - 2X_{t-1} + X_{t-2} = a_t - a_{t-1}$$

that is,

$$\nabla^2 X_t = \nabla a_t$$

which is close to a random walk as seen by the cancellation of the factor $(1-B)$. It can also be verified that, as $\omega_n\Delta \to 0$,

$$\sigma_a^2 \to 0$$

(c) Intermediate Values of $\omega_n\Delta$

The parameters ϕ_1 and ϕ_2 are independent of b_1; thus, as $\omega_n\Delta$ varies, they behave in the same manner as those given in Chapter 7. However, the behavior of θ_1 and σ_a^2, as seen from Eqs. (8.1.14) to (8.1.16), depends upon b_1. For clarity purposes, let us consider the overdamped and underdamped system separately.

(i) Overdamped System

Since the expression (8.1.14) for θ_1 involves P, we will first discuss the behavior of P with the changing value of ω_n and parameter b_1. From Eq. (8.1.15b),

$$P = \frac{P_1}{2P_2}$$

where

$$P_1 = b(1 + b_1^2 \omega_n^2) \sinh (2a\Delta)$$
$$- a(1 - b_1^2 \omega_n^2) \sinh (2b\Delta)$$
$$P_2 = a \sinh (b\Delta) \cosh (a\Delta)(1 - b_1^2 \omega_n^2)$$
$$- b \sinh (a\Delta) \cosh (b\Delta)(1 + b_1^2 \omega_n^2)$$

Since for all values of $\omega_n\Delta$, $b \sinh (2a\Delta) \geqslant a \sinh (2b\Delta)$

$$P_1 \geqslant 0 \qquad \text{for all} \qquad \omega_n\Delta \qquad \text{and} \qquad b_1 \qquad (8.5.4)$$

Also, if $\omega_n\Delta > 0$, then $[a \sinh (b\Delta) \cosh (a\Delta)] > [b \sinh (a\Delta) \cosh (b\Delta)]$. Thus, it follows that P_2 is positive for small b_1 and negative for large b_1. This brief analysis shows that P can be both positive and negative, depending on the values of $\omega_n\Delta$ and b_1. Hence, θ_1 can be positive and negative, and $1 \geqslant \theta_1 \geqslant -2 + \sqrt{3}$. The plots of θ_1 and σ_a^2 as a function of $\omega_n\Delta$ for various values of b_1 are shown in Fig. 8.3.

(ii) Underdamped Systems

Eq. (8.1.15b) specifies the value of P for an underdamped system. Since P contains sine and cosine functions that are oscillatory in nature and θ_1 is a finite function of P, θ_1 is oscillatory. The behaviors of θ_1 and σ_a^2 as a function of $\omega_n\Delta$ for various values of b_1 are shown graphically in Fig. 8.4.

8.6 AN ILLUSTRATIVE EXAMPLE—DIAMETER MEASUREMENTS OF A MACHINED PART

Recall that the results (8.2.3) to (8.2.4) for $b_1 = 0$ were illustrated by the IBM data for economic systems in Section 7.6.5. We now illustrate these results for $b_1 \neq 0$ to show how the model (8.1.12) tends to EWMA when the damping factor ζ becomes large compared to ω_n. As we remarked at the end of Section 8.4, such models are naturally appropriate in quality control. The present example uses quality control data from Wu and Dalal (1971), listed in Appendix I at the end of the book.

Figure 8.5 shows the last three digits of the micrometer readings in the quality control inspection of two different diameters (Diameter 1 and 2 are ¼ inch and ⅛ inch in nominal size) on the same machined part

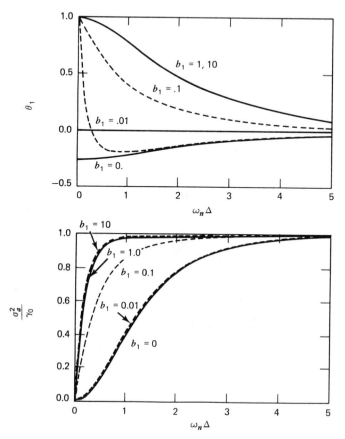

Figure 8.3   Behaviors of θ_1 and σ_a^2 for an AM(2,1) model with $\zeta = 1.2$.

manufactured on a single spindle automatic screw machine. The AM(2,1) model fitted to these series give the following values of parameters and 95% confidence limits.

Diameter 1

$$\phi_1 = 1.0144 \pm 0.0436$$

$$\phi_2 = -0.01305 \pm 0.04326$$

$$\theta_1 = 0.9523 \pm 0.1713$$

$$\text{mean} = 392.5613 \pm 1.6822$$

$$\text{Residual Variance} : 27.6361$$

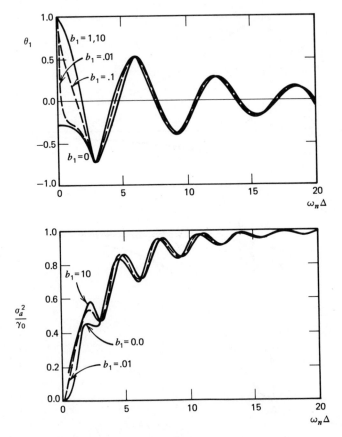

Figure 8.4   Behaviors of θ_1 and σ_a^2 for an AM(2,1) model with $\zeta = 0.1$.

Diameter 2

$$\phi_1 = 0.9987 \pm 0.0465$$

$$\phi_2 = -0.001 \pm 0.04586$$

$$\theta_1 = 0.8954 \pm 0.2345$$

$$\text{mean} = 394.2242 \pm 3.1988$$

$$\text{Residual Variance} : 143.3803$$

Therefore, the two discrete models are

$$X_t - 1.0144X_{t-1} + 0.01305X_{t-2} = a_t - 0.9523a_{t-1}$$

$$X_t - 0.9987X_{t-1} + 0.001X_{t-2} = a_t - 0.8954a_{t-1}$$

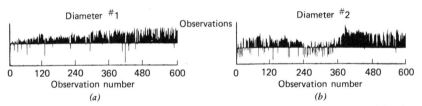

Figure 8.5   Quality control measurements of two diameters on a machined part. (Deviations from the target.)

and considering the confidence intervals on the parameters, these models may be represented by

$$\nabla X_t = a_t - 0.9523a_{t-1}$$

$$\nabla X_t = a_t - 0.8954a_{t-2}$$

which are of the form (8.2.1) and illustrate the results (8.2.3) to (8.2.4). The roots λ_i and μ_i are

Diameter 1

$$\lambda_1 = 1.0013 \quad \text{Unstable}$$

$$\lambda_2 = 0.01303$$

$$\mu_1 = 0.001325$$

$$\mu_2 = -4.3402$$

Diameter 2

$$\lambda_1 = 0.9977$$

$$\lambda_2 = 0.001002$$

$$\mu_1 = -0.0023$$

$$\mu_2 = -6.9054$$

Thus, the conditions $\lambda_1 \to 1$, $\lambda_2 \to 0$ or $\mu_1 \to 0$, $\mu_2 \to -\infty$ required for the limit of the discrete model to be EWMA are satisfied. (See Problem 8.8)

The underlying AM(2,1) models from Eqs. (8.1.13) to (8.1.15a) are

$$\frac{d^2 X}{dt^2} + 4.3389 \frac{dX}{dt} - 0.0057X = 13.5778 \frac{dZ}{dt} + Z$$

$$\frac{d^2 X}{dt^2} + 6.9077 \frac{dX}{dt} + 0.0159X = 4.8693 \frac{dZ}{dt} + Z$$

If we now interpret these AM(2,1) models as results of feedback, then comparing Eq. (8.4.5) with the above AM(2,1) equations, we have

Diameter 1

$$\alpha_0' + \alpha_f = \quad 4.3389$$

$$c + \alpha_0'\alpha_f = -0.0057$$

$$\alpha_f = \frac{1}{13.5778} = 0.0737$$

Solving these equations, we get

$$\alpha_0' = \quad 4.2652$$

$$\alpha_f = \quad 0.0737$$

$$c = -0.3201$$

so that the system equation for Diameter 1 corresponding to (8.4.3) to (8.4.5) are

$$\frac{dX}{dt} + 4.2652X = Z - Y$$

$$\frac{dY}{dt} + 0.0737Y = -0.3201X$$

Similarly, for

Diameter 2

$$\alpha_0' = \quad 6.7023$$

$$\alpha_f = \quad 0.2054$$

$$c = -1.3608$$

and the system equations are

$$\frac{dX}{dt} + 6.7023X = Z - Y$$

$$\frac{dY}{dt} + 0.2054Y = -1.3608X$$

The feedback given by these equations is not intentional but occurs in the machine tool itself, since no manual adjustments were made during the entire process.

To consider the implications of these models, we first note that μ_1 for Diameter 1 is slightly positive, whereas that for Diameter 2 is small but negative. This indicates that the process for Diameter 1 is unstable and therefore out of control, whereas for Diameter 2 it is almost on the boundary of stability, but still stable (in fact, asymptotically stable), and therefore in control. These facts can be qualitatively confirmed by visually inspecting the series in Fig. 8.5. Measurements of Diameter 1 clearly show a drift indicating out of control behavior, whereas Diameter 2 does not have such an unstable drift.

The system equations derived above also quantify the difference in stability or state of control of the two diameters. Diameter 1 has a lower parameter α_0', 4.2652 compared to 6.7023 for Diameter 2, and also a lower feedback parameter α_f, 0.0737 compared to 0.2054. Thus, the resistance to disturbances is much higher in the case of Diameter 2 as compared to Diameter 1.

The negative values of c in both cases show that the feedback, instead of aiding the system in curbing the disturbances and keeping the process under control, aids the disturbance and tends to throw the system out of control. This reflects the deteriorating condition of the cutting process, possibly caused by tool wear. In the absence of manual or automatic corrective feedback, the unintentional negative feedback leads to uncontrolled instability.

The negative c in this example relaxes a condition for the discrete model tending to EWMA, discussed in Section 8.4 with reference to relations (8.4.6). When c is a positive small number, Eq. (8.4.6) shows that $\alpha_0'\alpha_f$ must be small in order that $\mu_1 \to 0$ and $\mu_2 \to -\infty$. However, when c is negative, $\alpha_0'\alpha_f$ can be large as long as $c + \alpha_0'\alpha_f$ is small. For example, although $\alpha_0'\alpha_f$ is 1.3767 for Diameter 2, $c + \alpha_0'\alpha_f = 0.0159$ is still much smaller than $\alpha_0' + \alpha_f$ of 6.9077, leading to $\mu_1 = -0.0023$ and $\mu_2 = -6.9054$, and giving EWMA as the limiting case.

PROBLEMS

8.1 Verify the expressions (8.1.6), (8.1.15a–c), and (8.1.16a–c). [*Hint:* See Section 7.3.2.]

8.2 Find an expression for $\gamma(0)$ of an AM(2,1) model. For the same σ_z^2, is the variance larger or smaller compared to the A(2) model? Why? Give physical explanation in addition to mathematical reasoning.

8.3 Verify the limits of θ_1 and σ_a^2 given in Section 8.5.

8.4 Show that as $\zeta \to \infty$, the value of P given by Eq. (8.1.15a) can be approximated by

$$P \simeq \frac{\mu_2(1-\lambda_1^2)(1-b_1^2\mu_1^2)}{2\mu_1\,\lambda_1(1-b_1^2\mu_2^2)} - 1$$

[*Hint:* $\lambda_2 \to 0$ faster than $\mu_2 \to -\infty$]

8.5 For each of the models in the illustrative applications given in Section 8.6,

(a) compute θ_1 using the approximations derived in Problem 8.4 and see how they tally with the estimated values of θ_1;

(b) what values of the exponential smoothing parameter λ should be employed if one decides to use EWMA for prediction?

8.6 If the conditions required for EWMA are nearly satisfied, what intervals of values of λ should be employed for EWMA forecasting if the error variance is not to exceed 4% above the optimal value and the underlying model is (i) A(2), (ii) AM(2,1)?

8.7 Taking $\alpha_0' = 5, 10, 15$, give the approximate values of the feedback parameters c and α_f, both positive, that will justify the EWMA model for the sampled process within the confidence limits ± 0.01 of the parameter ϕ_1.

8.8 (a) Show that

$$b_1^2 = \frac{(\mu_2 D_1 - \mu_1 D_2)}{\mu_1 \mu_2 (\mu_1 D_1 - \mu_2 D_2)}$$

where

$$D_1 = (1 - \lambda_1^2)(1 + \lambda_2^2 + 2P\lambda_2)$$
$$D_2 = (1 - \lambda_2^2)(1 + \lambda_1^2 + 2P\lambda_1)$$

and

$$2P = -\left(\theta_1 + \frac{1}{\theta_1}\right)$$

(b) Verify the values of b_1 used in Section 8.6.

(c) What is the justification for choosing a positive value of b_1? Recalculate the parameters of the decomposed AM(2,1) model using negative b_1, and compare and contrast the resulting feedback system with the one given in section 8.6.

(d) What basic assumption is violated in calculating b_1 for Diameter 1? Why and where is it needed? Why do we get "reasonable" results in spite of this violation?

8.9 Via Eqs. (8.1.14) and (8.1.15c), show that for $\zeta < 1$

$$b_1^2 = \frac{(2PD_1 + N_2)^2 - [b(2PD_2 + N_1)]^2}{\omega_n^2 [2PD_1 + N_2 + b(2PD_2 + N_1)]^2}$$

where

$N_1 = \sinh 2a\Delta$

$N_2 = a \sin 2b\Delta$

$D_1 = a \sin b\Delta \cosh a\Delta$

$D_2 = \sinh a\Delta \cos b\Delta$

9

STOCHASTIC TRENDS AND SEASONALITY

In the preceding chapters, no explicit recognition was given to the trends and seasonal or periodic patterns, if present in the data. Nevertheless, it was found that no difficulty was encountered when data with such patterns was modeled by the methods of Chapter 4 and used for forecasting in Chapter 5.

For example, the IBM data in Fig. 2.3 clearly shows the series' tendency to remain constant at a given level, infrequently disturbed by its wandering off. In spite of this trend and the somewhat unstable or nonstationary appearance of the series, we did not take a difference or carry out other "preprocessing" of the data to "make the series stationary." The series was modeled without any difficulty in Chapter 2, as well as in Chapter 4. Rather than our guessing the trend and trying to remove it rightly or wrongly by preprocessing, the model itself told us about the existence of such a trend by returning a real characteristic root $\lambda_1 \simeq 1$, unambiguously exhibiting the random walk nature of the series. The model for the series was then simplified by the difference operator in Section 4.5.2 and this simplified model was used for forecasting in Chapter 5.

It would therefore seem that a large variety of apparent trends and seasonality patterns in the data can be modeled satisfactorily by the methods of Chapter 4. This is indeed true, as we will now show. After modeling, these patterns can be analyzed on the basis of the Green's function discussed in Chapter 3 or the limiting cases of continuous models† in Chapters 6 to 8.

If the roots λ_i or μ_i are very close to the limiting values required for trends and seasonality, then a more parsimonious representation of the model with a smaller number of parameters can be obtained, if desired, by using the known limiting values and reestimating only the remaining parameters. A special case of this technique was already encountered in Chapter 8. The discrete model with two parameters and the continuous model with three parameters were replaced by one-parameter EWMA models when the limiting conditions were satisfied. When such a parsimonious representation does not give a significantly increased sum of squares

---

†References to continuous models, parameters, roots μ_i and the Green's function $G(t)$ may be omitted without interrupting the flow.

over the full model, then it can be taken as an adequate model and used for forecasting, characterization, etc.

For the purpose of simplicity, we will first analyze the trend and seasonality via the characteristic roots and the Green's function in Section 9.1, together with illustrations of simulated series. Real examples of modeling and analysis will then be given in Section 9.2, with data from business, economics, hospital management, and transportation.

9.1 ANALYSIS OF STOCHASTIC TRENDS AND SEASONALITY

The key to the analysis of trends and seasonality is to remember the fact that the Green's function determines the overall pattern in the data. *The closer the roots are to the stability boundary, the more vividly the Green's function appears as the overall pattern in the data.* Therefore trends and seasonality patterns can be well represented by discrete models with roots λ_i closed to one in absolute value and by continuous models with real parts of the roots μ_i close to zero. The real roots represent constant, growth, or decay trends, and the complex roots represent seasonality with the period given by the imaginary part.

9.1.1 Stochastic Trends

Stochastic trends in the data arise when the roots are real. Let us consider a few examples of these trends. When the Green's function G_j tends to remain constant for all j, we get a constant trend. When G_j is proportional to j, we get a linear trend. When it varies with j as well as j^2, we have a quadratic trend, and in general, we can have a polynomial trend.

(a) Constant Trend

The constant trends, that is, an overall tendency in the data to remain at the same level, arise from a first order system. From Chapter 3 and Chapter 6 we know that the Green's function of the discrete and continuous first order models and their limiting forms are

$$G_j = \phi^j = \lambda_1^j \to G_j \equiv 1 \qquad \text{as} \qquad \phi = \lambda_1 \to 1$$

$$G(t) = e^{-\alpha_0 t} = e^{\mu_1 t} \to G(t) \equiv 1 \qquad \text{as} \qquad \alpha_0 = \mu_1 \to 0$$

For a first order system, both these limiting cases tend to random walk. Simulated data from the random walk model

$$(1 - B)X_t = a_t, \qquad \text{that is,} \qquad \nabla X_t = a_t$$

shown in Fig. 9.1 illustrates the constant trend. Note that although the series tends to remain at the same level, this level changes randomly since the trend is stochastic in nature.

Somewhat hazy patches of constant trends at different levels can arise

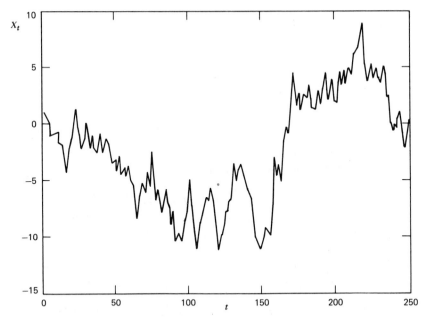

Figure 9.1   Simulated series from the model $\nabla X_t = a_t$.

from higher order systems with one root (discrete) close to one in absolute value and others relatively small. Two such examples can be seen from the papermaking input data and the IBM data in Figs. 2.1 and 2.3. The ARMA(2,1) models fitted to these in Tables 4.2 and 4.3 have the following parameters and roots

IBM: $\phi_1 = 1.05,$ $\phi_2 = -0.05,$ $\lambda_1 = 1,$ $\lambda_2 = 0.05$

Papermaking: $\phi_1 = 1.76,$ $\phi_2 = -0.76,$ $\lambda_1 = 1,$ $\lambda_2 = 0.76$

If a parsimonious representation is desired, such roots close to one from an adequate model found by the procedure of Chapter 4 can be replaced by the exact value one. This is equivalent to introducing the operator $\nabla = (1 - B)$ suggested by the adequate model. Such a procedure was implicit when we used the model

$$\nabla X_t = (1 - \theta_1 B)a_t$$

for the IBM data. After reestimating θ_1, and comparing with the ARMA(2,1) model, this model was found to be adequate in Chapter 4 and was used in Chapter 5 for forecasting. A similar procedure could also have been followed for the papermaking data.

If a continuous model is used instead of the discrete one, it can provide us with the valuable information whether the trend arises as a result of genuine limiting parameter values or simply because the sampling interval is too small. This is clear from the limiting cases discussed in Section 6.5 for a first order system, which are also applicable to higher order systems with only one real root close to the limiting value one.

(b)   Linear Trend

If two roots are close to one in absolute value for the discrete or with close to zero real parts for the continuous case, we get linear growth or decay trends. This is clear from the respective Green's functions and their limiting cases discussed in Section 3.1.8 and the results in Section 7.1.3 by which

$$G_j \rightarrow 1 + (1 - \theta_1)j \qquad \text{as} \qquad \lambda_1, \lambda_2 \rightarrow 1$$

$$G(t) \rightarrow C_1 + C_2 t \qquad \text{as} \qquad \mu_1, \mu_2 \rightarrow 0$$

Both these are straight lines that will give the corresponding trends in the data. In the discrete case the corresponding model is integrated random walk when θ_1 is zero.

For illustrative purposes a simulated set of data using the model

$$(1 - B)^2 X_t = (1 - \theta_1 B)a_t, \qquad \text{that is,} \qquad \nabla^2 X_t = (1 - \theta_1 B)a_t$$

is shown in Fig. 9.2 with $\theta_1 = -0.45$. Note that although there is a tendency in the data to behave linearly in straight line, the slope is slowly changing and, in fact, starts from the negative first and then becomes positive toward the end. Although the straight line behavior of the Green's function appears in the data, it is not exactly reproduced even when the roots are exactly one in absolute value since the trend is stochastic in nature. In particular, the sign of the Green's function is not preserved; this is because the a_t's that generate the series are of both positive and negative signs (their mean is zero), and hence the response generated by them can change signs.

In comparing Fig. 9.2 with the series in Fig. 9.1, which has been generated from the set of a_t's with the same variance, we can see that aside from the trend, there is a major difference between the two series. The small oscillations in Fig. 9.1 are smoothed out in Fig. 9.2. This happens because of the fundamental difference between the stability of the operators $(1 - B) = \nabla$ and $(1 - B)^2 = \nabla^2$. Both are not asymptotically stable, but $(1 - B)$ is stable, whereas $(1 - B)^2$ is unstable. This is reflected in the range of the two series generated from the set of a_t's with the same variance. The range in Fig. 9.1 is from -15 to $+10$; in Fig. 9.2 it is from -1400 to $+200$, which clearly brings out the difference between the stability of $(1 - B)$ and the instability of $(1 - B)^2$. Therefore, the scale in Fig. 9.2 is 64 times larger compared to Fig. 9.1, which masks the small oscillations, and the plot appears to be smooth. This tendency is even more marked in Fig. 9.3 that uses the operator $(1 - B)^3$ to be discussed later.

Again, when the adequate model obtained by the methods of Chapter 4 has two real roots close to one in absolute value and we want a parsi-

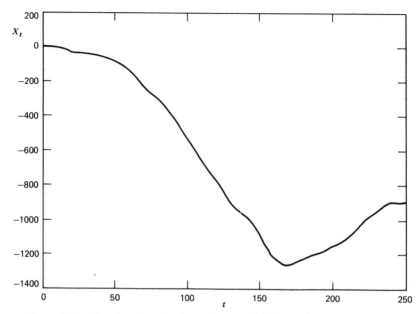

Figure 9.2   Simulated series from the model $\nabla^2 X_t = (1+0.45B)a_t$.

monious representation, we can reduce the number of parameters by refitting a model with the operator ∇^2. If the increase in the residual sum of
squares of the reestimated model is not significantly large compared with
the adequate model for two restricted parameters, then we can take the
parsimonious model as being adequate.

If a continuous model is used, the limiting cases discussed in Section
7.6 together with the understanding of the system are helpful in further
analyzing these trends. Such analysis can point out whether the linear trend
is due to excessively small sampling interval. In the latter case, one can
develop a better understanding of the system by either using smaller sampling intervals and the same number of observations or using the same
sampling interval with a larger number of observations.

These remarks become clearer if we consider the Green's function of
the second order discrete sampled system that has the form

$$G_j = g_1 e^{-(\zeta+\sqrt{\zeta^2-1})\omega_n \Delta j} + g_2 e^{-(\zeta-\sqrt{\zeta^2-1})\omega_n \Delta j}$$

If ω_n is very small so that both $\alpha_1 = 2\zeta\omega_n$ and $\alpha_0 = \omega_n^2$ are small, the roots
tend to absolute value one and we get a linear trend by the operator
$(1-2B+B^2)$. If the roots are complex, the Green's function is a sine wave
with very small frequency so that it stretches out very far and gives the
appearance of linear trends in the data. Similar is the case in which the
roots are real since the exponentially decaying Green's function decays

very slowly and gives rise to a linear trend. Thus, in these cases the trends are genuine in the sense that they arise from the limiting values of system parameters.

On the other hand, since the exponents in the Green's function involve $\omega_n \Delta$, the roots may be close to one because the sampling interval Δ is too small even though the parameters α_1, α_0 are not. This is like taking a large amount of data on a very small part of a sine wave or exponentially decaying Green's function of the second order system. This small part magnified in the data then appears to be a straight line. Such a spurious trend induced by the excessively small sampling interval can be detected and corrected only with the help of more data or physical knowledge of the system.

(c) Polynomial Trend

Generalizing the above discussion, we can easily see that the nth order polynomial trend can be represented by a model with $n+1$ discrete (continuous) real roots close to one in absolute value (zero real part). The respective Green's functions in such cases have the limiting forms

$$G_j = 1 + g_1 j + g_2 j^2 + \ldots + g_{n+1} j^n$$

$$G(t) = c_1 + c_2 t + c_3 t^2 + \ldots + c_{n+1} t^n$$

that will be reflected in a polynomial trend in the data.

When n such roots are found in an adequate model and we want a parsimonious representation, the operator ∇^n can be used. After reestimating the parameters, if the sum of squares is not significantly increased, we can take such an nth difference model as adequate. The class of processes for which the nth difference is stationary has been discussed by Yaglom (1952, 1955), and our treatment of such processes as limits of stationary processes rather than as nonstationary processes is in accord with Yaglom's corresponding treatment in frequency domain.

Such a treatment of stochastic trends as limits of stationary processes is also strongly supported by the limiting cases of first and second order systems discussed in Sections 6.5, 7.6, and 8.5. These results are readily generalized to a single root or a pair of roots in a higher order system approaching the limiting value one and the other roots relatively small. An analysis of the final adequate model or of the corresponding refitted model on the basis of such limiting cases can provide valuable insight into the system.

Figure 9.3 shows a plot of data simulated from the model

$$(1 - B)^3 X_t = (1 - \theta_1 B - \theta_2 B^2) a_t$$

with $\theta_1 = -0.45$, $\theta_2 = 0.45$ and the variance of a_t's the same as that used in Figs. 9.1 and 9.2. A quadratic or parabolic trend is clearly visible.

Comparison with Figs. 9.1 and 9.2 shows that the plot is relatively smoother. This is explained by the scale factors. The range in Fig. 9.1 with the operator $(1-B)$ is from -15 to $+10$; in Fig. 9.2 with the operator $(1-B)^2$ it is from -1400 to $+200$; and in Fig. 9.3 with the operator $(1-B)^3$ it is from $-120,000$ to 0. These ranges are a clear indication of increased instability of the underlying system.

Note that the plot in Fig. 9.3 looks like a deterministic function although generated by a stochastic process. It is thus possible to approximate deterministic functions by ARMA models over a finite amount of time. Two or more roots close to one in absolute value give an indication that such a deterministic component may be present in the data. One should then consider the physical understanding of the system to decide whether such a deterministic component is justified. If justified, the modeling methods outlined later in Chapter 10 should be employed.

9.1.2 Stochastic Seasonality

It is now easy to see that strongly seasonal or periodic trends in the data would arise when the roots are complex and of absolute value one in the discrete case or the real parts tend to zero in the continuous case. We will consider a few examples of such cases.

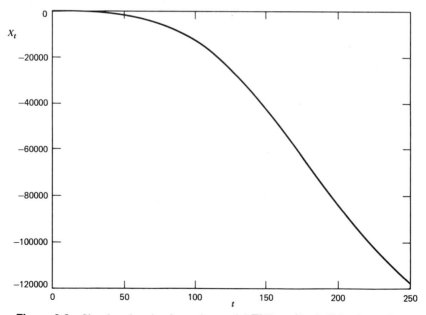

Figure 9.3   Simulated series from the model $\nabla^3 X_t = (1+0.45B-0.45B^2)a_t$.

(a) Periodicity of 12 (Yearly)

A very common seasonal pattern in the monthly data arises due to a yearly or 12-month period. Such a pattern would be clearly visible in the data when the absolute value of a complex conjugate pair of roots is nearly one and the imaginary part has a period of 12. Considering an ARMA(2,1) model, we have

$$\phi_2 = -\lambda_1\lambda_2 = -1$$

Since

$$\lambda_1,\lambda_2 = \frac{1}{2}(\phi_1 \pm \sqrt{\phi_1^2 + 4\phi_2})$$

$$= r(\cos\omega \pm i\sin\omega) \qquad (9.1.1)$$

$$r = \sqrt{-\phi_2}$$

$$\cos\omega = \frac{\phi_1}{2\sqrt{-\phi_2}}$$

For a period of 12

$$\cos\left(\frac{2\pi}{12}\right) = \frac{\sqrt{3}}{2}$$

and hence,

$$\phi_1 = \frac{\sqrt{3}}{2} \times 2\sqrt{-\phi_2} = \sqrt{3}\sqrt{-\phi_2} = \sqrt{3}$$

Thus,

$$\lambda_1,\lambda_2 = \frac{1}{2}(\sqrt{3} \pm i) \qquad (9.1.2a)$$

and the ARMA(2,1) model that gives such a seasonality is

$$(1 - \sqrt{3}B + B^2)X_t = (1 - \theta_1 B)a_t \qquad (9.1.2)$$

When the adequate ARMA(2,1) model is close to Eq. (9.1.2) or when a higher order model has a complex conjugate pair of roots very close to Eq. (9.1.2a), we can use the operator $(1 - \sqrt{3}B + B^2)$ to obtain a parsimonious representation accounting for the known seasonality of 12.

An example of an approximate period of 11 may be found in the sunspot data shown in Fig. 2.2. There are roughly $15\frac{1}{2}$ peaks in 176 years that gives a period of about 11 years. The adequate AR(2) model fitted to the data in Table 4.1 has parameters

$$\phi_1 = 1.34, \qquad \phi_2 = -0.65$$

for which

$$\cos\omega = \frac{1.34}{2\sqrt{0.65}}$$

which gives a period

$$\frac{2\pi}{\omega} \simeq 11 \text{ years}$$

and

$$\lambda_1, \lambda_2 \simeq 0.670 \pm i0.456$$

The reason why the periodic trend is not very regular is that ϕ_2 is not close to -1 so that there is considerable damping.

To illustrate the seasonality of 12, two sets of data simulated by the models

$$(1 - \sqrt{3}B + B^2)X_t = (1 + 0.45B)a_t$$

$$(1 - 1.65B + 0.903B^2)X_t = (1 + 0.45B)a_t$$

are shown in Figs. 9.4 and 9.5. Both these models have the same period of 12, but there is a slight damping in the second model since $|\phi_2| < 1$. The seasonal patterns can be clearly discerned from these plots.

(b) Periodicity of 3 (Quarterly)

A second common period in the monthly data is quarterly, which can be obtained by using $\omega = 2\pi/3$ in the preceding computations. This gives

$$\phi_2 = -1$$

$$\phi_1 = 2 \cos \frac{2\pi}{3} = -1$$

that is,

$$\lambda_1, \lambda_2 = \frac{1}{2}(-1 \pm i\sqrt{3}) \qquad (9.1.3)$$

and the corresponding ARMA(2,1) model

$$(1 + B + B^2)X_t = (1 - \theta_1 B)a_t \qquad (9.1.4)$$

Thus, when we get an ARMA(2,1) model with ϕ_1, ϕ_2 close to those in Eq. (9.1.4) or when a pair of complex conjugate roots have values close to Eq. (9.1.3) in a higher order model, we can introduce quarterly seasonality in the model by the operator $(1 + B + B^2)$ and reduce the number of parameters in the model. A set of simulated data using the model (9.1.4) with $\theta_1 = -0.45$ and another with the same periodicity as in (9.1.3) but with $|\phi_2| < 1$, providing a slight damping, are shown in Figs. 9.6 and 9.7.

(c) Arbitrary Period

It is now easy to see that the roots corresponding to an arbitrary periodicity p can be obtained from

$$\phi_2 = -1, \qquad \phi_1 = 2 \cos \left(\frac{2\pi}{p}\right)$$

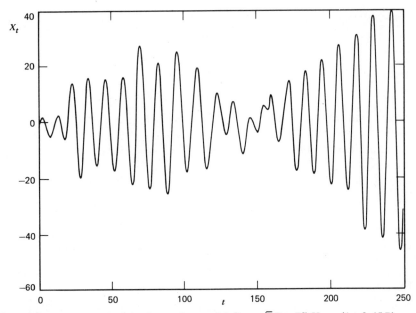

Figure 9.4   Simulated series from the model $(1 - \sqrt{3}B + B^2)X_t = (1 + 0.45B)a_t$.

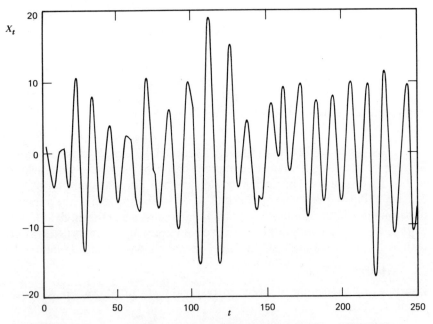

Figure 9.5   Simulated series from the model $(1 - 1.65B + 0.903B^2)X_t = (1 + 0.45B)a_t$.

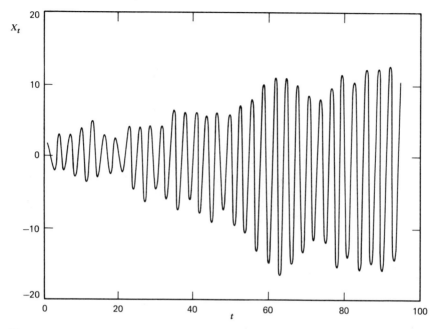

Figure 9.6   Simulated series from the model $(1 + B + B^2)X_t = (1 + 0.45B)a_t$.

and (9.1.5)

$$\lambda_1, \lambda_2 = \frac{1}{2}(\phi_1 \pm i\sqrt{\phi_1^2 + 4\phi_2})$$

When a pair of roots close to these are found in an adequate model, seasonality can be introduced by the operator $(1 - \phi_1 B - \phi_2 B^2)$ and the number of parameters reduced by two. If the increase in the sum of squares is not significant, such a parsimonious model can be taken to be adequate. Table 9.1 lists some typical roots and operators for ready reference.

9.1.3 Differencing or Seasonality Operators Before Modeling

In the preceding discussion, we have advocated the introduction of trend (differencing) or seasonality operators purely for the sake of parsimony *after* the adequate model has been found and its roots strongly suggest such operators. Throughout the book, we have discouraged the use of such operators on the data *before* modeling in order to reduce a seemingly nonstationary or unstable series to an apparently stationary stable series that is supposedly easier to model.

The procedure of simplifying the series of data by differencing or seasonality operators before modeling is often recommended in the liter-

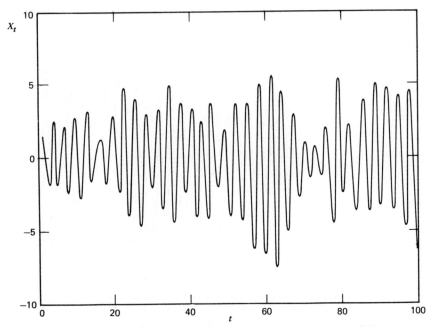

Figure 9.7 Simulated series from the model $(1 + 0.95B + 0.9B^2)X_t = (1 + 0.45B)a_t$.

ature. In such cases, modeling is based on identifying the model, either from the data or from the plots of sample autocorrelations, spectra, etc. When trends and seasonality are dominant in the data, the sample autocorrelations fail to damp out quickly, and the plots of spectra by the conventional methods are badly distorted, thus making it almost impossible to tentatively identify the model from their plots. The only way to get them to forms, from which a low order AR or MA model can be guessed, is to apply differencing or seasonality operators (which, in turn, have to be guessed from the data, sample autocorrelations, or spectra).

The danger of such indiscriminate operating or smoothing of the data simply for the sake of making it easier to analyze has been pointed out by Slutsky (1927). Such an operation itself may introduce spurious trends and periods in the resultant series that are absent in the original data. The final fitted model, although statistically adequate and apparently parsimonious, may give a completely distorted picture of the structure of the original series.

In particular, when there are deterministic trends present in the data, they will be completely masked and an irrelevant model fitted to represent them. As an example, a linear trend superimposed with an AR(2) model is shown in Fig. 9.8. The linear trend was $0.16t$ and the AR(2) data is the

Table 9.1   Some Typical Periodicities, Their Operators, and Characteristic Roots

Period p	$\phi_1 = 2\cos\left(\dfrac{2\pi}{p}\right)$	Operator $1 - \phi_1 B + B^2$	Roots λ_1, λ_2
2	-2	$1 + 2B + B^2$	$-1, \; -1$
3	-1	$1 + B + B^2$	$-0.500 \pm i0.866$
4	0	$1 + B^2$	$\pm i$
5	0.618	$1 - 0.618B + B^2$	$0.309 \pm i0.951$
6	1	$1 - B + B^2$	$0.500 \pm i0.866$
7	1.247	$1 - 1.247B + B^2$	$0.624 \pm i0.782$
8	1.414	$1 - 1.414B + B^2$	$0.707 \pm i0.707$
9	1.532	$1 - 1.532B + B^2$	$0.766 \pm i0.643$
10	1.618	$1 - 1.618B + B^2$	$0.809 \pm i0.588$
12	1.732	$1 - 1.732B + B^2$	$0.866 \pm i0.500$
14	1.802	$1 - 1.802B + B^2$	$0.901 \pm i0.434$
15	1.827	$1 - 1.827B + B^2$	$0.914 \pm i0.407$
20	1.902	$1 - 1.902B + B^2$	$0.951 \pm i0.309$
21	1.911	$1 - 1.911B + B^2$	$0.956 \pm i0.295$
24	1.932	$1 - 1.932B + B^2$	$0.966 \pm i0.259$
25	1.937	$1 - 1.937B + B^2$	$0.969 \pm i0.249$
28	1.950	$1 - 1.950B + B^2$	$0.975 \pm i0.222$
30	1.956	$1 - 1.956B + B^2$	$0.978 \pm i0.209$
38	1.968	$1 - 1.968B + B^2$	$0.984 \pm i0.178$
50	1.984	$1 - 1.984B + B^2$	$0.992 \pm i0.126$
60	1.989	$1 - 1.989B + B^2$	$0.995 \pm i0.105$
∞	2.000	$1 - 2.000B + B^2$	$1.00, \; 1.00$

same as that for the grinding wheel profile in Appendix I, with $\phi_1 = 0.76$ and $\phi_2 = -0.21$. If we "simplify" the data and "make it stationary" by differencing it and fit a model to the differenced series based on the estimated autocorrelation, we get an MA(3) model with

$$\theta_1 = 0.107, \qquad \theta_2 = 0.357, \qquad \theta_3 = 0.185$$

that has no relation to the actual AR(2) model whatsoever. Moreover, the valuable information contained in the linear trend in the original data is lost. (On the other hand, the methods of Chapter 10 recommended for such data almost exactly estimate the linear trend and the AR(2) parameters.)

No such simplification of the data to guess a model form is needed in our modeling procedure described in Chapter 4. The procedure is robust enough to model data with trends and seasonality. If the adequate model naturally indicates trends by real roots and seasonality by complex roots close to one in absolute value, we can then introduce these operators ensuring that the original series is not distorted.

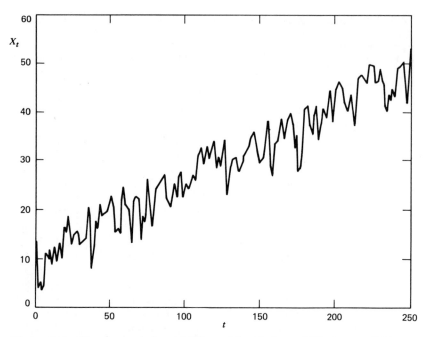

Figure 9.8   Linear trend plus AR(2) model with $\phi_1 = 0.76$, $\phi_2 = -0.21$.

When two or more roots are equal to or greater than one in absolute value with or without seasonality and the data has a smooth trend or a regular periodic behavior, the possibility of a deterministic component is indicated. If the physical understanding of the underlying system justifies such a deterministic component, one can use the methods discussed subsequently in Chapter 10.

9.2 EXAMPLES OF SERIES WITH STOCHASTIC TRENDS AND SEASONALITY

The examples of IBM stock prices, papermaking data, and sunspot series were considered in the last section to illustrate stochastic trends and seasonality. To illustrate the approach developed in the preceding section, let us now consider some more examples of actual data from business and economic series as well as hospital and airline management, where such trends and seasonality are quite prominent.

9.2.1 Modeling Investment and Money-Market Rate

Figure 9.9 shows the monthly averages of the weekly total investment in the U.S. treasury securities by the large commercial banks in New York

City. Fig. 9.10 shows the corresponding monthly averages of money market rates in weekly market yields of 3-month bills of U.S. Government securities. In all, there are 132 data points each, from January 1965 to December 1975. The second method of computing a_t's, starting with $a_1 = X_1$, was used in estimation as recommended in Section 4.5.

(a)   Money Rate

The results of modeling the money rate data by the procedure in Chapter 4 are given in Table 9.2. The ARMA(4,3) model is found to be adequate and the autoregressive roots are

$$\lambda_1 = -0.4582; \quad \lambda_2 = 0.9812; \quad \lambda_3, \lambda_4 = 0.0043 \pm i0.9653$$

None of these roots are close to the moving average roots:

$$\nu_1 = -0.6275, \quad \nu_{2,3} = -0.0269 \pm i1.0536$$

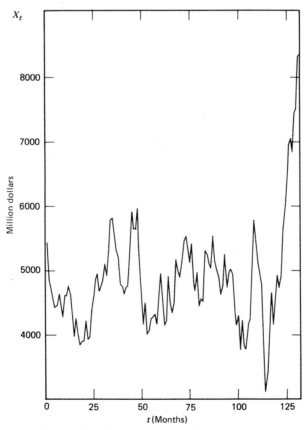

Figure 9.9   Monthly average investment.

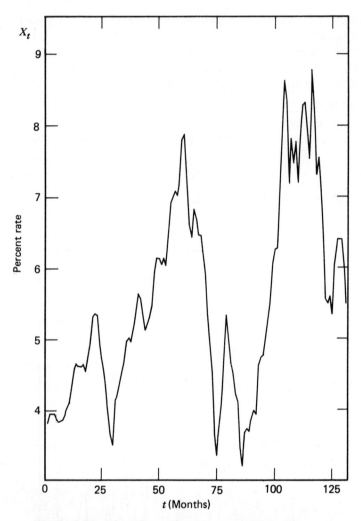

Figure 9.10   Monthly average money market rates.

Hence, none of the roots can be cancelled. The root $\lambda_2 = 0.9812$ indicates the possibility of a constant trend represented by $(1 - B)$. The complex pair of roots λ_3, λ_4 are nearly $\pm i$ that is a 4 monthly seasonal period since by Eq. (9.1.5)

$$\phi_2 = -\lambda_3\lambda_4 \simeq -1$$

and

$$\phi_1 = \lambda_3 + \lambda_4 = 0.0086 \simeq 0 = 2 \cos\left(\frac{2\pi}{4}\right)$$

Table 9.2 Modeling Monthly Money Market Rate Data

Parameters	ARMA Order			$(1-B)(1+B^2)$
	(2,1)	(4,3)	(6,5)	(1,3)
ϕ_1	0.941 ± 0.734	0.514 ± 0.293	1.458 ± 0.764	-0.694 ± 0.404
ϕ_2	0.0368 ± 0.726	-0.478 ± 0.305	-1.255 ± 1.24	
ϕ_3		0.491 ± 0.399	0.809 ± 1.42	
ϕ_4		0.419 ± 0.386	-0.0149 ± 1.45	
ϕ_5			-0.494 ± 1.14	
ϕ_6			0.456 ± 0.549	
θ_1	-0.276 ± 0.706	-0.681 ± 0.260	0.225 ± 0.729	-0.913 ± 0.311
θ_2		-1.145 ± 0.083	-0.856 ± 0.675	-1.097 ± 0.0278
θ_3		-0.697 ± 0.305	-0.158 ± 0.839	-0.841 ± 0.325
θ_4			0.0670 ± 0.751	
θ_5			-0.684 ± 0.539	
"Mean"	3.864 ± 0.726	3.962 ± 0.245	3.919 ± 0.212	0.1061 ± 0.126
Residual sum of squares	18.785	15.046	15.124	18.093
F		7.70	—	0.70^a

The adequate model is ARMA(4,3) with characteristic roots

$$\lambda_1 = -0.4582, \quad \lambda_2 = 0.9812, \quad \lambda_{3,4} = -0.0043 \pm i0.9653$$

$$\nu_1 = -0.6275, \quad \nu_{2,3} = -0.0269 \pm i1.0536$$

[a] Compared with (4,3).

which is also clear from the examination of Table 9.1. This 4 monthly period can be represented by the operator $(1 + B^2)$.

Using the combined operator $(1 - B)(1 + B^2)$ for the stochastic trend and seasonality, we can reduce the autoregressive parameters by 3 and fit an ARMA(1,3) model to the resultant data, that is, the model now is

$$(1 - B)(1 + B^2)(1 - \phi_1 B)(\dot{X}_t - \mu) = (1 - \theta_1 B - \theta_2 B^2 - \theta_3 B^3)a_t$$

The estimated parameter values for this model are shown in the last column of Table 9.2. The residual sum of squares is only slightly higher than that of the adequate (4,3) model. Therefore, the parsimonious model with trend and seasonality operators may also be considered adequate.

Either the full ARMA(4,3) or the parsimonious ARMA(1,3) with the operator $(1 - B)(1 + B^2)$ can be used for forecasting by the procedure discussed in Chapter 5. Both these models would give practically the same forecasts and probability limits. The fact that with the operator $(1 - B)(1 + B^2)$ the Green's function fails to die out does not cause an appreciable change in the error variances for short-term forecasts.

Table 9.3 Modeling Investment Data

(1) Parameters	(2) ARMA Order			(3) $(1+B^2)$ (2,3)	(4) $(1+B^2)$ (1,2)
	(2,1)	(4,3)	(6,5)		
ϕ_1	0.617 ± 1.31	0.0554 ± 0.144	0.0025 ± 0.406	0.00395 ± 0.112	0.963 ± 0.0074
ϕ_2	0.316 ± 1.27	-0.0829 ± 0.150	0.701 ± 1.14	0.837 ± 0.121	
ϕ_3		0.0346 ± 0.137	-0.624 ± 0.593		
ϕ_4		0.862 ± 0.118	0.437 ± 0.400		
ϕ_5			0.488 ± 0.889		
ϕ_6			-0.283 ± 0.827		
θ_1	-0.447 ± 1.25	-0.986 ± 0.145	-0.991 ± 0.362	-1.038 ± 0.0553	0.00710 ± 0.102
θ_2		-0.872 ± 0.141	-0.248 ± 0.798	-0.822 ± 0.0485	-0.868 ± 0.0896
θ_3		-0.968 ± 0.159	-1.030 ± 0.235	-0.937 ± 0.0869	
θ_4			-0.671 ± 0.203		
θ_5			0.0856 ± 0.898		
"Mean"	5462 ± 677	5354.8 ± 490	5238.9 ± 529	9988.0 ± 351	10314.1 ± 438
Residual sum of squares	17.919×10^6	15.743×10^6	15.198×10^6	16.791×10^6	18.881×10^6
F	4.28		1.07	4.13[a]	6.18[a]

The adequate model is ARMA(4,3) with characteristic roots:

$$\lambda_1 = 0.9652, \quad \lambda_2 = -0.9201, \quad \lambda_{3,4} = 0.0051 \pm 0.9852$$

$$\nu_1 = -1.0414, \quad \nu_{2,3} = 0.0276 \pm i0.9637$$

[a] Compared with ARMA(4,3).

(b) Investment

The results of modeling the investment data are shown in Table 9.3. The adequate model is ARMA(4,3) with autoregressive roots

$$\lambda_1 = 0.9652; \qquad \lambda_2 = -0.9201;$$

$$\lambda_3, \lambda_4 = 0.0051 \pm i0.9852$$

The complex pair of roots again indicates a 4 monthly period.

The results of exploring the possibility of introducing the seasonality operator $(1+B^2)$ to account for the 4 monthly period are given in the third column of Table 9.3. The tests of adequacy show that the parsimonious representation $(1+B^2)$ ARMA(2,3) is not adequate compared to the ARMA(4,3) model. If we try to cancel the root $\lambda_2 = -0.9201$ with $v_1 = -1.0414$ in addition to applying the operator $(1+B^2)$, the results given in the last column of Table 9.3 are obtained. Similar to the $(1+B^2)$ ARMA(2,3) model, this parsimonious model is also not found to be adequate. Therefore, the full ARMA(4,3) model is used for forecasting.

(c) Interpretation

The real significance of the trend and seasonality analysis far exceeds the parsimonious representation and the resultant simplicity of forecasting. With the help of characteristic roots λ_i, the trend and seasonality operators discussed in this chapter, and the form of the Green's functions studied in earlier chapters, we will try to briefly interpret the adequate ARMA(4,3) model for the money rate and the investment data.

We know from Chapter 3 that the Green's function of the above ARMA(4,3) model will consist of the sum of two exponentials corresponding to the two real roots λ_1 and λ_2 and a sine wave determined by the complex conjugate pair λ_3 and λ_4. With the 4 monthly seasonality analyzed above, the forms of the Green's function are shown in Fig. 9.11a, 9.11b and may be taken as

(a) $G_j = g_1(-0.4582)^j + g_2(0.9812)^j + A(0.9653)^j \cos\left(\dfrac{2\pi}{4}j+\beta\right)$

(b) $G_j = g_1(-0.9201)^j + g_2(0.9652)^j + A(0.9852)^j \cos\left(\dfrac{2\pi}{4}j+\beta\right)$

The long-term behavior of the Green's function and hence that of the overall series is determined by the autoregressive roots/parameters, whereas the short-term behavior up to three lags is also affected by the moving average roots/parameters.

In addition to the seasonality of four months, the series are dominated by a long-term factor leading to the near constant trend and is given by the exponential $(0.9812)^j$ for the money rate and $(0.9652)^j$ for the investment. This part of the Green's function effectively denotes high inertia or sluggishness or long memory similar to the case of random walk discussed in Section 2.2.6. Note that such inertia or sluggishness is more prominent for the money rate than for the investment since the exponential $(0.9812)^j$ dies slower than $(0.9652)^j$. This shows that a part of the money rate has a greater tendency to remain at a fixed level than the corresponding part of the investment.

The third part of the Green's function leads to the fluctuating tendency in the series due to the negative root. The root -0.9201 for the investment

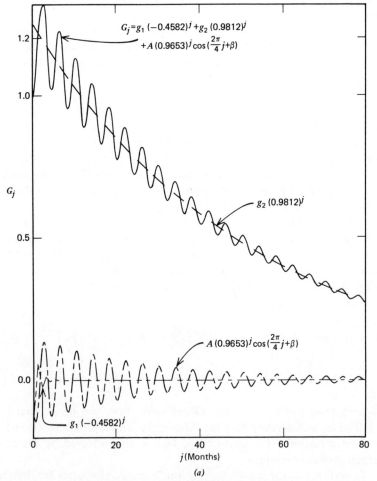

Figure 9.11a Green's function and its components for the money market rate data, ARMA(4,3) model. (Compare with Fig. 3.7)

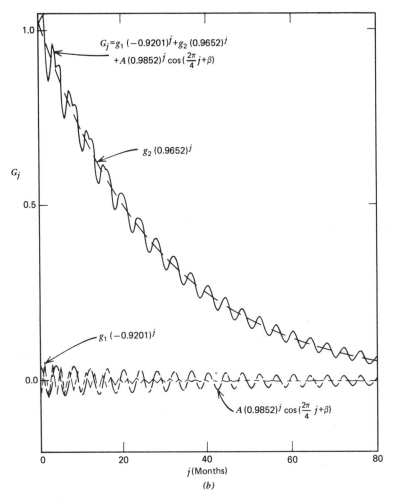

$$G_j = g_1 (-0.9201)^j + g_2 (0.9652)^j$$
$$+ A (0.9852)^j \cos(\tfrac{2\pi}{4} j + \beta)$$

$g_2 (0.9652)^j$

$g_1 (-0.9201)^j$

$A (0.9852)^j \cos(\tfrac{2\pi}{4} j + \beta)$

j(Months)

(b)

Figure 9.11b   Green's function and its components for the investment data, ARMA(4,3) model. (Compare with Fig. 3.7)

is larger, but since it can be approximately cancelled with the moving average part, its effect on the series will not be prominent. The smaller root -0.4582 for the money rate cannot be cancelled and will yield an alternately fluctuating component. This component could be due to the intelligent control of the money rate and investment by the banking institutions through their fiscal policies. The short-term effect of the three moving average parameters could also be due to such policies and the effect of the current market situation.

The above statements and inferences can be quantified and made precise in two ways. The first is to actually compute the coefficients g_1, g_2, A, and the phase angle β by the formulae given in Chapter 3, especially (3.1.17b), (3.1.17c) and (3.1.26). These coefficients quantify the relative influence of the three factors considered above and the effect of the moving average parameters in the overall behavior of the series. The second is to compute the variance components as discussed in Section 3.3.6. (For details, see Appendix A 9.1 and Problem 9.6.)

9.2.2 Modeling Hospital Patient Census

Figure 9.12 shows a plot of the daily in-patient census at the University of Wisconsin hospital, from April 21 to July 29, 1974. The data consisting of 100 observations are listed in Appendix I at the end of the book. The plot clearly indicates a cyclic trend of a weekly or 7-day period as would be expected. In other words, the observations that are seven days apart are similar. According to available literature on modeling seasonal series, one might use the operator $(1 - B^7)$, that is, take the seventh difference of the data before modeling since the operator $(1 - B^7)$ is expected to remove the nonstationarity that is present because of a 7-day period in the series. However, the results show that the use of the operator $(1 - B^7)$ is inappropriate for modeling hospital census data.

The results of modeling with initial a_t's set to zero are given in Table 9.4. Following the standard procedure, we find an ARMA(7,6) model adequate for the hospital census data. The characteristic roots (poles and zeros) of the ARMA(7,6) model are given in Table 9.5 and are also shown in Fig. 9.13 where they are compared with the roots of the operator $(1 - B^7)$. The polynomial $(1 - B^7)$ has seven roots of unity,

$$e^{i\frac{2\pi j}{7}}, \qquad j = 0, 1, 2, \ldots, 6$$

evenly spaced ($\theta = 2\pi/7 = 51.43°$) on the unit circle. It is seen from Fig. 9.13 that all the complex roots of the ARMA(7,6) model lie fairly close to the complex roots of unity, but the real root $\lambda_7 = 0.745$ is quite different from the real root of unity. The absence of the root $\lambda_7 = 1$ in the ARMA(7,6) model implies that the use of the operator $(1 - B^7)$ is not appropriate for the hospital census data.

The inappropriateness of the operator $(1 - B^7)$ is also clear from the fact that if the ARMA(7,6) model with this operator was appropriate, then we should have

$$\phi_1 = \phi_2 = \ldots = \phi_6 = 0, \qquad \phi_7 = 1$$

and the actual values in Table 9.4 are nowhere near these. On the other hand, to replace the roots $\lambda_7 = 1$ by $\lambda_7 = 0.745$ we note that

$$(1 - B^7) = (1 - B)(1 + B + B^2 + B^3 + B^4 + B^5 + B^6)$$

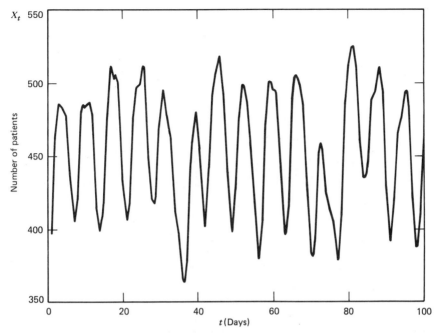

Figure 9.12   Daily inpatient census.

Therefore,

$$(1-0.745B)(1+B+B^2+B^3+B^4+B^5+B^6) = (1+0.255B+0.255B^2$$
$$+0.255B^3+0.255B^4+0.255B^5+0.255B^6-0.745B^7)$$

which gives

$$\phi_1 = \phi_2 = \ldots = \phi_6 = -0.255, \quad \phi_7 = 0.745$$

somewhat close to the values of the ARMA(7,6) model parameters in Table 9.4. It can also be readily confirmed by actually modeling the transformed data $(1-B^7)X_t = X_t - X_{t-7}$ that it does not lead to a simpler or better model than the ARMA(7,6) of Table 9.4.

Now let us examine three pairs of complex roots of the ARMA(7,6) model. The periods corresponding to these pairs of roots can be calculated using Eq. (9.1.1). For example, the pair of roots $(0.624 \pm i0.786)$ will give a period p equal to

$$\frac{2\pi}{\cos^{-1}\left(\dfrac{\lambda_1 + \lambda_2}{2\sqrt{\lambda_1\lambda_2}}\right)} = \frac{2\pi}{\cos^{-1}\left(\dfrac{0.624}{\sqrt{0.624^2+0.786^2}}\right)}$$

$$= 6.983 \simeq 7$$

Table 9.4   Modeling Hospital Census

Parameters	Order of the ARMA Model					
	(2,1)	(4,3)	(6,5)	(7,6)	(8,7)	(10,9)
ϕ_1	1.10 ± 0.18	1.30 ± 0.27	-0.62 ± 0.31	-0.27 ± 0.11	0.87 ± 0.18	-0.20 ± 0.33
ϕ_2	-0.74 ± 0.15	-0.84 ± 0.49	-0.37 ± 0.30	-0.24 ± 0.12	-0.09 ± 0.25	-0.10 ± 0.23
ϕ_3		-0.24 ± 0.48	-0.70 ± 0.20	-0.25 ± 0.10	-0.04 ± 0.24	0.60 ± 0.22
ϕ_4		0.23 ± 0.24	-0.80 ± 0.22	-0.28 ± 0.11	-0.14 ± 0.23	-0.31 ± 0.14
ϕ_5			-0.31 ± 0.29	-0.28 ± 0.09	0.004 ± 0.22	-0.14 ± 0.18
ϕ_6			-0.44 ± 0.23	-0.20 ± 0.11	0.07 ± 0.21	-0.14 ± 0.16
ϕ_7				0.72 ± 0.10	0.71 ± 0.20	0.86 ± 0.15
ϕ_8					-0.75 ± 0.15	0.05 ± 0.29
ϕ_9						-0.15 ± 0.22
ϕ_{10}						-0.60 ± 0.21
θ_1	0.005 ± 0.27	0.38 ± 0.19	-1.75 ± 0.30	-0.91 ± 0.22	0.08 ± 0.26	-0.95 ± 0.40
θ_2		0.08 ± 0.21	-1.48 ± 0.46	-0.98 ± 0.32	0.35 ± 0.28	-0.42 ± 0.46
θ_3		-0.85 ± 0.16	-1.39 ± 0.41	-1.24 ± 0.39	-0.06 ± 0.29	0.44 ± 0.41
θ_4			-1.31 ± 0.36	-1.30 ± 0.40	-0.06 ± 0.29	0.06 ± 0.43
θ_5			-0.51 ± 0.24	-1.53 ± 0.36	0.07 ± 0.28	-0.19 ± 0.41
θ_6				-1.26 ± 0.27	0.61 ± 0.28	0.38 ± 0.43
θ_7					0.82 ± 0.31	1.46 ± 0.47
θ_8						1.36 ± 0.62
θ_9						0.59 ± 0.44
Mean	454.0 ± 6.6	453.0 ± 8.2	453.4 ± 4.35	461.5 ± 6.5	452.6 ± 4.6	454.6 ± 1.3
Residual sum of squares	44547.9	26997.3	27136.8	11939.1	11330.8	10848.7

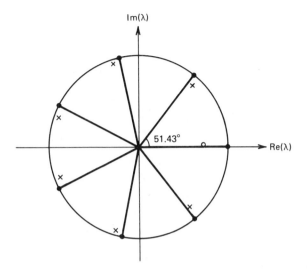

roots of unity $e^{\frac{i2\pi j}{7}}$; $j = 0, 1, \ldots, 6$

o   pole (autoregressive root)

×   zeros (moving average roots)

(For $j = 1$ to 6, autoregressive roots practically coincide with roots of unity)

Figure 9.13   Characteristic roots (poles and zeros) of the ARMA(7,6) model and seven roots of unity.

Similarly, the periods for the other two pairs $(-0.237 \pm i0.965)$ and $(-0.892 \pm i0.424)$ are found as 7/2 days and 7/3 days, respectively (Table 9.6). Thus, the roots $(-0.237 \pm i0.965)$ and $(-0.892 \pm i0.424)$ only reflect the higher harmonics of the fundamental period of 7 days.

In order to see the contribution of these harmonics to the hospital census data, the amplitude of these harmonics are computed using the Green's function as discussed in Chapter 3. Since the ARMA(7,6) model has seven characteristic roots, λ_i, $i = 1, 2, \ldots, 7$, the Green's function for this model may be represented from Eq. (3.1.24) as

$$G_j = g_1\lambda_1^j + g_2\lambda_2^j + \ldots + g_7\lambda_7^j$$

$$= \sum_{k=1}^{7} g_k\lambda_k^j$$

where the g_k are the strength of the mode represented by λ_k. As all the complex roots are close to 1 in absolute value (Table 9.6), the Green's

Table 9.5   The Characteristic Roots of the ARMA Models for the Hospital Census Data

Roots	Order of the ARMA Model				
	(2,1)	(4,3)	(6,5)	(7,6)	(8,7)
$\lambda_{1,2}$	$0.5511 \pm i0.6613$	$0.6821 \pm i0.7814$	$0.5968 \pm i0.7448$	$0.6243 \pm i0.7857$	$0.6256 \pm i0.7824$
$\lambda_{3,4}$			$-0.0463 \pm i0.7179$	$-0.2370 \pm i0.9654$	$-0.2376 \pm i0.9398$
$\lambda_{5,6}$			$-0.8603 \pm i0.4500$	$-0.8924 \pm i0.4237$	$-0.8586 \pm i0.4214$
λ_7		-0.4516		0.7449	$0.9053 \pm i0.2131$
λ_8		0.4984			
$\nu_{1,2}$		$0.6202 \pm i0.7751$	$0.2904 \pm i0.8655$	$0.6867 \pm i0.8721^*$	$0.6407 \pm i0.8187^*$
$\nu_{3,4}$			$-0.7967 \pm i0.4411$	$-0.2359 \pm i1.0050^*$	$-0.3088 \pm i0.8567$
$\nu_{5,6}$				$-0.9034 \pm i0.3759$	$-0.8507 \pm i0.3092^*$
ν_7	0.0053	-0.8605	-0.7338		1.1169^*

* Note that these roots are noninvertible.

function may be decomposed in undamped cosine waves as follows (see Chapter 3)

$$G_j = A_1 \cos (\omega_1 j + \beta_1) + A_2 \cos (\omega_2 j + \beta_2)$$
$$+ A_3 \cos (\omega_3 j + \beta_3) + g_7 \lambda_7 \qquad (9.2.3)$$
$$= \sum_{i=1}^{3} A_i \cos (\omega_i j + \beta_i) + g_7 \lambda_7$$

The relationship between the amplitude A_i and g_k or phase β_i and g_k is given by Eqs. (3.1.16b), (3.1.17a), (3.1.17b), and (3.1.17c). The values of g_k and A_i for the ARMA(7,6) model are calculated using Eqs. (3.1.16b), (3.1.17a), (3.1.17b), and (3.1.17c) and Eq. (9.2.3); and the results are given in Table 9.6. A sample calculation for the computation of g_k and A_i is given in Appendix A 9.1.

The results in the last column of Table 9.6 show that the amplitudes of the second and third harmonics are 0.23 and 0.22, respectively, in proportion to the fundamental frequency. Note that unlike the investment and money market rate models, further quantification of the relative influence of these seasonal factors cannot be done by means of variance components because the model is unstable. The model is also noninvertible.

However, it is possible to "deseasonalize" the series or obtain a "seasonally adjusted" series (as commonly practiced in business and economic analysis) by a far simpler and more natural procedure indicated by the model. For example, the 7-day seasonality can be removed by applying the (unstable and noninvertible) operator

$$\frac{[1 - (0.6243 + i0.7857)B][1 - (0.6243 - i0.7857)B]}{[1 - (0.6867 + i0.8721)B][1 - (0.6867 - i0.8721)B]}$$
$$= \frac{[1 - 1.2486B + 1.0071B^2]}{[1 - 1.3734B + 1.2321B^2]}$$

to X_t, that is, generating the seasonally adjusted series Y_t by

$$Y_t = X_t - 1.2486X_{t-1} + 1.0071X_{t-2} + 1.3734Y_{t-1} - 1.2321Y_{t-2}$$

Similar operators can be obtained for the other two seasonalities of 7/2 and 7/3 days.

If all three seasonalities are to be removed, then it is not hard to verify from the earlier discussion of the operator $(1 - B^7)$ and the roots in Table 9.5 that the desired operator, from the parameters of the ARMA(7,6) model:

$$\phi_1 = -0.2653, \quad \phi_2 = -0.2443, \quad \phi_3 = -0.2501, \quad \phi_4 = -0.2752,$$
$$\phi_5 = -0.2822, \quad \phi_6 = -0.1981; \quad \theta_1 = -0.9051, \quad \theta_2 = -0.9780,$$
$$\theta_3 = -1.2351, \quad \theta_4 = -1.2982, \quad \theta_5 = -1.5275, \quad \theta_6 = -1.2571$$

Table 9.6 Analysis of the Characteristic Roots of the ARMA(7,6) Model for Hospital Census Data

Roots		Absolute Value	Periodicity	Nature	Strength g_k	Amplitude A_i $(=2|g_k|)$	Ratio to Fundamental
λ_1,λ_2	$0.624 \pm i0.786$	1.0035	$6.983 \simeq 7$	Fundamental	$-0.105 \pm i0.102$	0.2916	1.00
λ_3,λ_4	$-0.237 \pm i0.965$	0.9941	$3.455 \simeq \frac{7}{2}$	Second harmonic	$-0.026 \pm i0.021$	0.0666	0.23
λ_5,λ_6	$-0.892 \pm i0.424$	0.9879	$2.329 \simeq \frac{7}{3}$	Third harmonic	$0.002 \pm i0.032$	0.0640	0.22
λ_7	0.7449	0.7449	—	Exponential	1.256	—	—

and $\lambda_7 = 0.7449$, is given by

$$\frac{(1+1.0102B+0.9892B^2+0.9950B^3+1.0201B^4+1.0271B^5+0.9430B^6)}{(1+0.9051B+0.9780B^2+1.2351B^3+1.2982B^4+1.5275B^5+1.2571B^6)}$$

Therefore the seasonally adjusted series Y_t can be obtained from the original series X_t by

$$Y_t = X_t + 1.0102X_{t-1} + 0.9892X_{t-2} + 0.9950X_{t-3} + 1.0201X_{t-4}$$
$$+ 1.0271X_{t-5} + 0.9430X_{t-6} - 0.9051Y_{t-1} - 0.9780Y_{t-2}$$
$$- 1.2350Y_{t-3} - 1.2982Y_{t-4} - 1.5275Y_{t-5} - 1.2571Y_{t-6}$$

Moreover, the seasonality-free series Y_t now has the simple AR(1) model

$$Y_t = 0.7449Y_{t-1} + a_t$$

with the variance of a_t same as before:

$$\sigma_a^2 = \frac{11939.1}{100-7} = 128.38$$

so that the variance of the seasonally adjusted series is

$$\frac{\sigma_a^2}{1-\phi_1^2} = \frac{128.38}{1-0.7449^2} = 288.41$$

compared to the variance of the original series 1699. It is thus seen that the seasonal components contribute nearly 83% to the variance of the original series. In actual computation the starting values in a short series may somewhat alter these numbers. (See Problem 9.7)

There are several possible explanations for the strong 7-day period observed in the series and incorporated in the model. The average length of stay is nearly seven days for almost all types of patients. Many services are cut back on weekends, for example, elective surgery is rarely performed on Saturday or Sunday. Doctors prefer to admit patients on Monday or Tuesday and many of the patients are then ready to go home by the weekend.

The above model can be useful to the hospital administration for forecasting the patient load and then adjusting the equipment and employee schedule to best accommodate the predicted demand. The model can also be used to ascertain the effect of policy changes introduced to keep the utilization of hospital resources at optimum levels.

9.2.3 Modeling Consumer and Wholesale Price Indexes

Figures 9.14 and 9.15 show the plots of the monthly U.S. consumer price index (CPI) and wholesale price index (WPI) from July 1953 to April 1970; the data are listed in Appendix I. The plot of the data clearly shows an

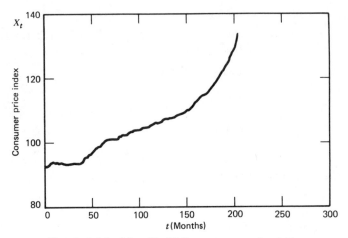

Figure 9.14   Monthly U.S. consumer price index.

exponential trend; in fact, the trend is so strong that it may be better modeled by a deterministic trend. We will nevertheless use the procedure of Chapter 4, starting the computation of a_i's with $a_1 = X_1$ as recommended in Section 4.5.

The results of modeling and the seasonality analysis are given in Tables 9.7 to 9.10. It is seen that the ARMA(4,3) is found to be adequate at a 5% level of significance for CPI, whereas AR(1) is adequate for WPI. Table 9.8 shows that the complex pair of autoregressive roots $-0.1783 \pm i0.9508$ for CPI is quite close to the pair of moving average

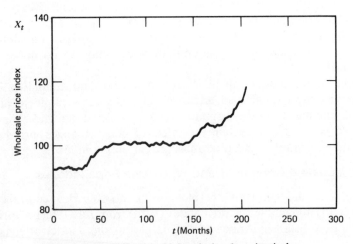

Figure 9.15   Monthly U.S. wholesale price index.

roots $-0.2140 \pm i0.9794$. However, cancelling the two roots on both sides takes us back to the fitted ARMA(2,1).

To get a simpler and/or more parsimonious model for CPI, we note from Table 9.8 that two real roots of the ARMA(4,3) model are close to one, which indicates the possibility of $\nabla = (1-B)$ or ∇^2 operators. The estimation results after introducing ∇ and ∇^2 are shown in Table 9.7. The requisite F-test in Table 9.7 indicates that we may replace the root 0.9113 by 1, but we cannot replace the other root 1.0156 since the reduction in the residual sum of squares from ∇^2 model to ARMA(4,3) is significant. For the WPI series the adequate model AR(1) is already simple.and no further simplification is necessary.

However, it should be noted from Table 9.7, that the calculated F-value of 3.60 is between the Table D values of $F_{0.95}(2,\infty) = 3.00$ and $F_{0.99}(2,\infty) = 4.61$. Moreover, since the mean for the $\nabla^2(2,3)$ model is practically zero, disregarding it gives an F-value of 2.40 that is insignificant compared to $F_{0.95}(3,\infty) = 2.60$. Additionally, the $\nabla^2(2,3)$ model has the narrowest confidence intervals; hence it may also be considered to be adequate.

The dominating root 1.0156 in CPI (Table 9.8) corresponding to the ARMA(4,3) model and 1.0139 in the WPI series (Table 9.10) indicate a strong exponential growth trend. Both the series are thus unstable and their Green's functions grow to infinity rather than decay to zero or remain constant. The series are therefore no longer stationary since the mean, variance and autocovariance will grow with lag and depend explicitly on origin. (The estimates of "mean" in Tables 9.7 and 9.9 are merely the estimates of starting values.) However, in spite of the nonstationary nature of the series, the stochastic models fitted above can be used for forecasting purposes by the procedure of Chapter 5. (See Problem 9.8.)

The two adequate models ARMA(4,3) and $\nabla^2(2,3)$ for CPI in Table 9.7 account for the instability/nonstationarity in two seemingly different ways: ARMA(4,3) by an exponential with small exponent and $\nabla^2(2,3)$ by an *unstable* operator ∇^2. However, this difference is only apparent and not real, because an exponential with a small exponent is practically a straight line, and the stochastic trend analysis in Section 9.1.1.b, showed that the ∇^2 operator yields a straight line trend! This may be confirmed by the respective Green's functions (see Problem 9.9) and will be further elaborated by introducing deterministic components in Chapter 10.

9.2.4 Modeling Airline Passenger Ticket Sales

Figure 9.16 shows the logged airliner passenger ticket sales from January 1949 to December 1960; the original data are listed in Appendix I. It should be noted that the logged series is being used here simply to compare with

Table 9.7 Modeling Consumer Price Index

Parameters	Order of the ARMA Model				
	(2,1)	(4,3)	(6,5)	∇ (3,3)	∇² (2,3)
ϕ_1	1.970 ± 0.083	1.570 ± 0.158	1.310 ± 3.02	0.660 ± 0.058	-0.361 ± 0.057
ϕ_2	-0.969 ± 0.085	-1.174 ± 0.155	-0.979 ± 2.29	-0.573 ± 0.072	-0.934 ± 0.057
ϕ_3		1.473 ± 0.140	1.539 ± 1.41	0.944 ± 0.058	
ϕ_4		-0.866 ± 0.141	-0.754 ± 1.65		
ϕ_5			0.101 ± 0.65		
ϕ_6			-0.214 ± 2.41		
θ_1	0.852 ± 0.140	0.374 ± 0.219	0.129 ± 3.00	0.468 ± 0.070	0.423 ± 0.078
θ_2		-0.662 ± 0.091	-0.818 ± 1.62	-0.622 ± 0.034	-0.634 ± 0.038
θ_3		0.806 ± 0.214	0.785 ± 1.07	0.901 ± 0.068	0.853 ± 0.076
θ_4			0.056 ± 0.913		
θ_5			0.249 ± 2.19		
"Mean"	93.48 ± 0.366	93.50 ± 0.334	93.50 ± 0.332	-0.029 ± 0.179	0.003 ± 0.004
RSS	7.3413	6.9118	6.8542	6.9797	7.1683
F		3.01	0.40	1.91[a]	3.60[a]

The adequate model is (4,3).

[a] Compared with (4,3).

Table 9.8 The Characteristic Roots of the ARMA Models for the Consumer Price Index Data

Roots	Order of the ARMA Model				
	(2,1)	(4,3)	(6,5)	∇ (3,3)	∇^2 (2,3)
λ_1	0.9542	0.9113	0.9381		
λ_2	1.0158	1.0156	1.0157	1.0136	
λ_3,λ_4		$-0.1783 \pm i0.9508$	$-0.1787 \pm i0.9542$	$-0.1768 \pm i0.9487$	$-0.1805 \pm i0.9492$
λ_5,λ_6			$0.1431 \pm i0.4670$		
ν_1	0.8523	0.8018	0.8443	0.8962	0.8523
ν_2,ν_3		$-0.2140 \pm i0.9794$	$-0.2138 \pm i0.9796$	$-0.2140 \pm i0.9795$	$-0.2148 \pm i0.9768$
ν_4,ν_6			$-0.1440 \pm i0.5221$		

Table 9.9 Modeling Wholesale Price Index

Parameters	Order of the ARMA Model		
	(1,0)	(2,1)	(4,3)
ϕ_1	1.014 ± 0.005	1.946 ± 0.180	1.524 ± 0.194
ϕ_2		-0.945 ± 0.183	-1.101 ± 0.169
ϕ_3			1.503 ± 0.157
ϕ_4			-0.924 ± 0.183
θ_1		0.879 ± 0.235	0.433 ± 0.244
θ_2			-0.581 ± 0.138
θ_3			0.859 ± 0.233
"Mean"	93.48 ± 0.664	93.50 ± 0.653	93.53 ± 0.652
Residual sum of squares	23.153	22.684	22.355
F		2.05	0.71

The adequate model is ARMA(1,0).

Table 9.10 The Characteristic Roots of the Autoregressive and Moving Average Models for the Wholesale Price Index Data

Roots	Order of the Model		
	(1,0)	(2,1)	(4,3)
λ_1	1.0139	1.0131	1.0132
λ_2		0.9328	0.9332
λ_3,λ_4			$-0.2112 \pm i0.9658$
ν_1		0.8792	0.8805
ν_2,ν_3			$-0.2238 \pm i0.9619$

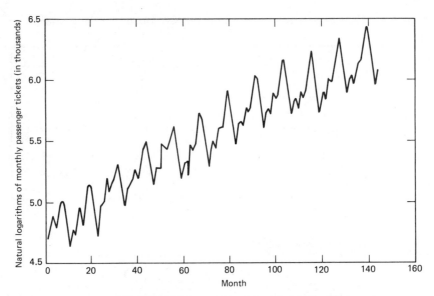

Figure 9.16 Logged airline ticket sales.

some of the existing models of the series and not in order to make the modeling simple. If the modeling strategy of Chapter 4 is followed, the original series is as easy to model as the logged one. The growth trend and the unstable nature of the series do not require a transformation "to make the series stationary," as was illustrated by the CPI and WPI series in Section 9.2.3.

Fitting ARMA$(2n, 2n-1)$ models for increasing order yields a significantly reduced sum of squares until $n = 7$, that is up to the ARMA(14,13) model. The estimation results of the ARMA(14,13) model are given in Table 9.11. It is also found that for the ARMA(14,13) model, the parameter ϕ_{14} is small ($\phi_{14} = -0.0767$); hence, the parameter ϕ_{14} is dropped and the ARMA(13,13) model is fitted to the data with the results given in Table 9.12. The residual autocorrelations and the $\pm 2/\sqrt{N}$ bounds are shown in Fig. 9.17.

The 13 autoregressive and moving average characteristic roots of the ARMA(13,13) model are listed in Table 9.13 together with the periods corresponding to the complex conjugate pairs calculated by Eq. (9.1.1). It is easy to see from the table that the first 12 estimated roots are practically the same as the 12 roots of unity, the theoretical values of which are listed in the left half of the table for comparison. The positions of these roots are indicated on the unit circle as shown in Fig. 9.18 on page 350. The angles and the complex number representations $e^{\pm i\theta}$ of these roots are given in columns 3 and 4 of Table 9.13.

Table 9.11 Logged Airline Data—Parameters and Characteristic Roots of the ARMA(14,13) Model

<table>
<tr><td colspan="2" align="center">Parameters</td></tr>
<tr>
<td>

$\phi_1 = 1.1452 \pm 0.0092$
$\phi_2 = -0.2476 \pm 0.0147$
$\phi_3 = 0.1606 \pm 0.0284$
$\phi_4 = -0.1221 \pm 0.0123$
$\phi_5 = 0.1186 \pm 0.0291$
$\phi_6 = 0.1186 \pm 0.0082$
$\phi_7 = 0.0712 \pm 0.0029$
$\phi_8 = -0.0812 \pm 0.0119$
$\phi_9 = 0.0996 \pm 0.0133$
$\phi_{10} = -0.0613 \pm 0.0339$
$\phi_{11} = 0.0447 \pm 0.0307$
$\phi_{12} = 0.9248 \pm 0.0058$
$\phi_{13} = -1.0161 \pm 0.0262$
$\phi_{14} = -0.0767 \pm 0.0557$

</td>
<td>

$\theta_1 = 0.5724 \pm 0.2050$
$\theta_2 = -0.0578 \pm 0.02850$
$\theta_3 = 0.1749 \pm 0.2240$
$\theta_4 = 0.0187 \pm 0.2700$
$\theta_5 = -0.1977 \pm 0.2780$
$\theta_6 = 0.2722 \pm 0.1980$
$\theta_7 = -0.1375 \pm 0.2240$
$\theta_8 = 0.0570 \pm 0.2800$
$\theta_9 = -0.0110 \pm 0.2300$
$\theta_{10} = 0.2738 \pm 0.2160$
$\theta_{11} = -0.1251 \pm 0.2070$
$\theta_{12} = 1.1077 \pm 0.2310$
$\theta_{13} = -0.4649 \pm 0.2170$

</td>
</tr>
</table>

Mean $= 5.8524 \pm 0.6370$

Position[a]	Characteristic Roots	Period	Angle[a]
R_0	$0.9929 \pm i0.0212$	294.31	1.22
R_1	$0.8683 \pm i0.4985$	12.0560	29.86
R_2	$0.4953 \pm i0.8689$	5.9686	60.32
R_3	$0.0063 \pm i0.9994$	4.0161	89.64
R_4	$-0.4951 \pm i0.8704$	3.0093	119.63
R_5	$-0.8589 \pm i0.5076$	2.4096	149.42
R_6	-0.9539		
	0.0816		

[a] For complex conjugate pairs, only one root $a + ib$ is considered. For a full configuration see Table 9.13 and Fig. 9.18.
Residual sum of squares $= 0.090698$.

Since the first pair of roots $0.9808 \pm i0.0211$ can be considered to be two roots equal to 1, the 12 autoregressive roots (poles) suggest the autoregressive operator $(1 - B^{12})$ and the thirteenth root leads to the operator $(1 - B)$. If indeed the autoregressive side were $(1 - B)(1 - B^{12})$, then the ARMA(13,13) model should have $\phi_1 = 1$, $\phi_2 = \phi_3 = \ldots = \phi_{11} = 0$, $\phi_{12} = 1$, $\phi_{13} = -1$. The estimated values of ϕ_i's in Table 9.12 are, in fact, very close to these values.

The first twelve moving average roots (zeros) and their periods given in Columns 8 and 9 are also close to the 12 roots of unity, although not as close as autoregressive roots. This suggests a moving average operator $(1 - \theta_{12} B^{12})$. Note that the operator $(1 - B^{12})$ cannot be used, not only because the values are somewhat different from the theoretical ones, but mainly because the operator $(1 - B^{12})$ would get cancelled on both sides

Table 9.12 Logged Airline Data—Parameters of the ARMA(13,13) Model

Parameters	
ϕ_1 = 1.0014 ± 0.0156	θ_1 = 0.4102 ± 0.1790
ϕ_2 = −0.0978 ± 0.0144	θ_2 = 0.0249 ± 0.1580
ϕ_3 = 0.0889 ± 0.0138	θ_3 = −0.0372 ± 0.1540
ϕ_4 = −0.0718 ± 0.0215	θ_4 = 0.1747 ± 0.0719
ϕ_5 = 0.0921 ± 0.0246	θ_5 = −0.1491 ± 0.1400
ϕ_6 = −0.0941 ± 0.0538	θ_6 = 0.0911 ± 0.2020
ϕ_7 = 0.0568 ± 0.0653	θ_7 = 0.0344 ± 0.2050
ϕ_8 = −0.0575 ± 0.0325	θ_8 = 0.0408 ± 0.1920
ϕ_9 = 0.0506 ± 0.0218	θ_9 = −0.1150 ± 0.1370
ϕ_{10} = 0.0013 ± 0.0277	θ_{10} = 0.3007 ± 0.1510
ϕ_{11} = −0.0136 ± 0.0301	θ_{11} = −0.0812 ± 0.1500
ϕ_{12} = 0.9921 ± 0.0257	θ_{12} = 1.249 ± 0.1630
ϕ_{13} = −0.9581 ± 0.0361	θ_{13} = −0.3814 ± 0.223
"Mean" = 6.2043 ± 0.4170	

Residual sum of squares = 0.091753.

and we would be left with an ARMA(1,1) model. On the other hand, the tests of adequacy have shown that the ARMA(14,13) model is better than the ARMA(2,1) model, and therefore, the ARMA(1,1) that will be obtained after cancelling $(1 - B^{12})$ on two sides cannot be considered an adequate model. The thirteenth moving average root leads to the operator $(1 - \theta_1 B)$.

It is thus seen that the fitted ARMA(13,13) model seems to suggest

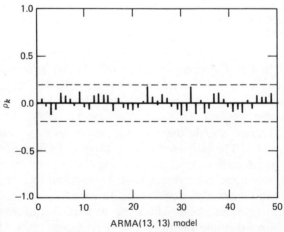

Figure 9.17 Residual autocorrelations of the ARMA(13,13) model.

Table 9.13 Theoretical and Estimated Characteristic Roots: Airline Data

| | | | | | | Estimated Roots | | |
| | | Theoretical Roots | | | Autoregressive | | Moving Average | |
(1) j	(2) Position[a]	(3) Angle[a]	(4) Value	(5) Period	(6) Value	(7) Period	(8) Value	(9) Period
0	R_0	0°	$e^0 = 1.00 \pm i0.00$	—	$0.9808 \pm i0.0211$	—	$1.0626 \pm i0.0000$	—
1	R_1	30°	$e^{\pm i\frac{2\pi}{12}} = 0.866 \pm i0.5$	12	$0.8674 \pm i0.5018$	11.980	$0.8905 \pm i0.5137$	12.008
2	R_2	60°	$e^{\pm i\frac{4\pi}{12}} = 0.5 \pm i0.866$	6	$0.4965 \pm i0.8707$	5.969	$0.5041 \pm i0.8800$	5.981
3	R_3	90°	$e^{\pm i\frac{6\pi}{12}} = 0.0 \pm i1.00$	4	$0.0047 \pm i1.0000$	4.012	$-0.0097 \pm i1.0075$	3.976
4	R_4	120°	$e^{\pm i\frac{8\pi}{12}} = -0.5 \pm i0.866$	3	$-0.4954 \pm i0.8690$	3.008	$-0.4626 \pm i0.8772$	3.008
5	R_5	150°	$e^{\pm i\frac{10\pi}{12}} = -0.866 \pm i0.5$	2.4	$-0.8662 \pm i0.5119$	2.409	$-0.8688 \pm i0.5119$	2.409
6	R_6	180°	$e^{\pm i\frac{12\pi}{12}} = -1.00 + i0.00$	2	$-0.9741 + i0.0000$	—	$-1.0648 + i0.0000$	—
							$0.3053 + i0.0000$	
Suggested Operators			$(1-B)(1-B^{12})$		$(1-\theta_1 B)(1-\theta_{12}B^{12})$			

[a] For complex conjugate pairs, only one root $a + ib$ is considered. See Fig. 9.18 for full configuration

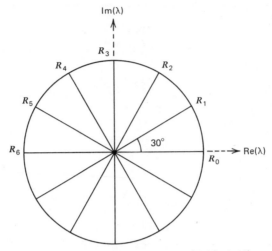

Figure 9.18   Positions of the characteristic roots (Table 9.13) on the unit circle.

a two-parameter parsimonious model for the logged airline data

$$(1 - B)(1 - B^{12})X_t = (1 - \theta_1 B)(1 - \theta_{12}B^{12})a_t \qquad (9.2.4)$$

The initial values of θ_1 and θ_{12} can be obtained from the characteristic roots of the moving average side given in Table 9.13. The initial value for θ_1 is 0.3053, the thirteenth moving average root that suggests the operator $(1 - \theta_1 B)$. For θ_{12} we take $0.98^{12} = 0.785$ since the parameter should be such that the absolute value of its twelveth root is slightly less than one. With these initial values, the following results are obtained:

$$\theta_1 = 0.40 \pm 0.08, \qquad \theta_{12} = 0.61 \pm 0.07$$

$$\text{Residual sum of squares} = 0.18252$$

Comparing this model with the ARMA(13,13) model by the F-criterion, we get

$$F = \frac{0.18252 - 0.09175}{0.09175} \times \frac{118}{25} = 4.63$$

and since $F_{0.95}(24,120) = 1.61$ and even $F_{0.999}(24,120) = 2.40$, the two-parameter model is not considered adequate. (All these residual sum of squares have been computed using zero initial a_t's.)

The two-parameter model (9.2.4) given above is practically the same as the one fitted by Box and Jenkins (1970); their slightly lower sum of squares is probably due to the refinement of back forecasting. They have suggested this model for the airline data based on the EWMA model

$$(1 - B)X_t = (1 - \theta_1 B)a_t$$

which is widely applicable, and one can use a similar model for the yearly seasonality

$$(1 - B^{12})X_t = (1 - \theta_{12}B^{12})a_t$$

Combining these two models gives the two-parameter model (9.2.4). After fitting, they conclude this model to be adequate on the basis of residual autocorrelations of a_t's, (Fig. 9.19), although a number of them appear rather large compared with their standard deviation $1/\sqrt{N} = 1/\sqrt{131} \simeq 0.09$, and the twenty-third autocorrelation is about 2.5 times the standard deviation.

The modeling of the airline data brings out several strong points of the ARMA($n,n-1$) modeling strategy. First of all, no trial and error was involved in modeling. The adequate ARMA(13,13) model was arrived at by a straightforward application of the modeling procedure given in Chapter 4. Even the parsimonious model (9.2.4) was clearly indicated by the parameter values of the ARMA(13,13) model and its characteristic roots given in Table 9.13. This shows that such parsimonious models are not "missed" by the ARMA($n,n-1$) strategy, but are naturally included in them, as was emphasized in Chapter 2, and can be logically arrived at as illustrated above.

Second, the ARMA($n,n-1$) strategy not only provides us with simpler models if we want, but also clearly informs us of the cost we have to pay for the simplicity of such parsimonious models. For the airline data, we can choose the two-parameter ARMA(13,13) model only at the cost of nearly an 80% increase in the one step ahead forecast error variance. Note that the one step ahead forecast error variance for the ARMA(13,13) model is 0.000637 (0.09175/144), whereas for the two-parameter model, it is 0.001267

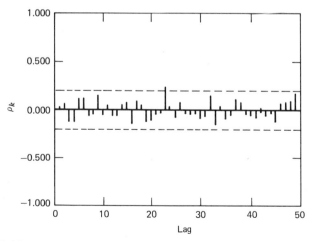

Figure 9.19 Residual autocorrelations of the two-parameter ARMA(13,13) model (9.2.4).

(0.18252/144). (Using unbiased estimates, these are 0.000784 and 0.001285 respectively.)

Third, the airline modeling shows that a higher order model does not necessarily mean estimation problems and bad estimates. ARMA(14,13) is a relatively high order model; rarely have such models been used in the existing literature. However, apart from the increase in computation, no serious estimation problems have been encountered. On the other hand, the estimates' closeness to the twelve roots of unity as well as their confidence intervals in Tables 9.11 to 9.13 show that they are quite good. This is statistically confirmed by the discussion at the end of Section A 2.2.3.

Note that in this chapter the logged series of the airline passenger ticket sales has been modeled. The original series will be used in Chapter 10, where a more comprehensive modeling procedure for nonstationary series will be discussed.

APPENDIX A 9.1

CALCULATION OF THE STRENGTH g_k AND AMPLITUDE A_i CORRESPONDING TO MODE λ_k FOR THE ARMA(7,6) MODEL

For the ARMA(7,6) model, the Green's function G_j can be written as (see Chapter 3)

$$G_j = \sum_{k=1}^{7} g_k \lambda_k^j$$

where g_k represents the strength of mode λ_k. When the number of g_k's is small, we can solve for them by first computing $G_0, G_1, \ldots, G_{n-1}$ from ϕ_i and θ_i by the implicit method of Section 3.1.5. When the number of g_k's is large, such as 7 in the present case, this method becomes cumbersome. However, from Eq. (3.1.26), g_k can be expressed as the function of autoregressive and moving average parameters and the characteristic roots $\lambda_i, i = 1, 2, \ldots, 7$:

$$g_k = \frac{\lambda_k^6 - \theta_1 \lambda_k^5 - \theta_2 \lambda_k^4 - \ldots - \theta_6}{(\lambda_k - \lambda_1)(\lambda_k - \lambda_2) \ldots (\lambda_k - \lambda_{k-1})(\lambda_k - \lambda_{k+1}) \ldots (\lambda_k - \lambda_7)}$$

$$k = 1, 2, \ldots, 7$$

Substituting $\theta_i, i = 1, 2, \ldots, 6$ from Table 9.4 and $\lambda_k, k = 1, 2, \ldots 7$ from Table 9.5

$$g_1 = \frac{\lambda_1^6 - \theta_1 \lambda_1^5 - \theta_2 \lambda_1^4 - \theta_3 \lambda_1^3 - \theta_4 \lambda_1^2 - \theta_5 \lambda_1 - \theta_6}{(\lambda_1 - \lambda_2)(\lambda_1 - \lambda_3)(\lambda_1 - \lambda_4)(\lambda_1 - \lambda_5)(\lambda_1 - \lambda_6)(\lambda_1 - \lambda_7)}$$

$$= 0.1046 + i0.1016$$

and g_2 is its complex conjugate

$$g_2 = 0.1046 - i0.1016$$

Similarly, g_3, g_4, g_5, g_6, and g_7 can be calculated and are given in Table 9.6. The corresponding amplitudes are simply twice the absolute values of

g_k by Eq. (3.1.16b). Thus, for (g_1, g_2)

$$A_1 = 2 \sqrt{(-0.1046)^2 + (0.1016)^2}$$
$$= 0.1458$$

Similarly, for (g_3, g_4) and (g_5, g_6) the amplitudes are computed as 0.03328 and 0.03202, respectively. The results of the amplitudes and their relative ratios are given in Table 9.6.

The phase angles β_i can be calculated using Eq. (3.1.17c). Thus, for (g_1, g_2)

$$\beta_1 = \tan^{-1} \left[\frac{\text{Imaginary Part of } g_1}{\text{Real Part of } g_1} \right]$$

$$= \tan^{-1} \left[\frac{0.1016}{-0.1046} \right] = 135.8°$$

Similarly, the phase angles β_2 and β_3 corresponding to (g_3, g_4) and (g_5, g_6) are found to be 140.3° and 86.2°.

PROBLEMS

9.1 An ARMA(2,1) model has periodicity of 6,
(a) Find the values of parameters ϕ_1 and ϕ_2.
(b) What is the effect of the parameter θ_1 on the seasonality?
(c) Find the natural frequency ω_n and the damping ratio ζ.

9.2 Given the ARMA(3,1) model

$$X_t = X_{t-3} + a_t + \theta_1 a_{t-1}$$

what kinds of trends are contained in the model?

9.3 A system has periodic trends of 6 and 10. If the adequate model for the system is ARMA(4,3), find the values of the autoregressive parameters, ϕ_i's, $i = 1,2,3,4$.

9.4 A set of daily data covering a period of two years gives an ARMA(4,1) model as adequate. The autoregressive roots are

$$\lambda_1, \lambda_2 = 0.625 \pm i0.780; \qquad \lambda_3, \lambda_4 = 0.978 \pm i0.208$$

(a) Determine the seasonality.
(b) Suggest the possible seasonality of operators and the form of the parsimonious model after using the seasonality operators.

9.5 An ARMA(7,1) model has characteristic roots λ's given by

$$\lambda_1 = 0.62, \quad \lambda_2 = 0.999, \quad \lambda_3 = 0.998,$$
$$\lambda_4, \lambda_5 = 0.5 \pm i0.866, \quad \lambda_6, \lambda_7 = -0.5 \pm i0.866$$

(a) Plot the characteristic roots on the complex plane.
(b) What kind of trend and seasonalities are contained in the model?
(c) Can the operator $(1 - B^N)$ be used for the system? If no, explain why not? If yes, what is the value of N?

9.6   The Green's function of the ARMA(4,3) model in Table 9.2 for the money market rate may be nearly represented by

$$G_j = g_1(-0.4582)^j + g_2(0.9812)^j$$
$$+ A(0.9653)^j \sin\left(\frac{2\pi j}{4} + \beta\right)$$

   (a)   Compute the coefficients g_1, g_2, A, and the phase angle β.
   (b)   Discuss the relative influence of three factors on the overall behavior of the series.
   (c)   If the autocovariance function for the above ARMA(4,3) model is represented by

$$\gamma_k = d_1(-0.4852)^k + d_2(0.9812)^k$$
$$+ d_3(-0.0043 + i0.9653)^k$$
$$+ d_4(-0.0043 - i0.9653)^k$$

   compute the variance components d_1, d_2, d_3, and d_4 corresponding to each characteristic root and discuss their relative effect. [*Hint:* See Eq. (3.3.26).]

9.7   (a)   Use the ARMA(8,7) model for the hospital census data to determine the operators for removing the seasonalities of 7, 7/2 and 7/3 days individually and together.

   (b)   Determine the model for the seasonally adjusted series and find its variance.

   (c)   Compute the seasonally and adjusted series, model them and compare against the theoretical results for (i)ARMA(7,6) and (ii)ARMA(8,7) models.

   (d)   Compare the four seasonal adjustment operators of ARMA(8,7) and ARMA(7,6) models. Which would you prefer and why?

9.8   The WPI and CPI series given in Appendix I are slightly larger than those used for modeling in Section 9.2.3. Use the remainder of the series at the end to compute a_i's and compare their variances with the estimates from the models. Also compute long term forecasts; assuming that the remainder is not given, compute the respective probability limits and plot with actual values to ascertain the effectiveness of forecasts.

9.9   (a)   Plot, compare and contrast the Green's functions of the three adequate models for the CPI and one adequate model for the WPI data using the implicit method of computing G_j.

   (b)   Determine the explicit expressions for the above Green's functions and comment on them with reference to the plots and unstable trends discussed at the end of Section 9.2.3.

10

DETERMINISTIC TRENDS AND
SEASONALITY: NONSTATIONARY SERIES

The stationary models employed in Chapter 9 were based on the assumptions that the first two moments—namely, the mean and the covariance—are independent of the time origin. These assumptions imply that the mean is fixed or constant, so that it may be subtracted and the series assumed to have zero mean, and that the autocovariance at a given lag depends only on the lag. What can be done if we know that the series is nonstationary, that is, its behavior depends on the time origin? Such a presence of nonstationary trends and seasonality may be indicated either from *a priori* knowledge or after fitting the models as illustrated in Sections 9.2.3 and 9.2.4.

We will show in this chapter that nonstationary trends and seasonal patterns in the data can be modeled by relaxing the first assumption of zero or fixed mean. The series will be decomposed into two parts. The first one, representing the mean of the series, accounts for the nonstationary trend by a *deterministic* function, which depends on the time origin. The second one is a stochastic part with zero mean so that it can be modeled by the methods of Chapter 4.

It is necessary to clarify the sense in which we use the word nonstationary series or trends. Some authors, notably Box and Jenkins (1970), use the word for discrete ARMA models with one or more roots with absolute value one, for example, random walk, integrated random walk, EWMA, etc. However, the probabilistic properties of these models, especially the correlation function, are still independent of the origin, although theoretically their variance is infinite. For this reason we treat these models as stationary or as limiting cases of stationary models with the roots tending to one. The data leading to these models can be analyzed by the methods of Chapter 4, as illustrated in Chapter 9.

We call a series of data nonstationary when its nature appears to be dependent on the time origin, following the terminology in system analysis, control theory, and stochastic processes. The models for such data therefore need to include functions that depend on the time origin. In this chapter, we will show that many of the nonstationary data can be modeled by explicitly including polynomial, exponential, or sinusoidal functions, dependent on the time origin, to represent the mean of the series.

Such nonstationary trends can also be modeled by first subjecting the series to transformations such as logarithmic and other operations such as differencing, either simple or seasonal, aimed at reducing the series to stationarity. The transformed stationary series is then modeled by an ARMA model. The pitfalls of such a modeling procedure are discussed in Section 9.1.3. We, however, avoid such transformations in our approach and put special emphasis on the system aspect of the data. For example, in Section 10.2 we will consider data with exponential trends. Although this trend can be suppressed by differencing and the resultant series can be fitted to a simple ARMA model, a very important piece of information about the exponential behavior and its physical interpretation could be lost. The distortion introduced by such an approach was illustrated in Section 9.1.3.

The procedure of modeling nonstationary series is developed in this chapter by considering data with different kinds of trends. The simplest linear trend is considered in Section 10.1 by means of data on material crack propagation. The exponential trends are treated in Section 10.2 using the data from the papermaking process and chemical relaxation. Seasonal series are dealt in Section 10.3, illustrated by international airline passenger data. The results are compared with those obtained by the method of transformations discussed in Chapter 9. Section 10.4 generalizes the procedure, relates it with the procedure for stationary series, and provides some practical hints for dealing with nonstationary data in general. The general procedure is illustrated in Section 10.5 by application to wholesale and consumer price indexes and to a wood surface profile in mechanical pulping. Special cases of this general procedure, leading to ARMA models with slight modifications, are pointed out in Problems 10.7 to 10.11.

10.1 LINEAR TRENDS

We will first consider the simplest nonstationarity that involves only a linear trend in the mean. This implies that the data appear to be scattered around a straight line. It was seen in Section 2.1.1 that when the data are independent, we can use the standard linear regression formulae to estimate the parameters and their confidence intervals. In this section, we will consider a method of modeling a linear trend when the data are no longer independent.

The method essentially consists of decomposing the model into two parts—deterministic and stochastic, removing the linear deterministic part that causes the nonstationarity in the data, and modeling the remaining stochastic part by the methods of the preceding chapters. For the purpose of removing the deterministic trend, and also for obtaining the initial values of the parameters of the final deterministic plus stochastic model, we will first fit the deterministic model alone. In this preliminary fitting of the deterministic part, we will obtain the estimates by the least squares method assuming uncorrelated errors. In particular, when the trend is linear, the linear regression results of Section 2.1.1 can be used to estimate the parameters and to obtain the residuals. The residuals can be examined together with their sample autocorrelations to check that no other nonstationary trends are present. The residuals can be modeled by the procedure in Chapter 4 to fit an ARMA model, which roughly represents the stochastic model. Then, the parameters of the final deterministic plus stochastic model can be estimated together, by starting with the estimates from the separate models as initial values.

10.1.1 Crack Propagation Data: Deterministic Part

To illustrate the method by a real-life data containing a linear trend, consider Fig. 10.1 that shows 90 coded observations (mean = 15.8772, variance = 7.7677) of the measured crack lengths of specimens subjected to Brittle Fracture tests. The crack length is measured after subjecting the specimen to a specific number of stress cycles, and the figure shows the plot of coded crack lengths versus the number of stress cycles. The numerical values of the series are given in Appendix I. The plot of the data shows a strong linear trend. The data points do not seem to be randomly scattered around the straight line but seem to have long drifts indicating a high positive

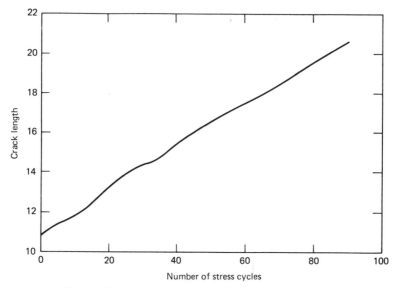

Figure 10.1   Observations of measured crack length.

correlation. Thus, the data appear to be dependent, which will be confirmed after modeling.

To model the deterministic part for preliminary analysis we use the method of Section 2.1.1 to fit a model

$$y_t = \beta_0 + \beta_1 t + \varepsilon_t \qquad (10.1.1)$$

where y_t is the observed crack length and t denotes the number of stress cycles, not the time. By linear regression formulae, the estimates of the parameters and their 95% confidence intervals are

$$\hat{\beta}_0 = 11.0278 \pm 0.0512$$

$$\hat{\beta}_1 = 0.10658 \pm 0.00986$$

Residual sum of squares = 1.3263

The residuals $\hat{\varepsilon}_t$ from the model (10.1.1) after using these estimated values are plotted in Fig. 10.2. The residuals have a tendency to increase once they start increasing and to decrease once they start decreasing. This behavior indicates a strong positive correlation in successive data points, and therefore the residuals are not independent.

We will therefore try to model the residuals following the procedure in Chapter 4. However, examining Figs. 10.1 and 10.2 for the original data and the residuals respectively, we see that the residuals are very small in numerical value compared to the data. Most of the residuals are less than

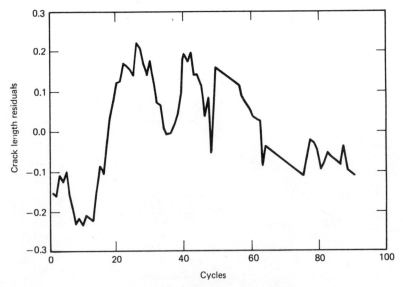

Figure 10.2 Crack length residuals.

2% of the corresponding data values. Therefore, if we are only interested in the linear trend and do not desire a precise fit, the improvement obtained by modeling the residuals and reestimating the final model may not be significant. On the other hand, if we want a precise fit and/or want to characterize the noise in the data or the deviations from the straight line trend, then further analysis would be fruitful. In fact, we will show later that the model for the stochastic part or the residuals can be given a valuable physical interpretation so that further analysis *is* physically meaningful.

10.1.2 Stochastic Part

In addition to examining the residuals, we also examine the sample autocorrelation of the residuals shown in Fig. 10.3. It is seen that the autocorrelations are asymptotically decaying to zero although initially they have large values. Thus, the residual series can be modeled by the procedure of Chapter 4, using the second method for computing a_t's starting with $a_1 = x_1$.

Table 10.1 shows the results printed by the computer program, similar to those given in Section 4.5. Although the ARMA(4,3) model is adequate, we select the simple AR(1) model, which is not significantly improved by the ARMA(2,1).

The continuous model then would be A(1). The parameter α_0 of this model can be directly obtained from the ϕ_1 of the AR(1) model by Eq. (6.4.7) as

$$\alpha_0 = -\ln(0.9299) = 0.07268$$

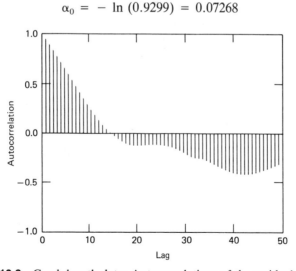

Figure 10.3   Crack length data: Autocorrelations of the residuals from the deterministic model.

Table 10.1 Results of Modeling Crack Length Residuals (Stochastic Part)

Parameters	Order of the ARMA Model		
	(2,1)	(4,3)	(1,0)
ϕ_1	0.4173 ± 1.240	0.2366 ± 0.454	0.9299 ± 0.079
ϕ_2	0.4940 ± 1.140	1.4668 ± 0.555	
ϕ_3		-0.1558 ± 0.377	
ϕ_4		-0.5870 ± 0.470	
θ_1	-0.4070 ± 1.320	-0.6981 ± 0.512	
θ_2		0.7507 ± 0.388	
θ_3		0.4157 ± 0.669	
Residual Sum of Squares	0.18650	0.15973	0.18986

considering the sampling interval as one stress cycle (which is "time" in the present case).

As an additional support for the AR(1), we examine the autocorrelations of its residuals plotted in Fig. 10.4. If we compare them with Fig. 10.3, we can see that no smooth pattern remains in Fig. 10.4, and therefore all the correlation seems to have been removed. In addition to their random nature, the autocorrelations are all small considering the 95% bounds $2/\sqrt{90} = 0.21$, except for lag 2, which slightly exceeds the bound. It

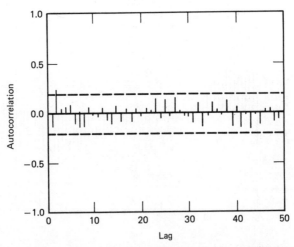

Figure 10.4 Crack length data: Autocorrelations of the residuals of the AR(1) model for the stochastic part.

therefore seems reasonable to consider the residuals from the AR(1) model as practically white noise and choose this model.

10.1.3 Complete Model: Deterministic Plus Stochastic

Combining the deterministic and stochastic parts, we get the complete model for the crack propagation data as

$$Y_t = \beta_0 + \beta_1 t + X_t \qquad (10.1.2)$$

where X_t is the stationary stochastic part that follows an AR(1) model in discrete t

$$X_t = \phi_1 X_{t-1} + a_t$$

Note that if we consider the data to be uniformly sampled from a continuous system, the equivalent continuous model is

$$Y(t) = \beta_0 + \beta_1 t + X(t) \qquad (10.1.3)$$

with

$$\frac{dX(t)}{dt} + \alpha_0 X(t) = Z(t)$$

For the purpose of estimation the discrete model may be written as

$$Y_t = \beta_0 + \beta_1 t + \phi_1 X_{t-1} + a_t \qquad (10.1.4)$$
$$X_t = Y_t - \beta_0 - \beta_1 t$$

or

$$Y_t = \beta_0 + \beta_1 t + \phi_1 [Y_{t-1} - \beta_0 - \beta_1(t-1)] + a_t$$
$$= \phi_1 Y_{t-1} + \beta_0(1-\phi_1) + \beta_1 t(1-\phi_1) + \beta_1\phi_1 + a_t$$

Since the derivative of Y_t with respect to either of the parameters β_0, β_1, or ϕ_1 is a function of the parameters itself, the reader may verify that Eq. (10.1.4) is a nonlinear regression model in its conditional aspect. Therefore, it is necessary to use the nonlinear least squares method to estimate its parameters. Starting with the estimates of β_0, β_1, and ϕ_1 in the separate models obtained above as the initial values, the nonlinear least squares routine minimizing the sum of squares of a_t's in the model (10.1.4) yields the following estimates and 95% confidence intervals:

$$\hat{\beta}_0 = 10.8896 \pm 0.0853$$
$$\hat{\beta}_1 = 0.1079 \pm 0.0024$$
$$\hat{\phi}_1 = 0.9592 \pm 0.0679 \quad (\hat{\alpha}_0 = 0.04166)$$

Residual sum of squares: 0.16938

The residual autocorrelations are shown in Fig. 10.5.

As expected, we see that these final values are not much different

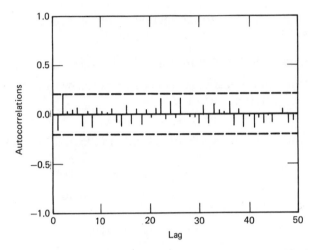

Figure 10.5 Crack length data: Autocorrelations of the residuals for the deterministic plus stochastic model.

from those obtained in the separate models (11.02783, 0.10658, and 0.9730, respectively) primarily because the deviations from the deterministic part or the "noise" are much smaller in magnitude. The residual sum of squares is also quite close to the one obtained from the stochastic part.

10.1.4 Physical Interpretation

The deterministic part of the model, which accounts for more than 82% of the variance of the crack length observation, represents the mode of crack propagation as a function of the number of stress cycles applied to the material. The model (10.1.2) shows that the mechanism of crack propagation can be represented by a simple linear model, at least over the range of stress cycles used in the data, that is, zero to 90 cycles. The crack length is directly proportional to the number of stress cycles the material has undergone.

A conjecture about the stochastic part of the model (10.1.2) can be made, based on the fact that this part represents a very small percentage of the actual observation. Let us assume that the stochastic part of the model arises from the measuring instrument. The instrument cannot respond immediately and will involve some time lag or inertia giving rise to instrument error or noise. We have seen in Chapter 6 that a system with inertia can be represented by an A(1) system. If this conjecture is indeed correct, then the preceding analysis gives us the time constant of the instrument as

$$\tau = \frac{1}{\alpha_0} = \frac{1}{0.04166} = 24.01$$

We can check the conjecture by measuring other sets of observations and seeing whether the stochastic part approximately gives the same model.

Accepting these interpretations, we can write the differential equations for crack propagation data. If we represent the measured crack length, actual crack length, and the instrument dynamics at t stress cycles by $Y(t)$, $C(t)$, and $X(t)$ respectively, then the data may be represented by the following set of equations:

$$\text{Observations: } Y(t) = C(t) + X(t) \qquad (10.1.5a)$$

$$\text{Crack length: } \frac{dC(t)}{dt} = \beta_1 = 0.108, \qquad (10.1.5b)$$

$$C(0) = \beta_0 = 10.89$$

$$\text{Instrument: } \frac{dX(t)}{dt} + 0.04166X(t) = Z(t), \qquad (10.1.5c)$$

$$\sigma_z^2 = 0.00195$$

10.1.5 Calibration of Measuring Instruments

Engineers and scientists are often faced with the problem of instrument calibration. The static constant bias in the instrument can be readily found and in many cases easily corrected. However, the calibration of the dynamics is more difficult. In general, one is forced to carry out the calibration of the dynamics under laboratory conditions, using the instrument to measure a fixed measurement. Such a calibration under laboratory conditions may not hold under actual working conditions.

The above analysis provides a possible method for calibrating the dynamics of the instrument under working conditions. If the above analysis is repeated for more sets of data, one can get a fairly good idea of the dynamic behavior of the instrument. Of course, the assumption inherent in such a procedure, that the stochastic part in the data is due to the measuring instrument *alone*, will have to be checked and confirmed by looking into the physical circumstances and any additional information available.

10.2 EXPONENTIAL TRENDS

We have seen in the last section that when the data clearly indicate the nature of the deterministic part and the noise or the disturbance added to the deterministic part is relatively small, the data can be modeled by a simple extension of the modeling strategy discussed in the earlier chapters. In this section we will consider more complex trends, namely the exponential ones, either individually or a mixture of several in the presence of a large proportion of noise.

Such exponential trends very often arise in the data obtained from systems in the physical sciences and engineering. Many of these systems can be well-represented by differential equations with constant coefficients, but the order of the differential equations, values of the constant coefficients, and the nature of noise are generally unknown. When such a system has a step or impulse input, the response is a sum of exponentials (see Chapters 6 and 7) with added noise. In this section we will consider the case of real roots; complex roots leading to a sinusoidal response will be considered in the next section. The problem of estimating the parameter values, when the equations or transfer functions for the system and the noise are known, is sometimes referred to as "system identification" and is a special case of the one discussed here.

When the amount of disturbance is relatively small and white or uncorrelated, the problem can be solved with relative ease either approximately by graphical methods or more accurately by least squares methods. The procedure given in this section is applicable even in the case of "colored" or correlated noise possibly present in large proportion. Of course, a large noise or disturbance has the inevitable consequence of large uncertainty associated with the parameter estimates (especially for the deterministic component), reflected in wide confidence intervals. When the noise is present in large proportion, it becomes important to characterize the noise itself by a mathematical model besides finding a good representation for the system. The first example in this section has therefore been chosen with the feature of large correlated noise.

The basic modeling strategy remains the same as before fitting models of successively higher order until no further improvement can be obtained. However, in the case of linear trends, since the deterministic component was fairly well known, the strategy was applied only to the stochastic part after separating the deterministic one in the preliminary analysis. In the present case, since the deterministic component may be a sum of many exponentials, we will apply the strategy to *both* the deterministic and the stochastic components. Thus, the model will be assumed to be the sum of n exponentials for the deterministic part, plus ARMA($n, n-1$) for the stochastic part. The models for successively larger values of $n = 1, 2, 3,$ . . . will be fitted until the reduction in the residual sum of squares is small and the confidence intervals on the parameters indicate that the lower order model is adequate.

10.2.1 Basis Weight Response to Step Input

Figure 10.6 shows a plot of 100 data points of the observed basis weights in response to a step input in the stock flow rate of a papermaking process. The numerical values of the data, taken at one-second intervals, are given in Appendix I. A brief description of the papermaking process together with a schematic diagram is presented in Section 11.1.1. The modeling and analysis of this data provide a technique of system identification from on-line measurements as discussed in Pandit, Goh, and Wu (1976).

During the operation of the papermaking machine, there are variations of paper basis weight not only in the machine direction but also across the machine direction. The beta-ray gage recording the basis weight moves back and forth across the machine direction, forming a zig-zag measurement path, as shown in Fig. 10.7, due to the movement of the paper in the machine direction. Under steady state conditions, when weight variation in the cross-direction is negligible, the record looks similar to the sketch shown in Fig. 10.8a, but if there is a significant cross-machine variation, for example, if the paper is thicker at the center, the measurements exhibit a pattern such as the one shown in Fig. 10.8b.

In the particular papermaking machine under study, there was a strong cross-directional variation superimposed as noise over the dynamic response. It is therefore important to characterize not only the dynamics but also the noise. The dynamics is controlled by means of the stock flow rate (time-wise adjustments), whereas the cross-directional variation can be controlled by adjusting the openings on various sections of the headbox

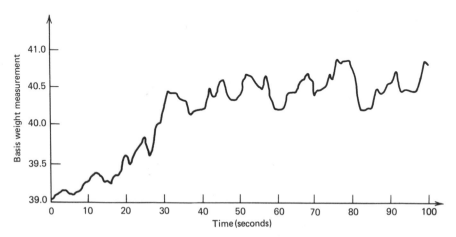

Figure 10.6 Basis weight measurement data showing the machine response to a step increase in stock flow.

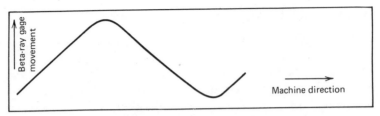

Figure 10.7 Measurement path of beta-ray gage.

slice lip (spacewise adjustments). A comprehensive characterization of the dynamics, as well as the noise dominated by cross-directional variation, is necessary for successful control of the papermaking process.

10.2.2 First Order Dynamics

Since the observed data shown in Fig. 10.6 is the result of a step increase in the input stock flow, under the assumption of the first order dynamics with superimposed noise or disturbance, the model for the basis weight may be written as

$$y_t = A_0 + g\left(1 - e^{-\frac{t}{\tau}}\right) + \varepsilon_t \qquad (10.2.1)$$

where y_t is the measured basis weight at time t seconds, g is the steady state gain, τ is the time constant in seconds, and ε_t represents the unknown disturbance effects. Although a part of the disturbance may come from the input, machine-directional variation, or the instrument, it is believed that a major portion arises from the cross-directional variation.

In the preliminary analysis, primarily aimed at obtaining the initial values for the final adequate model to be considered later, we assume ε_t to be uncorrelated and estimate the values A_0, g, and τ by the nonlinear

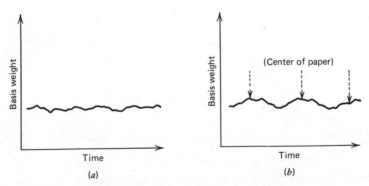

Figure 10.8 Measurement data patterns.

least squares method. The initial values of these parameters required for the least squares computer routine minimizing the sum of squares of the ε_t's are easily obtained from the data. The initial value of A_0 can be taken as the smallest observed value about 39. In Fig. 10.6, the basis weight measurement data after 30 seconds fluctuate around 40.6, which is the sum of A_0 and g; therefore, g was taken as $40.6 - 39 = 1.6$. At $t = \tau$, the average response should reach the value $A_0 + (1 - .37)g \simeq 40$; therefore, from the plot, $\tau \simeq 27$.

Starting with these initial values, the nonlinear least squares routine gives the following estimates and 95% confidence intervals:

$$\hat{A}_0 = 38.7586 \pm 0.1840$$

$$\hat{g} = 1.9321 \pm 0.1661$$

$$\hat{\tau} = 27.1719 \pm 6.2593$$

Residual sum of squares: 4.0293

The plot of the residuals and their autocorrelations is shown in Figures 10.9 and 10.10. From both these plots it is clear that the residuals are stationary and highly correlated; in fact, the plot in Fig. 10.10 shows a decaying pattern characteristic of a stationary stochastic process. If the model for the residual is AR(1), then the initial value for the parameter ϕ_1 will be $\hat{\phi}_1 = \hat{\rho}_1 = 0.755$.

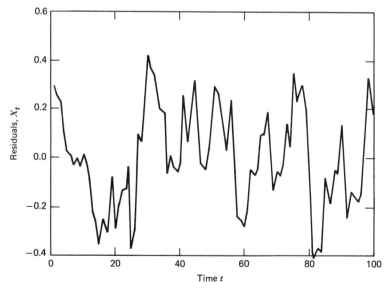

Figure 10.9   Basis weight residuals from a first order deterministic model (10.2.1).

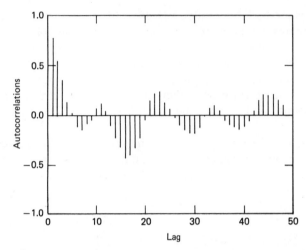

Figure 10.10 Papermaking data: Autocorrelations of the residuals after the deterministic model.

Rather than analyzing the residuals separately and finding a model for them, we will directly formulate a combined model consisting of a deterministic and stochastic part as

$$y_t = A_0 + g\left(1 - e^{-\frac{t}{\tau}}\right) + X_t \qquad (10.2.2)$$
$$X_t = \phi_1 X_{t-1} + a_t$$

where the stochastic part X_t is assumed to follow an AR(1) model. The parameters of this model can be estimated by nonlinear least squares minimizing the sum of squares of a_t's recursively computed from

$$y_t = A_0 + g(1 - e^{-\frac{t}{\tau}}) + \phi_1 X_{t-1} + a_t \qquad (10.2.2a)$$

Taking the estimated values given above for the model (10.2.1) as the initial values of A_0, g, and τ and the initial value of ϕ_1 as the one lag autocorrelation of the residuals $\hat{\phi}_1 = \hat{\rho}_1 = 0.755$, the nonlinear least squares computer routine yields the following estimates and 95% confidence intervals:

$$\hat{A}_0 = 38.9974 \pm 0.2682$$
$$\hat{g} = 1.7992 \pm 0.4228$$
$$\hat{\tau} = 35.3912 \pm 23.5606$$
$$\hat{\phi}_1 = 0.7887 \pm 0.1270$$

Residual sum of squares: 1.6639

Comparing these with the estimates for the model (10.2.1), we see that these estimates are considerably different, especially for the parameters g and τ. This difference is primarily due to the large proportion of noise in the data leading to significant errors in estimates when the residuals are incorrectly assumed to be uncorrelated. The large proportion of the noise also results in more uncertainty regarding the dynamic parameters g and τ as reflected in their wide confidence intervals.

A plot of the residual a_t's of the model (10.2.2) is shown in Fig. 10.11. This plot indicates that the residuals have no trend and the model seems to be adequate. To check the adequacy of the noise model, we should, however, fit a higher order model, for example, the ARMA(2,1), and second or higher order dynamics for the deterministic part to see if it provides a better fit. (Such a second order model, in fact, gives an insignificant reduction in the residual sum of squares. See Problem 10.4.)

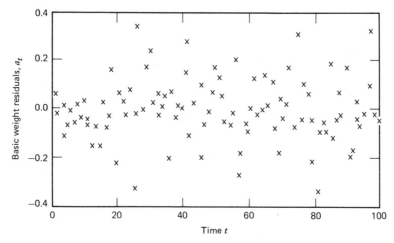

Figure 10.11   Plot of the residuals of an AR(1) for stochastic and first order deterministic.

10.2.3 Differential Equations

Since the sampling interval is one second, model (10.2.2) implies that the differential equation for the stochastic part $X(t)$ may be written as [see Eq. (6.4.7)]

$$\frac{dX(t)}{dt} + \alpha_0 X(t) = Z(t) \qquad (10.2.3)$$

with $\alpha_0 = -\ln(\phi_1) = -\ln(0.7887) = 0.2374$. The deterministic part is the step response of the basis weight $B(t)$ measured along the path in Fig. 10.7. Its differential equation is (see Section 6.1)

$$\tau \frac{d}{dt}[B(t) - A_0 - g] + [B(t) - A_0 - g] = 0 \qquad (10.2.4)$$

where

$$\tau = 35.3912$$

$$B(0) = A_0 = 38.9974$$

$$B(\infty) = A_0 + g = (38.9974 + 1.7992) = 40.7966$$

If we assume that the primary source of correlation in the noise is the cross-machine variation, Eq. (10.2.3) may be considered to represent the dynamics of the paper profile in the cross-machine direction, whereas Eq. (10.2.4) represents it in the machine direction. The time constant for the cross-machine variation represented by Eq. (10.2.3) is

$$\tau_X = \frac{1}{0.2374} = 4.2128$$

which is much smaller than that of the machine-directional variation represented by Eq. (10.2.4).

10.2.4 Possibility of a Single Time Constant

Although the two time constants τ and τ_X for the machine and cross-directional variation are considerably different, the confidence interval on τ is very large 35.3912 ± 23.5606. Therefore, there is a possibility that the two time constants could be the same. If they are the same, then the model (10.2.2) takes the form

$$y_t = A_0 + g(1 - \phi_1^t) + X_t$$

$$X_t = \phi_1 X_{t-1} + a_t$$

To explore this possibility, the sum of squares of a_t's from

$$y_t = A_0 + g(1 - \phi_1^t) + \phi_1 X_{t-1} + a_t$$

corresponding to (10.2.2a) can be minimized to estimate the parameters A_0, g, and ϕ_1. The results of estimation are as follows:

$$\hat{A}_0 = 39.0486 \pm 0.2939$$

$$\hat{g} = 1.4915 \pm 0.7360$$

$$\hat{\phi}_1 = 0.9522 \pm 0.0521$$

Residual sum of squares: 1.7944

$$\tau = -\frac{1}{\ln(0.9522)} = 20.42$$

The residual sum of squares is increased considerably from the value 1.6639 for the model (10.2.2). To see whether this increase is significantly large for the reduction of one parameter, we apply the F-criterion on the basis of a local linear hypothesis.

$$F = \frac{1.7944 - 1.6639}{1} \div \frac{1.6639}{100 - 3}$$

$$= 7.4$$

Since $F_{0.99}(1,100) \simeq 6.7$, it is seen that the increase is significant even at a 1% level. Thus, the data gives strong evidence against the hypothesis of the same time constant. We tentatively conclude that the model (10.2.2) and the corresponding differential equations (10.2.3) and (10.2.4) provide the correct representation of the system.

10.2.5 Chemical Relaxation—Estimation of Time Constant

As a second illustration of exponential trends, we consider a set of data collected for the evaluation of the relaxation time of a fast chemical reaction. A temperature-jump system is used to study the relaxation process of the concentration change of a fast chemical reaction. An electric spark provides a sudden rise in temperature, which disrupts the equilibrium state of the reacting solution and initiates the relaxation process. An absorption spectrometry system is installed to indirectly measure the concentration change of the reaction solution. The underlying chemistry and the experimental procedure are briefly discussed in Chu (1972). Fig. 10.12 shows a recording of the data from the system; the numerical values of 100 obser-

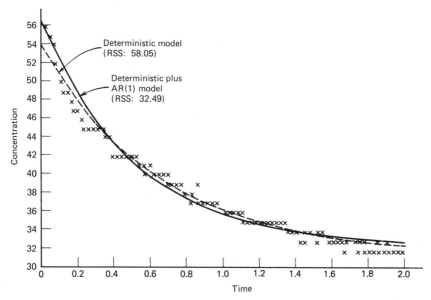

Figure 10.12 Chemical relaxation data.

vations of this reaction data taken at a sampling interval of 0.02 seconds are given in Appendix I.

The procedure for modeling the relaxation data is exactly the same as that for the preceding basis weight data, although the model form is slightly different. Therefore, only the final results are given below.

As a preliminary step toward fitting a single exponential plus AR(1) model, the following model was fitted:

$$y_t = A_0 + A_1 e^{-\frac{t}{\tau}} + \varepsilon_t \tag{10.2.5}$$

where $(A_0 + A_1)$ is the initial concentration, A_0 is the final equilibrium concentration, and τ is the relaxation time constant. Finding the initial values from the data as described before, the nonlinear least squares routine minimizing the sum of squares of ε_t's gives the following results:

$$\hat{A}_0 = 31.347 \pm 0.539$$
$$\hat{A}_1 = 22.575 \pm 0.599$$
$$\hat{\tau} = 0.6447 \pm 0.0520$$

Residual sum of squares: 58.05

The autocorrelations of the residual ε_t's are shown in Fig. 10.13 and the fit is shown by a dotted line in Fig. 10.12.

Taking the estimated values as the initial values of A_0, A_1, τ and the first autocorrelation as that of ϕ_1, the following deterministic plus stochastic model was fitted:

$$y_t = A_0 + A_1 e^{-\frac{t}{\tau}} + X_t \tag{10.2.6}$$
$$X_t = \phi_1 X_{t-1} + a_t$$

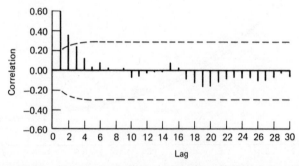

Figure 10.13 Chemical relaxation data: Plot of the autocorrelation of deterministic model residuals.

The estimation results are

$$\hat{A}_0 = 32.198 \pm 1.24$$

$$\hat{A}_1 = 24.109 \pm 1.50$$

$$\hat{\tau} = 0.5248 \pm 0.0934$$

$$\hat{\phi}_1 = 0.7863 \pm 0.1272$$

Residual sum of squares: 32.49

The autocorrelations of the residual a_t's are shown in Fig. 10.14. They seem to indicate that the model (10.2.6) is adequate, which may be further confirmed by fitting a single or double exponential model with the ARMA(2,1) stochastic part. The fit is shown by a solid line in Fig. 10.12.

Since the sampling interval Δ is 0.02, we can write the differential equation for the uniformly sampled system represented by Eq. (10.2.6) after computing the time constant for the stochastic part. The parameter α_0 of the corresponding A(1) system is

$$\alpha_0 = -\frac{1}{\Delta} \ln(\phi_1) = -\frac{1}{0.02} \ln(0.7863) = 12.021$$

If we denote the concentration at time t above the equilibrium level $A_0 = 32.198$ by $C(t)$, then the differential equations for the system are (see Chapter 6)

$$\frac{dC(t)}{dt} + \frac{1}{\tau} C(t) = A_1 \delta(t) \tag{10.2.7}$$

$$\frac{dX(t)}{dt} + \alpha_0 X(t) = Z(t) \tag{10.2.8}$$

$$y(t) - A_0 = y'(t) = C(t) + X(t)$$

where $y(t)$ denotes the observed value at t.

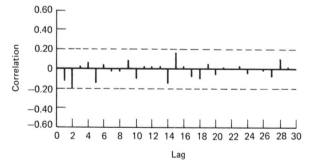

Figure 10.14 Chemical relaxation data: Plot of the autocorrelation of residual of an AR(1) and deterministic model.

The differential equation (10.2.7) says that the change of concentration from the equilibrium level is the response of a first order system to impulse input. This is in accord with the physical nature of the reaction process described at the beginning of this section. The sudden change in the temperature gives rise to an instantaneous change in the concentration acting like an impulse. Since the reaction is first order, the concentration returns to the equilibrium position following Eq. (10.2.7), of which the deterministic part of the model (10.2.6) is the solution.

10.2.6 Another Differential Equation for Relaxation

In order to interpret the disturbance or the stochastic part $X(t)$, it is necessary to ascertain whether $X(t)$ can be accounted for by the reaction itself. Does it arise from the reaction by some random changes in the input so that it is an integral part of the relaxation process, or is it added to the relaxation response by instrument error, etc., at the output? In the present case, since the measurement system is sophisticated and accurate, the first possibility is more likely. We will now show that the data also seems to support the first possibility.

If the stochastic part arises from the reaction itself, then the first order differential equation for $X(t)$ should have the same time constant as that of the relaxation reaction. Therefore from Eqs. (10.2.7) and (10.2.8), the model for the observation $y'(t)$ (measured from the equilibrium concentration A_0) should be

$$\frac{dy'(t)}{dt} + \frac{1}{\tau} y'(t) = A_1 \delta(t) + Z(t) \tag{10.2.9}$$

so that the response would be a sum of the response to the deterministic impulse $A_1 \delta(t)$ at the start of the reaction and the stochastic "impulses" $Z(t)$ occurring throughout the reaction. By the theory of Chapter 6, it follows that if the system (10.2.9) is sampled at uniform intervals Δ, it will have the representation

$$y_t = A_0 + A_1 e^{-\frac{t\Delta}{\tau}} + X_t \tag{10.2.10}$$

$$X_t = e^{-\frac{\Delta}{\tau}} X_{t-1} + a_t$$

The estimates of the *three* parameters A_0, A_1, and τ can be obtained by minimizing the sum of squares of a_t's recursively computed from

$$y_t = A_0 + A_1 e^{-\frac{t\Delta}{\tau}} + e^{-\frac{\Delta}{\tau}} X_{t-1} + a_t \tag{10.2.10a}$$

The results of estimation are as follows:

$$\hat{A}_0 = 33.3395 \pm 3.2216$$

$$\hat{A}_1 = 23.8367 \pm 3.2984$$

$$\hat{\tau} = 0.3953 \pm 0.1669$$

Residual sum of squares: 34.19

It is thus seen that the residual sum of squares of the model (10.2.10) with three parameters is only slightly larger than that of the four-parameter model (10.2.6). If the F-criterion is used on the basis of the local linear hypothesis used in nonlinear least squares, we get

$$F = \frac{34.19 - 32.49}{32.49} \div \frac{1}{100 - 4}$$

$$= 5.02$$

whereas $F_{0.95}(1,120) = 3.92$ and $F_{0.99}(1,120) = 6.85$. Therefore, the data does not have strong evidence to reject the model (10.2.10), which may therefore be also considered as adequate. Statistically, it is difficult to choose between the models (10.2.10) and (10.2.6). The choice will have to be made on the basis of other evidence based on a physical understanding of the process.

10.3 PERIODIC TRENDS: SEASONALITY

In the preceding Section 10.2 we considered exponential trends with models in which the exponents were real. In this section we will allow some or all of the exponents to be complex or, more precisely, complex conjugate since the response in the form of the data is real. The resulting deterministic plus stochastic models are useful in dealing with a continually increasing or decreasing nonstationary data with seasonal or periodic trends. Strict sine cosine waves or their combinations superimposed with correlated noise form a special case of these models. Therefore, data with these features can also be modeled by the method of this section.

Sometimes the data fluctuate around a fixed mean and the periodic tendency is present but not explicitly clear. An overall period or periods are discernible (for example, by counting average number of peaks), but the periodicity appears to be more or less changing from one part of the data to another, that is, it exhibits a stochastic tendency. In such case, the series should be considered stationary and modeled by methods of Chapter 9; the method of nonstationary trends discussed in this section will not give satisfactory results. It is only when the periodic tendency repeats regularly that a deterministic trend is called for and then the method of this section is needed; although, even in such a case, methods of Chapter 9 would give satisfactory results.

In all three examples considered in Sections 10.1 and 10.2, only one nonstationary trend was present. This is usually not the case with seasonal series, particularly those observed in business and economics. In addition to the main (monthly or weekly) period, its harmonics are also present. Therefore, it is often necessary to remove quite a few trends before the data can be considered stationary as judged by their plot and the plot of autocorrelations.

The growth (or decay) trends in the data can be removed in the preliminary analysis by the models, as in Section 10.2, containing exponentials with real roots. The periodic trends can then be added to the model by pairs of complex conjugate exponentials, with the imaginary parts of the roots containing the dominant frequency and its multiples. However, it is advisable to use the equivalent representation by a product of an exponential with a real exponent and sine function with a known or unknown period and phase angle. After the addition of further periods and trends fails to result in a significant improvement in the residual sum of squares, the residuals can be checked for stationarity and an ARMA/AM model fitted to them. Finally, the parameters of the combined model can be estimated. As before, we will explain and illustrate the procedure by means of an example.

10.3.1 International Airline Passenger Data

Figure 10.15 shows the plot of monthly totals of international airline passengers that was used in Section 9.2.4 for modeling after logarithmic transformation. The data show a regular periodic tendency; a major peak occurs in the months of July and August and a smaller one in the month of March every year. There is an overall growth tendency on which these periodic trends are superimposed. It seems, therefore, logical to fit a model with the deterministic part consisting of exponentials—with real exponents for the growth trends and complex conjugate exponentials for the periodic trends. A general model of this kind may be formulated as

$$y_t = \sum_{j=1}^{s} A_j e^{K_j t} + X_t \qquad (10.3.1)$$

where X_t is an ARMA process for discrete time or an AM process for continuous time.

In the present case, since we know that the complex conjugate exponents K_j's will have imaginary parts with frequency in radians corresponding to the dominant period (12 months) and its harmonics, we con-

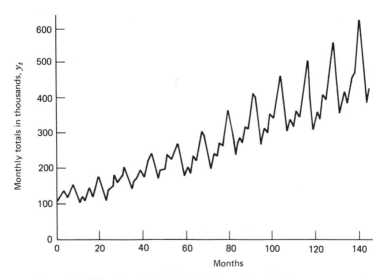

Figure 10.15   Monthly totals of international airline passengers.

sider the following special case of an equivalent representation of Eq. (10.3.1):

$$y_t = \sum_{j=1}^{\ell} R_j e^{r_j t} + \sum_{j=1}^{i} B_j e^{b_j t} \sin(j\omega t + \psi_j) + X_t \qquad (10.3.2)$$

where ℓ is the number of real exponents corresponding to growth trends, i is the number of pairs of complex conjugate roots corresponding to periodic trends, and $\ell + 2i = s$. R_j, r_j, B_j, b_j, and ψ_j are the unknown parameters to be estimated, and ω is the dominant frequency in radians per unit t. The parameters B_j and ψ_j denote the amplitude and phase of the periodic trends. The term $e^{b_j t}$ takes care of the growth trend of the periodic term and its harmonics. The significance of these parameters will be discussed later in Section 10.3.4.

However, in using a nonlinear least squares routine for estimation, the parameter ψ_j is often difficult to estimate. Therefore, we will use the following reparameterized form of the model (10.3.2)

$$y_t = \sum_{j=1}^{\ell} R_j e^{r_j t} + \sum_{j=1}^{i} B_j e^{b_j t} [C_j \sin(j\omega t) + \sqrt{1 - C_j^2} \cos(j\omega t)] + X_t$$

$$(10.3.3)$$

where $C_j = \cos \psi_j$.

10.3.2 Exponential Growth Trend

In the preliminary analysis aimed at finally fitting a deterministic plus stochastic model (10.3.3), we begin by first fitting the exponential growth trend and then add the periodic trends, one by one. The first model to be fitted is therefore

$$y_t = R_1 e^{r_1 t} + \varepsilon_t \qquad (10.3.4)$$

The initial guesses for the parameters R_1 and r_1 can be obtained from the plot of the data by using the facts: $e^0 = 1$ and $e^1 \simeq 2.7$. For example, when $t = 0$, $y_0 \simeq R_1$, the value of R_1 from Fig. 10.15 is approximately 130. Similarly, when $t = 1/r_1$, $y_t \simeq R_1 e^1 = 130\,(2.7) = 351$ and Fig. 10.15 gives an average value of $t = 100$ months. This provides an initial estimate of $r_1 = 1/t_1 = 0.01$. With these initial estimates, the nonlinear least squares routine minimizing the sum of squares of ε_t's in Eq. (10.3.4) yields the results given in the second column of Table 10.2.

The residuals from the model (10.3.4) are shown in Fig. 10.16. The exponential growth present in the original data has been removed and the residuals only have periodic trends with increasing amplitude. Therefore, it seems that a single exponential with real roots is enough and we need not add more exponentials with real roots, unless warranted at a later stage in the preliminary analysis.

10.3.3 Addition of Periodic Trends

Taking $\ell = 1$ in the model (10.3.3), we now start adding the periodic trends. The first model to be entertained is therefore

$$y_t = R_1 e^{r_1 t} + B_1 e^{b_1 t} \left[C_1 \sin(\omega t) + \sqrt{1 - C_1^2} \cos(\omega t) \right] + \varepsilon_t \quad (10.3.5)$$

where

$$\omega = \frac{2\pi}{12}$$

is the dominant frequency in radians per month for the major peaks clearly visible in the residuals of the model (10.3.4) as also in the original data.

The initial estimates of the additional parameters B_1, b_1, and C_1 were roughly obtained from the plot of the residuals in Fig. 10.16 as $\hat{B}_1 = -20$, $\hat{b}_1 = 0.01$, $\hat{C}_1 = 0.7$. The results of estimation with 95% confidence bounds are shown in the third column of Table 10.2. Note that the residual sum of squares is drastically reduced from 296,250 to 95,783 by inclusion of only one period of 12 months. This is also clear from Figs. 10.15 and 10.16 that show that the yearly period can account for a large part of the variation in the series. Also, the estimate of the exponent is significantly large and its confidence interval does not include zero. This means that the amplitude

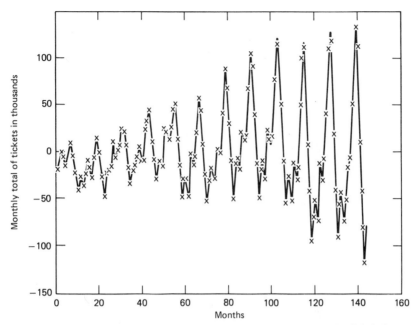

Figure 10.16 Plot of residuals from the deterministic model (10.3.4).

of the yearly oscillations are growing as postulated by the model. The plot in Fig. 10.16 again confirms this observation.

Since the improvement in the sum of squares is large, we successively fit the models

$$
y_t = \sum_{j=1}^{\ell} R_j e^{r_j t} + \sum_{j=1}^{i} B_j e^{b_j t} [C_j \sin (j\omega t) \\
+ \sqrt{1 - C_j^2} \cos (j\omega t)] + \varepsilon_t \tag{10.3.6}
$$

for values of $\ell = 1$ and $i = 2, 3, \ldots$ until the reduction in the residual sum of squares is insignificantly small and the statistical tests show the adequacy of the model. The estimation results are shown in columns 4, 5, 6 and 7 for the values of $i = 2, 3, 4$, and 5, respectively. Adding the sixth period does not result in a significant improvement and therefore we stop at $i = 5$.

The introduction of five periods in the model accounts for yearly, half-yearly, 4 monthly, quarterly, and 2-2/5 monthly periods. It is therefore natural that periods smaller than 2-2/5 monthly may not be necessary. Residuals from the model (10.3.6) with $\ell = 1, i = 5$ are shown in Fig. 10.17. No further periodic trends are visible, which confirms that the introduction of further periodic components is not necessary. The fit of this deterministic

Table 10.2 Deterministic and Stochastic Models for International Airline Data [mean = 280.3, variance = 14,392].

Model:

$$Y_t = \sum_{j=1}^{\ell=1} R_j e^{r_j t} + \sum_{j=1}^{i=5} B_j e^{b_j t} [C_j \sin(j\omega t) + \sqrt{1 - C_j^2}\, \cos(j\omega t)] + X_t$$

$$X_t = \sum_{s=1}^{n} \phi_s X_{t-s} - \sum_{w=1}^{m} \theta_w a_{t-w} + a_t$$

(i,n,m)	(0,0,0)	(1,0,0)	(2,0,0)	(3,0,0)	(4,0,0)	(5,0,0)	(5,3,3)
Total No. of Parameters $2\ell + 3i + n + m$	2	5	8	11	14	17	23
(1) Parameters	(2)	(3)	(4)	(5)	(6)	(7)	(8)
R_1	130.83±10.14	131.07±5.85	130.43±4.03	130.63±3.71	130.63±3.51	130.62±3.36	127.48±0.45
r_1	0.0095±0.0007	0.0095±0.0005	0.0095±0.0003	0.0095±0.0002	0.0095±0.0002	0.0095±0.0003	0.0096±0.0007
B_1		−14.83±6.56	−16.16±4.83	−15.78±4.36	−15.77±4.132	−15.79±3.96	−15.89±4.063
b_1		0.0136±0.0039	0.0126±0.0027	0.0129±0.0025	0.0129±0.0024	0.0128±0.0022	0.0128±0.0020
C_1		0.426±0.106	0.433±0.076	0.433±0.069	0.433±0.065	0.434±0.063	0.436±0.059
B_2			−8.88±4.961	−9.31±4.708	−9.396±4.49	−9.414±4.30	−9.503±2.40

b_2			0.0121 ± 0.0050	0.0115 ± 0.0046	0.0114 ± 0.0038	0.0113 ± 0.0041	0.0112 ± 0.0022
C_2			-0.997 ± 0.016	-0.996 ± 0.015	-0.996 ± 0.012	-0.996 ± 0.012	-0.996 ± 0.006
B_3				4.529 ± 5.225	4.267 ± 4.76	4.187 ± 4.50	3.914 ± 1.652
b_3				0.0095 ± 0.0107	0.0103 ± 0.0103	0.0105 ± 0.0098	0.0113 ± 0.0039
C_3				-0.431 ± 0.340	-0.433 ± 0.316	-0.432 ± 0.300	-0.411 ± 0.122
B_4					4.320 ± 5.536	4.379 ± 5.328	5.357 ± 2.642
b_4					0.0072 ± 0.0124	0.0071 ± 0.0118	-0.006 ± 0.0472
C_4					0.912 ± 0.191	0.903 ± 0.191	0.423 ± 0.425
B_5						2.289 ± 4.114	2.383 ± 1.785
b_5						0.0121 ± 0.0160	0.0115 ± 0.0068
C_5						0.995 ± 0.049	0.994 ± 0.025
ϕ_1							-0.087 ± 0.163
ϕ_2							-0.140 ± 0.110
ϕ_3							0.892 ± 0.112
θ_1							-0.716 ± 0.234
θ_2							-0.955 ± 0.227
θ_3							0.180 ± 0.256
Residual sum of squares	296,250	95,783	44,753	36,897	32,354	28,908	8,091

component is pictorially represented in Fig. 10.18 by a solid line, the original data are shown by cross marks. This figure also confirms the adequacy of the deterministic part of the model.

10.3.4 Stochastic Part and the Combined Model

The residuals from the model (10.3.6) with $\ell = 1$, $i = 5$, shown in Fig. 10.17, appear to be stationary series without persistent nonstationary trends. The autocorrelations of these residuals are shown in Fig. 10.19. The asymptotic decay of the autocorrelations also indicates that the data are a stationary time series.

Thus, the residuals can be considered as the stochastic part of the final model and the modeling procedure of Chapter 4 is applicable. This procedure indicates that a significant reduction in the residual sum of squares is obtained in going from ARMA(2,1) to ARMA(4,3), but the reduction is insignificant comparing the ARMA(4,3) with ARMA(6,5). However, the parameter ϕ_4 of the ARMA(4,3) is small and after refitting the ARMA(3,3) turns out to be the adequate model.

We are now in a position to formulate the final adequate combined model with a deterministic plus stochastic part and estimate all the param-

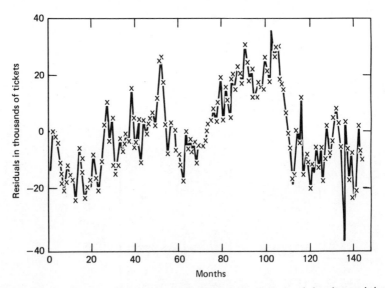

Figure 10.17   International airline data: Plot of residuals of the deterministic model (10.3.6) with $\ell = 1$, $i = 5$.

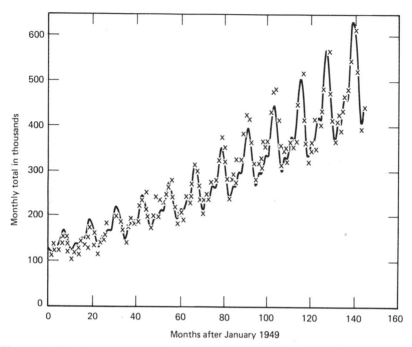

Figure 10.18 International airline data: Comparison of the deterministic trend and actual data.

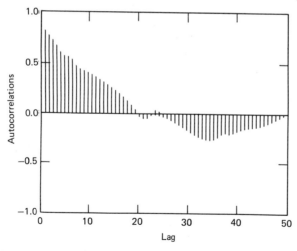

Figure 10.19 International airline data: Plot of autocorrelations of residuals from the deterministic model.

eters simultaneously. The combined model is

$$X_t - \phi_1 X_{t-1} - \phi_2 X_{t-2} - \phi_3 X_{t-3} = a_t - \theta_1 a_{t-1} - \theta_2 a_{t-2} - \theta_3 a_{t-3}$$

$$y_t = R_1 e^{r_1 t} + \sum_{j=1}^{5} B_j e^{b_j t} \left[C_j \sin\left(\frac{2j\pi t}{12}\right) \right.$$

$$\left. + \sqrt{1 - C_j^2} \cos\left(\frac{2j\pi t}{12}\right) \right] + X_t \quad (10.3.7)$$

The last column of Table 10.2 shows the results of minimizing the sum of squares of a_t's in the model (10.3.7), starting with the initial values that are the estimated values of the separate models fitted earlier. It can be seen from the table that the introduction of the stochastic part represented by an ARMA(3,3) drastically reduces the sum of squares from 28,908 to 8,091. On the other hand, it can be verified that the introduction of one more period in place of the stochastic part by making $i = 6$ reduces the sum of squares only to 28,849. This strengthens our decision to stop at $i = 5$ in the deterministic part. Since the preceding analysis of the residuals from the deterministic part (10.3.6) had clearly indicated that the ARMA(3,3) model is adequate, we conclude that the model 10.3.7 is the final adequate model.

The residual a_t's from the model (10.3.7) are shown in Fig. 10.20.

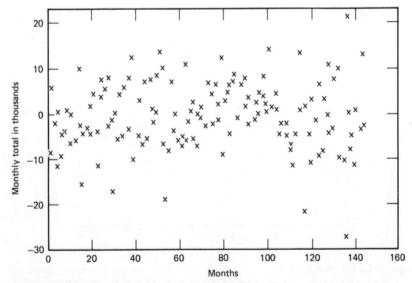

Figure 10.20 Plot of residuals from the ARMA(3,3) and the deterministic model.

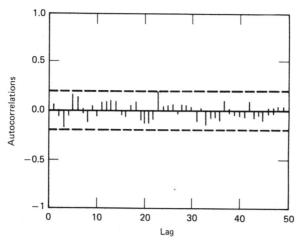

Figure 10.21 International airline data: Plot of autocorrelations of residuals from the ARMA(3,3) and the deterministic model.

They appear to be scattered and do not show any trends characteristic of periodic components or autocorrelation in the data. The plot of the autocorrelations of a_t's given in Fig. 10.21 also confirms that the a_t's are uncorrelated, all the $\hat{\rho}_k$'s are within the $\pm 2/\sqrt{N} = 0.167$ bounds. Thus, there are no indications of the inadequacy of the model (10.3.7).

The fit of the combined model (10.3.7) is graphically represented in Figs. 10.22a to 10.22n. Figure 10.22a shows the plot of the original airline passenger ticket sales. A combined deterministic and stochastic model fit is shown in Fig. 10.22g. In essence, the fit is decomposed into three parts. First, an exponential trend $R_1 e^{r_1 t}$ is shown in Fig. 10.22h. The second part consists of five periods, that is, yearly, half yearly, 4 monthly, quarterly and 2-2/5 monthly, that are represented in Figs. 10.22i to 10.22m, respectively. The summation of the exponential trend (curve h) and the first period (curve i) is shown in Fig. 10.22b. Figure 10.22c depicts the summation of the exponential curve (h) and two periods (i) and (j). Similarly, the summation of the exponential curve (h) with three, four, and five periods, is shown in Figs. 10.22d, 10.22e, and 10.22f, respectively. The third part represents the stochastic model (Fig. 10.22n). The summation of Figs. 10.22f and 10.22n actually yields the curve in (g). The residuals from the model (10.3.7) have already been shown in Fig. 10.20.

In order to clearly see the significance of various parameters, the first two periodic terms are plotted on a large scale and shown in Figs. 10.23 and 10.24. The terms $|B_1|$ and $|B_2|$ represent the starting amplitudes; the parameters b_1 and b_2 are the exponents of the exponential growth terms.

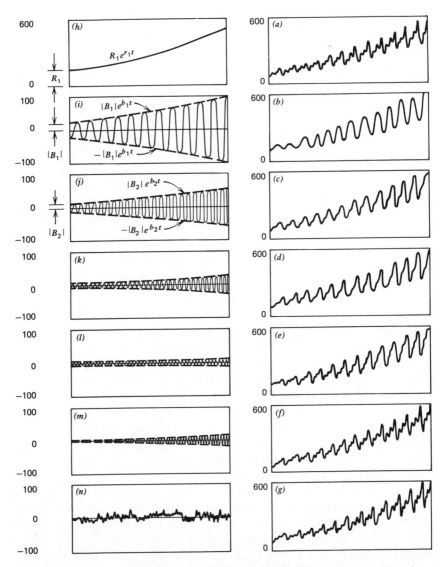

Figure 10.22 A graphical representation of the fit of deterministic and stochastic models to the airline data.

The large value of $b_1 = 0.013$ in Fig. 10.23 indicates a faster rate of increase for the amplitude of the periodic term than in Fig. 10.24 where $b_2 = 0.01$.

The parameter C, which is actually the cosine of the phase angle ψ, controls the starting point of the periodic cycle. For example, in Fig. 10.24, the absolute value of the starting point is closer to the datum line (zero

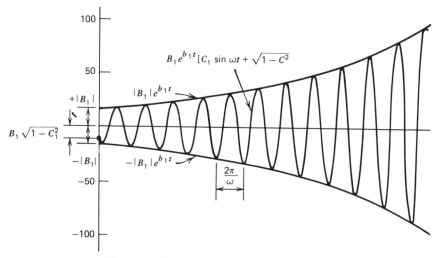

Figure 10.23 Plot of the first periodic term
$-15.89e^{0.013t}[0.436 \sin \omega t + \sqrt{1-0.436^2} \cos \omega t]$.

line) than in Fig. 10.23. The first two periodic terms without the exponential growth in their amplitude are further amplified and shown in Figs. 10.25 and 10.26. For the first periodic term (the value of parameter $C_1 = 0.436$ and the frequency ω), the two components $C_1 \sin \omega t$ and $\sqrt{1 - C_1^2} \cos \omega t$ are plotted to yield a curve with the starting point at a

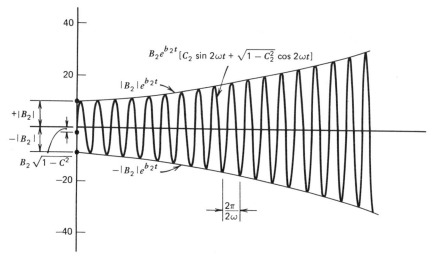

Figure 10.24 Plot of the second periodic term
$-9.5e^{0.01t}[-.996 \sin 2\omega t + \sqrt{1+0.996^2} \cos 2\omega t]$.

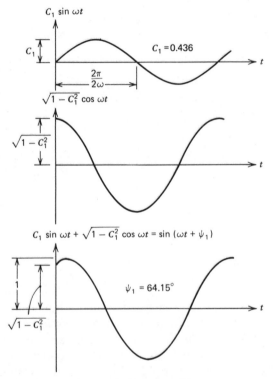

Figure 10.25   Plot of the first periodic term without the exponential growth,
[$\sin(\omega t + 64.15°)$].

distance $\sqrt{1 - C_1^2}$ from the datum line and showing a phase shift of 64.15°.
For the second periodic term (with parameter $C_2 = -0.996$ and frequency
2ω), Fig. 10.26 shows the position of the starting point closer to the datum
line, that is, at a distance $\sqrt{1 - C_2^2}$ showing a phase shift of 174.87°.

10.3.5 Comparison With a Multiplicative Model

Recall that in Chapter 9 the following multiplicative model was fitted to
the airline data considered above after taking logarithms

$$(1 - B)(1 - B^{12})w_t = (1 - 0.40B)(1 - 0.61B^{12})a_t \quad (10.3.8)$$

where

$$w_t = \ell n \, y_t$$

This model reduces the unstable series to a stable one by three transformations: the first is logarithmic, the second is the first differencing $\nabla =$

$(1 - B)$, and the third is a seasonal 12th differencing $(1 - B^{12})$. Such a transformed series is then modeled by a stable moving average model of the 13th order given by the right-hand side of (10.3.8).

Although both the models (10.3.7) and (10.3.8) are statistically adequate, there is a fundamental difference in their structure and assumptions that should be carefully noted. The main feature of the airline passenger data is the growth trend coupled with periodicity. The model (10.3.7) accounts for these trends by means of deterministic components while the remaining part is described by a *stable* stochastic model. On the other hand, the model (10.3.8) assumes that these trends are stochastic in nature and accounts for the entire series by means of an *unstable* stochastic model after logarithmic transformation. As we will now see, this fundamental difference in the structure leads to a large difference in their prediction errors.

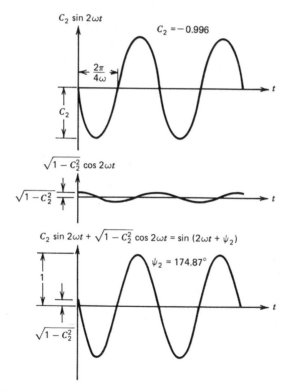

Figure 10.26 Plot of the second periodic term without the exponential growth, $[\sin(2\omega t + 174.87°)]$.

The model (10.3.8) has only two parameters, whereas (10.3.7) has 23. If we are interested in the data analysis only for the purpose of prediction or forecasting, is it worthwhile to increase the number of parameters by 21 and use the model (10.3.7) rather than (10.3.8)? To answer this question, we have computed ℓ step ahead forecast error variances assuming each of these models to be true for $\ell = 1$ to 36, that is, one month to three years ahead forecasts. The mean squared errors of prediction for the two models are plotted in Fig. 10.27. In order to clearly see the difference in the mean squared errors of the two models, the first 14 months are also plotted on a large scale. (The estimated parameters are assumed known for both as usual.)

It can be seen from this plot that the prediction errors for the model (10.3.8) are consistently larger for all lags. For one step ahead forecasts, the mean squared error (σ_a^2) is about 76% larger for (10.3.8) compared to the model (10.3.7), that is, 99.06 versus 56.18; for $\ell = 12$ or one year ahead forecasts, it is larger by about 165% (498.91 versus 188.23). This difference grows very rapidly for higher lags and for $\ell = 36$ or a three year ahead forecast, the squared error by (10.3.8) is about 14 times as large as that of (10.3.7). The other feature that is apparent from the plot

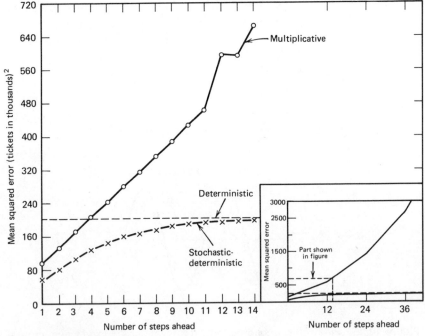

Figure 10.27 International airline data: Comparison of forecasts.

is a jump in the mean squared error value of the multiplicative model after every 12 months. It can be shown that these jumps arise because of the nature of the Green's function of the multiplicative model.†

This enormous difference in long-term prediction errors is inherent in the structure and assumptions of the two models emphasized above. The deterministic part of the model (10.3.7) can be predicted without error so that its mean squared error for long-term forecasts is limited to the error of its stochastic part. Since the stochastic part of the model (10.3.7) is stable, its variance that is the upper bound for its long-term mean squared prediction error is finite. For the airline data, the variance of the stochastic part is (28908/144) = 200.75 (Table 10.2), and the mean squared prediction error for arbitrarily large lags can never exceed this limit theoretically.

On the other hand, there is only one unstable model in (10.3.8) to account for the entire series. Because of its instability, the variance for the model (10.3.8) is theoretically infinite, and therefore, as the lag increases, the long-term mean squared prediction error tends to infinity, by its very assumptions. Thus, in practice, one can expect the long-term prediction errors for the model (10.3.8) to increase without any bound.

When the models (10.3.7) or (10.3.8) are used for forecasting long periods, there is a possibility that the model parameters would change over long periods of time. This would introduce additional error in forecasting that was not considered above. This error could be larger for the model (10.3.7) since it has a larger number of parameters. However, one would generally update the parameter values as the new data comes in and thus try to reduce this type of error. Also, we have taken the estimated param-

---

† It is somewhat difficult to compare the mean squared error of the model (10.3.8) for the *logged* series with that of the model (10.3.7) for the *original* series, especially since the series is nonstationary. For illustrative purposes, we have converted the error variances for the model (10.3.8) by assuming lognormal distribution for the original series with mean μ_x = 280.3. Then the expression for the ℓ step ahead forecast error variance $\sigma_x^2(\ell)$ for the original series in terms of $\sigma_w^2(\ell)$ for the logged series w_t is given by

$$\sigma_x^2(\ell) = \mu_x^2 \left\{ \exp \left[\sigma_w^2(\ell)\right] - 1 \right\}$$

The graphs for the multiplicative model in Fig. 10.27 have been based on this expression. Actually the series mean is increasing because of instability/nonstationarity, and the values plotted in Fig. 10.27 would be underestimates. Also, σ_a^2 for the w_t series has been taken as 0.18252/144 = 0.00126, rather than the slightly larger unbiased estimate 0.18252/129 = 0.00141.

It is interesting to note that by a similar computation, for the ARMA(13,13) model of Section 9.2.4,

$$\sigma_w^2(1) = \sigma_a^2 = 0.09102/144 = 0.000632$$
$$\sigma_x^2(1) = (280.3)^2 \left\{ \exp (0.000632) - 1 \right\} = 49.677$$

which compares favorably with the value 56.18 for the 23 parameter model (10.3.7).

eters as known for both models. As pointed out in Section 2.1.1, relaxing this assumption would inflate error variances in both models.

Apart from improved prediction, the model (10.3.7) has many advantages in characterizing the system. Since no transformation has been used, its parameters can be related to the system characteristics. For the airline data the yearly, half-yearly, and quarterly trends can be traced to the corresponding travel patterns. The exponential and periodic parameters can be used for a meaningful comparison of the two systems, of the same system under different periods or different management policies, and so on. All this information is lost in a model such as (10.3.8).

10.4 GENERAL NONSTATIONARY MODELS

We will now generalize the considerations developed in the preceding sections of this chapter in the context of specific examples. These general considerations will reveal that the basic modeling strategy for nonstationary data is essentially the same as that for the stationary data developed in Chapter 4. Moreover, some practical aspects in dealing with nonstationary data will be discussed in light of earlier sections and the general model considered here.

10.4.1 A General Model for Nonstationary Series

Following Eq. (10.3.1), we may write a general model for a series nonstationary in the mean as

$$y_t = \sum_{j=0}^{s} A_j e^{K_j t} + X_t \tag{10.4.1}$$

where X_t is the response of an ARMA$(n, n-1)$ system for discrete time

$$(1 - \phi_1 B - \phi_2 B^2 - \ldots - \phi_n B^n) X_t$$
$$= (1 - \theta_1 B - \theta_2 B^2 - \ldots - \theta_{n-1} B^{n-1}) a_t \tag{10.4.2}$$

or of an AM$(n, n-1)$ system for continuous time

$$(D^n + \alpha_{n-1} D^{n-1} + \ldots + \alpha_0) X(t)$$
$$= (1 + b_1 D + \ldots + b_{n-1} D^{n-1}) Z(t) \tag{10.4.3}$$

The deterministic part of the "exponential" model (10.4.1) can also be expressed as the response of a stable or unstable system governed by difference/differential equations with constant coefficients. (See Problems 10.7 to 10.12 for special cases.)

When all the A_j's are zero, or when only one A_j is nonzero and the corresponding K_j is zero, the model (10.4.1) represents a zero or constant mean series. In this case the data fluctuates around a fixed level. Such a series is stationary and can be modeled by methods of Chapter 4.

When the exponents K_j are real, very small, but nonzero, the model represents polynomial trends. This is clear since the exponential function is a limit of such a polynomial and with small exponents the higher order terms are negligible. With real, large, and positive or negative exponents, it can represent increasing or decreasing exponential trends.

With complex conjugate exponents K_j with negative real parts, the model (10.4.1) can represent damped sine cosine trends, with zero real parts exact sine cosines buried in noise and with positive real parts sine cosine trends with growing amplitude. As explained in Section 10.3, it is preferable to use the equivalent form (10.3.3) whenever periodic trends are found in the data.

It can be seen that the exponential model (10.4.1) is a very powerful device capable of representing correlated or uncorrelated data with a wide variety of nonstationary trends. The corresponding differential equations for the deterministic and stochastic parts can provide an insight into the underlying system and its coefficients can be related to the system characteristics.

When very little is known about the underlying system, the model (10.4.1) and the corresponding differential equations can help in conjecturing theories of its behavior. When some conjectures are available, they can be tested against the evidence of the model arrrived at from the data. When the deterministic model for the system is fairly well-known, it can be used in the model (10.4.1) replacing the sum of exponentials.

If the model (10.4.1) is used to represent the data from an inherently nonlinear system, it provides the closest linear approximation over the range of the data values. A further analysis with more data may reveal the actual nonlinear representation of the system.

10.4.2 Modeling as Decomposition

The concept of orthogonal or Wold's decomposition of a stationary time series was introduced in Chapter 3 (See Section 3.1.4). The essence of this decomposition is expressed by the Green's function or the autocovariance function, either of which provides a complete *deterministic* characterization of a zero mean stationary stochastic process. It was then shown in Chapter 4 that the most natural form of the Green's function or the autocovariance function is a sum of exponentials ($e^{\mu t} = \lambda^t$), which is equivalent to an ARMA/AM model for the corresponding time series. Thus, modeling a zero or constant mean time series by an ARMA/AM model is the same as decomposing or expressing its deterministic characteristic, the Green's function or the autocovariance function, into a sum of exponentials.

For the nonstationary series considered in this chapter the mean is no longer zero or constant but is a (deterministic) function of time. In modeling nonstationary time series, it is therefore logical to express the mean into a sum of exponentials, as in the model (10.4.1), and use the ARMA/AM model (10.4.2)/(10.4.3) for the remaining zero mean stochastic part. Thus, modeling a nonstationary series by the exponential model (10.4.1) is equivalent to decomposing or expressing the mean and the autocovariance (or Green's function) of the series each into a sum of exponentials. Just as the stochastic part can be represented by a difference/differential equation with white noise input in the form of an ARMA/AM model, the deterministic part can also be represented by a difference/differential equation with impulse, step, or periodic input.

There is, however, an important difference in the sums of exponentials used for the deterministic and stochastic parts. In the Green's function or the autocovariance function of the stochastic part, the exponents are restricted to a nonpositive real part so that the stochastic system is either asymptotically stable or stable. On the other hand, the exponents in the representation for the mean in the deterministic part can have positive real values and thus represent an unstable system. In fact, this was the case for the airline passenger data considered in Section 10.3. There is also a possibility that the exponents in the deterministic and stochastic parts are equal so that each is respectively the response to a deterministic input and white noise input from the *same* system. This possibility was illustrated by the chemical relaxation example in Section 10.2, and is further elaborated in Problems 10.7 to 10.12.

Decomposition into exponentials is a common method employed in linear system analysis. Fourier series or transform is a special case of such a decomposition in which the exponents are restricted to imaginary values so that one essentially gets a decomposition into sine-cosine waves. The Laplace transform commonly used by control engineers is a similar decomposition into damped sine-cosine by means of exponents with a *fixed* negative real part.

The exponents in the Green's function or the autocovariance function of the stochastic part in (10.4.1) are also complex with negative real parts, so that it is also a decomposition into damped sine-cosines. However, in this case the real part may change from one exponent to another so that this decomposition is a slight generalization of the Laplace transform. On the other hand, the exponents in the deterministic part may take any complex value with positive or negative real parts changing from one exponent to another. Thus, modeling by (10.4.1) may be termed a generalized

Laplace transform of a nonstationary time series, elaborated upon in Pandit (1973).

10.4.3 Some Practical Aspects

The procedure of modeling for (10.4.1) is basically the same as that used in the examples considered earlier. The dominant linear or exponential trends are taken out first by least squares and the residuals are examined for further trends. Such an examination of residuals can provide information about further trends and rough initial estimates of the corresponding parameters. This process is also continued for periodic trends until the data appear stationary. Then, the ARMA/AM model may be fitted to the stationary residuals. Finally, the combined deterministic plus stochastic model is estimated simultaneously.

It is thus seen that unlike the procedure for stationary series, the modeling procedure for nonstationary series does make use of the plot of the residuals and no "automatic" routine is provided that directly yields the final adequate model from the data. This difference is inherent in the nature of the two classes: the class of stationary series is by nature a definite, limited class amenable to mathematical analysis, on the other hand, the class of nonstationary series is only negatively defined and therefore indefinite and unlimited. The lack of a positive definition and restriction makes it difficult to analyze the modeling problem theroetically, as in the case of the stationary series, and devise a routine modeling strategy applicable in all cases.

However, once the "kind" of nonstationarity is discerned from the data or the residuals, modeling can be executed without much difficulty. For example, once we know from the data that it involves one or two exponentials and/or a periodic component and its harmonics, a computer routine can be written for modeling. The examples considered in this chapter illustrate the principal "types" of nonstationarity for which such a routine may be written.

In the preliminary analysis, the procedure for modeling the deterministic part of (10.4.1) involves successively adding the exponentials, one by one. Such a gradual addition of two or three parameters at a time considerably reduces the problems and difficulties in estimation. Fairly accurate values of the parameters of a single exponential or periodic component can be obtained from the residuals, and therefore the convergence of the minimization routine with respect to these parameters is quite fast. Moreover, since the final values of the remaining parameters in an earlier model are used as the initial values, the minimization routine has to essentially work only with the additional two or three parameters.

For example, in fitting a 17-parameter deterministic component for the airline passenger data, the initial estimates of the first 14 parameters taken from the preceding model are almost the same as the final estimates, so the routine has to effectively minimize only with the three additional parameters of the last 2-2/5 month period. It can be verified from Table 10.2 that there is relatively little change in the old parameters when the new components are added. For example, R_1 changes from 130.83 to 127.48, r_1 from 0.0095 to 0.0096, and so on.

The same comment holds for the final deterministic plus stochastic model estimation. In spite of a significantly large number of parameters, no estimation difficulties are encountered since the initial values obtained from the previous models are very close to the final values of the combined model. Therefore, convergence occurs after only a few minor adjustments in the estimates. Of course, the computer time required for each iteration increases very fast with the addition of more parameters.

10.5 MORE ILLUSTRATIVE EXAMPLES

We will now consider two more examples in which the model (10.4.1) or its modification based on some known forms can be used along the guidelines developed in Section 10.4. Unless a specific form of the deterministic part is known, the general procedure is to successively fit the model (10.4.1) of higher order. Once the adequate model is obtained, it may be tallied and, if necessary, modified on the basis of any *a priori* information that may give some clues about the model form. When a specific deterministic form is known, it can then replace the exponential deterministic form in (10.4.1). The first example does not use any *a priori* information, whereas in the second example, the form of the deterministic part is known.

10.5.1 Consumer and Wholesale Price Index

These data shown in Figs. 9.14 and 9.15 clearly exhibit exponential trends and it was observed at the beginning of Section 9.2.3 that it would be better to model them by deterministic trends. This observation was also confirmed by the stochastic trend analysis of the adequate ARMA models in Section 9.2.3, which resulted in dominant real roots 1.0156 and 1.0139 for the CPI and the WPI respectively (See Tables 9.8 and 9.10). The exponential growth of the resulting Green's function again underscores the possibility of an unstable/nonstationary trend represented by a deterministic function that depends on the time origin.

It is therefore appropriate to fit the model (10.4.1) with $s = 2$ and $K_0 = 0$. In the preliminary analysis aimed at obtaining the initial values of the model, we fit only the deterministic component with

$$y_t = A_0 + A_1 e^{K_1 t} + \varepsilon_t$$

by minimizing the sum of squares of ε_t. Note that this fitting is, strictly speaking, not correct since ε_t is actually correlated. The initial values of A_0, A_1, and K_1 are easily obtained from the graph by using the properties of the exponential function.

Consumer Price Index

The least square estimates and the 95% confidence intervals are as follows:

$$\hat{A}_0 = 89.076 \pm 1.47$$

$$\hat{A}_1 = 4.941 \pm 1.02$$

$$\hat{K}_1 = 0.010495 \pm 0.00092$$

The CPI data residuals ε_t from this fitting are shown in Fig. 10.28. No additional exponential or seasonal deterministic trends are visible, which can be further confirmed by fitting double exponential and showing that the residual sum of squares is not significantly improved.

We can now apply the procedure of Chapter 4 to the residuals in Fig. 10.28. The results of modeling are given in Table 10.3. (The second method of computing a_t's starting with $a_1 = X_1$ has been used.) The ARMA(6,5) model is considered to be adequate at the 5% level. For the ARMA(6,5) model (Table 10.4), the damped frequency of the root $-0.3641 \pm i0.8467$ gives a period $p = 3.18$, whereas the undamped root $0.0094 \pm i0.9993$ is almost the same as $\pm i$, which is a 4 monthly seasonality by Table 9.1.

Table 10.3 Models for CPI Residuals from First Order Exponential

Para- meters	Order of the ARMA Model				
	(1,0)	(2,1)	(4,3)	(6,5)	(8,7)
ϕ_1	1.005 ± 0.018	2.003 ± 0.025	1.317 ± 0.081	1.214 ± 0.085	3.012 ± 0.102
ϕ_2		-1.006 ± 0.024	-0.443 ± 0.149	-1.419 ± 0.150	-4.612 ± 0.321
ϕ_3			0.940 ± 0.150	2.250 ± 0.179	5.898 ± 0.522
ϕ_4			-0.820 ± 0.082	-1.269 ± 0.175	-6.455 ± 0.636
ϕ_5				0.984 ± 0.147	5.323 ± 0.618
ϕ_6				-0.853 ± 0.086	-3.750 ± 0.478
ϕ_7					2.439 ± 0.279
ϕ_8					-0.859 ± 0.085
θ_1		0.948 ± 0.054	0.236 ± 0.007	0.183 ± 0.076	1.911 ± 0.099
θ_2			-0.239 ± 0.007	-1.259 ± 0.074	-2.584 ± 0.162
θ_3			1.240 ± 0.006	1.137 ± 0.108	3.491 ± 0.193
θ_4				-0.282 ± 0.075	-3.516 ± 0.208
θ_5				0.938 ± 0.087	2.572 ± 0.192
θ_6					-1.899 ± 0.132
θ_7					0.945 ± 0.075
Residual sum of squares	8.857	7.408	6.801	6.384	6.179
F		9.63	4.30	3.09	1.53

The adequate model is (6,5).

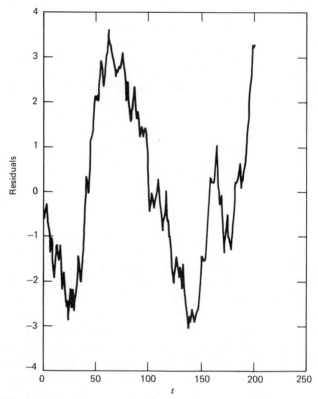

Figure 10.28 CPI data: Residuals from first order exponential.

The combined model (10.4.1) with one exponential and $K_0 = 0$ with X_t given by an ARMA(6,5) model can be estimated together. The initial values of this model can be taken from the final model separately fitted above. The results of the combined model estimation are shown in Table 10.5 and the residual autocorrelations are plotted in Fig. 10.29.

Table 10.4 The Characteristic Roots of the Models Given in Table 10.3

Roots	ARMA Model				
	(1,0)	(2,1)	(4,3)	(6,5)	(8,7)
λ_1, λ_2	1.0053	$1.0015 \pm i0.0519$	$1.0023 \pm i0.0457$	$1.0016 \pm i0.0519$	$1.0011 \pm i0.0519$
λ_3, λ_4			$-0.3438 \pm i0.8344$	$-0.3641 \pm i0.8467$	$-0.3716 \pm i0.8405$
λ_5, λ_6				$0.0094 \pm i0.9993$	$0.0089 \pm i0.09984$
λ_7, λ_8					$0.8673 \pm i0.5132$
ν_1		0.9484	1.0077	0.9483	0.9491
ν_2, ν_3			$-0.3857 \pm i0.9313$	$-0.3859 \pm i0.9311$	$-0.3856 \pm i0.9312$
ν_4, ν_5				$0.0030 \pm i0.9866$	$0.0032 \pm i0.9845$
ν_6, ν_7					$0.8635 \pm i0.5157$

Table 10.5 Combined Estimation of Stochastic and Deterministic Parts for the CPI Data

Deterministic Parameters	Stochastic			
	Autoregressive		Moving Average	
	Parameters	Roots	Parameters	Roots
A_0: 89.78 ± 2.55	ϕ_1: 1.24 ± 0.083	$-0.387 \pm i0.851$	θ_1: 0.219 ± 0.075	1.0206
A_1: 3.915 ± 2.47	ϕ_2: -1.34 ± 0.145	$1.001 \pm i0.040$	θ_2: -1.16 ± 0.080	$-0.399 \pm i0.928$
K_1: 0.012 ± 0.003	ϕ_3: 2.20 ± 0.168	$0.008 \pm i0.996$	θ_3: 1.25 ± 0.059	$-0.0015 \pm i0.980$
	ϕ_4: -1.22 ± 0.163		θ_4: -0.19 ± 0.082	
	ϕ_5: 0.98 ± 0.141		θ_5: 1.00 ± 0.085	
	ϕ_6: -0.87 ± 0.082			

Residual sum of squares: 5.8455

Note that the estimated value of $K_1 = 0.012$ closely agrees with the characteristic root 1.0156 for the stochastic model in Table 9.8 since $e^{0.012} = 1.0121$. The other roots are fairly close to those of the stochastic part alone, and hence the seasonalities in these roots discussed above are unaffected.

Wholesale Price Index

For the WPI series, the least squares estimates of the single exponential deterministic model alone, with their 95% confidence interval, are

$$\hat{A}_0 = 93.131 \pm 1.59$$

$$\hat{A}_1 = 2.155 \pm 1.05$$

$$\hat{K}_1 = 0.01122 \pm 0.0022$$

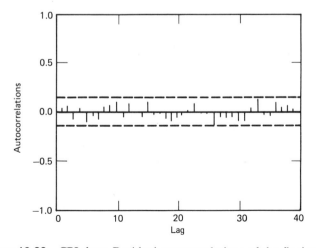

Figure 10.29 CPI data: Residual autocorrelations of the final model.

The residuals from this model are shown in Fig. 10.30. No additional deterministic trends are visible; hence the procedure of Chapter 4 can be
applied with the results in Tables 10.6 and 10.7. The ARMA(4,3) model
is found to be adequate. It is further verified from the fit of the ARMA(4,3)
model that since the added pair of roots (Table 10.7) gets almost cancelled
on both sides, it gives only a marginal improvement over the ARMA(2,1).

The estimation results for the combined model (10.4.1) for the WPI
with one exponential and $K_0 = 0$ with X_t given by an ARMA(2,1) model
are given in Table 10.8; using ARMA(4,3) for X_t reduces the residual sum
of squares to only 22.644. The residual autocorrelations are shown in
Fig. 10.31. The estimated values are quite close to those obtained from
the separate estimation, although the confidence intervals of the deterministic parameters are now wider.

Comparing the final results for the CPI and WPI in Tables 10.5 and
10.8, we can see that exponents K_1 are almost equal, the confidence in-

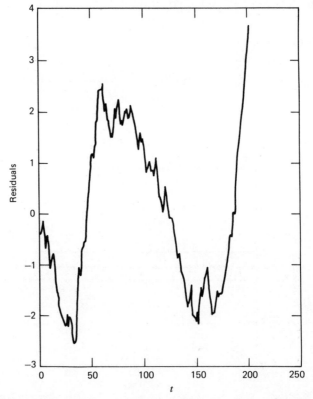

Figure 10.30 WPI data: Residuals from first order exponential.

Table 10.6   Models for WPI Residuals from First Order Exponential

Parameters	Order of the ARMA Model		
	(1,0)	(2,1)	(4,3)
ϕ_1	0.9872 ± 0.026	1.9863 ± 0.025	-0.0536 ± 0.078
ϕ_2		-0.9885 ± 0.025	1.8561 ± 0.079
ϕ_3			0.0491 ± 0.079
ϕ_4			-0.8712 ± 0.080
θ_1		1.0055 ± 0.004	-1.2127 ± 0.004
θ_2			0.6386 ± 0.003
θ_3			0.8618 ± 0.004
Residual sum of squares	26.082	24.274	22.850
F		3.66	3.01

The adequate model is the ARMA(4,3).

tervals for the deterministic parameters overlap and the characteristic roots of the ARMA(2,1) model for the WPI are also present in the ARMA(6,5) of the CPI. Similarly, in the purely stochastic models in Section 9.2.3, the only root 1.014 of the adequate AR(1) model for the WPI was also found to be present in the ARMA(4,3) model for the CPI. Thus, the dynamics of the CPI is that of the WPI plus some other modes given by the additional characteristic roots not present in the WPI. This result is intuitively appealing since the consumer prices will always be determined based on the wholesale price, but will in addition be affected by factors relevant to retail trading. In particular, the above analysis shows that the the CPI is significantly affected by the approximately 3 monthly and 4 monthly seasonal effects, whereas the WPI is not. Such a dynamic dependence of the CPI on the WPI suggests that these two should be modeled together as multiple series.

The question of whether one should use the stochastic models obtained in Section 9.2.3 or the deterministic plus stochastic models in the present section now arises. A purely statistical answer to this question can be given

Table 10.7   The Characteristic Roots of the ARMA Models for the WPI

Roots	ARMA Models		
	(1,0)	(2,1)	(4,3)
λ_1, λ_2	0.9872	$0.9932 \pm i0.0457$	$0.9515 \pm i0.0513$
λ_3, λ_4			$-0.9783 \pm i0.0482$
ν_1		1.0055	0.8255
ν_2, ν_3			$-1.0191 \pm i0.0736$

Table 10.8   Combined Estimation of Stochastic and Deterministic Parts for the WPI Data

Deterministic Parameters	Stochastic		
	Autoregressive		Moving Average
	Parameters	Roots	Parameter
A_0: 92.46 ± 3.77	ϕ_1: 1.99 ± 0.028	0.99 ± i0.043	θ_1: 1.017 ± 0.024
A_1: 2.28 ± 3.52	ϕ_2: −0.99 ± 0.028		
K_1: 0.011 ± 0.008			

Residual sum of squares: 22.840.

on the basis of the F-criterion. For the CPI the adequate ARMA(4,3) model in Table 9.7 has 8 parameters and RSS of 6.9118, whereas the stochastic plus deterministic model in Table 10.5 has 14 parameters and RSS of 5.8455. With 202 observations the F-test gives

$$F = \frac{6.9118 - 5.8455}{5.8455} \times \frac{188}{6} = 5.716$$

whereas $F_{0.99}(6,\infty) = 2.80$. Therefore, the reduction of RSS by the deterministic plus stochastic model is statistically significant at 1% level. Similarly, for the WPI series, the reduction in the F value with five additional parameters is also significant. Therefore, the models of this section are significantly better than those of Section 9.2.3 although the deterministic parameters in Table 10.5 have rather large confidence intervals. From forecasting considerations, it can be seen that the models of this section

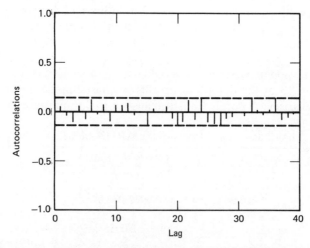

Figure 10.31   WPI data: Residual autocorrelations from the final model.

will reduce the one step ahead forecast error by about 16% (since the one step ahead forecast, that is, the variance of the residuals σ_a^2, reduces from 0.0344 to 0.0289) and the reduction will be substantially larger for long-term forecasts as illustrated in Section 10.3.5.

However, the decision in choosing the model should be made not only on the basis of a statistical check, but also by taking into account the economic theory regarding the price behavior. The presence of the deterministic part, supported by the F-criterion, should also be confirmed by the knowledge of the price behavior. If there are theories supporting both types of models, then one may use the above analysis to choose between them.

10.5.2 Wood Surface Profile

We will now consider an example in which the form of the deterministic function is known *a priori* and is not in the form of a straight line, exponential or sinusoidal. Figure 10.32 shows the profile of a block of wood after being subjected to a grinding process. The numerical data are listed in the Appendix I and are taken from Dornfeld (1976). The configuration and the position of the wood surface from which the data have been collected are shown in Fig. 10.33.

As illustrated in Fig. 10.33, the surface profile (Fig. 10.32), in addition to showing the actual profile variation, has an inherent curvature due to the radius of the grinding stone used in the experiments. Thus, the deterministic component in this case will be the arc of a circle. To remove

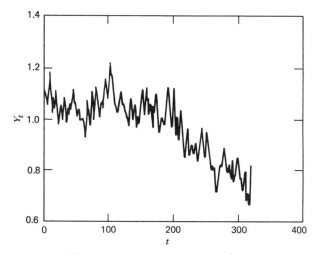

Figure 10.32   Wood surface profile.

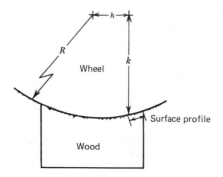

Figure 10.33   Physical significance of the deterministic parameters h, k, and R.

such a deterministic component from the data, three parameters h, k, and R (Fig. 10.33) corresponding to the offset from the center of the grinding stone to the first height location in the x-direction, the y-direction, and the radius of the profile are to be estimated. The reason for introducing these three parameters to the deterministic function is because we do not exactly know the center of the grinding stone. We only know the profile heights measured with respect to some reference axes. In addition, the location of the grinding stone center is not fixed and the profile radius varies with grinding stone condition and grinding parameters. Therefore, the offset parameters together with the radius of the profile are estimated to take into account the proper deterministic trend in the data.

The basic equation for estimating the deterministic portion of the data is

$$(x - h)^2 + (y' - k)^2 = R^2 \tag{10.5.1}$$

where

x = the sampling distance along the surface;
y' = the profile height at x because of a deterministic trend;
h = the x-direction offset of the surface profile from the theoretical center of the profile arc;
k = the y-direction offset corresponding to h;
R = the theoretical radius of the profile arc.

Equation (10.5.1) can be reformulated as

$$y' = k + \sqrt{R^2 - (x - h)^2}$$

Here, the actual profile height y can be decomposed into two parts. The first part consists of the profile height due to the curvature and offset, y', and the second part shows the actual profile variation, ε, that is,

$$y = k + \sqrt{R^2 - (x - h)^2} + \varepsilon \tag{10.5.2}$$

To model the deterministic part, we use the method of least squares to fit a model

$$y_t = k + \sqrt{R^2 - (t - h)^2} + \varepsilon_t \qquad (10.5.3)$$

where

t = observation number and represents the sampling distance (x/Δ where Δ = the sampling interval in inches).

y_t = the profile height at t.

For simplicity, let us take the initial value of R as equal to the wheel radius, that is, $R = 6$ inches, and estimate the parameters h and k from the model

$$y_t = k + \sqrt{36 - (t - h)^2} + \varepsilon_t \qquad (10.5.4)$$

Note that if R is completely unknown, we could also estimate it as an additional parameter, although estimation may become a little more difficult. The estimation results with 95% confidence limits are as follows:

$$\hat{h} = -4.3349 \pm 0.0470$$

$$\hat{k} = -4.0722 \pm 0.0535$$

Residual sum of squares = 1.4318

The modeling results of the residuals ε_t from Eq. (10.5.4) using the above estimated values are given in Table 10.9. It is seen that the ARMA(3,2)

Table 10.9 Wood Surface Profile—Models for Residuals from the Circular Arc

Parameters	Order of the ARMA Model			
	(2,1)	(4,3)	(6,5)	(3,2)
ϕ_1	0.864 ± 0.706	0.885 ± 0.721	0.207 ± 0.765	1.795 ± 0.431
ϕ_2	-0.120 ± 0.589	-0.015 ± 0.943	-0.856 ± 0.341	-1.421 ± 0.665
ϕ_3		-0.080 ± 0.875	-0.590 ± 0.635	0.544 ± 0.334
ϕ_4		0.098 ± 0.518	-0.552 ± 0.473	
ϕ_5			0.531 ± 0.371	
ϕ_6			0.934 ± 0.524	
θ_1	-0.138 ± 0.707	-0.188 ± 0.710	-0.822 ± 0.741	0.813 ± 0.491
θ_2		0.128 ± 0.621	0.317 ± 0.686	-0.347 ± 0.482
θ_3		0.233 ± 0.576	$0.637 \pm .0428$	
θ_4			-0.783 ± 0.424	
θ_5			-0.403 ± 0.450	
Residential Sum of Squares	0.6000	0.5321	0.5037	0.5696

The adequate model is ARMA(3,2).

Table 10.10 Wood Surface Profile: Deterministic Plus Stochastic Model

Deterministic Parameters	Stochastic Parameters	
h: -3.773 ± 0.0068	ϕ_1:	1.609 ± 0.2920
	ϕ_2:	-0.7413 ± 0.3065
k: -4.166 ± 4.988	ϕ_3:	0.1157 ± 0.1281
	θ_1:	0.6273 ± 0.2736
R: 5.651 ± 3.630	θ_2:	0.1629 ± 0.4819

Residual sum of squares: 0.2561.

model is found to be adequate. Now, we can formulate a combined model

$$y_t = k + \sqrt{R^2 - (t - h)^2} + X_t$$

$$X_t = \phi_1 X_{t-1} + \phi_2 X_{t-2} + \phi_3 X_{t-3} + a_t - \theta_1 a_{t-1} - \theta_2 a_{t-2} \quad (10.5.5)$$

for which the estimation results are given in Table 10.10. It is interesting to note that the estimated value of the radius is 5.651 inches rather than the nominal wheel radius of 6 inches. This estimate of the mean radius could form a characterization of the wood profile under different conditions of speed, feed, pressure, types of wood, etc. The dynamics of the grinding process will be reflected in the parameters and characteristic roots of the ARMA(3,2) model, which can be used to quantify the pulping process.

PROBLEMS

10.1 For the airline data, compare the periods in the stochastic part as well as the deterministic part in the model (10.3.7) and Table 10.2 with those obtained in Section 9.2.4. Give your interpretations.

10.2 (a) In Section 10.5.1, the exponentially increasing rate of the Consumer Price Index (CPI) Data and the Wholesale Price Index (WPI) Data are determined by the term Ae^{Kt}. Compare the increasing rates of these two sets of data from Ae^{Kt} for small t and large t.

   (b) Compare, by means of their characteristic roots,
   (i) the ARMA(8,7) model in Table 10.3 with the ARMA(6,5) model in Table 9.7 for the CPI data;
   (ii) the ARMA(4,3) model in Table 10.6 with the ARMA(4,3) model in Table 9.9 for the WPI. Are there any complex conjugate roots giving almost the same periodicity? Explain your results.

10.3 Using the CPI or WPI data as an example, compare and contrast between the stochastic model in Chapter 9 and the deterministic

+ stochastic model in Chapter 10. Give your opinion on which model is preferred and why.

10.4 The second order deterministic plus stochastic model mentioned in Section 10.2.2 would take the form

$$y_t = A_0 + g[1 - \frac{\tau_1}{(\tau_1 - \tau_2)} e^{-\frac{t}{\tau_1}} + \frac{\tau_2}{(\tau_1 - \tau_2)} e^{-\frac{t}{\tau_2}}] + X_t$$

$$X_t - \phi_1 X_{t-1} - \phi_2 X_{t-1} = a_t - \theta_1 a_{t-1}$$

This model, fitted to the basis weight data, gives the following results:

$$\hat{A}_0 = 38.9277, \ \hat{g} = 1.6485, \ \hat{\tau}_1 = 11.6708,$$

$$\hat{\tau}_2 = 13.5005, \ \phi_1 = 1.6873, \ \phi_2 = -0.7499$$

$$\hat{\theta}_1 = 0.9788, \text{ residual sum of squares} = 1.5774$$

(a) Compute the one and two step ahead forecast error variances for the combined model of (i) first order given by (10.2.2a) and (ii) second order given above.

(b) Find the percentage improvement in the forecast error from (i) to (ii).

(c) Use the F-test to show that the improvement from the first order to the second order model is insignificant.

(d) Part (c) implies that the noise given by the stochastic part is essentially represented by an AR(1) model. Why then are the parameters ϕ_2 and θ_2 given above so large, and why do their 95% confidence intervals -0.7499 ± 0.1434 and 0.9788 ± 0.1609, respectively, not include zero?

10.5 Given $Y(t) = C \sin \omega t + \sqrt{1 - C^2} \cos \omega t$, sketch $Y(t)$ and find the phase angle ψ for

(i) $C = 0.7$

(ii) $C = -0.3$

10.6 Given $Y(t) = Be^{bt} [C \sin \omega t + \sqrt{1 - C^2} \cos \omega t]$ where

$$B = 10$$

$$b = -0.05$$

$$C = 0.5$$

sketch $Y(t)$, showing correctly

(1) the initial value $Y(0)$;

(2) the period of $Y(t)$;

(3) the exponentially decaying envelope.

10.7 As discussed in Section 10.4.2, when the exponents in the deterministic part are equal to those in the Green's function of the stochastic part, special cases of the models (10.4.1 and 10.4.2) can be written with fewer number of parameters; these can be checked as illustrated for the first order case in Sections 10.2.4 and 10.2.6. In the discrete time case, this special case of models (10.4.1) and (10.4.2) can be written as

$$(1 - \phi_1 B - \phi_2 B^2 - \ldots - \phi_n B^n) y_t$$
$$= (1 - \theta_1 B - \theta_2 B^2 - \ldots - \theta_m B^m)(f_t + a_t)$$

where f_t usually contains suitable combinations of Kronecker delta

$$\delta_t = 1, \quad t = 0$$
$$= 0, \quad t \neq 0$$

and the discrete step function

$$s_t = 1, \quad t \geq 0$$
$$= 0, \quad t < 0$$

to initiate the trend from a desired origin.

(a) Show that the model in Section 10.2.4 can be represented by this model with $n = 1$, $m = 0$ as:

$$(1 - \phi_1 B) y_t = f_t + a_t$$

where

$$f_t = A_0 \delta_0 + g(1 - \phi_1) s_{t-1}$$

(Note that, for a *finite* geometric series, the sum:

$$1 + \phi + \phi^2 + \ldots + \phi^{t-1} = \frac{1 - \phi^t}{1 - \phi})$$

(b) Show that the model (10.2.10) can also be represented by the one given in (a) with $\phi_1 = e^{-\Delta/\tau}$; find the corresponding f_t.

(c) Show that further specializing to

$$(1 - B) y_t = A_1 s_{t-1} + a_t$$

gives a straight line trend buried in random walk. What is the slope of this straight line? What is the Y-intercept? How would you modify the model if you want the Y-intercept (initial condition) to be Y_0?

(d) Show that the model in (c) is suggested by the results for the Wholesale Price Index (WPI) in Table 10.8. Find the initial values of A_0 and A_1 using Fig. 9.15 and the data at the end. Compare this model with those in Chapters 9 and 10, and discuss your preference.

10.8 The appearance of the model in Problem 10.7(c) can be simplified if it is assumed that $y_0 = A_0$ is given, and we start from $t = 1$. Then

$$(1 - B)y_t = A_1 + a_t, \qquad t \geq 1$$

or

$$(1 - B)y_t = \varepsilon_t, \qquad t \geq 1$$

$$E(\varepsilon_t) = A_1, \qquad E(\varepsilon_t \varepsilon_{t-k}) = \delta_k \sigma_a^2$$

This is the same as a random walk with a constant added to the a_t's or a random walk excited by white noise with nonzero mean. Such models are often used in economic literature because of the ease of their estimation.

(a) Show that only the first and the last observations are needed to estimate A_1.

(b) Estimate A_1 and the residual sum of squares of a_t's for the WPI data given at the end of the book.

(c) Estimate the residual of sum of squares using the model in problem (10.7) with the initial values found in (10.7.d) and compare it with the one obtained in (b) above. Which is smaller? Why?

(d) What procedure would you suggest to estimate A_0 and A_1 so as to minimize the residual sum of squares?

(e) Compare the residual sum of squares in (b to d) above with those in Chapters 9 and 10. Comment on their differences. Which model would you prefer to use and why?

10.9 (a) Show that the stable ARMA(2,1) model with a constant added to a_t's:

$$(1 - \phi_1 B - \phi_2 B^2)y_t = (1 - \theta_1 B)(A_0 \delta_0 + A_1 s_{t-1} + a_t)$$

or the simplified form, given $y_0 = A_0$, $y_1 = A_0 + A_1$,

$$(1 - \phi_1 B - \phi_2 B^2)y_t = (1 - \theta_1 B)(A_0 + a_t), \qquad t \geq 2$$

is equivalent to the following special case of (10.4.1) and (10.4.2):

$$y_t = A_0 + A_1 \left[\frac{g_1(1 - \lambda_1^t)}{1 - \lambda_1} + \frac{g_2(1 - \lambda_2^t)}{1 - \lambda_2} \right] + X_t$$

$$(1 - \phi_1 B - \phi_2 B^2)X_t = (1 - \theta_1)a_t$$

where, as usual $g_1 = (\lambda_1 - \theta_1)/(\lambda_1 - \lambda_2)$, $g_2 = (\lambda_2 - \theta_1)/(\lambda_2 - \lambda_1)$ and $(1 - \phi_1 B - \phi_2 B^2) = (1 - \lambda_1 B)(1 - \lambda_2 B)$.

(b)   Specializing further, show that
$$(1 - \lambda B)\nabla y_t = (1 - \theta_1 B)(A_0 \delta_0 + A_1 s_{t-1} + a_t)$$
or, given $y_0 = A_0$, $y_1 = A_0 + A_1$,
$$(1 - \lambda B)\nabla y_t = (1 - \theta_1 B)(A_1 + a_t), \qquad t \geq 2$$
is equivalent to the straight line trend with a transient exponential (assuming $|\lambda| < 1$) buried in an ARMA(2,1) noise:

$$y_t = A_0 + A_1 \left[\frac{(\lambda - \theta_1)(\lambda' - 1)}{(1 - \lambda)^2} + \left(\frac{1 - \theta_1}{1 - \lambda} \right) t \right] + X_t$$
$$(1 - \lambda B)\nabla X_t = (1 - \theta_1 B)a_t$$

10.10 To further illustrate the implications of the model in Problem 10.7(a to b) note the similarity between the chemical relaxation data in Fig. 10.12 and the papermaking data in Fig. 2.1.

(a)   If we decide to fit the model (10.2.10) with $\Delta = 1$ to the papermaking data, what would be the initial values of A_0, A_1, τ, and ϕ_1? (Recall that $e^{-t/\tau} = e^{-1}$ when $t = \tau$.)

(b)   Now consider the first model fitted to these data in Chapter 2, taking $\overline{X} = 32.02$ as the estimate of the mean. In terms of the original data $\dot{X}_t$ (*before* subtracting the average $\overline{X}$) this model can be written as
$$(1 - 0.898B)(\dot{X}_t - 32.02) = a_t$$
Using the results of Problem 10.7a and 10.7b with $y_t \equiv \dot{X}_t$, find the initial values of A_0, A_1, τ, and ϕ_1 and compare them with the ones obtained in (a) above.

10.11 (a)   When the mean was estimated as an additional parameter, the AR(1) model for the papermaking data in Section 4.5.3 was found to be
$$(1 - 0.98B)(\dot{X}_t - 33.4) = a_t$$
Show that this model in fact contains a weak deterministic trend and give an equation for the trend.

(b)   The adequate model for the papermaking data in Section 4.5.3 is
$$(1 - 0.76B)\nabla X_t = (1 - 0.94B)a_t$$
Are there any deterministic trends contained in this model? How does it account for the trend in the data?

(c)   Discuss how and why the models in Problems 10.10(b), 10.11(a), and (10.11b) successively account for the dynamics in the data.

10.12   Nelson (1973) has fitted the following models to the quarterly data on the Gross National Product (GNP) and the actual Expenditures of Producers' Durables (EDP) in the United States in billions of current dollars from 1947–01 through 1966–04 (80 observations):

$$\text{GNP: } (1 - 0.62B)\nabla X_t = 2.69 + a_t, \qquad \sigma_a^2 = 22.66$$

$$\text{EPD: } \nabla X_t = a_t + 0.35a_{t-1} + 0.52, \qquad \sigma_a^2 = 1.12$$

Determine the parameters of all the components of deterministic trends contained in these models. The data are listed at the end of the book. Do you agree with these models? Can you improve them? How?

11

MULTIPLE SERIES: OPTIMAL CONTROL AND FORECASTING BY LEADING INDICATOR

In earlier chapters, we were concerned with a single series of data, considered as a realization of a stationary stochastic system. Such data can always be represented by a model of the ARMA form. The ARMA models can be extended to vector systems to show that two or more sets of data, treated as the realization of a vector stationary stochastic system, can be similarly represented by vector models. In this chapter, we will consider simple systems relating two or more, that is, multiple series of data.

The most prominent applications involving multiple sets of data arise in control systems, although there are other important applications such as economic forecasting, system simulation, etc. In this chapter we will be primarily concerned with models for discrete systems and developing strategies for their optimal forecasting and control in the sense of minimum mean squared error. The models employed here can be readily used in other related applications.

The control systems can be formulated in two equivalent but distinct ways. The first one corresponds to the classical control theory, employs block diagrams and transfer functions, and utilizes transform techniques such as Fourier, Laplace, or Z-transforms. In this formulation, an almost exclusive emphasis is placed upon those specific control systems classified as servomechanisms. In addition, attention is directed in most cases to systems of the single-input–single-output variety.

The second formulation corresponds to modern control theory, employs the state space technique, and utilizes methods such as mathematical programming suited to computer calculation. The state space technique essentially consists of treating all the inputs and outputs (as well as their derivatives and differences if necessary) as the "states" of the system and forming a single vector with each of the states as its elements. This formulation is particularly suitable for multiple-input–multiple-output systems. The advantages are its comprehensiveness and the ease of theoretical and computational treatment even in the case of complicated systems. Kalman filtering is a well-known example of this formulation of modern control theory. Both these formulations together with the importance of transition from classical to modern control theory may be found in most control texts.

412

For discrete stochastic systems, the autoregressive moving average models may be formulated in either framework, following classical control theory in the form of transfer function models or modern control theory in the form of vector ARMA models. Each of these formulations have their respective advantages and disadvantages. The transfer function methodology is intuitively appealing and easily introduced and grasped by means of block diagrams, especially for single-input–single-output systems. It is, however, cumbersome in modeling and becomes very complicated for a large number of inputs and outputs. The vector models lose their intuitive appeal, but they are easy to treat and particularly simple for the purpose of modeling; they are almost indispensible for systems with a large number of inputs and outputs.

A comprehensive treatment of the vector models requires matrix methods and is therefore beyond the scope of this book. [See Pandit (1973)] However, a special case of the vector models can be treated by the same theory as the one developed in earlier chapters. The special form of these models are often adequate to deal with one output and one or two input systems. In this chapter we will consider modeling such systems by the procedure developed in Chapter 4 and the forecasting and control of systems by the procedure in Chapter 5. Extensions to multi-input–single-output systems will be indicated whenever relevant.

Section 11.1 introduces several forms of models for the multiple series by considering a real papermaking system for basis weight control. The relationships between these different forms are discussed. The extension of the procedure in Chapter 4 for modeling multiple series is given in Section 11.2 with examples of papermaking data. Section 11.3 deals with the minimum mean squared error control strategy and shows the derivation of the control equation by the theory of Chapter 5. The full theory of forecasting by conditional expectation and error variances by the Green's function is used in Section 11.4 for the purpose of forecasting by leading indicators in business and economic series.

11.1 TRANSFER FUNCTION AND ARMAV MODELS

We will introduce and develop the single-input–single output models by considering a real control system. This is the papermaking system schematically represented in Fig. 11.1. A brief description of this process is given before formulating its control model.

11.1.1 The Papermaking Process

The Fourdrinier papermaking process shown in Fig. 11.1 starts with a mixture of water and wood fibers (pulp) in the mixing box. The gate

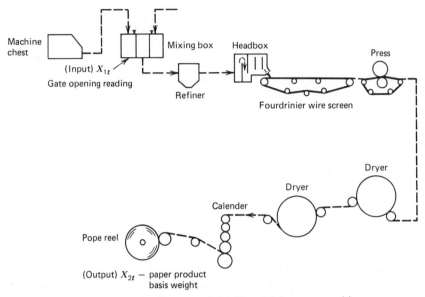

Figure 11.1   Schematic diagram of the Fourdrinier papermaking process.

opening in the mixing box can be controlled to allow a greater or smaller flow of the thick stock (a mixture of water and fiber) entering the headbox. A turbulence is created in the headbox by means of suspended plates to improve the consistency of the pulp. The pulp then descends on a moving wire screen, as a jet from the headbox nozzles. Water is continuously drained from the wet sheet of paper so formed on the wire screen. The paper sheet then passes through press roles, driers, and calender roles to be finally wound.

It is important to produce paper of as uniform a thickness as possible since irregularities on the surface such as ridges and valleys cause trouble in later operations such as winding, coating, printing, etc. This uniformity is measured by what is called a basis weight, the weight of dry paper per unit area. It may be measured directly or by means of a beta-ray gauge that makes use of the beta-ray absorption properties of the paper material. The regulation of paper basis weight is one of the major goals of the paper control system.

The basis weight is affected by variables such as stock consistency, stock flow, headbox turbulence and slice opening, drier steam pressure, machine speed, etc. However, thick stock flow is often used as the main control input measured by the gate opening in the mixing box. For the sake of simplicity, we will consider the gate opening denoted by X_{1t} as the only input, which is to be manipulated so that the output basis weight is maintained at the given target value depending upon the grade of the paper being made. We will denote the output by X_{2t}, which is the deviation of the basis weight at time t from the target value. The aim of the control system is to keep the output X_{2t} at zero.

11.1.2 Transfer Function Model

To formulate a model for the control system is therefore to establish a relation between the input X_{1t} and the output X_{2t}. If we consider the output basis weight X_{2t} alone, we can write an ARMA model for it expressing its dependence on the past values. The simplest such model will be AR(1). If we take the autoregressive parameter as 0.7 for illustrative purposes, we can write the model as

$$X_{2t} = 0.7X_{2t-1} + a_{2t}$$

However, the output X_{2t} does not depend on its past values alone but also depends on the input gate opening X_{1t}. Therefore, the dependence on X_{1t} must also be added on the right-hand side of this model.

Now the input gate opening X_{1t} cannot immediately affect the output X_{2t}. In other words, the lag between the output and input cannot be zero; it has to be at least one for a discrete system. Therefore, the simplest dependence of X_{2t} on the input that can be added in the above model is on the term X_{1t-1}. If we take the corresponding coefficient as, say 0.25, we can write the simple input-output relationship as

$$X_{2t} = 0.7X_{2t-1} + 0.25X_{1t-1} + a_{2t} \qquad (11.1.1)$$

When the model (11.1.1) is adequate, it says that the output basis weight at time t can be decomposed into three parts: one dependent on its own preceding value X_{2t-1}, one dependent on the input gate opening X_{1t-1} at the preceding time, and one independent of these two denoted by a_{2t}. Thus, Eq. (11.1.1) is the simplest input-output model corresponding to AR(1) of the univariate case in Chapter 2.

Using the B operator, we can also write the model (11.1.1) as

$$(1 - 0.7B)X_{2t} = 0.25BX_{1t} + a_{2t}$$

that is,

$$X_{2t} = \frac{0.25B}{(1 - 0.7B)} X_{1t} + \frac{1}{(1 - 0.7B)} a_{2t} \qquad (11.1.1a)$$

$$\underset{\text{Transfer}}{} \qquad \underset{\text{Noise}}{}$$

Function

In this form the model is called a transfer function model and is often schematically represented by a block diagram shown in Fig. 11.2. The

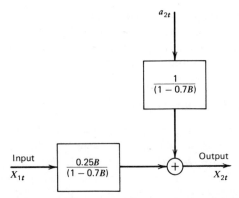

Figure 11.2 Block diagram for the transfer function model (11.1.1a) without feedback.

transfer function is represented by the operator

$$\frac{0.25B}{(1 - 0.7B)}$$

which transfers the input X_{1t} to the output. To this transferred input the "noise" represented by

$$\frac{1}{(1 - 0.7B)} a_{2t}$$

is added, giving the actual output X_{2t}.

11.1.3 Transfer Function versus State Variable Approach

The transfer function methodology in classical control theory arose mainly because the form of the transfer function was generally known or guessed fairly well by physical considerations. The discrepancy between the output given by the transfer function alone and the actual output was then termed noise. In many cases, the physical source of noise can be clearly identified as separate from the system represented by the transfer function. When the transfer function is derived from physical reasoning and the noise source entering the system after the transfer from the input to output is clearly identified, the transfer function methodology and its block diagram representation in Fig. 11.2 are intuitively appealing.

In general, however, the noise or disturbance as well as the transfer of input into output occur or are distributed all over the system; hence, the transfer and noise models and their representation in Fig. 11.2 are fictitious. For this reason and also for other theoretical and computational advantages, the modern control theory uses state space or state variable approach in preference to the transfer function methodology of classical control theory. In the state variable approach the variables of interest in the system are just considered as states and an equation (generally differ-ence or differential equation) relating them is taken as the system equation. The state variables may be observable (such as input, output, etc.) or unobservable (such as disturbances or derivatives of input-output, etc.).

The ARMA or AM models for univariate time series discussed in earlier chapters are actually state variable models. For example, the discrete time AR(1) model relates the observed states X_t and X_{t-1} with unobserved state a_t; the continuous time A(1) model relates the unobserved states $dX(t)/dt$ and $Z(t)$ with the observed state $X(t)$. Therefore, a natural extension of the ARMA models to control systems involving one or more inputs and one output is a model such as (11.1.1). This model can always be converted into a transfer function form such as (11.1.1a), if desired. However, it will be subsequently seen that the extended ARMA models such as (11.1.1) are quite sufficient to deal with control systems. In fact, for both the modeling and derivation of optimal control strategies, it is far simpler to use the form (11.1.1) than the form (11.1.1a).

Although every model such as (11.1.1) can be reduced to transfer function form, there is a difference between such a reduced model and one originally formulated in the transfer function form. It is seen in the model (11.1.1a) that both the transfer function and the noise part have the same autoregressive operator $(1-0.7B)^{-1}$. This may not necessarily be the case for a transfer function model formulated a priori. When a transfer function model is being formulated a priori, if it is known that the transfer function and the noise have different modes of dynamics, then the transfer function model may be of the form, say,

$$X_{2t} = \frac{0.2B}{(1-0.7B)} X_{1t} + \frac{1}{(1-0.5B)} a_{2t} \qquad (11.1.2a)$$

which is the same as the model

$$(1-0.7B)(1-0.5B)X_{2t} = 0.2B(1-0.5B)X_{1t} + (1-0.7B)a_{2t}$$

that is,

$$X_{2t} = 1.2X_{2t-1} - 0.35X_{2t-2} + 0.2X_{1t-1}$$

$$- 0.1X_{1t-2} + a_{2t} - 0.7a_{2t-1} \qquad (11.1.2)$$

Therefore, when converting an autoregressive moving average model (11.1.2) into a transfer function model, one must cancel nearly equal roots to write the model in its simplest form.

11.1.4 Large Delay or Dead Time

In developing the input-output relation by the model (11.1.1), we empha-sized that for a physical system, the input cannot immediately affect the output. Therefore, it is generally necessary to include at least one lag, and the output X_{2t} can be affected earliest by X_{1t-1}.

However, in practice, the lag could be much larger than one. That is, it may take several units of time before the input can affect the output. Such a lag is called delay or dead time because it delays the effect of input on the output. If we denote the lag by L, then Eq. (11.1.1) may be reformulated as

$$X_{2t} = 0.25X_{1t-L} + 0.7X_{2t-1} + a_{2t} \qquad (11.1.3)$$

with the corresponding transfer function model

$$X_{2t} = \frac{0.25B^L}{(1-0.7B)} X_{1t} + \frac{1}{(1-0.7B)} a_{2t} \qquad (11.1.3a)$$

It is clear that Eq. (11.1.1) is a special case of Eq. (11.1.3) for $L=1$, the minimum value of L.

11.1.5 ARMAV Models

So far we have not considered the nature of the input X_{1t}. In the transfer function form or the autoregressive moving average form as discussed above, X_{1t} can be any discrete input, either deterministic or stochastic. The deterministic inputs commonly used are step, ramp, or sinusoidal. An example of modeling with step input was presented in Sections 10.2.1 to 10.2.5. However, here we will be primarily concerned with stochastic or random inputs. When the input is a stationary stochastic process or time series, the input-output model can be formulated as a vector system, which we will now discuss.

Suppose the input X_{1t} is a stochastic process or time series and there is no feedback, that is X_{1t} does not depend on X_{2t} at any t, then the input X_{1t} can be represented by an ARMA model involving dependence on its past values only. For simplicity we will take this model to be also AR(1) with, say, $\phi = 0.8$. Let the output-input model be (11.1.1). Then, the complete control system may be represented by a pair of models

$$X_{1t} = 0.8\,X_{1t-1} \qquad\qquad\qquad + a_{1t} \qquad (11.1.4a)$$

$$X_{2t} = 0.25\,X_{1t-1} + 0.7\,X_{2t-1} + a_{2t}$$

which is the same as the vector model

$$\mathbf{X}_t = \boldsymbol{\phi}\,\mathbf{X}_{t-1} + \mathbf{a}_t \qquad (11.1.4)$$

where

$$\mathbf{X}_t = \begin{bmatrix} X_{1t} \\ X_{2t} \end{bmatrix}, \qquad \boldsymbol{\phi} = \begin{bmatrix} \phi_{11} & \phi_{12} \\ \phi_{21} & \phi_{22} \end{bmatrix} = \begin{bmatrix} 0.8 & 0 \\ 0.25 & 0.7 \end{bmatrix}$$

$$\mathbf{a}_t = \begin{bmatrix} a_{1t} \\ a_{2t} \end{bmatrix}$$

The model (11.1.4) is a discrete Autoregressive Vector model of order one, which we will denote by ARV(1). The scalar form of this ARV(1) model is given by Eq. (11.1.4a).

It should be noted that the absence of feedback, or the absence of dependence of the output X_{2t} on the input X_{1t}, is reflected by the fact that the parameter ϕ_{12} is zero. If such a feedback is present, for example, the transfer function models are of the form, represented in Fig. 11.3,

$$X_{1t} = \frac{0.1}{(1-0.8B)} X_{2t-1} + \frac{1}{(1-0.8B)} a_{1t}$$

$$X_{2t} = \frac{0.25}{(1-0.7B)} X_{1t-1} + \frac{1}{(1-0.7B)} a_{2t} \qquad (11.1.5a)$$

then the corresponding vector model is

$$X_{1t} = 0.8 \, X_{1t-1} + 0.1 \, X_{2t-1} + a_{1t}$$

$$X_{2t} = 0.25 X_{1t-1} + 0.7 \, X_{2t-1} + a_{2t} \qquad (11.1.5)$$

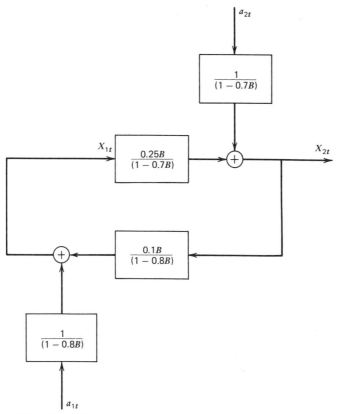

Figure 11.3   Block diagram for the transfer function model (11.1.5a) with feedback.

which is the same as Eq. (11.1.4a) with the only difference being that the value of ϕ_{12} has changed from zero to 0.1. It is thus seen that the feedback is easily incorporated in the vector ARMA models with no change in their form, but it changes the structure of the transfer function model considerably.

The ARV(1) model (11.1.4) can be developed by exactly the same reasoning as in Chapter 2 applied to the vector $\mathbf{X}_t$ rather than to the scalar X_t. When both the input and the output are discrete white noise sequences, we get an "ARV(0)" model

$$\mathbf{X}_t = \mathbf{a}_t$$

which says that vectors $\mathbf{X}_t$ at different t are uncorrelated or independent. If they are dependent, the simplest way to express this dependence is by the ARV(1) model

$$\mathbf{X}_t = \phi_1 \mathbf{X}_{t-1} + \mathbf{a}_t \qquad (11.1.6)$$

which is an orthogonal decomposition of $\mathbf{X}_t$ into two parts, one dependent on the preceding $\mathbf{X}_{t-1}$ and the other, $\mathbf{a}_t$, independent of $\mathbf{X}_{t-1}, \mathbf{X}_{t-2}, \ldots$, as well as $\mathbf{a}_{t-1}, \mathbf{a}_{t-2}, \ldots$. If, however, $\mathbf{a}_t$ of the ARV(1) model depends upon $\mathbf{X}_{t-2}$ and/or $\mathbf{a}_{t-1}$, we get an ARMAV(2,1) model

$$\mathbf{X}_t = \phi_1 \mathbf{X}_{t-1} + \phi_2 \mathbf{X}_{t-2} + \mathbf{a}_t - \theta_1 \mathbf{a}_{t-1} \qquad (11.1.7)$$

In fact, continuing the discussion in Chapter 2 in this way for bold face X_t's, we arrive at the ARMAV($n, n-1$) model or, in general, the ARMAV(n, m) model

$$\mathbf{X}_t = \phi_1 \mathbf{X}_{t-1} + \phi_2 \mathbf{X}_{t-2} + \ldots + \phi_n \mathbf{X}_{t-n}$$
$$+ \mathbf{a}_t - \theta_1 \mathbf{a}_{t-1} - \ldots - \theta_m \mathbf{a}_{t-m} \qquad (11.1.8)$$

It should be noted that the basic assumption on $\mathbf{a}_t$ in the structure of the ARMAV models does not guarantee that $a_{1t}, a_{2t}, \ldots$ at simultaneous time t are independent or uncorrelated. Therefore, $E[a_{1t}a_{2t}]$ may or may not be zero. In general, for a bivariate series, the variance-covariance matrix of $\mathbf{a}_t$ can be written as

$$\gamma_a = \begin{bmatrix} \gamma_{a_{11}} & \gamma_{a_{12}} \\ \gamma_{a_{21}} & \gamma_{a_{22}} \end{bmatrix}$$

where $\gamma_{a_{11}}$ and $\gamma_{a_{22}}$ denote the variances of a_{1t} and a_{2t}, and $\gamma_{a_{12}}$ and $\gamma_{a_{21}}$ represent their cross-dependence or covariance.

It is important to note the change of notation for the variances of the residuals. When $\gamma_{a12} = \gamma_{a21} = 0$, so that a_{1t} and a_{2t} are uncorrelated or independent, we use $\sigma_{a_1}^2$ and $\sigma_{a_2}^2$ or $\gamma_{a_{11}}$ and $\gamma_{a_{22}}$ interchangeably; when they are correlated or dependent, we use $\gamma_{a_{11}}$ and $\gamma_{a_{22}}$, exclusively to emphasize their bivariate nature.

A thorough understanding of the ARMAV model structure and modeling procedure, etc. is beyond the scope of this book. However, the usual assumption in control theory that the input noise (or the input itself) is independent of the output noise implies that modeling and control can be accomplished by considering the two models separately, as in Eqs. (11.1.4a) and (11.1.5), rather than together in their vector form (11.1.6) and (11.1.7). The assumption of independent noises, in turn, means that a_{1t} is independent of a_{2t}, so that the matrix γ_a and the θ_i matrices in (11.1.7) and (11.1.8) are diagonal. This will be shown in the next section 11.1.6 by relating ARMAV models with their simplified scalar forms used in this chapter.

However, the assumption of independent noises is so common that the reader can skip Section 11.1.6 and proceed to Section 11.2, keeping in mind the assumption that a_{1t}, a_{2t}, etc. are assumed independent at the same time. No continuity will be lost if all the subsequent transformations to ARMAV models are skipped.

*11.1.6 A Special Form of the ARMAV Models

A simpler form can be deduced from the general ARMAV model first by a simple transformation and next by assuming that the input and output noises are uncorrelated. This form of the vector autoregressive moving average model can greatly simplify the estimation of the parameters of the multiple input-one output systems.

The input and output noises in a bivariate autoregressive moving average model can be made uncorrelated by making γ_a diagonal, that is,

$$E[a_{1t}a_{2t}] = 0$$

For illustration, let us rewrite an ARV(1) model, using *'s to emphasize that a_{1t}^* and a_{2t}^* are *correlated*

$$\mathbf{X}_t = \boldsymbol{\phi}_1^* \mathbf{X}_{t-1} + \mathbf{a}_t^* \qquad (11.1.6a)$$

In expanded form it can be written as

$$\begin{bmatrix} X_{1t} \\ X_{2t} \end{bmatrix} = \begin{bmatrix} \phi_{111}^* & \phi_{121}^* \\ \phi_{211}^* & \phi_{221}^* \end{bmatrix} \begin{bmatrix} X_{1t-1} \\ X_{2t-1} \end{bmatrix} + \begin{bmatrix} a_{1t}^* \\ a_{2t}^* \end{bmatrix} \qquad (11.1.6b)$$

where ϕ_{111}^*, ϕ_{121}^*, ϕ_{211}^*, and ϕ_{221}^* are the elements of the autoregressive parameter matrix $\boldsymbol{\phi}_1^*$. Note that the first two subscripts of the parameter elements designate the row and column of the matrix; the last subscript denotes the subscript of the parameter matrix $\boldsymbol{\phi}_1^*$, that is, 1. The variance-covariance matrix of a_{1t}^* and a_{2t}^*, γ_a^*, is a symmetric matrix and with possibly nonzero $\gamma_{a21}^* = \gamma_{a12}^*$. Transformation from correlated a_{1t}^* and a_{2t}^* to uncorrelated a_{1t} and a_{2t} may be represented by

$$\mathbf{a}_t = \boldsymbol{\phi}_0 \mathbf{a}_t^* \qquad (11.1.9a)$$

or

$$\begin{bmatrix} a_{1t} \\ a_{2t} \end{bmatrix} = \begin{bmatrix} 1 & \phi_{120} \\ 0 & 1 \end{bmatrix} \begin{bmatrix} a_{1t}^* \\ a_{2t}^* \end{bmatrix}$$

where ϕ_0 is a transformation matrix. The variance-covariance matrix γ_a is written as

$$\begin{aligned}
\gamma_a &= E[\mathbf{a}_t \mathbf{a}_t'] \\
&= E \begin{bmatrix} 1 & \phi_{120} \\ 0 & 1 \end{bmatrix} \begin{bmatrix} a_{1t}^* \\ a_{2t}^* \end{bmatrix} (a_{1t}^* a_{2t}^*) \begin{bmatrix} 1 & 0 \\ \phi_{120} & 1 \end{bmatrix} \\
&= \begin{bmatrix} 1 & \phi_{120} \\ 0 & 1 \end{bmatrix} E \left[\begin{pmatrix} a_{1t}^* \\ a_{2t}^* \end{pmatrix} (a_{1t}^* a_{2t}^*) \right] \begin{bmatrix} 1 & 0 \\ \phi_{120} & 1 \end{bmatrix} \\
&= \begin{bmatrix} 1 & \phi_{120} \\ 0 & 1 \end{bmatrix} \begin{bmatrix} \gamma_{a11}^* & \gamma_{a12}^* \\ \gamma_{a21}^* & \gamma_{a22}^* \end{bmatrix} \begin{bmatrix} 1 & 0 \\ \phi_{120} & 1 \end{bmatrix} \\
&= \begin{bmatrix} \gamma_{a11}^* + \phi_{120}\gamma_{a21}^* & \gamma_{a12}^* + \phi_{120}\gamma_{a22}^* \\ \gamma_{a21}^* & \gamma_{a22}^* \end{bmatrix} \begin{bmatrix} 1 & 0 \\ \phi_{120} & 1 \end{bmatrix} \\
&= \begin{bmatrix} (\gamma_{a11}^* + \phi_{120}\gamma_{a21}^*) + \phi_{120}(\gamma_{a12}^* + \phi_{120}\gamma_{a22}^*) & \gamma_{a12}^* + \phi_{120}\gamma_{a22}^* \\ \gamma_{a21}^* + \phi_{120}\gamma_{a22}^* & \gamma_{a22}^* \end{bmatrix}
\end{aligned}$$

$$(11.1.9b)$$

Since γ_a is a diagonal matrix representing the cross-dependence of a_{1t} and a_{2t} as zero, the cross terms in Eq. (11.1.9b) should be zero, that is,

$$\gamma_{a12}^* + \phi_{120}\gamma_{a22}^* = 0$$

and

$$\gamma_{a21}^* + \phi_{120}\gamma_{a22}^* = 0$$

Since $\gamma_{a12}^* = \gamma_{a21}^*$ these equations give

$$\phi_{120} = -\frac{\gamma_{a12}^*}{\gamma_{a22}^*} = -\frac{\gamma_{a21}^*}{\gamma_{a22}^*} \qquad (11.1.9c)$$

Substituting $\phi_{120} = -(\gamma_{a12}^*/\gamma_{a22}^*)$ in the element (1,1) of γ_a matrix [Eq. (11.1.9b)] gives

$$\left(\gamma_{a11}^* - \frac{\gamma_{a12}^*}{\gamma_{a22}^*} \cdot \gamma_{a21}^* \right) + \left(-\frac{\gamma_{a12}^*}{\gamma_{a22}^*} \right) \left(\gamma_{a12}^* - \frac{\gamma_{a12}^*}{\gamma_{a22}^*} \cdot \gamma_{a22}^* \right) = \left(\gamma_{a11}^* - \frac{\gamma_{a12}^{*\,2}}{\gamma_{a22}^*} \right)$$

Thus, the γ_a matrix is

$$\gamma_a = \begin{bmatrix} \left(\gamma_{a11}^* - \dfrac{\gamma_{a12}^{*\,2}}{\gamma_{a22}^*} \right) & 0 \\ 0 & \gamma_{a22}^* \end{bmatrix} = \phi_0 \gamma_a^* \phi_0' \qquad (11.1.9)$$

Explicitly, the $\boldsymbol{\phi}_0$ matrix is

$$
\boldsymbol{\phi}_0 = \begin{bmatrix} 1 & -\dfrac{\gamma^*_{a12}}{\gamma^*_{a22}} \\ 0 & 1 \end{bmatrix} \tag{11.1.9d}
$$

and the uncorrelated a_{1t} and a_{2t} can be expressed as

$$
a_{1t} = a^*_{1t} - \frac{\gamma^*_{a12}}{\gamma^*_{a22}} a^*_{2t}
$$

$$
a_{2t} = a^*_{2t}
$$

If we multiply Eq. (11.1.6b) by $\boldsymbol{\phi}_0$, we get

$$
\begin{bmatrix} 1 & \phi_{120} \\ 0 & 1 \end{bmatrix} \begin{bmatrix} X_{1t} \\ X_{2t} \end{bmatrix} = \begin{bmatrix} \phi_{111} & \phi_{121} \\ \phi_{211} & \phi_{221} \end{bmatrix} \begin{bmatrix} X_{1t-1} \\ X_{2t-1} \end{bmatrix} + \begin{bmatrix} a_{1t} \\ a_{2t} \end{bmatrix} \tag{11.1.6c}
$$

where $\phi_{120} = -(\gamma^*_{a12}/\gamma^*_{a22})$ and ϕ_{111}, ϕ_{121}, ϕ_{211}, ϕ_{221} are transformed parameters, with $\boldsymbol{\phi}_1 = \boldsymbol{\phi}_0\boldsymbol{\phi}^*_1$, $\mathbf{a}_t = \boldsymbol{\phi}_0\mathbf{a}^*_t$, and a_{1t} and a_{2t} are now uncorrelated. Thus, the series X_{1t} and X_{2t} can be expressed as

$$
X_{1t} + \phi_{120}X_{2t} = \phi_{111}X_{1t-1} + \phi_{121}X_{2t-1} + a_{1t}
$$

$$
X_{2t} = \phi_{211}X_{1t-1} + \phi_{221}X_{2t-1} + a_{2t}
$$

Note that since each series includes only its own a_{it}, the model is just a conditional regression and the conditional least squares estimates of the parameters can be obtained separately as if by a univariate series rather than a bivariate series.

Next, let us consider an ARMAV(2,1) model:

$$
X_t = \boldsymbol{\phi}^*_1 X_{t-1} + \boldsymbol{\phi}^*_2 X_{t-2} + \mathbf{a}^*_t - \boldsymbol{\theta}^*_1\mathbf{a}^*_{t-1} \tag{11.1.7a}
$$

Using a similar transformation as given for ARV(1), that is, premultiplying (11.1.7a) by $\boldsymbol{\phi}_0$, we get

$$
\boldsymbol{\phi}_0 X_t = \boldsymbol{\phi}_0\boldsymbol{\phi}^*_1 X_{t-1} + \boldsymbol{\phi}_0\boldsymbol{\phi}^*_2 X_{t-2} + \boldsymbol{\phi}_0\mathbf{a}^*_t - \boldsymbol{\phi}_0\boldsymbol{\theta}^*_1\mathbf{a}^*_{t-1} \tag{11.1.7b}
$$

Now substituting

$$
\boldsymbol{\phi}_1 = \boldsymbol{\phi}_0\boldsymbol{\phi}^*_1
$$

$$
\boldsymbol{\phi}_2 = \boldsymbol{\phi}_0\boldsymbol{\phi}^*_2
$$

$$
\mathbf{a}_t = \boldsymbol{\phi}_0\mathbf{a}^*_t \tag{11.1.7c}
$$

$$
\mathbf{a}_{t-1} = \boldsymbol{\phi}_0\mathbf{a}^*_{t-1}
$$

and

$$
\boldsymbol{\theta}_1 = \boldsymbol{\phi}_0\boldsymbol{\theta}^*_1\boldsymbol{\phi}_0^{-1}
$$

in Eq. (11.1.7b) yields

$$
\boldsymbol{\phi}_0 X_t = \boldsymbol{\phi}_1 X_{t-1} + \boldsymbol{\phi}_2 X_{t-2} + \mathbf{a}_t - \boldsymbol{\theta}_1\mathbf{a}_{t-1} \tag{11.1.7d}
$$

For a bivariate model, Eq. (11.1.7d) can be written in expanded form as

$$
\begin{bmatrix} 1 & \phi_{120} \\ 0 & 1 \end{bmatrix} \begin{bmatrix} X_{1t} \\ X_{2t} \end{bmatrix} = \begin{bmatrix} \phi_{111} & \phi_{121} \\ \phi_{211} & \phi_{221} \end{bmatrix} \begin{bmatrix} X_{1t-1} \\ X_{2t-1} \end{bmatrix}
$$
$$
+ \begin{bmatrix} \phi_{112} & \phi_{122} \\ \phi_{212} & \phi_{222} \end{bmatrix} \begin{bmatrix} X_{1t-2} \\ X_{2t-2} \end{bmatrix} + \begin{bmatrix} a_{1t} \\ a_{2t} \end{bmatrix} - \begin{bmatrix} \theta_{111} & \theta_{121} \\ \theta_{211} & \theta_{221} \end{bmatrix} \begin{bmatrix} a_{1t-1} \\ a_{2t-1} \end{bmatrix} \quad (11.1.7e)
$$

In scalar form, say,

$$
X_{2t} = \phi_{211}X_{1t-1} + \phi_{221}X_{2t-1} + \phi_{212}X_{1t-2}
$$
$$
+ \phi_{222}X_{2t-2} + a_{2t} - \theta_{211}a_{1t-1} - \theta_{221}a_{2t-1} \quad (11.1.10)
$$

Since each series includes both a_{1t} and a_{2t}, its parameters cannot be estimated by minimizing the sum of squares of either a_{1t} or a_{2t} alone. However, since a_{1t} and a_{2t} are already independent, this difficulty can be resolved and the parameters estimated separately, if $\theta_{121} = \theta_{211} = 0$, that is, if θ_1, is diagonal.

To repeat, the basic assumption states that thé input noise is independent of output noise. This assumption implies that a_{1t} is independent of $a_{2t}, a_{2t-1}, a_{2t-2}, \ldots$, that is,

$$
E[a_{1t}a_{2t-k}] = 0, \qquad k = 0, \pm 1, \pm 2, \ldots
$$

and that the terms θ_{121} and θ_{211} in Eq. (11.1.7e) are zero. Thus, Eq. (11.1.10) becomes

$$
X_{2t} = \phi_{211}X_{1t-1} + \phi_{221}X_{2t-1} + \phi_{212}X_{1t-2}
$$
$$
+ \phi_{222}X_{2t-2} + a_{2t} - \theta_{221}a_{2t-1} \quad (11.1.10a)
$$

which does not include $a_{1t}, a_{1t-1}, \ldots$, and therefore the estimation of the parameters can be performed separately by minimizing the sum of squares of the residuals a_{2t}'s alone. It is easy to see that Eq. (11.1.10a) has the same form as Eqs. (11.1.1) to (11.1.3). In this chapter, we will only be using these special forms of the ARMAV model. These may be thought of as extended ARMA models, as developed through Section 11.1.3, keeping in mind that a_{1t} is assumed to be independent of a_{2t}.

11.2 MODELING AND ILLUSTRATIONS

The special form of the ARMAV models can be obtained by the same modeling strategy as the univariate ARMA models. Such a simpler treatment of modeling more or less corresponds to Chapter 2. The only additional computation needed is that for the initial values of the coefficients due to the additional series and checking the assumption of a_{it} independent of a_{jt}, $i \neq j$, using plots of cross-correlations of the residuals. Generally, with a limited number of series, say two or three, the simplified ARMAV form of the modeling is useful. With a large number of series and/or when the independent noise assumption is not satisfied, then the general AR-MAV modeling should be used.

11.2.1 Modeling Procedure

The modeling procedure for the simplified ARMAV models can be outlined along the same lines as the ARMA$(n, n-1)$ models in Section 4.5.

Step 1

Obtain the initial values of the parameters by the inverse function method as given in Appendix A 11.1. The initial values also provide an indication of the presence of initial delay or dead time. For example, if the initial values of the parameters, say, ϕ_{211}, ϕ_{212} of one input-one output system are small (confidence intervals include zero), then the model for the output X_{2t} could start with lag 3. However, it should be noted that the choice of the initial delay L is tentative, and if there is doubt, the parameters should be included in the final estimation.

Step 2

By the nonlinear least squares method fit the models for input and output separately, in a form modified for lag, if necessary, starting from the initial values in Step 1. Increase the value of n until the reduction in the sum of squares is insignificant as discussed in Sections 4.4.2 and 4.4.3. (The first method of setting $a_t = 0$, $t = 1$ to n and starting the computation of a_t's from $t = n + 1$ is preferable.)

Step 3

Drop the small parameter values to reduce the order and/or increase the lag or dead time and refit the model. Check the refitted model against the one in Step 2 and determine the adequate model.

Step 4

Compute the auto and cross-correlation estimates of a_{it}, a_{jt}, using the *average substracted* residuals in the following

$$\hat{\gamma}_{a_{ijk}} = \frac{1}{N} \sum_{t=k+1}^{N} a_{it} a_{jt-k}, \ \hat{\gamma}_{a_{ij(-k)}} = \hat{\gamma}_{a_{jik}}, \ k \geqslant 0 \qquad (11.2.1)$$

$$\hat{\rho}_{ijk} = \frac{\hat{\gamma}_{a_{ijk}}}{\sqrt{\hat{\gamma}_{a_{iio}} \hat{\gamma}_{a_{jjo}}}}, \qquad k = 0, \pm 1, \pm 2, \ldots \qquad (11.2.2)$$

If the $\hat{\rho}_{ijk}$'s lie within the $\pm 2/\sqrt{N}$† band (or the "unified" correlations are less than two in magnitude), the series may be considered independent. If not, repeat in Steps 2 and 3 with a greater number of parameters. If the

†These are slightly conservative, actually one can use $1.96/\sqrt{N-k}$.

cross-correlations are still not within the $\pm\,2/\sqrt{N}$ band, then the general ARMAV models are perhaps needed.

11.2.2 One Input–One Output Papermaking System: Comparison with Deterministic Inputs

In a project undertaken by a papermaking plant, an open-loop (that is, without feedback) test with a step input was used to study the dynamic relation between the input stock flow and output basis weight. The basis weight data from this test was used in Sections 10.2.1 to 10.2.4 to determine the dynamics and estimate the parameters. However, an open loop normal operating record was obtained before the application of the step input. A record of the input X_{1t} (stock flow, gal/min) and the corresponding output basis weight (lbs/3,300 sq ft) X_{2t} is shown in Fig. 11.4; the numerical data are given in Appendix I at the end of the book. The data have been adjusted for a lag of 50 seconds computed from the machine speed.

The dynamics obtained from a normal operating record is more desirable, being more realistic compared to that under external inputs. How-

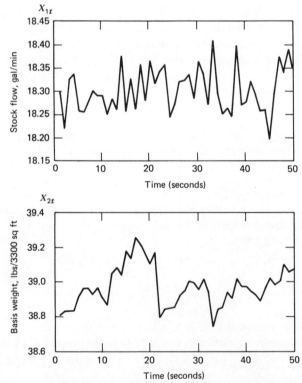

Figure 11.4 Input X_{1t} and output X_{2t} for the papermaking process.

ever, the external inputs such as step or impulse are still commonly used because of the difficulties of obtaining the dynamics from the normal operating data. An attempt by Goh (1973) to model this data by empirical cross-correlation methods did not succeed because of large disturbance.

Modeling of this data by the procedure of Section 11.2.1 will now be illustrated. We first fit a first order model to the output X_{2t} series. As in Step 1, the initial values of the parameters ϕ_{211} and ϕ_{221} are computed using the procedure illustrated in Appendix A 11.1. The initial values also indicate that there is no significant delay between input and output, and hence the lag was chosen as one. The nonlinear least squares method gives the results for the output X_{2t} and the input X_{1t} presented in Tables 11.1 and 11.2, respectively.

It is seen from Table 11.1 that the first order model fitted to the output-input model is

$$X_{2t} = 0.247X_{1t-1} + 0.696X_{2t-1} + a_{2t}, \quad \gamma_{a22} = 0.00624$$

The improvement in the residual sum of squares from the first order to second order is significant since

$$F = \frac{0.3060 - 0.1950}{3} \div \frac{0.1950}{50 - 5} = 8.538$$

compared with $F_{0.95}(3,45) = 2.84$. Comparing the (2,1) model with the (3,2) gives $F = 3.14$. On the other hand, ARMA(4,3) model (not given

Table 11.1 Modeling Papermaking Process Output, Basis Weight, X_{2t} (Mean = 38.969, Variance = .0134)

Parameters	Model Order		
	(1,0)	(2,1)	(3,2)
ϕ_{211}	0.247 ± 0.461	0.275 ± 0.457	0.454 ± 0.422
ϕ_{212}		-0.593 ± 0.471	-0.416 ± 0.378
ϕ_{213}			-0.536 ± 0.428
ϕ_{221}	0.696 ± 0.195	1.453 ± 0.245	0.470 ± 0.308
ϕ_{222}		-0.523 ± 0.241	0.923 ± 0.251
ϕ_{223}			-0.518 ± 0.267
θ_{221}		1.125 ± 0.239	0.029 ± 0.325
θ_{222}			1.304 ± 0.349
Residual sum of squares	0.3060	0.1950	0.1593
γ_{a22}	0.00624	0.00406	0.00339
F		8.538	3.140

The adequate model is (3,2).
Number of observations = 50.

Table 11.2 Modeling Papermaking Process Input; Gate Opening X_{1t} (Mean = 18.302, Variance = 0.0024)

Parameters	Model Order			
	(1,0)	(2,1)	(3,2)	(4,3)
ϕ_{120}	0.104 ± 0.168	0.121 ± 0.164	0.123 ± 0.150	0.112 ± 0.113
ϕ_{111}	0.002 ± 0.276	-0.290 ± 0.569	0.300 ± 0.228	0.265 ± 0.537
ϕ_{112}		0.023 ± 0.277	-0.553 ± 0.056	-0.455 ± 0.392
ϕ_{113}			-0.053 ± 0.191	-0.059 ± 0.241
ϕ_{114}				0.022 ± 0.155
ϕ_{121}	0.191 ± 0.164	0.006 ± 0.176	0.151 ± 0.212	0.126 ± 0.144
ϕ_{122}		0.260 ± 0.173	0.136 ± 0.220	0.199 ± 0.155
ϕ_{123}			-0.0121 ± 0.163	-0.036 ± 0.055
ϕ_{124}				-0.013 ± 0.058
θ_{111}		-0.221 ± 0.643	0.591 ± 0.319	0.494 ± 0.195
θ_{112}			-1.451 ± 0.291	-1.523 ± 0.103
θ_{113}				-0.089 ± 0.453
Residual sum of squares	0.1051	0.08464	0.03525	0.02892
γ_{a11}	0.00214	0.00176	0.00075	0.00063
F	—	3.552	19.14	2.845

The adequate model is (3,2).

in Table 11.2) has the residual sum of squares 0.1512, giving F $= 0.693$. Therefore the (3,2) model is taken as an adequate model. (However, since $F_{0.99}(3,45) = 4.31$, the (2,1) model may also be considered adequate if 0.01 level of significance is used.) Note that the average has been subtracted from both the series although the respective means could have been estimated as additional parameters. Thus the adequate model for the output series X_{2t} is

$$X_{2t} = 0.454X_{1t-1} - 0.416X_{1t-2} - 0.536X_{1t-3}$$
$$+ 0.470X_{2t-1} + 0.923X_{2t-2} - 0.518X_{2t-3}$$
$$+ a_{2t} - 0.029a_{2t-1} - 1.304a_{2t-2}, \gamma_{a22} = 0.00399$$

For the input X_{1t} series, the tests of adequacy indicate that the adequate model is (3,2), that is,

$$X_{1t} = -0.123X_{2t} + 0.300X_{1t-1} - 0.553X_{1t-2} - 0.053X_{1t-3}$$
$$+ 0.151X_{2t-1} + 0.136X_{2t-2} - 0.012X_{2t-3} + a_{1t} - 0.591a_{1t-1}$$
$$+ 1.451a_{1t-2}, \gamma_{a11} = 0.000615$$

compared to a white noise model $(X_{1t} = a_{1t})$ with variance $= 0.0024$ (Table 11.2) for an ideal step input with added white noise.

As a final step in modeling (Step 4 of Section 11.2.1), we check the basic assumption of uncrosscorrelated a_{1t} and a_{2t} necessary for these models. The sample auto- and cross-correlations of a_{1t} and a_{2t} from the models (2,1) and (3,2) for the X_{2t} and X_{1t} series, respectively, are shown in Fig. 11.5. They are all within the $\pm 2/\sqrt{N}$ bands.

Deterministic Inputs

We have thus shown that the extended ARMA model can be obtained from the normal operating data without disturbing the system. The transfer function of the system can be readily obtained from the ARMA model as discussed in Section 11.1.

The transfer functions are often obtained following the classical control theory by means of deterministic inputs such as impulse, step, sinusoidal, etc. The response of these inputs can be used to obtain the model by graphical methods such as Bode plots, spectrum, etc. These methods cannot be effectively applied when there is considerable noise in the system. The modeling of such noisy data by the deterministic plus ARMA approach was illustrated in Sections 10.3.2 and 10.3.4.

Can we use the model obtained from the deterministic input response data to formulate an extended ARMA or transfer function model? Strictly speaking, the answer is "no." The dynamics under actual operating conditions may differ from that under controlled conditions with deterministic inputs. However, one can get a rough idea about the transfer function

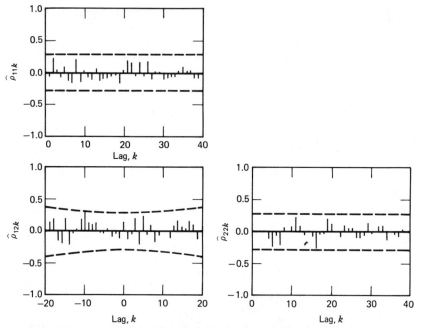

Figure 11.5 Sample autocorrelation functions (diagonal plots) and sample cross-correlation function (off diagonal plot) of the residuals.

from the deterministic models, which may be later refined based on the operating data as illustrated above. Therefore, we will illustrate how to obtain the transfer function from the deterministic models by means of the papermaking data and the model in Section 10.2.2.

The general method for obtaining a discrete transfer function from a continuous deterministic input response consists of first finding the continuous transfer function and then using Z-transforms. The general formulae for this purpose are given in Appendix A 11.2 together with an illustrative example. For a first order system under consideration, it can be simply obtained by considering the gain g as illustrated below.

The gain g is defined as the eventual response of a system (measured from the initial equilibrium position) to unit step input. For a first order discrete system with deterministic input Y_{1t} and output Y_{2t}

$$Y_{2t} = \frac{\phi_{211}}{(1 - \phi_{221}B)} Y_{1t-1}$$

$$= \phi_{211}(1 + \phi_{221}B + \phi_{221}^2 B^2 + \ldots)Y_{1t-1}$$

If the input is a unit step, then

$$Y_{1t-k} \equiv 1, \qquad k = 1, 2, 3, \ldots$$

and therefore the output g is

$$g = \phi_{211}(1 + \phi_{221} + \phi_{221}^2 + \ldots)$$

$$= \frac{\phi_{211}}{1 - \phi_{221}} \tag{11.2.3}$$

It is not difficult to see from the above expansion that the gain g for *any* discrete transfer functions can be obtained by simply substituting $B = 1$.

Now recall from Section 10.2.2 that the adequate model for the step response of the papermaking system was found to be

$$Y_t = A_0 + g(1 - e^{-(t/\tau)}) + X_t$$

$$X_t = \phi_1 X_{t-1} + a_t$$

$$\hat{g} = 1.7992 \pm 0.4228$$

$$\tau = 35.3912 \pm 23.5606$$

$$\hat{\phi}_1 = 0.7887 \pm 0.1270$$

The autoregressive parameter of the discrete transfer function that represents the dynamics is given by

$$\phi_{221} = e^{-\alpha_0 \Delta} = e^{-(\Delta/\tau)} = e^{-1/35.3912} = 0.9189$$

since the sampling interval $\Delta = 1$. The gain for the discrete as well as the continuous system are the same and therefore by Eq. (11.2.3)

$$\phi_{211} = g(1 - \phi_{221}) = 1.7992(1 - 0.9189) = 0.144$$

Hence, the discrete transfer function is

$$\frac{0.14 B}{(1 - 0.92 B)}$$

as opposed to the one obtained from the first order model fitted to the operating data as

$$\frac{0.25 B}{(1 - 0.7B)}$$

which do not coincide even after taking into consideration the confidence interval that is between 0.501 and 0.891 for ϕ_{221} in Table 11.1.

On the other hand, the noise, which was of course treated as stochastic in both Sections 10.2.2 and 11.2.2, gave the estimates $\hat{\phi}_1 = 0.7887 \pm 0.1270$ and $\phi_{221} = 0.696 \pm 0.195$, respectively, which are quite close. The first order model was found to be adequate with deterministic inputs in Section 10.2.2, whereas ARMA(3,2) was adequate in Section 11.2.2.

Deterministic Inputs versus Operating Data

It is thus possible to obtain the transfer function either by means of deterministic inputs, using the procedure of Chapter 10 combined with Appendix A 11.2, or from the operating data, using the extended ARMA modeling procedure of Section 11.2.1. From the point of view of understanding the dynamics as well as prediction and control, the modeling of Section 11.2.1 from the operating data has the following advantages:

1   The normal operation of the system need not be disturbed just for the purpose of analysis.
2   The dynamics is obtained under realistic conditions and therefore would provide more precise prediction and control.
3   By continuously monitoring the data it is possible to detect changes in the system over time and to take appropriate corrective action when necessary.

Apart from these advantages, there are some inherent difficulties in the deterministic input approach. It is difficult to decide the scale of such inputs that would provide an accurate and reliable estimate of the dynamics. If inputs that are too weak are used, the response may be submerged in the noise, making it difficult to estimate the transfer function. If inputs that are too large are used to provide a sufficiently high signal to noise ratio, the test may yield a distorted picture of the actual dynamics under actually smaller normal inputs. Moreover, the input noise may be added to the response and may inflate and distort the so-called output noise.

These difficulties are somewhat illustrated by the papermaking example considered above. The deterministic input indicates a first order dynamics although the dynamics under operating conditions is second or third order. The confidence intervals are relatively large and the variance of a_t's is inflated from 0.0062 to 0.0166 comparing the first order model in both cases of operating data and deterministic inputs respectively.

*Transformation to ARMAV

Once the separate adequate models for X_{1t} and X_{2t} are estimated, the corresponding bivariate ARMAV model in the form of (11.1.6a) can be determined by back transformation [see Eq. (11.1.7c)]:

$$\mathbf{\phi}_1^* = \mathbf{\phi}_0^{-1} \mathbf{\phi}_1$$

$$= \begin{bmatrix} 1 & -0.123 \\ 0 & 1 \end{bmatrix} \begin{bmatrix} 0.300 & 0.151 \\ 0.454 & 0.470 \end{bmatrix} = \begin{bmatrix} 0.244 & 0.093 \\ 0.454 & 0.470 \end{bmatrix}$$

$$\mathbf{\phi}_2^* = \mathbf{\phi}_0^{-1} \mathbf{\phi}_2$$

$$= \begin{bmatrix} 1 & -0.123 \\ 0 & 1 \end{bmatrix} \begin{bmatrix} -0.553 & 0.136 \\ -0.416 & 0.923 \end{bmatrix} = \begin{bmatrix} -0.502 & 0.023 \\ -0.416 & 0.923 \end{bmatrix}$$

$$\mathbf{\phi}_3^* = \mathbf{\phi}_0^{-1} \mathbf{\phi}_3$$

$$= \begin{bmatrix} 1 & -0.123 \\ 0 & 1 \end{bmatrix} \begin{bmatrix} -0.053 & -0.012 \\ -0.536 & -0.518 \end{bmatrix} = \begin{bmatrix} -0.013 & -0.052 \\ -0.536 & -0.518 \end{bmatrix}$$

$$\theta_1^* = \Phi_0^{-1}\theta_1\Phi_0$$

$$= \begin{bmatrix} 1 & -0.123 \\ 0 & 1 \end{bmatrix} \begin{bmatrix} 0.591 & 0 \\ 0 & 0.029 \end{bmatrix} \begin{bmatrix} 1 & 0.123 \\ 0 & 1 \end{bmatrix}$$

$$= \begin{bmatrix} 0.591 & 0.069 \\ 0 & 0.029 \end{bmatrix}$$

$$\theta_2^* = \Phi_0^{-1}\theta_2\Phi_0$$

$$= \begin{bmatrix} 1 & -0.123 \\ 0 & 1 \end{bmatrix} \begin{bmatrix} -1.451 & 0 \\ 0 & 1.304 \end{bmatrix} \begin{bmatrix} 1 & 0.123 \\ 0 & 1 \end{bmatrix}$$

$$= \begin{bmatrix} -1.451 & -0.019 \\ 0 & 1.304 \end{bmatrix}$$

and the ARMAV model in expanded form is written as

$$\begin{bmatrix} X_{1t} \\ X_{2t} \end{bmatrix} = \begin{bmatrix} 0.244 & 0.093 \\ 0.454 & 0.470 \end{bmatrix} \begin{bmatrix} X_{1t-1} \\ X_{2t-1} \end{bmatrix}$$

$$+ \begin{bmatrix} -0.502 & 0.023 \\ -0.416 & 0.923 \end{bmatrix} \begin{bmatrix} X_{1t-2} \\ X_{2t-2} \end{bmatrix}$$

$$+ \begin{bmatrix} 0.013 & 0.052 \\ -0.536 & -0.518 \end{bmatrix} \begin{bmatrix} X_{1t-3} \\ X_{2t-3} \end{bmatrix}$$

$$+ \begin{bmatrix} a_{1t}^* \\ a_{2t}^* \end{bmatrix} - \begin{bmatrix} 0.591 & 0.069 \\ 0 & 0.029 \end{bmatrix} \begin{bmatrix} a_{1t-1}^* \\ a_{2t-1}^* \end{bmatrix}$$

$$- \begin{bmatrix} -1.451 & -0.021 \\ 0 & 1.304 \end{bmatrix} \begin{bmatrix} a_{1t-2}^* \\ a_{2t-2}^* \end{bmatrix} \qquad (11.2.4)$$

with the variance-covariance matrix from Eq. (11.1.9) as

$$\gamma_a^* = \Phi_0^{-1}\gamma_a\Phi_0^{-1'}$$

$$\gamma_a^* = \begin{bmatrix} 1 & -0.123 \\ 0 & 1 \end{bmatrix} \begin{bmatrix} 0.00075 & 0 \\ 0 & 0.00339 \end{bmatrix} \begin{bmatrix} 1 & 0 \\ -0.123 & 1 \end{bmatrix}$$

$$= \begin{bmatrix} 0.000801 & -0.000417 \\ -0.000417 & 0.003390 \end{bmatrix}$$

11.2.3 Two Inputs–One Output System With Dead Time: Application to the Paper Pulping Digester

The simplest one-input–one-output example was considered in Section 11.2.2. Since the data was adjusted for the initial delay or dead time of 50 seconds calculated from machine speed, the complications of estimating the dead time outlined in Step 1 of Section 11.2.1 did not arise. In this section, we will consider an example in which such dead time exists but its exact amount is unknown, and moreover, in addition to the one input that is manipulated for control purposes, there is one more input that is measured but not manipulated. The data for this example are taken from a continuous pulping digester in a paper mill.

The schematic diagram showing the continuous pulping digester process is given in Fig. 11.6. Wood chips are fed through a low pressure feeder (at about 16 psig) to a steam vessel to drive off air to facilitate liquor penetration. The chips are then mixed with circulating liquor and forced into the digester by a high pressure (at about 160 psig) feeder. Recycling white liquor for cooking is added and after about 45 minutes in the impregnation zone in the upper part of the digester, heat is added to raise

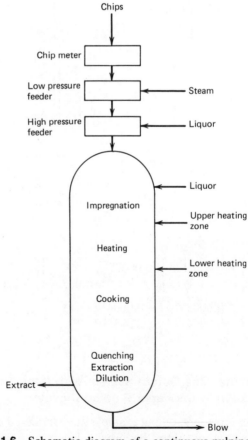

Figure 11.6   Schematic diagram of a continuous pulping digester.

the temperature for the delignification process. The main temperature control is on the lower heating zone. After about two hours, the resulting pulp leaves the cooking zone to be quenched by black liquor and diluted; then, it is removed from the bottom of the digester about four hours later. The wood lignin is removed in the cooking process and the extent of delignification is indicated by the K number of the pulp obtained in a laboratory test; a higher K number corresponds to a higher quantity of residual lignin and vice versa.

The objective of the digester control effort is to keep the K number as close to a target value as possible so that a consistent quality of paper product can be maintained. Of the factors that affect the K number, the temperature at the lower heating zone immediately above the cooking zone is considered most important and is the main input variable used for control purposes. The overall process throughout time is about eight hours; therefore, there is a long lag time between the application of temperature adjustment and the resulting K number. The chemical mechanism of the process is so complex that control schemes based on analytical designs are extremely difficult to devise and implement; therefore, the time series approach is used in an attempt to handle the modeling and control problem of the digester process.

One could use the temperature and K number series for a one-input–one-output model to control the K number. Since the measurements of another variable—blow to feed (B/F) ratio—were available, it was decided to explore the possibility of improving the control by incorporating this information in the model. It should be noted that this variable B/F ratio is only a "secondary input" or "disturbance" since it can be measured but not manipulated for control purposes.

The set point temperature, blow to feed (B/F) ratio, and the K number of the output pulp were recorded every hour, sets of 110 observations (from more than four and a half days of operation) for softwood cooking were obtained and had the following characteristics:

X_{1t}: temperature with a mean of 341°F and a variance of 4.29.
X_{2t}: B/F ratio with a mean of 54.1 and a variance of 6.51.
X_{3t}: K number with a mean of 18.31 and a variance of 9.81.

The data of these three series, taken from Goh (1973), are given in Appendix I and plotted in Figure 11.7.

Physically, since the input temperature is a manipulable variable, it should not be affected by other variables. The second input B/F ratio is mainly related to adjustments for the chip level. Since these two inputs are adjusted for different purposes, they need not be dependent on each other. Therefore, the two series X_{1t} and X_{2t} can be modeled separately by

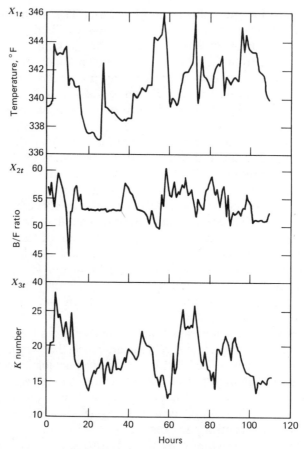

Figure 11.7 Pulping digester data.

ARMA models. However, for illustrative purposes, we will ignore this information and use the extended ARMA or the special formulation of the ARMAV models for X_{1t} and X_{2t}. If indeed these series are independent of each other and there is no feedback, the models should reduce to ARMA.

Determination of Dead Time or Initial Delay

As indicated in Step 1 of the modeling procedure in Section 11.2.1, if the dead time or initial delay is present but its exact number of lags are unknown, they can be determined from inverse function coefficients, which can be estimated from parameters of a high order autoregressive model. These coefficients are needed for the computation of the initial values as discussed in Appendix A 11.1. As in the case of the ARMA(n,m) models discussed in Chapter 4, it usually suffices to compute $m+n$ coefficients.

However, when an initial delay or dead time L is present, the first L coefficients are theoretically zero and it is necessary to compute at least $L + m + n$ coefficients to get reasonably good estimates of L and the initial values of parameters.

When L is unknown, the quantity $L + m + n$ cannot be determined exactly. One way to overcome this difficulty is to plot the parameter estimates of the autoregressive models until a stable picture of the dead time emerges. Once it is seen that the estimate of the initial delay or dead time does not change even after increasing the order of the model, one can stop, choose the value of L, and compute the initial values from the autoregressive model.

Theoretically, the parameters of an autoregressive model of a large enough order should be zero up to the lag L. However, the estimated values will not be exactly zero, but they can be considered to be zero if their confidence intervals include zero. Note that in the case of autoregressive models, the estimates as well as their confidence intervals can be computed by linear least squares explicitly and no recursive routine is required. We can plot the estimates and their confidence intervals for increasing autoregressive order until a stable pattern emerges. Then, we take L as the smallest value for which the confidence interval does not include zero. A possible mistake in the determination of L can be corrected in the final estimation and checked by the F-test along with the other small parameters whose confidence intervals include zero (see Step 3 of the modeling procedure in Section 11.2.1).

Since the dead time is most important for the output series, consider the K number X_{3t} first. The estimates $\hat{\phi}_{31k}$ and $\hat{\phi}_{32k}$, $k = 1, 2, \ldots$ as denoted by crosses and their 95% confidence intervals are shown in Figs. 11.8a and 11.8b for the models of orders 7, 8, and 9. These plots indicate that the dead time between X_{3t} and X_{1t} appears to be six and between X_{3t} and X_{2t} is one. Note that X_{3t} is an output variable and therefore must have at least one lag with X_{1t} and X_{2t}; therefore, ϕ_{310} or ϕ_{320} are neither estimated nor plotted in Figs. 11.8a and 11.8b.

Next consider the input temperature X_{1t}. Being a manipulated variable, it may depend upon both X_{2t} and X_{3t} at the same time. Therefore, X_{1t} may have zero lag with X_{2t} and X_{3t}, and the model parameters should start from ϕ_{120} and ϕ_{130}. The estimated values and 95% confidence intervals of the parameters ϕ_{12k} and ϕ_{13k} for the models or order 7, 8, and 9 are plotted in Figs. 11.8c and 11.8d. It is seen that the confidence intervals of $\hat{\phi}_{120}$ and $\hat{\phi}_{130}$, $\hat{\phi}_{121}$ and $\hat{\phi}_{131}$, etc. include zero. Hence, from Figs. 11.8c and 11.8d and from the physical conditions of the process, it is assumed that there is practically no dependence of X_{1t} on X_{2t} and X_{3t}, and the series X_{1t} is modeled by itself as an ARMA model.

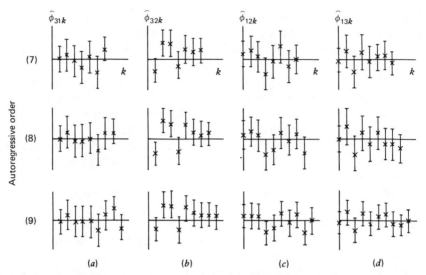

Figure 11.8(a–d) Autoregressive parameter estimates and 95% confidence intervals for the model of output series X_{3t} and input series X_{1t}.

Finally, for the second input X_{2t}, the plots of $\hat{\phi}_{21k}$ and $\hat{\phi}_{23k}$ together with their confidence intervals are shown in Figs. 11.8e and 11.8f. Again, in accordance with the model (11.1.10), $\hat{\phi}_{21k}$ starts from $k = 1$ and $\hat{\phi}_{23k}$ from $k = 0$, which is physically reasonable. From the plot of the autoregressive model of order 9 in Fig. 11.8e, we take the dead time between X_{2t} and X_{1t} as two. On the other hand, Fig. 11.8f shows that the confidence interval of $\hat{\phi}_{230}$ itself does not include zero. Therefore, there is no lag between X_{2t} and X_{3t}, and hence the model should include the parameter $\hat{\phi}_{230}$.

The tentative dead times may be summarized as follows:

$$X_{3t} - X_{1t} : \quad 6.$$
$$X_{3t} - X_{2t} : \quad 1.$$
$$X_{2t} - X_{1t} : \quad 2.$$
$$X_{2t} - X_{3t} : \quad 0.$$

X_{1t}—no dependence on X_{2t} and X_{3t}.

These dead times will either be confirmed or rejected in the final estimation. As explained earlier, the physical evidence indicates that not only X_{1t} but also X_{2t} should not depend on other series. We will, however, retain the dependence of X_{2t} on X_{1t} and X_{3t}, with lags two and zero obtained

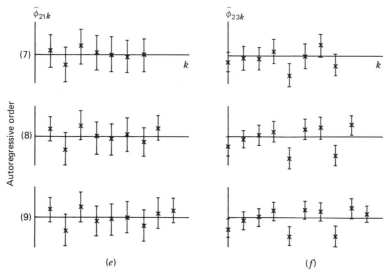

Figure 11.8(e–f)   Autoregressive parameter estimates and 95% confidence intervals for the X_{2t} series.

above, to illustrate how the errors in initial lag estimation are corrected in the final estimation and, also, how ARMA models emerge as special cases of the extended ARMA or the ARMAV models.

Modeling

The results of modeling the X_{3t} series with the above dead times are shown in Table 11.3. The facts that for the adequate (2,1) model the values of ϕ_{316} and ϕ_{321} are not very small, and that their confidence intervals do not include zero, confirm the choice of six lags for X_{3t} with X_{1t} and one lag for X_{3t} with X_{2t}. The total time for the digester process is about eight hours; hence, the six-hour dead time between the temperature manipulation and the resultant K number of the output pulp is physically reasonable.

Since X_{1t} does not depend upon the other series, it can be modeled by the ARMA methods of Chapter 4 with the results shown in Table 11.4. The adequate model turns out to be AR(1). The adequacy of the AR(1) model is also confirmed by the confidence intervals for ARMA(2,1) and ARMA(3,2) models in addition to the F-values.

Table 11.5 gives the results of modeling the B/F ratio X_{2t}. It is seen that all the parameters, except ϕ_{221} for the (1,0) model and ϕ_{222} and θ_{221} for the (2,1) model, are small and their confidence intervals include zero. This indicates that the series X_{2t} does not depend upon X_{1t} and X_{3t}, as conjectured by the physical evidence earlier. Hence, using the method of

Table 11.3   Modeling Paper Pulping Digester Output—K Number X_{3t}

Parameters	Model Order		
	(1,0)	(2,1)	(3,2)
ϕ_{316}	-0.105 ± 0.161	-0.286 ± 0.220	-0.282 ± 0.205
ϕ_{317}		0.264 ± 0.221	-0.056 ± 0.216
ϕ_{318}			0.237 ± 0.220
ϕ_{321}	0.024 ± 0.135	-0.167 ± 0.149	-0.164 ± 0.157
ϕ_{322}		0.318 ± 0.160	0.180 ± 0.162
ϕ_{323}			0.325 ± 0.169
ϕ_{331}	0.764 ± 0.116	1.231 ± 0.251	-0.005 ± 0.308
ϕ_{332}		-0.356 ± 0.212	0.308 ± 0.288
ϕ_{333}			0.251 ± 0.238
θ_{331}		0.635 ± 0.256	-0.768 ± 0.335
θ_{332}			-0.430 ± 0.321
Residual sum of squares	326.9	248.1	230.1
F		8.178	1.935

The adequate model is (2,1).

Chapter 4, ARMA models are fitted to the secondary input X_{2t} alone, which gives ARMA(2,1) as the adequate model

$$X_{2t} = 0.11X_{2t-1} + 0.44X_{2t-2} + a_{2t} + 0.36a_{2t-1}, \quad (11.2.5)$$

$$\sigma_{a2}^2 = 4.01$$

As a final check of adequacy, the plots of the auto- and cross-correlations of a_{1t}, a_{2t}, and a_{3t} from the three adequate models in Tables 11.3 to 11.5 are shown in Fig. 11.9. $\hat{\rho}_{ijk}$ are plotted in their matrix order: $\hat{\rho}_{11k}$

Table 11.4   Modeling Paper Pulping Digester Main Input—Temperature X_{1t}

Parameters	Model Order		
	(1,0)	(2,1)	(3,2)
ϕ_{111}	0.784 ± 0.116	0.784 ± 1.036	0.124 ± 0.721
ϕ_{112}		0.067 ± 0.831	-0.105 ± 0.720
ϕ_{113}			0.562 ± 0.432
θ_{111}		0.144 ± 1.032	-0.538 ± 0.768
θ_{112}			-0.587 ± 0.646
Residual sum of squares	178.4	172.7	169.5
F		1.774	0.982

The adequate model is (1,0).

Table 11.5   Modeling Paper Pulping Digester Secondary Input—Blow to Feed Ratio X_{2t}

Parameters	Model Order	
	(1,0)	(2,1)
ϕ_{230}	-0.101 ± 0.207	-0.083 ± 0.231
ϕ_{211}	-0.014 ± 0.190	0.019 ± 0.277
ϕ_{212}		-0.011 ± 0.276
ϕ_{221}	0.557 ± 0.159	0.054 ± 0.582
ϕ_{222}		0.425 ± 0.309
ϕ_{231}	-0.029 ± 0.207	0.013 ± 0.191
ϕ_{232}		-0.031 ± 0.208
θ_{221}		-0.417 ± 0.645
Residual sum of squares	459.9	420.5
F		2.391

at the top, $\hat{\rho}_{21k}$ in the second row, and $\hat{\rho}_{31k}$, $\hat{\rho}_{32k}$, and $\hat{\rho}_{33k}$ in the third row. These plots show that a_{1t}, a_{2t}, and a_{3t} are neither autocorrelated nor cross-correlated. Therefore, the basic assumption for the special formulation of the ARMAV modeling is satisfied and the fitted models are entertained.

It should be noted that modeling all three series is necessary to check the justification for using the special form of the ARMAV models and to obtain a comprehensive knowledge of the system including feedback and the interactions between the inputs. However, if one knows by prior modeling or physical knowledge that the special form of the ARMAV model is justified and the model is to be used only for control purposes, it suffices to get the model for the output alone rather than models for all the input and output series. Furthermore, it will be seen in the next section that when the dead time of the additional input X_{2t}, which is not manipulated, is greater than or equal to the dead time of the manipulated input X_{1t}, the derivation of the optimal control equation does not require the models for the inputs.

In summary, the adequate models for the series X_{1t}, X_{2t}, and X_{3t} are listed below:

$$X_{1t} = 0.784X_{1t-1} + a_{1t}, \qquad \gamma_{a_{11}} = \sigma_{a_1}^2 = 1.64$$

$$X_{2t} = 0.11X_{2t-1} + 0.44X_{2t-2} + a_{2t} + 0.36a_{2t-1},$$

$$\gamma_{a_{22}} = \sigma_{a_2}^2 = 4.01 \tag{11.2.6}$$

$$X_{3t} = -0.286X_{1t-6} + 0.264X_{1t-7} - 0.167X_{2t-1} + 0.318X_{2t-2}$$

$$+ 1.231X_{3t-1} - 0.356X_{3t-2} + a_{3t} - 0.635a_{3t-1},$$

$$\gamma_{a_{33}} = 2.41$$

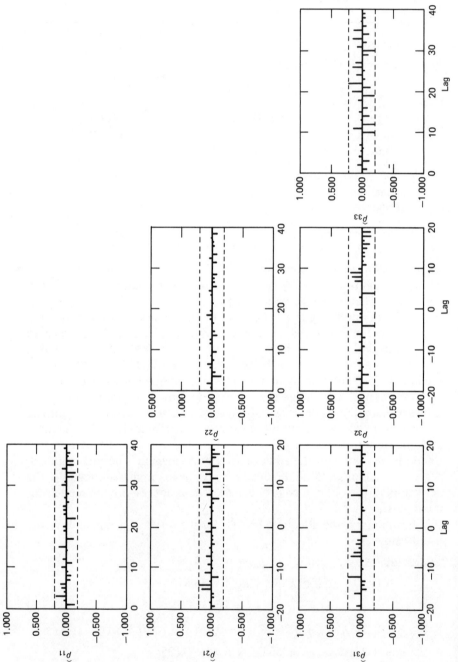

Figure 11.9 Sample autocorrelation functions (diagonal plots) and sample cross-correlation functions (off diagonal plots) of the residuals.

*Transformation to ARMAV

If we consider the results of the (2,1) model for illustrative purposes from Table 11.5 for the X_{2t} series, the ϕ_0 matrix can be written as

$$\phi_0 = \begin{bmatrix} 1 & 0 & 0 \\ 0 & 1 & -0.083 \\ 0 & 0 & 1 \end{bmatrix}$$

Thus, similar to the one input-one output models for papermaking data, a combined three series ARMAV model can be obtained from three separate models by back transformation using $\phi_1^* = \phi_0^{-1}\phi_1$, $\phi_2^* = \phi_0^{-1}\phi_2$, $\phi_6^* = \phi_0^{-1}\phi_6$ and $\theta_1^* = \phi_0^{-1}\theta_1$ from Eq. (11.1.7c):

$$\begin{bmatrix} X_{1t} \\ X_{2t} \\ X_{3t} \end{bmatrix} = \begin{bmatrix} 0.78 & 0 & 0 \\ 0.019 & 0.040 & 0.15 \\ 0 & -0.167 & 1.231 \end{bmatrix}\begin{bmatrix} X_{1t-1} \\ X_{2t-1} \\ X_{3t-1} \end{bmatrix}$$
$$+ \begin{bmatrix} 0 & 0 & 0 \\ -0.011 & 0.451 & -0.061 \\ 0 & 0.318 & -0.356 \end{bmatrix}\begin{bmatrix} X_{1t-2} \\ X_{2t-2} \\ X_{3t-2} \end{bmatrix} \tag{11.2.7}$$
$$+ \begin{bmatrix} 0 & 0 & 0 \\ -0.024 & 0 & 0 \\ -0.286 & 0 & 0 \end{bmatrix}\begin{bmatrix} X_{1t-6} \\ X_{2t-6} \\ X_{3t-6} \end{bmatrix} + \begin{bmatrix} 0 & 0 & 0 \\ 0.021 & 0 & 0 \\ 0.264 & 0 & 0 \end{bmatrix}\begin{bmatrix} X_{1t-7} \\ X_{2t-7} \\ X_{3t-7} \end{bmatrix}$$
$$+ \begin{bmatrix} a_{1t}^* \\ a_{2t}^* \\ a_{3t}^* \end{bmatrix} - \begin{bmatrix} 0 & 0 & 0 \\ 0 & -0.417 & 0.087 \\ 0 & 0 & 0.635 \end{bmatrix}\begin{bmatrix} a_{1t-1}^* \\ a_{2t-1}^* \\ a_{3t-1}^* \end{bmatrix}$$

with the variance-covariance matrix as

$$\gamma_a^* = \begin{bmatrix} 1.64 & 0 & 0 \\ 0 & 3.905 & 0.193 \\ 0 & 0.193 & 2.41 \end{bmatrix}$$

Note that because of the lag between X_{3t} and X_{1t}, the order of the complete ARMAV model is (7,1), whereas the order of the models for individual series when formulated as the special form of the ARMAV model are low. Consequently, the effort involved in estimating the parameters of the simplified form of the ARMAV model is far less than in the case of a complete ARMAV model. In fact, it is easy to see that for the ARMAV model, the number of parameters to be estimated increases considerably with the increase of the number of input-output series.

11.3 OPTIMAL CONTROL

Once the models are obtained from the data as discussed and illustrated in Section 11.2, they can be used for prediction or forecasting and control.

No additional concepts are involved in forecasting; the procedure developed in Chapter 5 can be applied in a straightforward manner. The forecasts are obtained by conditional expectation, and the forecast errors are obtained by means of the Green's function.

Equating the forecasts to zero provides us with the minimum mean squared error control equation. The output turns out to be the same as the forecast error, which has minimum variance as explained in Chapter 5. Hence, for a given model, manipulating the input following the control equation based on such forecasts yields control that is optimal in the sense of a minimum mean squared error.

11.3.1 Minimum Mean Squared Error Control Strategy

Let us now consider the problem of deriving the optimal control strategy for a given model. The goal of the control effort is to keep the output at the target value, which may be taken as zero, when the output is measured from the target value as the zero. However, because of the noise and disturbance in the system, it is seldom possible to maintain the output exactly at zero level. The best one can do is to keep the deviations or errors from the zero target values as small as possible. Since these deviations that form the output are random variables with zero mean, the measure of their smallness is given by their variance. Therefore, the optimal control is defined as those adjustments in the manipulable input values that yield minimum variance or minimum mean squared error of the output.

We already know from Chapter 5 that the minimum mean squared error forecast that yields minimum variance of the prediction error can be obtained by taking conditional expectation. Since there is always some lag $L \geq 1$ between the manipulated input and the output to be controlled, the earliest time at which the input X_{1t} can affect the output is $t + L$. Therefore, the minimum mean squared error control strategy is to adjust the input X_{1t} such that the forecast of X_{2t+L} made at time t, $\hat{X}_{2t}(L)$, is zero (or target value). Thus, the optimal control strategy may be simply written as

$$\hat{X}_{2t}(L) = 0 \qquad (11.3.1)$$

which may be rigorously proved using orthogonal decomposition [see Pandit (1973) or Pandit and Wu (1977)].

Since the observation X_{2t+L} is

$$X_{2t+L} = \hat{X}_{2t}(L) + e_{2t}(L)$$

it is clear that *after* implementing the control equation (11.3.1), the output is given by

$$X_{2t+L} = e_{2t}(L) \qquad (11.3.2)$$

which has the smallest variance that can be achieved based on observations at time t. This optimally controlled output, or output error or deviation from the mean (target value), is of course the same as the L step ahead forecast error, and by Chapter 3 [see Eq. (5.2.11)], *it has a MA(L−1) model.*

In this section, the development of the optimal control equation and an assessment of its effectiveness are presented in four convenient steps:

(i)   Obtain the output model before control.
(ii) Derive a control equation based on the minimum mean square error forecast.
(iii) Obtain the model after implementing the control equation.
(iv) Evaluate the efficiency of control.

For illustration purposes, four examples are chosen, namely: (i) The first order model with lag 1, (ii) the second order model with lag 1, (iii) the first order model with lag 2, and (iv) the second order model with lag 2. General formulae are given after the examples.

11.3.2 Illustrative Example 1—First Order Model with Lag 1

(i)   Model Before Control

As the simplest case, let us consider the model (11.1.1), which is in fact a first order model fitted to the papermaking data in Section 11.2, with rounded-off parameter values

$$X_{2t} = 0.25X_{1t-1} + 0.7X_{2t-1} + a_{2t} \tag{11.3.3}$$

with $\gamma_{a22} = 0.0062$.

(ii)   Control Equation

For deriving the control equation, we only need the above model for the output X_{2t}; the one for X_{1t} is not needed. Since the earliest input term on the right-hand side is X_{1t-1}, the lag $L = 1$. Hence, taking the one step ahead conditional expectation by the usual rules in Chapter 5 (now exjended to X_{1t} as well) and using the control strategy $\hat{X}_{2t}(L) = 0$, we get

$$\hat{X}_{2t}(1) = 0.25X_{1t} + 0.7X_{2t} = 0 \tag{11.3.4a}$$

that is,

$$0.25X_{1t} = -0.7X_{2t}$$

or

$$X_{1t} = -2.8X_{2t} \tag{11.3.4}$$

which gives the desired *control equation.*

(iii) Model after Control

Substituting the control equation (11.3.4a) in the model (11.3.3), we get the model for the output *after* implementing the control strategy as

$$X_{2t} = a_{2t} \qquad (11.3.5)$$

which is the optimally controlled output. This is the one step ahead forecast or control error as the lag L is one.

The control equation (11.3.4) can be readily explained in the context of the papermaking process if Eq. (11.3.3) is taken as its adequate model. This equation says that if the output basis weight increases to an amount X_{2t} from its target value, the gate opening should be set at $-2.8X_{2t}$. For example, if the target value were 38.97 lbs/3300 sq ft (50 g/m^2), and the observed output measured 39.07 lbs/3300 sq ft (50.5 g/m^2), the gate opening should be changed to read -0.28 (-1.4) on its scale, calibrated with zero as the value corresponding to the target output.

Since such a calibration may be difficult, the control strategy is often expressed in the difference form as

$$X_{1t} - X_{1t-1} = -2.8X_{2t} + 2.8X_{2t-1}$$

that is,

$$\nabla X_{1t} = -2.8\nabla X_{2t}$$

This equation states that if the output deviation *increases* by, say, 0.5, the input scale reading should be *reduced* by 1.4 from its preceding position. This is intuitively meaningful. Since the paper thickness increases, we have to reduce the gate opening so that less stock flows in and eventually reduces the thickness measured by the basis weight. If the thickness decreases, then the gate opening has to be enlarged as the equation indicates. (The equation can be readily implemented by the PID controller, see Appendix A 11.3.)

(iv) Control Efficiency

The equation for the optimally controlled output given by (11.3.5) shows that the output is now an independent series and cannot be further improved since it cannot be predicted at one lag. The minimum variance of such an optimally controlled output is $\gamma_{a22} = 0.0062$. The variance of X_{2t} data, that is, the variance of the output before control is $\gamma_{220} = 0.0134$ (Table 11.1). Hence, the optimal control strategy can reduce the variance by 53%. Actually, as we know from Section 11.2, the first order model is not adequate for the papermaking data, and therefore further improvement can be made in the controlled output by using higher order models.

11.3.3 Illustrative Example 2—Second Order Model with Lag 1

(i) Model Before Control

Let us now consider the second order model fitted to the papermaking data in Section 11.2, with rounded-off parameter values to illustrate how the control equation can be derived (Table 11.1).

$$X_{2t} = 0.28X_{1t-1} - 0.59X_{1t-2} + 1.45X_{2t-1}$$
$$- 0.52X_{2t-2} - 1.13a_{2t-1} + a_{2t} \quad (11.3.6)$$

with $\gamma_{a22} = 0.0041$.

(ii) Control Equation

Since $L = 1$, the control equation is

$$\hat{X}_{2t}(1) = 0.28X_{1t} - 0.59X_{1t-1} + 1.45X_{2t}$$
$$- 0.52X_{2t-1} - 1.13a_{2t} = 0 \quad (11.3.7a)$$

that is,

$$0.28X_{1t} = -1.45X_{2t} + 0.59X_{1t-1} + 0.52X_{2t-1} + 1.13a_{2t} \quad (11.3.7)$$

The control equation (11.3.7) cannot be directly implemented since it involves a_{2t} that is unknown. However, this problem can be solved after determining the equation of the controlled output.

(iii) Model After Control

Substituting (11.3.7a) in (11.3.6), we get the equation for the *optimally controlled output* as

$$X_{2t} = a_{2t} \quad (11.3.8)$$

Now, if the a_{2t} in Eq. (11.3.7) is replaced by using Eq. (11.3.8), then the *control equation* becomes

$$0.28X_{1t} = -0.32X_{2t} + 0.59X_{1t-1} + 0.52X_{2t-1}$$

that is,

$$X_{1t} = -1.14X_{2t} + 2.11X_{1t-1} + 1.86X_{2t-1}$$

or in the difference form

$$\nabla X_{1t} = -1.14\nabla X_{2t} + 2.11\nabla X_{1t-1} + 1.86\nabla X_{2t-1}$$

(iv) Control Efficiency

The variance of the optimally controlled output is now $\gamma_{a22} = 0.0041$ and comparing with the variance before control $\gamma_{220} = 0.0134$, we see that the control equation reduces the variance by about 69%. This is 16% more than that given by the first order model in Section 11.3.2.

11.3.4 Illustrative Example 3—First Order Model with Lag 2

(i)   Model

Derivation of the control equation becomes complicated when the dead time or initial delay L is greater than one. To illustrate its effect, let us again consider the first order model (11.3.3) but with a dead time 2 instead of one so that the model is

$$X_{2t} = 0.25X_{1t-2} + 0.7X_{2t-1} + a_{2t} \qquad (11.3.9)$$

(ii)   Control Equation

Since the dead time is two, the control equation is

$$\hat{X}_{2t}(2) = 0$$
$$= 0.25X_{1t} + 0.7\hat{X}_{2t}(1)$$

that is,

$$0.25X_{1t} = -0.7\hat{X}_{2t}(1) \qquad (11.3.10a)$$

where $\hat{X}_{2t}(1)$ is the one step ahead forecast of the *optimally controlled output*.

Unlike the case of initial delay $L=1$ (Section 11.3.2), the substitution of the control equation (11.3.10a) in Eq. (11.3.9) does not give the optimally controlled output in a straightforward manner. Nevertheless, from Eq. (11.3.9)

$$\hat{X}_{2t}(1) = 0.25X_{1t-1} + 0.7X_{2t} \qquad (11.3.11)$$

Hence, by substituting Eq. (11.3.11) into Eq. (11.3.10a),

$$0.25X_{1t} = -0.7(0.25X_{1t-1} + 0.7X_{2t})$$

and the *control equation* is

$$X_{1t} = -0.7X_{1t-1} - 1.96X_{2t} \qquad (11.3.10)$$

In the difference form, the control equation is

$$\nabla X_{1t} = -0.7\nabla X_{1t-1} - 1.96\nabla X_{2t}$$

(iii)   Model After Control

To see the optimally controlled output after the control Eq. (11.3.10) is implemented, we rewrite Eq. (11.3.9) as

$$(1 - 0.7B)X_{2t} = a_{2t} + 0.25X_{1t-2}$$

then

$$X_{2t} = \frac{1}{1 - 0.7B} a_{2t} + \frac{0.25}{1 - 0.7B} X_{1t-2} \qquad (11.3.12)$$

$$= (1 + 0.7B + 0.7^2B^2 + \ldots)a_{2t} + \frac{0.25}{1 - 0.7B} X_{1t-2}$$

The two step ahead observation X_{2t+2} is

$$X_{2t+2} = \underbrace{a_{2t+2} + 0.7a_{2t+1}}_{} + 0.7^2a_{2t} + 0.7^3a_{2t-1} + \dots$$

$$e_{2t}(2) \qquad + \underbrace{\frac{0.25}{1 - 0.7B} X_{1t}}_{}$$

$$\hat{X}_{2t}(2)$$

or

$$X_{2t+2} = e_{2t}(2) + \hat{X}_{2t}(2)$$

Since from the control equation the two step ahead forecast $\hat{X}_{2t}(2) = 0$, the *optimally controlled output* is the two step ahead forecast error

$$X_{2t+2} = e_{2t}(2)$$
$$= a_{2t+2} + 0.7a_{2t+1}$$

that is,

$$X_{2t} = (1 + 0.7B)a_{2t} \qquad (11.3.13)$$

Therefore, the output after control is a MA(1) model.

(iv) Control Efficiency

The variance of the optimally controlled output X_{2t} given by Eq. (11.3.13) can be computed as

$$V[X_{2t}] = \text{Var}[a_{2t} + 0.7a_{2t-1}]$$
$$= (1 + 0.7^2)\gamma_{a22}$$
$$= 1.49\gamma_{a22}$$

The output variance is, thus, increased by 49% for the dead time $L=2$ compared with the $L=1$ case. It will be shown later that as L increases, the output variance increases and reaches to the same as that without any control.

11.3.5 Illustrative Example 4—Second Order Model with Lag 2

(i) Model Before Control

The control equation of the first order model with lag 2 can be obtained by successive substitution. However, when the model includes the moving average terms and has a lag larger than one, a simple substitution will not work. It then requires a combination of the approaches illustrated in examples 2 and 3 to get the control equation. For illustration, let

$$X_{2t} = 0.28X_{1t-2} - 0.59X_{1t-3} + 1.45X_{2t-1}$$
$$- 0.52X_{2t-2} - 1.13a_{2t-1} + a_{2t} \qquad (11.3.14)$$

(ii) Control Equation

As before, the control equation is

$$\hat{X}_{2t}(2) = 0 \tag{11.3.15a}$$
$$= 0.28X_{1t} - 0.59X_{1t-1} + 1.45\hat{X}_{2t}(1) - 0.52X_{2t}$$

If $\hat{X}_{2t}(1)$ is substituted into Eq. (11.3.15a), we get

$$0.28X_{1t} - 0.59X_{1t-1} + 1.45(0.28X_{1t-1} - 0.59X_{1t-2}$$
$$+ 1.45X_{2t} - 0.52X_{2t-1} - 1.13a_{2t}) - 0.52X_{2t} = 0$$

The above control equation involves an a_{2t} term that cannot be implemented directly. Similar to Example 2, we need to represent a_{2t} in terms of the observation X_{2t}.

(iii) Model After Control

Similar to Eq. (11.3.12), the output X_{2t} can be written as

$$X_{2t} = \frac{1 - 1.13B}{1 - 1.45B + 0.52B^2} a_{2t} + \frac{0.28 - 0.59B}{1 - 1.45B + 0.52B^2} X_{1t-2}$$
$$= (a_{2t} + G_1a_{2t-1} + G_2a_{2t-2} + \ldots)$$
$$+ \frac{0.28 - 0.59B}{1 - 1.45B + 0.52B^2} X_{1t-2}$$

Thus,

$$X_{2t+2} = \underbrace{a_{2t+2} + G_1a_{2t+1}}_{} + \underbrace{G_2a_{2t} + \ldots + \frac{0.28 - 0.59B}{1 - 1.45B + 0.52B^2} X_{1t}}_{}$$
$$= \qquad e_{2t}(2) \qquad + \qquad\qquad \hat{X}_{2t}(2)$$

where

$$G_1 = \phi_1 - \theta_1$$
$$= 1.45 - 1.13 = 0.32$$
$$G_2 = \phi_1 G_1 + \phi_2 = 1.45(0.32) - 0.52$$
$$= -0.056$$

etc. Since $\hat{X}_{2t}(2) = 0$ by the control strategy, the output after implementing the control is $e_{2t}(2)$, that is,

$$X_{2t} = (1 + G_1B)a_{2t} \tag{11.3.16}$$
$$= (1 + 0.32B)a_{2t}$$

This confirms the previous conclusion that after control, the output becomes a MA(1) model.

From Eq. (11.3.16)

$$a_{2t} = \frac{1}{(1 + 0.32B)} X_{2t}$$

and

$$\hat{X}_{2t}(1) = 0.32a_{2t}$$

$$= 0.32 \cdot \frac{1}{(1 + 0.32B)} X_{2t}$$

$$= \frac{0.32}{(1 + 0.32B)} X_{2t} \qquad (11.3.17)$$

Now, substituting Eq. (11.3.17) into Eq. (11.3.15a), the *control equation* becomes

$$0.28X_{1t} - 0.59X_{1t-1} + 1.45 \cdot \frac{0.32}{1 + 0.32B} X_{2t} - 0.52X_{2t} = 0$$

or

$$(1 + 0.32B)(0.28 - 0.59B)X_{1t} = -1.45(0.32)X_{2t} + 0.52(1 + 0.32B)X_{2t}$$

After simplification

$$X_{1t} = 1.87X_{1t-1} - 1.06X_{2t} + 0.59X_{2t-1} \qquad (11.3.15)$$

(iv)   Control Efficiency

The variance of the optimally controlled output X_{2t} can be computed from Eq. (11.3.16) as

$$\mathrm{Var}[X_{2t}] = \mathrm{Var}[a_{2t} + G_1 a_{2t-1}] = (1 + G_2^2)\gamma_{a22} = 1.1\gamma_{a22}$$

11.3.6 Effect of Large Lag

It should be noted that the model of the output after implementing the control equation is $\mathrm{MA}(L-1)$ if the lag is L, that is,

$$X_{2t} = (1 + G_1 B + G_2 B^2 + \ldots G_{L-1}B^{L-1})a_{2t} \qquad (11.3.18)$$

and

$$\mathrm{Var}[X_{2t}] = (1 + G_1^2 + G_2^2 + \ldots G_{L-1}^2)\gamma_{a22} \qquad (11.3.19)$$

As L increases, the output variance increases, which is intuitively clear since it takes a longer time for the manipulated input to affect the output. Moreover as $L \to \infty$, the right-hand side of Eq. (11.3.18) becomes the same as the noise in the output before applying the control strategy, with variance $\sum_{j=0}^{\infty} G_j^2 \gamma_{a22}$. Thus, Eq. (11.3.19) shows that in the limit as $L \to \infty$, the variance of the output series after control will tend to one without control.

Hence, for large lags the optimal control is practically the same as no control. For the papermaking process, this says that the longer it takes to realize the effect of change in the stock flow on the output basis weight, the larger the variance of the optimally controlled output will be. In other words, the optimal control will deteriorate as the lag L increases. This is understandable since nothing can be done about the disturbances that come in the lag time. In fact, it is these disturbances that form the minimum mean squared error of prediction or control and inevitably appear as the output. No further improvement is possible unless the system design is altered to change the lag and/or model form including parameter values.

11.3.7 General Results: One-Input One-Output System

The control equation, the optimally controlled output and its variance can now be readily derived for the general model with an initial lag or dead time L using the following steps:

Step 1

Obtain the L step ahead forecast for the output using the conditional expectation procedure as developed in Chapter 5 and equate it to zero. This gives the basis for developing the optimal control equation.

Step 2

Find the optimally controlled output by obtaining the L step ahead forecast error $e_t(L)$ from the orthogonal decomposition of the model using the method of Chapter 5. In essence, this is a MA$(L-1)$ model.

Step 3

Solve for the optimal control equation in terms of input and output alone by using the optimally controlled output model and express the control strategy in the difference form.

Step 4

Obtain the variance of the optimally controlled output.

The output model for one input-one output system with initial lag or dead time L is

$$X_{2t} = \phi_{21L}X_{1t-L} + \phi_{21L+1}X_{1t-L-1} + \ldots + \phi_{21n}X_{1t-n}$$
$$\phi_{221}X_{2t-1} + \phi_{222}X_{2t-2} + \ldots + \phi_{22n}X_{2t-n} \qquad (11.3.20)$$
$$- \theta_{221}a_{2t-1} - \theta_{222}a_{2t-2} - \ldots - \theta_{22m}a_{2t-m} + a_{2t}$$
$$n \geq L > 0$$

where n and m represent the autoregressive and moving average order of the model.

The control equation is obtained by the strategy

$$\hat{X}_{2t}(L) = 0$$

as

$$
\begin{aligned}
\hat{X}_{2t}(L) = {} & \phi_{21L}X_{1t} + \phi_{21L+1}X_{1t-1} \\
& + \ldots + \phi_{21n}X_{1t+L-n} \\
& + \phi_{221}\hat{X}_{2t}(L-1) + \phi_{222}\hat{X}_{2t}(L-2) \\
& + \ldots + \phi_{22n}X_{2t+L-n} \qquad\qquad (11.3.21a) \\
& - \theta_{22L}a_{2t} - \theta_{22L+1}a_{2t-1} \\
& - \ldots - \theta_{22m}a_{2t+L-m} \\
= {} & 0
\end{aligned}
$$

with the a_{2t} terms absent when $L > m$.

The optimally controlled output after implementing this control equation is

$$X_{2t} = (1 + G_1B + G_2B^2 + \ldots + G_{L-1}B^{L-1})a_{2t} \qquad (11.3.22)$$

where G_j is the Green's function of the ARMA(n,m) model for X_{2t} in (11.3.20) and can be computed by comparing the coefficients of equal powers of B in (See Section 3.1.5)

$$
\begin{aligned}
(1 - \phi_{221}B - \phi_{222}B^2 - \ldots - {} & \phi_{22n}B^n)(G_0 + G_1B + G_2B^2 \\
+ \ldots) \equiv {} & (1 - \theta_{221}B - \theta_{222}B^2 - \ldots - \theta_{22m}B^m) \qquad (11.3.23)
\end{aligned}
$$

The control equation (11.3.21a) can be expressed in terms of X_{1t} and X_{2t} alone by using the output model (11.3.22). From Eq. (11.3.22)

$$
\begin{aligned}
\hat{X}_{2t}(L-j) &= \left(\sum_{i=1}^{j} G_{L-i}B^{j-i} \right) a_{2t} \\
&= \frac{\left(\sum_{i=1}^{j} G_{L-i}B^{j-i} \right)}{\left(\sum_{k=0}^{L-1} G_kB^k \right)} X_{2t}, \quad j = 1,2,\ldots,L-1, \\
& \qquad\qquad\qquad\qquad\qquad\qquad (11.3.24)
\end{aligned}
$$

and substituting Eqs. (11.3.22) and (11.3.24) in Eq. (11.3.21a), we have the control equation

$$\left(1 + \frac{\phi_{21L+1}}{\phi_{21L}} B + \frac{\phi_{21L+2}}{\phi_{21L}} B^2 + \ldots + \frac{\phi_{21n}}{\phi_{21L}} B^{n-L}\right)(1 + G_1 B$$

$$+ \ldots + G_{L-1}B^{L-1})X_{1t} = -\sum_{j=1}^{L-1} \sum_{i=1}^{j} \frac{\phi_{22j}}{\phi_{21L}} (G_{L-i}B^{j-i})X_{2t} \quad (11.3.21)$$

$$- \sum_{k=0}^{n-L} \frac{\phi_{22L+k}}{\phi_{21L}} (1 + G_1 B + \ldots + G_{L-1}B^{L-1})X_{2t-k}$$

$$+ \sum_{i=0}^{m-L} \frac{\theta_{22L+i}}{\phi_{21L}} X_{2t-i}$$

It is left as an exercise for the reader to verify that the results of Sections 11.3.2 to 11.3.6 can be obtained from the above general results by proper substitutions.

The variance of the optimally controlled output is obtained from Eq. (11.3.22) as

$$V[X_{2t}] = \text{Var}\{[1 + G_1 B + G_2 B^2 + \ldots + G_{L-1}B^{L-1}]a_{2t}\} \quad (11.3.25)$$
$$= [1 + G_1^2 + G_2^2 + \ldots + G_{L-1}^2]\gamma_{a22}$$

Only in exceptionally simple cases, such as the one in Section 11.3.2, is the explicit form of the control equation (11.3.21) simple and easily implemented by means of a nomogram. For most practical cases, however, the explicit form (11.3.21) becomes very cumbersome and it is much easier and computationally more efficient to use the form (11.3.21a) together with the recursive relations for the conditional forecasts required therein. These forecasts and the a_{2t}'s in Eq. (11.3.21a) can be obtained from Eq. (11.3.22) by the usual rules of conditional expectation. Thus,

$$\hat{X}_{2t}(\ell) = G_\ell a_{2t} + G_{\ell+1}a_{2t-1} + \ldots + G_{L-1}a_{2t-L+\ell+1} \quad (11.3.26)$$
$$\ell = 1, 2, \ldots, L-1$$

where

$$a_{2t} = X_{2t} - G_1 a_{2t-1} - \ldots - G_{L-1}a_{2t-L+1}$$

Equations (11.3.21a) and (11.3.26) provide a complete recursive sequence for optimal control, which is computationally equivalent to the explicit equations (11.3.21) but is easier to use.

11.3.8 Improved Control by Additional Inputs

We have seen in the last section that when the initial delay or dead time of the manipulated input variable is large, the optimal control strategy cannot be sufficiently effective. Directly reducing the lag or dead time by the alteration of the system design is often very difficult in practice. However, this can be indirectly achieved by identifying additional inputs, measuring them, and then improving the control by utilizing these measurements, especially when the lag of the manipulated variable is larger than this additional input. In this section, we will derive the optimal control equation in such a case for which there are two inputs: one that is measured and manipulated and another that is measured but not manipulated.

Such an added input aimed at improving the control was already encountered in the digester applications of Section 11.2.3. Let us therefore consider the simple first order model fitted to this data from Table 11.3 for illustrative purposes only.

(i)   Model Before Control

$$X_{3t} = -0.10X_{1t-6} + 0.02X_{2t-1} + 0.76X_{3t-1} + a_{3t} \quad (11.3.27)$$

with $\gamma_{a_{33}} = (327/109) = 3.0$

(ii)   Control Equation

In Eq. (11.3.27) $L=6$, and taking the conditional expectation gives the optimal control equation

$$\hat{X}_{3t}(6) = 0 = -0.10X_{1t} + 0.02\hat{X}_{2t}(5) + 0.76\hat{X}_{3t}(5) \quad (11.3.28a)$$

(iii)   Model After Control

To find $\hat{X}_{3t}(5)$ we have to first find the optimally controlled output, which is the same as the six step ahead forecast error $e_{3t}(6)$. This error is most easily obtained from the Green's function form of the model (11.3.27), written for X_{3t+6} as

$$X_{3t+6} = \frac{-0.10}{(1-0.76B)}X_{1t} + \frac{0.02}{(1-0.76B)}X_{2t+5}$$

$$+ \frac{1}{(1-0.76B)}a_{3t+6} \quad (11.3.29)$$

Before we take the conditional expectation on this form to get $e_{3t}(6)$, we have to substitute X_{2t+5} in terms of a_{2t} using the adequate ARMA model for X_{2t} from Eq. (11.2.5) as

$$(1-0.11B-0.44B^2)X_{2t} = (1+0.36B)a_{2t} \quad (11.3.30)$$

with $\gamma_{a_{22}} = \sigma_{a_2}^2 = 4.01$.

Substituting in (11.3.29), we can get the Green's function expansion as

$$X_{3t+6} = -0.10 \sum_{j=0}^{\infty} 0.76^j X_{1t-j} + \sum_{j=1}^{\infty} G_{32j}a_{2t+6-j}$$

$$+ \sum_{j=0}^{\infty} G_{33j}a_{3t+6-j} \quad (11.3.29a)$$

and hence,

$$e_{3t}(6) = \sum_{j=1}^{5} G_{32j}a_{2t+6-j} + \sum_{j=0}^{5} G_{33j}a_{3t+6-j} \qquad (11.3.31)$$

where G_{32j} is given by

$$\frac{0.02}{(1-0.76B)} X_{2t-1} = \frac{0.02(1+0.36B)}{(1-0.76B)(1-0.11B-0.44B^2)} a_{2t-1}$$

$$= \left(\sum_{j=1}^{\infty} G_{32j}B^j \right) a_{2t}$$

and hence can be computed by equating the coefficients in

$$(G_{321} + G_{322}B + G_{323}B^2 + \ldots)(1-0.76B) \qquad (11.3.32)$$
$$(1-0.11B-0.44B^2) \equiv 0.02(1+0.36B)$$

and

$$G_{33j} = 0.76^j, \qquad j = 0, 1, 2, \ldots \qquad (11.3.33)$$

It follows from Eq. (11.3.32) that

$G_{321} = 0.02$

$G_{322} = 0.02(0.76 + 0.11 + 0.36) = 0.025$

$G_{323} = 0.02(-0.76 \times 0.11 + 0.44) + (0.76 + 0.11)0.025$

$\quad\;\; = 0.029$

$G_{324} = 0.029(0.76 + 0.11) + 0.025(-0.76 \times 0.11 + 0.44)$

$\quad\;\;\; - 0.02 \times 0.76 \times 0.44 = 0.028$

$G_{325} = 0.028(0.76 + 0.11) + 0.029(-0.76 \times 0.11 + 0.44)$

$\quad\;\;\; - 0.025 \times 0.76 \times 0.44 = 0.026$

Hence, after implementing the control equation, the optimally controlled output, which is the same as $e_{3t}(6)$, may be written as

$$X_{3t+6} = 0.02a_{2t+5} + 0.025a_{2t+4} + 0.029a_{2t+3}$$
$$+ 0.028a_{2t+2} + 0.026a_{2t+1} + \sum_{j=0}^{5}(0.76)^j a_{3t+6-j} \qquad (11.3.31a)$$

The terms $\hat{X}_{2t}(5)$ and $\hat{X}_{3t}(5)$ in the optimal control equation (11.3.28a) may now be obtained from (11.3.30) and (11.3.31a). Thus, the optimal control equation takes the form

$$X_{1t} = 0.2\hat{X}_{2t}(5) + 7.6\hat{X}_{3t}(5) \qquad (11.3.28)$$

where

$$\hat{X}_{2t}(\ell) = 0.11\hat{X}_{2t}(\ell-1) + 0.44\hat{X}_{2t}(\ell-2), \quad \ell = 5, 4, 3$$

$$\hat{X}_{2t}(2) = 0.11\hat{X}_{2t}(1) + 0.44X_{2t}$$

$$\hat{X}_{2t}(1) = 0.11X_{2t} + 0.44X_{2t-1} + 0.36a_{2t}$$

$$a_{2t} = X_{2t} - 0.11X_{2t-1} - 0.44X_{2t-2} - 0.36a_{2t-1}$$

and

$$\hat{X}_{3t}(5) = 0.026a_{2t} + (0.76)^5 a_{3t}$$

$$a_{3t} = X_{3t} - 0.02a_{2t-1} - 0.025a_{2t-2} - 0.029a_{2t-3}$$

$$- 0.028a_{2t-4} - 0.026a_{2t-5} - \sum_{j=1}^{5} (0.76)^j a_{3t-j}$$

By carrying out the algebra similar to the one used in Section 11.3.7 in deriving Eq. (11.3.21), we can express the control equation (11.3.28) completely in terms of the present and the past values of the output X_{3t} and the inputs X_{1t} and X_{2t}. However, the resulting equation is quite cumbersome. On the other hand, the recursive relations given above are computationally equivalent, but much easier to deal with. These equations completely specify the amount of optimally controlled adjustment. If necessary, Eq. (11.3.28) can be expressed in the difference ∇X_{1t} by applying the operator ∇ on both sides as illustrated before in Sections 11.3.2 to 11.3.4.

(iv) Control Efficiency

It follows from Eq. (11.3.31a) that the (minimum) variance of the optimally controlled output is

$$\text{Var}(X_{3t}) = (0.02^2 + 0.025^2 + 0.029^2 + 0.028^2 + 0.026^2)\gamma_{a_{22}}$$

$$+ \left(\sum_{j=0}^{5} (0.76)^{2j} \right)\gamma_{a_{33}}$$

$$= 0.00332 \times 4.01 + 2.27951 \times 3 = 6.85$$

since a_{1t} and a_{2t} are uncorrelated for an adequate model. The variance before control is 9.72; therefore, the optimal control reduces the variance by

$$\frac{9.81 - 6.85}{9.81} \times 100 = 30.17\%$$

Note that since the model is not actually adequate the control is not really optimal. In fact, the small coefficient 0.02 of the secondary input X_{2t} in Eq. (11.3.27) shows that it contributes very little to control, which is also verified by the calculated G_{32j} and its contribution to the variance above.

11.3.9 Optimal Control of a Two-Input One-Output Paper Pulping Digester

The simple first order model was chosen in Section 11.3.8 for illustrative purposes. The adequate model for the digester was found to be (2,1) that may be written from Table 11.3 or Eq. (11.2.6) as

$$X_{3t} = -0.29X_{1t-6} + 0.26X_{1t-7} - 0.17X_{2t-1} + 0.32X_{2t-2}$$
$$+ 1.23X_{3t-1} - 0.36X_{3t-2} + a_{3t} - 0.64a_{3t-1} \quad (11.3.34)$$

with $\sigma_{a3}^2 = \gamma_{a_{33}} = 2.41$.

Hence, the optimal control equation is

$$X_{3t}(6) = 0 = -0.29X_{1t} + 0.26X_{1t-1} - 0.17\hat{X}_{2t}(5)$$
$$+ 0.32\hat{X}_{2t}(4) + 1.23\hat{X}_{3t}(5) - 0.36\hat{X}_{3t}(4)$$

The optimal control scheme can now be specified as

$$X_{1t} = 0.90X_{1t-1} - 0.59\hat{X}_{2t}(5) + 1.10\hat{X}_{2t}(4) + 4.24\hat{X}_{3t}(5)$$
$$- 1.24\hat{X}_{3t}(4)$$
$$\hat{X}_{2t}(\ell) = 0.11\hat{X}_{2t}(\ell-1) + 0.44\hat{X}_{2t}(\ell-2), \quad \ell = 5, 4, 3$$
$$\hat{X}_{2t}(2) = 0.11\hat{X}_{2t}(1) + 0.44\hat{X}_{2t}$$
$$\hat{X}_{2t}(1) = 0.11X_{2t} + 0.44X_{2t-1} + 0.36a_{2t}$$
$$a_{2t} = X_{2t} - 0.11X_{2t-1} - 0.44X_{2t-2} - 0.36a_{2t-1}$$
$$\hat{X}_{3t}(5) = 0.358a_{2t} + 0.112a_{3t}$$
$$\hat{X}_{3t}(4) = 0.311a_{2t} + 0.371a_{2t-1} + 0.212a_{3t} + 0.163a_{3t-1}$$
$$a_{3t} = X_{3t} + 0.18a_{2t-1} - 0.049a_{2t-2} - 0.181a_{2t-3} - 0.311a_{2t-4}$$
$$- 0.371a_{2t-5} - 0.59a_{3t-1} - 0.393a_{3t-2} - 0.282a_{3t-3} -$$
$$0.212a_{3 \cdot -4} - 0.163a_{3t-5}$$

The (minimum variance) of the optimally controlled output is

$$\text{Var}(X_{3t}) = \sum_{j=1}^{5} G_{32j}^2 \sigma_{a2}^2 + \sum_{j=0}^{5} G_{33j}^2 \sigma_{a3}^2$$
$$= [(-0.17)^2 + 0.031^2 + 0.166^2 + 0.306^2 + 0.358^2]4.01$$
$$+ (1 + 0.59^2 + 0.366^2 + 0.238^2 + 0.161^2 + 0.112^2)2.41$$
$$= 0.2792 \times 4.01 + 1.5771 \times 2.41$$
$$= 4.92$$

and reduces the output variance by

$$\frac{9.81 - 4.92}{9.81} \times 100 = 49.85\%$$

*11.3.10 General Results: Optimal Control of Multi-Input One-Output System

We can now generalize the results of the last two sections and derive the optimal control equations for a multi-input–one-output system. To take into account the different lags for the inputs, let us generalize Eq. (11.2.6) for the output X_{pt} to the form

$$\Phi_{pp}(B)X_{pt} = -B^L\Phi_{p1}(B)X_{1t} - \sum_{j=2}^{q'} B^{\ell_j}\Phi_{pj}(B)X_{jt}$$

$$- \sum_{j=q'+1}^{p-1} B^{\ell_j}\Phi_{pj}(B)X_{jt} + H_{pp}(B)a_{pt} \quad (11.3.35)$$

where, $\ell_p = 0$, $\ell_1 = L$, and

$$\ell_p, \ell_{p-1}, \ldots, \ell_{q'+2}, \ell_{q'+1} < L \leq \ell_2, \ell_3, \ldots, \ell_{q'}$$

and $\Phi_{pj}(B)$, $H_{pp}(B)$ are suitable operators in B. X_{1t} is the manipulated input; $X_{2t}, X_{3t}, \ldots, X_{p-1t}$ are the inputs that are observed but not manipulated. Of these, $X_{2t}, X_{3t}, \ldots, X_{q't}$ have dead times ℓ_j greater than or equal to the main dead time $L = \ell_1$ of the manipulated input X_{1t}; and $X_{q'+1t}, X_{q'+2t}, \ldots, X_{p-1t}$ have dead times ℓ_j less than L.

Taking the conditional expectation of X_{pt+L} gives the optimal control equation

$$\hat{X}_{pt}(L) = 0 = -\Phi_{p1}(B)X_{1t} - \sum_{j=2}^{q'} \Phi_{pj}(B)X_{jt+L-\ell_j}$$

$$+ \sum_{j=q'+1}^{p-1} [\phi_{pj0}\hat{X}_{jt}(L-\ell_j) + \phi_{pj1}\hat{X}_{jt}(L-\ell_j-1)$$

$$+ \ldots + \phi_{pj(L-\ell_j-1)}\hat{X}_{jt}(1) + \phi_{pj(L-\ell_j)}X_{jt} \quad (11.3.36)$$

$$+ \phi_{pj(L-\ell_j+1)}X_{jt-1} + \ldots]$$

$$+ \phi_{pp1}\hat{X}_{pt}(L-1) + \phi_{pp2}\hat{X}_{pt}(L-2) + \ldots$$

$$- \theta_{ppL}a_{pt} - \theta_{ppL+1}a_{pt-1} - \ldots$$

The conditional expectations of X_{jt}, $j=q'+1, q'+2, \ldots, p-1$ required in the above equation can be obtained by applying the standard rules to input models that may be written in the operator form

$$\sum_{k=1}^{p} \Phi_{jk}(B)X_{kt} = H_{jj}(B)a_{jt} \quad (11.3.37)$$

$$j = q'+1, q'+2, \ldots, p-1$$

Note that, for convenience of notation, lags ℓ_{jk} of X_{kt} relative to X_{pt} are included in $\Phi_{jk}(B)$.

To find the conditional expectation of X_{pt}, we have to obtain the optimally controlled output. The Green's function form of Eq. (11.3.35) for this purpose may be written as

$$X_{pt+L} = -\Psi_1(B)X_{1t} - \sum_{k=2}^{q} B^{\ell_k - L}\Psi_k(B)X_{kt}$$

$$+ \sum_{k=q+1}^{p} \sum_{j=\ell_k}^{\infty} G_{pkj}a_{kt+L-j} \quad (11.3.38)$$

where ℓ_k is the smallest lag of a_{kt} relative to X_{pt} expressed in terms of a_{1t}, $a_{2t}, \ldots, a_{pt}$ alone as discussed later; q is such that $\ell_k < L$ for all $k > q$, that is, $\ell_p = 0, \ell_{p-1}, \ldots, \ell_{q+2}, \ell_{q+1} < L = \ell_1 \leq \ell_2, \ell_3, \ldots, \ell_q; \Psi_1(B)$, $\Psi_2(B), \ldots, \Psi_q(B)$ are possibly infinite B operators that we do not need explicitly since they remain unchanged under conditional expectation. Therefore, the optimally controlled output after implementing the control equation is

$$X_{pt} = e_{t-L}(L) = \sum_{k=q+1}^{p} \sum_{j=\ell_k}^{L-1} G_{pkj}a_{kt-j} \quad (11.3.39)$$

that can be used to recursively obtain the conditional expectations of X_{pt} in the optimal control equation (11.3.36).

The Green's function required in Eq. (11.3.39) can be found by solving Eqs. (11.3.35) and (11.3.37) as simultaneous equations treating $\Phi_{jk}(B)$ and $H_{jj}(B)$ as constant coefficients. Writing these equations together

$$\sum_{k=1}^{r} \Phi_{jk}(B)X_{kt} = H_{jj}(B)a_{jt}, \quad j = q+1, q+2, \ldots, p-1$$

$$B^L\Phi_{p1}(B)X_{1t} + \sum_{k=2}^{p-1} B^{\ell_k}\Phi_{pk}(B)X_{kt} + \Phi_{pp}(B)X_{pt} = H_{pp}(B)a_{pt}$$

by Cramer's rule, it can be shown that

$$\frac{\begin{vmatrix} \Phi_{(q+1)(q+1)}(B) & \Phi_{(q+1)(q+2)}(B) & \cdots & H_{(q+1)(q+1)}(B)a_{(q+1)t} \\ \Phi_{(q+2)(q+1)}(B) & \Phi_{(q+2)(q+2)}(B) & \cdots & H_{(q+2)(q+2)}(B)a_{(q+2)t} \\ & \cdots & \cdots & \cdots \\ B^{\ell_q+1}\Phi_{p(q+1)}(B) & B^{\ell_q+2}\Phi_{p(q+2)}(B) & \cdots & H_{pp}(B)a_{pt} \end{vmatrix}}{\begin{vmatrix} \Phi_{(q+1)(q+1)}(B) & \Phi_{(q+1)(q+2)}(B) & \cdots & \Phi_{(q+1)p}(B) \\ \Phi_{(q+2)(q+1)}(B) & \Phi_{(q+2)(q+2)}(B) & \cdots & \Phi_{(q+2)p}(B) \\ & \cdots & \cdots & \cdots \\ B^{\ell_q+1}\Phi_{p(q+1)}(B) & B^{\ell_q+2}\Phi_{p(q+2)}(B) & \cdots & \Phi_{pp}(B) \end{vmatrix}}$$

$$= \sum_{k=q+1}^{p} \left(\sum_{j=\ell_k}^{\infty} G_{pkj}B^j \right) a_{kt} \quad (11.3.40)$$

Equating equal powers of B for an a_{kt} in Eq. (11.3.40) gives the G_{pkj} required in Eq. (11.3.39). Once we know G_{pkj}, Eq. (11.3.39) yields the (minimum) variance of the optimally controlled output

$$\text{Var}(X_{pt}) = \sum_{k=q+1}^{p} \left(\sum_{j=\ell_k}^{L-1} G_{pkj}^2 \right) \sigma_{ak}^2 \qquad (11.3.41)$$

The choice of q is often clear from the context, as illustrated earlier and also in the special cases discussed below. In general, however, q can be determined by first obtaining ℓ_k from the operators in the left-hand side of Eq. (11.3.40) evaluated at $q = 0$. Then q is the index such that $\ell_k < L$ from all $k > q$. Note that (11.3.40) is equal to X_{pt} when $q = 0$. Some reordering of models may be necessary to use Eq. (11.3.40) for obtaining G_{pkj}.

Special Case: $p=2$

To illustrate the implications of the general equations derived above, consider their special case $p=2$ so that $q' = q=1$. The Eq. (11.3.35) takes the form

$$\Phi_{22}(B)X_{2t} = -B^L\Phi_{21}(B)X_{1t} + H_{22}(B)a_{2t} \qquad (11.3.42)$$

and Eq. (11.3.36) becomes

$$\Phi_{21}(B)X_{1t} = \phi_{221}\hat{X}_{2t}(L-1) + \phi_{222}\hat{X}_{2t}(L-2) + \cdots \qquad (11.3.43)$$
$$- \theta_{22L}a_{2t} - \theta_{22L+1}a_{2t-1} - \cdots$$

that are the same as Eqs. (11.3.20) and (11.3.21a). Since there is only one input, Eq. (11.3.37) is not needed to get the conditional expectations. Equation (11.3.38) can be obtained immediately from Eq. (11.3.42) as

$$X_{2t+L} = -\frac{\Phi_{21}(B)}{\Phi_{22}(B)} X_{1t} + \frac{H_{22}(B)}{\Phi_{22}(B)} a_{2t} \qquad (11.3.44)$$

Note that in this simple case $\Psi_1(B) = \Phi_{21}(B)\Phi_{22}^{-1}(B)$.
   The optimally controlled output follows from Eq. (11.3.39) as

$$X_{2t} = e_{t-L}(L) = \sum_{j=0}^{L-1} G_{22j}a_{2t-j} \qquad (11.3.45)$$

which is the same as Eq. (11.3.22). G_{22j} follows from Eq. (11.3.40) as

$$\frac{H_{22}(B)}{\Phi_{22}(B)} a_{2t} \equiv \left(\sum_{j=0}^{\infty} G_{22j}B^j \right) a_{2t} \qquad (11.3.46)$$

Equation (11.3.46) is also clear from Eq. (11.3.44) and is the operator equivalent of Eq. (11.3.23). The conditional expectation in Eq. (11.3.43) can now be obtained in a straightforward way from Eq. (11.3.45) with

$$a_{2t} = X_{2t} - \sum_{j=1}^{L-1} G_{22j}a_{2t-j}$$

and the control equations can be implemented recursively. This recursive procedure is, of course, equivalent to Eq. (11.3.21) as pointed out in Section 11.3.7.

Special Case: $p=3$

To illustrate the usefulness of Eq. (11.3.40), let us consider $p=3$. The Eq. (11.3.35) becomes

$$\Phi_{33}(B)X_{3t} = -B^L\Phi_{31}(B)X_{1t} - B^{\ell_2}\Phi_{32}(B)X_{2t} + H_{33}(B)a_{3t} \quad (11.3.47)$$

Case (i): $L \leq \ell_2$

Then $q'=2$ and the control equation follows from Eq. (11.3.36) as

$$\begin{aligned} X_{3t}(L) &= 0 \\ &= -\Phi_{31}(B)X_{1t} - \Phi_{32}(B)X_{2t+L-\ell_2} + \phi_{331}\hat{X}_{3t}(L \\ &\quad -1) + \phi_{332}\hat{X}_{3t}(L-2) + \ldots - \theta_{33L}a_{3t} \\ &\quad - \theta_{33L+1}a_{3t-1} - \cdots \end{aligned} \quad (11.3.48)$$

Since the conditional expectations of X_{2t} are not involved, Eq. (11.3.37) is again not required. The optimally controlled output follows from Eq. (11.3.39) as (assuming $q=2$)

$$X_{3t} = \sum_{j=0}^{L-1} G_{33j}a_{3t-j} \quad (11.3.49)$$

where, by Eq. (11.3.40)

$$\frac{H_{33}(B)}{\Phi_{33}(B)}a_{3t} = \left(\sum_{j=0}^{\infty} G_{33j}B^j\right)a_{3t} \quad (11.3.50)$$

Case (ii): $L > \ell_2$

Then $q' = q=1$ and the control equation follows from Eq. (11.3.36) as

$$\begin{aligned} X_{3t}(L) = 0 &= -\Phi_{31}(B)X_{1t} + \phi_{320}\hat{X}_{2t}(L-\ell_2) \\ &\quad + \phi_{321}\hat{X}_{2t}(L-\ell_2-1) \end{aligned}$$

$$+ \ldots + \phi_{32(L-\ell_2-1)}\hat{X}_{2t}(1)$$
$$+ \phi_{32(L-\ell_2)}X_{2t} + \ldots \tag{11.3.51}$$
$$+ \phi_{331}\hat{X}_{3t}(L-1) + \phi_{332}\hat{X}_{3t}(L-2)$$
$$+ \ldots + \phi_{33(L-1)}\hat{X}_{3t}(1)$$
$$+ \phi_{33L}X_{3t} + \ldots$$
$$- \theta_{33L}a_{3t} - \theta_{33L+1}a_{3t-1} - \ldots$$

The conditional forecasts of X_{2t} required in this equation can be obtained from Eq. (11.3.37) with $p=3$, that is,

$$\Phi_{21}(B)X_{1t} + \Phi_{22}(B)X_{2t} + \Phi_{23}(B)X_{3t} = H_{22}(B)a_{2t} \tag{11.3.52}$$

by the usual rules of conditional expectations.

The optimally controlled output is now obtained from Eq. (11.3.39)

$$X_{3t} = \sum_{j=\ell_2}^{L-1} G_{32j}a_{2t-j} + \sum_{j=0}^{L-1} G_{33j}a_{3t-j} \tag{11.3.53}$$

which can be used recursively to compute the conditional expectations of X_{3t} required in Eq. (11.3.51) together with a_{3t} and a_{2t} computed from Eqs. (11.3.53) and (11.3.52), respectively. Note that since the lag of the additional input X_{2t} is smaller than L, Eq. (11.3.52) *is* needed and the optimally controlled output contains a_{2t}, as opposed to $L \leq \ell_2$ case above, in which Eq. (11.3.49) does not contain a_{2t} and the Eq. (11.3.52) for X_{2t} is not required.

To compute G_{32j} and G_{33j}, we use Eq. (11.3.40) that takes the form

$$\frac{\begin{vmatrix} \Phi_{22}(B) & H_{22}(B)a_{2t} \\ B^{\ell_2}\Phi_{32}(B) & H_{33}(B)a_{3t} \end{vmatrix}}{\begin{vmatrix} \Phi_{22}(B) & \Phi_{23}(B) \\ B^{\ell_2}\Phi_{32}(B) & \Phi_{33}(B) \end{vmatrix}} = \sum_{j=\ell_2}^{\infty} G_{32j}a_{2t-j}$$
$$+ \sum_{j=0}^{\infty} G_{33j}a_{3t-j} \tag{11.3.54}$$

which reduces to the two operator identities for a_{3t} and a_{2t}

$$\Phi_{22}(B)H_{33}(B) \equiv [\Phi_{22}(B)\Phi_{33}(B)$$
$$- B^{\ell_2}\Phi_{32}(B)\Phi_{23}(B)] \sum_{j=0}^{\infty} G_{33j}B^j \tag{11.3.54a}$$

$$- B^{\ell_2}\Phi_{32}(B)H_{22}(B)$$

$$\equiv [\Phi_{22}(B)\Phi_{33}(B) - B^{\ell_2}\Phi_{32}(B)\Phi_{23}(B)] \sum_{j=\ell_2}^{\infty} G_{32j}B^j \quad (11.3.54b)$$

These identities can also be verified by the substitution of Eq. (11.3.52) in Eq. (11.3.47), which is quite simple when $\Phi_{23}(B)\equiv0$, that is, when the input X_{2t} has no feedback. It is left as an exercise to check that the results of Sections 11.3.8 and 11.3.9 then follow easily by proper substitution. When $\Phi_{23}(B)$ is not identically zero, then the substitution procedure becomes cumbersome even for $p=3$, and much more so for larger p; it is then that the usefulness of Eq. (11.3.40) becomes clear.

11.4 FORECASTING BY LEADING INDICATOR

In many business and economic systems, one is often more interested in forecasting than control. For a single time series, the procedure of forecasting as described in Chapter 5 can be used. However, in many cases, the forecast of a series of interest can be improved by using information from a related series. Such a related series is then called a "leading indicator." Treating the leading indicator series as X_{1t} or X_{it} and the desired series as X_{2t} or X_{pt}, we can obtain an extended ARMA (that is, a simplified ARMAV) model for them by the procedure outlined in Section 11.2. The procedure in this section can then be used for forecasting X_{2t} or X_{pt} by the leading indicator X_{1t} or X_{it}.

11.4.1 Forecasting Using Conditional Expectation

As we know, in an input-output system, the output series X_{2t} depends upon its own dynamics as well as the past values of the input series, that is, X_{1t-1}, X_{1t-2}, etc. For illustration purposes, let us consider an extended first order autoregressive model for the X_{2t} series

$$X_{2t} = \phi_{211}X_{1t-1} + \phi_{221}X_{2t-1} + a_{2t} \quad (11.4.1)$$

Then using the rules for the conditional expectation, we obtain the forecasts from the model as

$$\hat{X}_{2t}(1) = \phi_{211}X_{1t} + \phi_{221}X_{2t}$$
$$\hat{X}_{2t}(\ell) = \phi_{211}\hat{X}_{1t}(\ell-1) + \phi_{221}\hat{X}_{2t}(\ell-1), \quad \ell \geq 2 \quad (11.4.2)$$

Equation (11.4.2) shows that except for lead time $\ell=1$, the forecasts for the series X_{1t} are also needed to compute the forecasts of X_{2t}.

Case (i)

Let us consider X_{1t} as a white noise series; it then has the model

$$X_{1t} = a_{1t}$$

and

$$\hat{X}_{1t}(\ell) = 0, \qquad \ell \geqslant 1 \tag{a1.4.3}$$

Equation (11.4.2) reduces to

$$\hat{X}_{2t}(1) = \phi_{211}X_{1t} + \phi_{221}X_{2t}$$
$$\hat{X}_{2t}(\ell) = \phi_{221}\hat{X}_{2t}(\ell-1), \qquad \ell \geqslant 2 \tag{11.4.4}$$

It is thus seen that the forecasts beyond the first step for $\ell > 1$ are not at all affected by the leading indicator. This is obvious since the leading indicator X_{1t}, being a white noise series, cannot provide any forecasts that can be used in improving the forecasts of the required series X_{2t} after lead time one.

Case (ii)

If X_{1t} also has an extended first order autoregressive model

$$X_{1t} = \phi_{111}X_{1t-1} + \phi_{121}X_{2t-1} + a_{1t}$$

then using the rules for conditional expectation,

$$\begin{aligned}
\hat{X}_{1t}(\ell-1) &= \phi_{111}\hat{X}_{1t}(\ell-2) + \phi_{121}\hat{X}_{2t}(\ell-2) \\
&= \phi_{111}[\phi_{111}\hat{X}_{1t}(\ell-3) + \phi_{121}\hat{X}_{2t}(\ell-3)] \\
&\quad + \phi_{121}\hat{X}_{2t}(\ell-2) \\
&\quad \vdots \\
&= \phi_{111}^{\ell-1}X_{1t} + \phi_{121}[\hat{X}_{2t}(\ell-2) \\
&\quad + \phi_{111}\hat{X}_{2t}(\ell-3) + \ldots + \phi_{111}^{\ell-2}X_{2t}]
\end{aligned} \tag{11.4.5}$$

Substituting the conditional expectation from Eq. (11.4.5) in Eq. (11.4.2), we get the expression for forecasts as

$$\begin{aligned}
\hat{X}_{2t}(1) &= \phi_{211}X_{1t} + \phi_{221}X_{2t} \\
\hat{X}_{2t}(\ell) &= \phi_{211}[\phi_{111}^{\ell-1}X_{1t} + \phi_{121}\{\hat{X}_{2t}(\ell-2) \\
&\quad + \ldots + \phi_{111}^{\ell-2}X_{2t}\}] \\
&\quad + \phi_{221}\hat{X}_{2t}(\ell-1), \qquad \ell \geqslant 2 \\
&= \phi_{211}\phi_{111}^{\ell-1}X_{1t} + \phi_{221}\hat{X}_{2t}(\ell-1) \\
&\quad + \phi_{211}\phi_{121}[\hat{X}_{2t}(\ell-2) + \phi_{111}\hat{X}_{2t}(\ell-3) \\
&\quad + \ldots + \phi_{111}^{\ell-2}X_{2t}]
\end{aligned} \tag{11.4.6}$$

Note that the leading indicator X_{1t} affects the forecasts of the desired series X_{2t} for lead times $\ell \geq 1$. The significance of the results will be later illustrated by a simple numerical example.

11.4.2 Conditional Expectation from Orthogonal Decomposition

Conditional expectation of the orthogonal decomposition for the desired series can be used, as in Chapter 5, to find the probability limits on forecasts. For example, the orthogonal decomposition of the X_{2t} series (11.4.1) is given by

$$X_{2t} = (1 - \phi_{221}B)^{-1}(\phi_{211}X_{1t-1} + a_{2t}) \tag{11.4.7}$$

Using $X_{1t} = a_{1t}$ for Case (i), we get

$$X_{2t} = (1 - \phi_{221}B)^{-1}(\phi_{211}a_{1t-1} + a_{2t})$$
$$= (1 + \phi_{221}B + \phi_{221}^2 B^2 + \ldots)(\phi_{211}a_{1t-1} + a_{2t})$$
$$= \sum_{j=0}^{\infty} G_j(\phi_{211}a_{1t-j-1} + a_{2t-j}) \tag{11.4.8}$$

where G_j is the Green's function of the AR(1) model, $X_{2t} = \phi_{221}X_{2t-1} + a_{2t}$, and can be obtained by the methods of Chapter 3. Furthermore, from Eq. (11.4.8),

$$X_{2t+\ell} = \sum_{j=0}^{\infty} G_j(\phi_{211}a_{1t+\ell-j-1} + a_{2t+\ell-j})$$
$$= (\phi_{211}a_{1t+\ell-1} + a_{2t+\ell})$$
$$+ G_1(\phi_{211}a_{1t+\ell-2} + a_{2t+\ell-1})$$
$$+ G_2(\phi_{211}a_{1t+\ell-3} + a_{2t+\ell-2})$$
$$+ \ldots + G_{\ell-1}(\phi_{211}a_{1t} + a_{2t+1}) \tag{11.4.9}$$
$$+ G_\ell(\phi_{211}a_{1t-1} + a_{2t})$$
$$+ G_{\ell+1}(\phi_{211}a_{1t-2} + a_{2t-1}) + \ldots$$

Using the rules of conditional expectation,

$$\hat{X}_{2t}(\ell) = G_{\ell-1}\phi_{211}a_{1t} + G_\ell(\phi_{211}a_{1t-1} + a_{2t})$$
$$+ G_{\ell+1}(\phi_{211}a_{1t-2} + a_{2t-1}) + \ldots, \qquad \ell \geq 1 \tag{11.4.10}$$

It is easy to verify that Eqs. (11.4.4) and (11.4.10) are equivalent.

11.4.3 Forecast Errors and Probability Limits

The forecast errors can be obtained from the orthogonal decomposition of the model (11.4.9) in the same way as in Chapter 5, that is, for $\ell = 1$,

$$
\begin{aligned}
e_{2t}(1) &= X_{2t+1} - \hat{X}_{2t}(1) \\
&= [(\phi_{211}a_{1t}+a_{2t+1}) + G_1(\phi_{211}a_{1t-1}+a_{2t}) \\
&\quad + G_2(\phi_{211}a_{1t-2}+a_{2t-1}) + \ldots] \\
&\quad - [\phi_{211}a_{1t}+G_1(\phi_{211}a_{1t-1}+a_{2t}) + G_2(\phi_{211}a_{1t-2}+a_{2t-1}) + \ldots] \\
&= a_{2t+1}
\end{aligned}
$$

In general, from Eqs. (11.4.9) and (11.4.10)

$$
\begin{aligned}
e_{2t}(\ell) &= X_{2t+\ell} - \hat{X}_{2t}(\ell) \tag{11.4.11} \\
&= (\phi_{211}a_{1t+\ell-1}+a_{2t+\ell}) + G_1(\phi_{211}a_{1t+\ell-2}+a_{2t+\ell-1}) \\
&\quad + \ldots + G_{\ell-2}(\phi_{211}a_{1t+1}+a_{2t+2}) + G_{\ell-1}a_{2t+1}, \ell \geq 2
\end{aligned}
$$

Since a_{1t} and a_{2t} are assumed independent, the variances of the forecasts (actually of the forecast errors) are given by

$$
\begin{aligned}
V[e_{2t}(1)] &= \gamma_{a22} \\
V[e_{2t}(\ell)] &= (1+G_1^2+G_2^2 + \ldots + G_{\ell-2}^2)(\phi_{211}^2\gamma_{a11}+\gamma_{a22}) \tag{11.4.12} \\
&\quad + G_{\ell-1}^2\gamma_{a22}, \qquad \ell \geq 2
\end{aligned}
$$

Probability Limits on the Forecasts

The probability limits on the forecasts specify how close the true value will be to the forecasts. Thus, 95% probability limits on the forecast of X_{2t+1} made at time t are

$$
\hat{X}_{2t}(1) \pm 1.96[V(e_{2t}(1))]^{1/2}
$$

that is,

$$
(\phi_{211}X_{1t}+\phi_{221}X_{2t}) \pm 1.96(\gamma_{a22})^{1/2}
$$

Similarly, the 95% probability limits on the forecasts of $X_{2t+\ell}$, $\ell \geq 2$ made at time t are

$$
\hat{X}_{2t}(\ell) \pm 1.96[V(e_{2t}(\ell))]^{1/2}
$$

that is,

$$
\begin{aligned}
\hat{X}_{2t}(\ell) \pm 1.96[(1+G_1^2+G_2^2 + \ldots \\
+ G_{\ell-2}^2)(\phi_{211}^2\gamma_{a11}+\gamma_{a22}) + G_{\ell-1}^2\gamma_{a22}]^{1/2} \tag{11.4.13}
\end{aligned}
$$

11.4.4 An Illustrative Example

Let us consider an example of the first order models for papermaking data as discussed in Section 11.2 with rounded-off parameter values (Tables 11.1 and 11.2) for illustrative purposes only.

$$X_{1t} = -0.10X_{2t} + 0.19X_{2t-1} + a_{1t}, \quad \gamma_{a_{11}} = 0.0021 \quad (11.4.14)$$

$$X_{2t} = 0.25X_{1t-1} + 0.7X_{2t-1} + a_{2t}, \quad \gamma_{a_{22}} = 0.0062$$

where there is a small but significant feedback, or in economic terms, the leading indicator X_{1t}, in turn, depends on the past values of the desired series X_{2t}. The orthogonal decomposition of the model for the X_{2t} series is written as

$$
\begin{aligned}
X_{2t} &= 0.25(-0.10X_{2t-1}+0.19X_{2t-2}+a_{1t-1}) \\
&\quad + 0.7X_{2t-1} + a_{2t} \\
&= -0.025X_{2t-1} + 0.048X_{2t-2} \\
&\quad + 0.7X_{2t-1} + a_{2t} + 0.25a_{1t-1} \quad (11.4.15) \\
(1 &- 0.675B - 0.048B^2)X_{2t} = 0.25a_{1t-1} + a_{2t} \\
X_{2t} &= (1 - 0.675B - 0.048B^2)^{-1}(0.25a_{1t-1}+a_{2t}) \\
&= \sum_{j=0}^{\infty} G_j(0.25a_{1t-j-1} + a_{2t-j})
\end{aligned}
$$

where G_j is the Green's function of the AR(2) model

$$X_{2t} = 0.675X_{2t-1} + 0.048X_{2t-2} + a_{2t}$$

that is,

$$G_0 = 1$$

$$G_1 = \phi_1 = 0.675$$

$$G_2 = \phi_1 G_1 + \phi_2 G_0 = 0.675^2 + 0.048 = 0.504$$

$$\vdots$$

Using the conditional expectation of the model (11.4.14), we get the expressions for the forecasts as (recall that X_{1t} and X_{2t} are the values *after* subtracting the mean)

$$\hat{X}_{2t}(1) = 0.25X_{1t} + 0.7X_{2t} \quad (11.4.16)$$

$$\hat{X}_{2t}(\ell) = 0.25\hat{X}_{1t}(\ell-1) + 0.7\hat{X}_{2t}(\ell-1), \quad \ell \geq 2$$

or using Eq. (11.4.15),

$$\hat{X}_{2t}(\ell) = G_{\ell-1}(0.25a_{1t})$$

$$+ \sum_{j=\ell}^{\infty} G_j(0.25a_{1t+\ell-1-j}+a_{2t+\ell-j}), \qquad \ell \geq 1 \quad (11.4.17)$$

The forecast errors can be obtained by using Eq. (11.4.11) as

$$e_{2t}(1) = a_{2t+1}$$

$$e_{2t}(\ell) = (0.25a_{1t+\ell-1}+a_{2t+\ell}) + 0.675(0.25a_{1t+\ell-2}+a_{2t+\ell-1})$$

$$+ \ldots + G_{\ell-2}(0.25a_{1t+1}+a_{2t+2}) + G_{\ell-1}a_{2t+1}, \qquad \ell \geq 2$$

and the 95% probability limits on the forecasts are

$$\hat{X}_{2t}(1) \pm 1.96(\gamma_{a22})^{1/2} \tag{11.4.18}$$

$$\hat{X}_{2t}(\ell) \pm 1.96[V(e_{2t}(\ell))]^{1/2}$$

where

$$V[e_{2t}(\ell)] = (1+0.675^2+0.504^2 + \ldots$$

$$+ G_{\ell-2}^2(0.0625\gamma_{a11}+\gamma_{a22}) + G_{\ell-1}^2\gamma_{a22}$$

For example, the 95% probability limits on two step ahead forecasts are

$$\hat{X}_{2t}(2) \pm 1.96[\{0.0625(0.0021) + 0.0062\} + 0.675^2(0.0062)]^{1/2}$$

$$= \hat{X}_{2t}(2) \pm 1.96\sqrt{0.0091561} = \hat{X}_{2t}(2) \pm 0.1875$$

Figure 11.10 depicts the forecasts and the 95% probability limits for lead time $\ell = 1, 2, \ldots, 7$ made at origin $t = 43$. It is seen that the probability limits of the forecasted values include all the actual observations.

To illustrate the advantage of using a leading indicator in forecasting, suppose that only the output series X_{2t} is available. An AR(1) model was fitted to this data following the Chapter 2 methodology.

$$X_{2t} = 0.7168X_{2t-1} + a_{2t} \qquad \text{with} \qquad \sigma_{a2}^2 = 0.00674$$

Table 11.6 shows the estimated variances of forecast errors made with and without the leading indicator. As expected, the leading indicator improves the accuracy of the forecasts since the error variances are smaller. The improvement in the error variances is maximum at lag 2 and reduces for large lags.

11.4.5 Usefulness of the Leading Indicator

It appears that if the modeling procedure of Chapter 4 is used, one can always find a scalar ARMA model of sufficiently high order for the desired

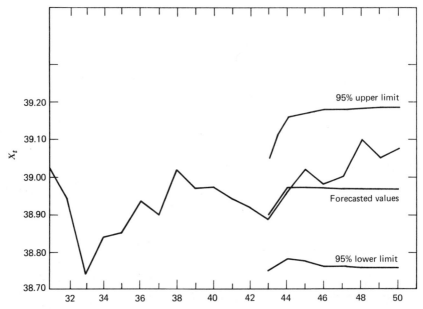

Figure 11.10 Forecasting by leading indicator for papermaking data.

series alone which can provide forecasts at least as good as, if not better than, those provided by using the leading indicator. Sometimes, it may be necessary to disregard the F-criterion and use a scalar ARMA model of higher order although it does not provide a significant improvement in the sum of squares at a given level of probability such as 95%.

Table 11.6 Comparison of Forecast Error Variances for First Order Models

Lead Time ℓ	$V[e_t(\ell)]$ Without Leading Indicator	With Leading Indicator	Percentage Improvement With Leading Indicator
1	0.00674	0.00610	9.50
2	0.01020	0.00901	11.66
3	0.01197	0.01060	11.44
4	0.01287	0.01146	10.95
5	0.01334	0.01193	10.57
6	0.01358	0.01218	10.30
7	0.01370	0.01232	10.07
8	0.01376	0.01239	9.95
9	0.01379	0.01243	9.86
10	0.01386	0.01245	9.85

Such a possibility that the addition of one or more associated series may fail to improve the forecasts of a given series is quite understandable. The forecasts and their effectiveness for a given series are dependent upon the dynamics, memory, or correlation in that series alone; it is immaterial whether this dynamics is the result of association with some related series. Therefore, unless the scalar model for that series alone fails to fully capture the dynamics (due to inefficient modeling, estimation, etc.), a multiple series model using leading indicators may not improve the forecasts. As we have stressed in Chapter 4, one of the strong points of the $ARMA(n, n-1)$ modeling strategy is that it can capture the dynamics as fully as we desire.

In practice, of course, forecasting by a leading indicator is well worth exploring. If the number of observations in a given series is too small, the additional information provided by the leading indicator may improve the parameter estimates, the effect of which has not been considered above. Second, contrary to the stationarity assumption, the probabilistic structure of the data may change over time; such a change may again be more effectively detected with the help of a leading indicator. The choice between forecasting by a single series and forecasting by the addition of a leading indicator can therefore be made only on the basis of practice. The above discussion is aimed not so much at discouraging the use of a leading indicator, but at discouraging the often false expectation that the leading indicator will always improve the forecasts!

Irrespective of the improvement in forecasts, the simplified ARMAV models with leading indicators are important in clarifying the structure of the underlying economic system. The parameters of the various indices on the series of interest may be useful in policy decisions, control and manipulation of the system. The dynamic relationship between the series and its leading indicator can be analyzed along the same lines as in Chapter 3.

11.4.6 General Results: Forecasting By a Single Leading Indicator

Let X_{2t} be the main series of interest and X_{1t} be the leading indicator. To ensure that a_{1t} is independent of a_{2t}, the zero lag parameter may be included either in the X_{1t} model as ϕ_{120} [illustrated in Eq. (11.4.14)] or in the X_{2t} model as ϕ_{210}. For the sake of definiteness we will include it in the X_{2t} model as ϕ_{210}; this implies that the series of interest is affected by the leading indicator at the same time, but the feedback effect of the series on the leading indicator requires at least one lag. Then the models are

$$X_{2t} = \phi_{21L}X_{1t-L} + \phi_{21L+1}X_{1t-L-1} + \ldots + \phi_{21n}X_{1t-n}$$
$$+ \phi_{221}X_{2t-1} + \phi_{222}X_{2t-2} + \ldots + \phi_{22n}X_{2t-n} - \theta_{221}a_{2t-1} \quad (11.4.19)$$
$$- \theta_{222}a_{2t-2} - \ldots - \theta_{22m}a_{2t-m} + a_{2t}$$
$$0 \leqslant L \leqslant n$$

$$X_{1t} = \phi_{121}X_{2t-1} + \phi_{122}X_{2t-2} + \ldots + \phi_{12n}X_{2t-n}$$
$$+ \phi_{111}X_{1t-1} + \phi_{112}X_{1t-2} + \ldots + \phi_{11n}X_{1t-n} - \theta_{111}a_{1t-1} \quad (11.4.20)$$
$$- \theta_{112}a_{1t-2} - \ldots - \theta_{11m}a_{1t-m} + a_{1t}$$

For the forecasts at lead times $\ell \leq L$, the model (11.4.20) is not needed since the forecasts involve only the known present and past values of X_{1t}. Thus,

$$\hat{X}_{2t}(\ell) = \phi_{21L}X_{1t-L+\ell} + \phi_{21L+1}X_{1t-L-1+\ell} + \ldots$$
$$+ \phi_{21n}X_{1t-n+\ell} + \phi_{221}\hat{X}_{2t}(\ell-1) + \phi_{222}\hat{X}_{2t}(\ell-2) + \ldots \quad (11.4.21)$$
$$+ \phi_{22\ell-1}\hat{X}_{2t}(1) + \phi_{22\ell}X_{2t} + \phi_{22\ell+1}X_{2t-1} + \ldots$$
$$+ \phi_{22n}X_{2t-n} - \theta_{22\ell}a_{2t} - \theta_{22\ell+1}a_{2t-1} + \ldots + \theta_{22m}a_{2t-m}$$
$$\ell \leq L$$

$$e_{2t}(\ell) = \sum_{j=0}^{\ell-1} G_{22j}a_{2t+\ell-j} \quad (11.4.22)$$

where G_{22j} is the Green's function of the ARMA(n,m) model in (11.4.19) and can be computed by comparing the coefficients of equal powers of B in

$$(1 - \phi_{221}B - \phi_{222}B^2 - \ldots - \phi_{22n}B^n)(1 + G_{221}B + G_{222}B^2 + \ldots)$$
$$\equiv (1 - \theta_{221}B - \theta_{222}B^2 - \ldots - \theta_{22m}B^m) \quad (11.4.23)$$

and the error variances are

$$V[e_t(\ell)] = \gamma_{a22} \sum_{j=0}^{\ell-1} G_{22j}^2 \quad (11.4.24)$$

For the case $\ell > L$, we can write Eqs. (11.4.19) and (11.4.20) in operator form as

$$(1 - \phi_{221}B - \phi_{222}B^2 - \ldots - \phi_{22n}B^n)X_{2t} = (\phi_{21L}B^L + \phi_{21L+1}B^{L+1}$$
$$+ \ldots + \phi_{21n}B^n)X_{1t} + (1 - \theta_{221}B - \ldots - \theta_{22m}B^m)a_{2t} \quad (11.4.19a)$$

$$(1 - \phi_{111}B - \phi_{112}B^2 - \ldots - \phi_{11n}B^n)X_{1t} = (\phi_{121}B + \phi_{122}B^2$$
$$+ \ldots + \phi_{12n}B^n)X_{2t} + (1 - \theta_{111}B - \ldots - \theta_{11m}B^m)a_{1t} \quad (11.4.20a)$$

Substituting for X_{1t} from Eq. (11.4.20a) in (11.4.19a), we get

$$
\begin{aligned}
&[(1 - \phi_{221}B - \phi_{222}B^2 - \ldots - \phi_{22n}B^n) \\
&\cdot(1 - \phi_{111}B - \phi_{112}B^2 - \ldots - \phi_{11n}B^n) \\
&- (\phi_{21L}B^L + \phi_{21L+1}B^{L+1} + \ldots \\
&+ \phi_{21n}B^n)(\phi_{121}B + \phi_{122}B^2 + \ldots + \phi_{12n}B^n)]X_{2t} \quad (11.4.25) \\
&= (1 - \phi_{111}B - \phi_{112}B^2 - \ldots - \phi_{11n}B^n) \\
&\cdot(1 - \theta_{221}B - \ldots - \theta_{22m}B^m)a_{2t} \\
&+ (\phi_{21L}B^L + \phi_{21L+1}B^{L+1} + \ldots + \phi_{21n}B^n) \\
&\cdot(1 - \theta_{111}B - \ldots - \theta_{11m}B^m)a_{1t}
\end{aligned}
$$

Now the forecasts can be obtained by conditional expectation on this form and the forecast errors from its orthogonal decomposition in the usual way. [See Eqs. (11.4.29) and (11.4.30) in the next section.] Alternatively, the forecasts may be obtained by using conditional expectation on *both* (11.4.19) and (11.4.20), using relations similar to (11.4.21) for $\ell > L$.

*11.4.7 General Results: Forecasting by Multiple Leading Indicators

No conceptual difficulties arise in forecasting by more than one leading indicator. However, the algebra of substitution becomes quite cumbersome and it is much easier to use the operator methods with Cramer's rule. Consider X_{pt} as the series of interest and $X_{1t}, X_{2t}, \ldots, X_{p-1t}$ as the leading indicators. Then, the operator form equations, similar to Eq. (11.3.37), can be succinctly written as

$$
\sum_{k=1}^{p} \Phi_{jk}(B)X_{kt} = H_{jj}(B)a_{jt}, \qquad j = 1, 2, \ldots, p \quad (11.4.26)
$$

The forecasts can be obtained by taking the conditional expectation of the pth equation. The required expectations on $X_{1t}, X_{2t}, \ldots, X_{p-1t}$ can be obtained from the other respective equations. Note that following Eqs. (11.4.19) and (11.4.20), the full form of these equations is

$$
X_{jt} + \sum_{k=1}^{j-1} \phi_{jk0}X_{kt} = \sum_{k=1}^{p}\sum_{i=1}^{n} \phi_{jki}X_{k,t-i} - \sum_{i=1}^{m} \phi_{jji}a_{j,t-i} + a_{jt} \quad (11.4.26a)
$$
$$
j = 1, 2, \ldots, p
$$

which may be compared with (11.2.6). It is clear that the conditional expectations are easy to get by the usual rules.

The ℓ step ahead forecast error in the series of interest X_{pt} is given by

$$e_{pt}(\ell) = \sum_{k=1}^{p} \sum_{j=0}^{\ell-1} G_{pkj} a_{kt+\ell-j} \qquad (11.4.27)$$

where the Green's functions G_{pkj} are defined by [compare with Eq. (11.3.40)]

$$\begin{vmatrix} \Phi_{11}(B) & \Phi_{12}(B) & \cdots & \Phi_{1(p-1)}(B) & H_{11}(B)a_{1t} \\ \Phi_{21}(B) & \Phi_{22}(B) & \cdots & \Phi_{2(p-1)}(B) & H_{22}(B)a_{2t} \\ \cdots & \cdots & \cdots & \cdots & \cdots \\ \Phi_{p1}(B) & \Phi_{p2}(B) & \cdots & \Phi_{p(p-1)}(B) & H_{pp}(B)a_{pt} \end{vmatrix}$$

$$\begin{vmatrix} \Phi_{11}(B) & \Phi_{12}(B) & \cdots & \Phi_{1(p-1)}(B) & \Phi_{1p}(B) \\ \Phi_{21}(B) & \Phi_{22}(B) & \cdots & \Phi_{2(p-1)}(B) & \Phi_{2p}(B) \\ \cdots & \cdots & \cdots & \cdots & \cdots \\ \Phi_{p1}(B) & \Phi_{p2}(B) & \cdots & \Phi_{p(p-1)}(B) & \Phi_{pp}(B) \end{vmatrix}$$

$$= \sum_{k=1}^{p} \left(\sum_{j=0}^{\infty} G_{pkj} B^j \right) a_{kt} \qquad (11.4.28)$$

Equating equal powers of B for an a_{kt} in Eq. (11.4.28) gives G_{pkj} required in (11.4.27). The variance of the ℓ step ahead forecast is then given by

$$\text{Var}[e_t(\ell)] = \sum_{k=1}^{p} \sum_{j=0}^{\ell-1} G_{pkj}^2 \, \sigma_{ak}^2 \qquad (11.4.29)$$

from which the probability limits on the forecasts can be obtained.

To illustrate the use of Eq. (11.4.28), consider the special case $p=2$. Then Eq. (11.4.28) takes the form

$$\frac{\begin{vmatrix} \Phi_{11}(B) & H_{11}(B)a_{1t} \\ \Phi_{21}(B) & H_{22}(B)a_{2t} \end{vmatrix}}{\begin{vmatrix} \Phi_{11}(B) & \Phi_{12}(B) \\ \Phi_{21}(B) & \Phi_{22}(B) \end{vmatrix}} = \sum_{k=1}^{2} \left(\sum_{j=0}^{\infty} G_{2kj} B^j \right) a_{kj}$$

Hence, we have the operator identities

$$\Phi_{11}(B) H_{22}(B) \equiv \left(\sum_{j=0}^{\infty} G_{22j} B^j \right)$$

$$\cdot [\Phi_{11}(B)\Phi_{22}(B) - \Phi_{21}(B)\Phi_{12}(B)] \quad (11.4.30)$$

$$- \Phi_{21}(B) H_{11}(B) \equiv \left(\sum_{j=0}^{\infty} G_{21j} B^j \right)$$

$$\cdot [\Phi_{11}(B)\Phi_{22}(B) - \Phi_{21}(B)\Phi_{12}(B)]$$

Appendix A 11.1

INITIAL VALUES FOR THE SPECIAL FORM OF THE ARMAV MODELS

The initial values of the parameters for the ARMAV models and their special cases can be obtained from their inverse function coefficients. A procedure to get the initial guess values based on the inverse function coefficients for the ARMA model parameters has been described in Chapter 4. In this section, the same approach is extended to find the initial values of the parameters for the input and output series models. We will use the bivariate model to illustrate the method

It follows from Section 11.1.6 that the bivariate model (one input-one output system) may be written as

$$X_{1t} + \phi_{120} X_{2t} = \phi_{111} X_{1t-1} + \phi_{121} X_{2t-1} + \phi_{112} X_{1t-2} + \phi_{122} X_{2t-2}$$

$$+ \ldots + \phi_{12n} X_{2t-n} + a_{1t} - \theta_{111} a_{1t-1} - \ldots - \theta_{11m} a_{1t-m}$$

$$X_{2t} = \phi_{211} X_{1t-1} + \phi_{221} X_{2t-1} + \ldots + \phi_{21n} X_{1t-n} + \phi_{22n} X_{2t-n}$$

$$+ a_{2t} - \theta_{221} a_{2t-1} - \ldots - \theta_{22m} a_{2t-m}$$

where n and m are the autoregressive and moving average order, respectively. Using the B operator

$$(1 - \phi_{111} B - \ldots - \phi_{11n} B^n) X_{1t} + (\phi_{120} - \phi_{121} B - \ldots - \phi_{12n} B^n) X_{2t}$$

$$= (1 - \theta_{111} B - \ldots - \theta_{11m} B^m) a_{1t}$$

$$(A\ 11.1.1)$$

and

$$(-\phi_{211} B - \ldots - \phi_{21n} B^n) X_{1t} + (1 - \phi_{221} B - \ldots - \phi_{22n} B^n) X_{2t}$$

$$= (1 - \theta_{221} B - \ldots - \theta_{22m} B^m) a_{2t}$$

$$(A\ 11.1.2)$$

The inverse function expansion of these models may be written as

$$a_{1t} = (1 - I_{111}B - I_{112}B^2 - \ldots)X_{1t}$$
$$+ (I_{120} - I_{121}B - I_{122}B^2 - \ldots)X_{2t} \qquad \text{(A 11.1.3)}$$
$$a_{2t} = (-I_{211}B - I_{212}B^2 - \ldots)X_{1t}$$
$$+ (1 - I_{221}B - I_{222}B^2 - \ldots)X_{2t} \qquad \text{(A 11.1.4)}$$

Substituting for a_{1t} and a_{2t} from Eqs. (A 11.1.3) and (A 11.1.4) in Eqs. (A 11.1.1) and (A 11.1.2),

$$(1 - \phi_{111}B - \ldots - \phi_{11n}B^n)X_{1t}$$
$$+ (\phi_{120} - \phi_{121}B - \ldots - \phi_{12n}B^n)X_{2t} \qquad \text{(A 11.1.5)}$$
$$= (1 - \theta_{111}B - \ldots - \theta_{11m}B^m)[(1 - I_{111}B - \ldots)X_{1t}$$
$$+ (I_{120} - I_{121}B - \ldots)X_{2t}]$$

and

$$(-\phi_{211}B - \phi_{212}B^2 - \ldots - \phi_{21n}B^n)X_{1t}$$
$$+ (1 - \phi_{221}B - \phi_{222}B^2 - \ldots - \phi_{22n}B^n)X_{2t}$$
$$= (1 - \theta_{221}B - \ldots - \theta_{22m}B^m)[(-I_{211}B - I_{212}B^2 - \ldots)X_{1t}$$
$$+ (1 - I_{221}B - \ldots)X_{2t}]$$

$$\text{(A 11.1.6)}$$

Equating the coefficients of equal powers of B in Eq. (A 11.1.5) we get two sets of equations for each X_{1t} and X_{2t}

$$\phi_{111} = \theta_{111} + I_{111}$$
$$\phi_{112} = \theta_{112} - \theta_{111}I_{111} + I_{112}$$
$$\vdots \qquad \qquad \text{(A 11.1.5a)}$$
$$\phi_{11j} = \theta_{11j} - \theta_{111}I_{11j-1} - \theta_{112}I_{11j-2} - \ldots$$
$$- \theta_{11j-1}I_{111} + I_{11j}$$

and

$$\phi_{120} = I_{120}$$
$$\phi_{121} = \theta_{111}I_{120} + I_{121}$$
$$\phi_{122} = \theta_{112}I_{120} - \theta_{111}I_{121} + I_{122} \qquad \text{(A 11.1.5b)}$$
$$\vdots$$
$$\phi_{12j} = \theta_{11j}I_{120} - \theta_{111}I_{12j-1} - \ldots - \theta_{11j-1}I_{121} + I_{12j}$$

Similarly, from Eq. (A 11.1.6), we get simultaneous equations for the remaining parameters

$$\phi_{211} = I_{211}$$

$$\phi_{212} = I_{212} - \theta_{221}I_{211}$$

$$\vdots$$

$$\phi_{21j} = I_{21j} - \theta_{221}I_{21j-1} - \ldots - \theta_{22j-1}I_{211}$$

(A 11.1.6a)

and

$$\phi_{221} = \theta_{221} + I_{221}$$

$$\phi_{222} = \theta_{222} - \theta_{221}I_{221} + I_{222}$$

$$\vdots$$

$$\phi_{22j} = \theta_{22j} - \theta_{221}I_{22j-1} - \ldots - \theta_{22j-1}I_{221} + I_{22j}$$

(A 11.1.6b)

for all j with the assumption that θ_{11j} and $\theta_{22j} = 0$ for $j > m$; and ϕ_{11j}, ϕ_{12j}, ϕ_{21j}, and $\phi_{22j} = 0$ for $j > n$ for the (n,m) model. In particular, for $j > \max(n,m)$

$$(1 - \theta_{111}B - \theta_{112}B^2 - \ldots - \theta_{11m}B^m)I_{1ij} = 0, \quad i = 1,2$$

and

(A 11.1.7)

$$(1 - \theta_{221}B - \theta_{222}B^2 - \ldots - \theta_{22m}B^m)I_{2ij} = 0, \quad i = 1,2$$

It is thus clear that the initial values of the parameters ϕ's and θ's can be computed from Eqs. (A 11.1.5) to (A 11.1.7) if the values of I_{1ij} and I_{2ij}, $i = 1,2$, $j = 1,2, \ldots$, $\max(n,m)$ are known. The values of the inverse function coefficients for the series X_{1t} and X_{2t} can be determined from the parameters of the pure autoregressive model of order p. For example, let us consider the autoregressive model of order p for the series X_{1t}

$$X_{1t} = -\phi_{120}X_{2t} + \phi_{111}X_{1t-1} + \phi_{121}X_{2t-1}$$

$$+ \ldots + \phi_{11p}X_{1t-p} + \phi_{12p}X_{2t-p} + a_{1t}$$

or

$$a_{1t} = X_{1t} + \phi_{120}X_{2t} - \phi_{111}X_{1t-1} - \phi_{121}X_{2t-1}$$

$$- \ldots - \phi_{11p}X_{1t-p} - \phi_{12p}X_{2t-p} \quad \text{(A 11.1.8)}$$

Comparing Eq. (A 11.1.8) and (A 11.1.3), we have $I_{1ij} = \phi_{1ij}$, $i = 1,2$ and $j = 1,2, \ldots, p$, $I_{1ij} = 0$, $j > p$. The choice of order p and the estimation of I_{1ij} are carried out in a similar way, as discussed in Chapter 4.

Once the estimates I_{1ij} and I_{2ij}, $i = 1,2$, $j = 1,2, \ldots, p$ are available, they can be substituted in the preceding equations (A 11.1.5) to (A 11.1.7), and then one can make use of only those many equations as the number of desired initial values and solve these simultaneous linear equations to get the initial values. As in the case of AR models (see Chapter 4), one can also use more equations and obtain the estimates of initial values by linear least squares.

In solving by the simultaneous equation method, some redundancy arises since the θ_{iij}'s can be obtained from any one set of Eq. (A 11.1.7), that is, equations either with I_{i1j} or I_{i2j}. Rather than making an arbitrary choice, a sensible method is to add the corresponding equations and then solve the resultant m simultaneous equations. Substituting these in the first n equations, one can recursively compute ϕ_{ijk} as in the ARMA case.

A 11.1.2 An Illustrative Example

Let us consider a second order bivariate model

$$X_{1t} = -\phi_{120}X_{2t} + \phi_{111}X_{1t-1} + \phi_{121}X_{2t-1} + \phi_{112}X_{1t-2}$$
$$+ \phi_{122}X_{2t-2} + a_{1t} - \theta_{111}a_{1t-1} \tag{A 11.1.9}$$

$$X_{2t} = \phi_{211}X_{1t-1} + \phi_{221}X_{2t-1} + \phi_{212}X_{1t-2}$$
$$+ \phi_{222}X_{2t-2} + a_{2t} - \theta_{221}a_{2t-1} \tag{A 11.1.10}$$

As discussed earlier, we first fit a pure autoregressive model of order $(2+1)=3$ and obtain the estimates of I_{ikj}, $i,k=1,2$, $j=1,2,3$. Then using Eqs. (A 11.1.5) to (A 11.1.7), we get

$$\theta_{111} = \frac{I_{113} + I_{123}}{I_{112} + I_{122}}$$

$$\phi_{120} = I_{120}$$

$$\phi_{111} = I_{111} + \theta_{111}$$

$$\phi_{121} = I_{121} + \theta_{111}I_{120} \tag{A 11.1.11}$$

$$\phi_{112} = I_{112} - \theta_{111}I_{111}$$

$$\phi_{122} = I_{122} - \theta_{111}I_{121}$$

$$\theta_{221} = \frac{I_{213} + I_{223}}{I_{212} + I_{222}}$$

$$\phi_{211} = I_{211}$$

$$\phi_{221} = I_{221} + \theta_{221} \tag{A 11.1.12}$$

$$\phi_{212} = I_{212} - \theta_{221}I_{211}$$

$$\phi_{222} = I_{222} - \theta_{221}I_{221}$$

For the papermaking data, the third order autoregressive model gives

$I_{120} = 0.1293$, $I_{111} = -0.0195$, $I_{121} = 0.0280$

$I_{112} = 0.0424$, $I_{122} = 0.3093$, $I_{211} = 0.3012$, $I_{113} = -0.0386$

$I_{123} = -0.2192$, $I_{221} = 0.6125$, $I_{212} = -0.2517$, $I_{222} = -0.0387$

$I_{213} = -0.2355$, $I_{223} = -0.0259$

Substituting these values in Eq. (A 11.1.11) and (A 11.1.12) gives

$$\theta_{111} = -0.733, \quad \phi_{120} = 0.129, \quad \phi_{111} = -0.752$$

$$\phi_{112} = 0.028, \quad \phi_{121} = -0.0666, \quad \phi_{122} = 0.3298$$

$$\theta_{221} = 0.90, \quad \phi_{211} = 0.3012, \quad \phi_{221} = 1.5125$$

$$\phi_{212} = -0.5228, \quad \phi_{222} = -0.5899$$

Note that the method described here to obtain the initial values of the parameters refers to the bivariate model of order (n,m). However, it is easy to extend this method to a general multivariate model $[(p-1)$ inputs and one output systems].

Appendix A 11.2

CONTINUOUS TO DISCRETE TRANSFER FUNCTION

The procedure for obtaining the discrete transfer function for "pulsed" inputs (that is, inputs that are assumed to be constant between the sampling intervals) from the known transfer functions for the continuous time inputs may be found in standard control theory books for *deterministic* sampled data systems, see, for example, Kuo (1970). The relevant formulae in our notation are given below.

Suppose that the differential equation relating the input $Y_1(t)$ with the output $Y_2(t)$ is

$$(D^n + \alpha_{n-1}D^{n-1} + \ldots + \alpha_0)Y_2(t)$$
$$= (\beta_m D^m + \beta_{m-1}D^{m-1} + \ldots + 1)cY_1(t) \quad \text{(A 11.2.1)}$$
$$m < n$$

and μ_i are the characteristic roots

$$(D^n + \alpha_{n-1}D^{n-1} + \ldots + \alpha_0)$$
$$= (D-\mu_1)(D-\mu_2)\ldots(D-\mu_n) \quad \text{(A 11.2.2)}$$

so that the continuous time transfer functions are

$$T(D) = \frac{c \sum_{i=0}^{m} \beta_i D^i}{\sum_{i=0}^{n} \alpha_i D^i}, \qquad \beta_0 = 1 = \alpha_n \tag{A 11.2.3}$$

$$= \frac{c \sum_{i=0}^{m} \beta_i D^i}{\prod_{i=1}^{n} (D - \mu_i)}$$

Then, the required discrete time transfer function for sampling interval Δ to be used for control is given by

$$T(B) = (1 - B) \sum_{i=0}^{n} \frac{c_i}{1 - \lambda_i B} \tag{A 11.2.4}$$

where

$$\lambda_i = e^{\mu_i \Delta}, \qquad i = 0, 1, \ldots n; \qquad \mu_0 = 0 \tag{A 11.2.5}$$

and the coefficients c_i are obtained from the partial fraction decomposition

$$\sum_{i=0}^{n} \frac{c_i}{(D - \mu_i)} = \frac{c \sum_{i=0}^{m} \beta_i D^i}{\prod_{i=0}^{n} (D - \mu_i)} = \frac{T(D)}{D} \tag{A 11.2.6a}$$

that is, for distinct roots

$$c_0 = \frac{c}{\prod_{i=1}^{n} (-\mu_i)} \tag{A 11.2.6}$$

$$c_k = \frac{c \sum_{i=0}^{m} \beta_i \mu_k^i}{\mu_k \prod_{\substack{i=1 \\ i \neq k}}^{n} (\mu_k - \mu_i)}, \qquad k = 1, 2, \ldots n$$

If there is a lag of L sampling intervals in the continuous time transfer function, $T(B)$ in Eq. (A 11.2.4) is multiplied by B^L.

As an illustration, consider the first order example used in Section 11.2.2. The step response for this example was found in Section 10.3.2 to be

$$g(1 - e^{-t/\tau}) = g(1 - e^{-\alpha_0 t})$$

It is easily seen from Chapter 6 that the differential equation that gives such a response is

$$(D + \alpha_0)Y_2(t) = g\alpha_0 Y_1(t)$$

so that the continuous time transfer function is

$$T(D) = \frac{g\alpha_0}{(D + \alpha_0)}$$

By Eqs. (A 11.2.5) and (A 11.2.6)

$$\lambda_0 = 1, \quad \lambda_1 = \phi_{221} = e^{-\alpha_0\Delta}$$

$$\frac{T(D)}{D} = \frac{g\alpha_0}{D(D+\alpha_0)} = g\left[\frac{1}{D} - \frac{1}{D + \alpha_0}\right]$$

and by Eq. (A 11.2.4)

$$T(B) = g(1 - B)\left[\frac{1}{(1-B)} - \frac{1}{(1 - \phi_{221}B)}\right]$$

$$= \frac{g(1-\phi_{221})B}{(1 - \phi_{221}B)}$$

which agrees with the results of Section 11.2.2.

Appendix A 11.3

RELATION TO PROPORTIONAL PLUS INTEGRAL PLUS DERIVATIVE (PID) CONTROL

The conventional feedback control strategy consists of comparing the measured system output variable with the desired (or set point) value, and adjusting the input by means of the controller to drive the error toward zero. If $m(t)$ denotes the deviation of the controller output from the steady state, then the operation of the widely used three-mode or three-term controller is represented by the following equation:

$$m(t) = K_p e(t) + K_i \int e(t) \, dt + K_d \frac{de(t)}{dt} \qquad \text{(A 11.3.1)}$$

where $e(t)$ is the error or the deviation of the measured output variable from the desired or target (set point) value. K_p, K_i, and K_d are constants that can be adjusted on the controller, the first term being proportional (P) to the error, the second and the third use integral (I) and derivative (D) of the error. Hence the name PID controller.

The discrete form of equation (A 11.3.1) can be obtained by replacing the integral by summation and the derivative by the difference. In our usual notation of X_{1t} and X_{2t} for the deviations of the system input (or controller output) and system output, respectively, from mean, target, or sets point values, sampling interval Δ, difference operator $\nabla = (1-B)$, and summation operator $1/\nabla = (1-B)^{-1}$, the discrete form of equation (A 11.3.1) becomes

$$X_{1t} = K_p X_{2t} + K_i \Delta \Sigma X_{2t} + \frac{K_d}{\Delta} \nabla X_{2t}$$

or

$$X_{1t} = K_p X_{2t} + \frac{K_i \Delta}{(1-B)} X_{2t} + \frac{K_d}{\Delta} (1-B) X_{2t}$$

Multiplying by $(1-B)$ on both sides, we can write the control equation in our usual notation as

$$X_{1t} = \left(K_p + K_i \Delta + \frac{K_d}{\Delta} \right) X_{2t} + X_{1t-1}$$

$$- \left(K_p + \frac{2 K_d}{\Delta} \right) X_{2t-1} + \frac{K_d}{\Delta} X_{2t-2} \quad (\text{A } 11.3.2)$$

Note the formal similarity of this equation with the control equation (11.3.15), showing that equation (A 11.3.2) is a special case of the general one-input–one-output control equation (11.3.21). When $K_i = K_d = 0$, one gets proportional only—P—control as illustrated by Eq. (11.3.4) for the papermaking data. Similarly, when $K_d = 0$ one gets proportional plus integral—PI—control, which is most prevalent in industry.

    With the increasing use of digital computers or microprocessors, Eq. (A 11.3.2) rather than (A 11.3.1) is used for process control. However, such PID control assumes a specific form of the model and may lead to suboptimal control when this model form is not appropriate.

PROBLEMS

11.1 For a two input-one-output Paper Pulping Digester system, the adequate models are found as follows:

$$X_{1t} = 0.78 X_{1t-1} + a_{1t} \tag{1}$$

$$X_{2t} = 0.11 X_{2t-1} + 0.44 X_{2t-2} + 0.36 a_{2t-1} + a_{2t} \tag{2}$$

$$X_{3t} = 1.23 X_{3t-1} - 0.36 X_{3t-2} - 0.29 X_{1t-6} + 0.26 X_{1t-7}$$
$$- 0.17 X_{2t-1} + 0.32 X_{2t-2} - 0.64 a_{3t-1} + a_{3t} \tag{3}$$

(a) Find the transfer function model for the output X_{3t}.

(b) Draw the block diagram for the complete transfer function model.

(c)   If the model of the input series X_{1t} is found as
$$X_{1t} = 0.78X_{1t-1} + 0.2X_{3t-1} + a_{1t}$$
and the models for X_{2t} and X_{3t} do not change, show the changes in the block diagram for the transfer function model and discuss it.

11.2   In the simplified form of the ARMAV models, it is assumed that the off diagonal terms of the moving average matrices are zero (for example, Eq. 11.1.10). In a physical situation, what is really implied by this assumption? Use a bivariate example to support your answer.

11.3   To compute the initial values of the parameters of a (2,1) model for a control system, a third order model was fitted for the input-output series and the estimates of the parameters are found as

$$I_0 = \begin{bmatrix} 1 & 0.1 \\ 0 & 1 \end{bmatrix}, \qquad I_1 = \begin{bmatrix} 0.1 & 0.2 \\ 0.05 & 0.3 \end{bmatrix},$$

$$I_2 = \begin{bmatrix} 0.4 & 0.6 \\ 0.5 & 0.9 \end{bmatrix}, \qquad I_3 = \begin{bmatrix} 0.3 & 0.4 \\ 0.14 & 0.7 \end{bmatrix}$$

Compute the initial values of the parameters of the input and output series for this model. Assume the lag between the output-input is one, $L=1$, and the system has feedback.

11.4   Verify Eq. (11.2.7) using the estimated values of the parameters of adequate models for two input-one output paper pulping digester data as given in Tables 11.3 to 11.5.

11.5   (a)   Suppose you are a consultant to a chemical firm. If the manager of that firm wishes to know about process control only from the analysis and not from the engineering viewpoint, what is the first step you would suggest?

(b)   Suppose the management informs you that the output data of the chemical process, when modeled, yield an MA(2) model. Can you make any improvement in the sense of optimum control? Explain your answer.

11.6   For a two input-one output ARMAV model with a given γ_a^* matrix, derive the expressions for the matrices ϕ_0 and γ_a.

11.7   Suppose the models for a one input-one output system are obtained as

$$X_{1t} = 0.25X_{1t-1} + a_{1t}$$
$$X_{2t} = 0.7X_{1t-2} - 0.25X_{1t-3} + 0.9X_{2t-1} - 0.2X_{2t-2}$$
$$+ 0.3a_{2t-1} + a_{2t}$$

(a) Derive the optimal control equation using the minimum mean squared error strategy.

(b) Obtain the optimally controlled output and its variance in terms of error variance γ_{a22}.

(c) If the lag in the above system is increased to four, compute the percentage increase in the variation of the optimally controlled output.

11.8 The output X_{2t} and the input X_{1t} of a system are related by the following model:

$$X_{2t} = X_{1t-3} - 0.5X_{1t-4} + 1.1X_{2t-1} - 0.3X_{2t-2} + a_{2t} - 0.2a_{2t-1}$$

(a) Derive the optimal control equation from the first principles.

(b) It is possible to reduce the dead time from three to two by suitable design changes estimated to cost $1,000,000 ($10^6$) every year. If the yearly savings from the reduction in the optimal output variance is estimated as $100,000 ($10^5$) per one percent, would you advise the management to make the design changes? If yes, specify the yearly net savings; if no, specify the yearly net loss resulting from the changes.

11.9 Verify the results of Section 11.3.9.

11.10 In a certain economic system, it was decided to improve the forecasts of the main series X_{2t} by using the information from its leading indicator. The models fitted to two series are

$$X_{1t} = 0.1X_{2t} + a_{1t}, \quad \gamma_{a11} = 1.0$$
$$X_{2t} = 0.7X_{2t-1} - 0.12X_{2t-2} + 1.1X_{1t-2}$$
$$- 0.3X_{1t-3} + a_{2t}, \quad \gamma_{a22} = 1.5$$

Suppose $X_{1t} = 0.5$, $X_{1t-1} = -0.5$, $X_{1t-2} = 1.5$, $X_{2t} = 0.6$, $X_{2t-1} = 0.8$.

(a) Find the forecasts for the main series through $\ell = 5$.

(b) Compute the corresponding forecast errors variances.

11.11 For the pulping digester data, assume that only 100 observations are given. Using the models obtained in the text, find

(a) forecasts of the last 10 observations;

(b) the corresponding forecast errors;

(c) the corresponding 95% probability limits.

11.12 Consider Fig. 11.10. The forecasts seem to follow the series quite well up to the first lag and then move away. Why does this happen? Can you improve the forecasts? If yes, give the new forecasts, probability limits and plot them as in Fig. 11.10 to show the improvement. If no, explain why not.

DATA LISTING

TABLE A1　Observed Values of Adjustments in the Gate Opening in a Papermaking Process Taken Every Twenty Minutes.

TABLE A2　Yearly Average Sunspot Numbers from 1749 Through 1924.

TABLE A3　IBM Stock Prices (daily) from May 17, 1961 Through November 2, 1962.

TABLE A4　Response Data of a Mechanical Vibratory System of Mass, Spring, and Dashpot.

TABLE A5　Grinding Wheel Profile Data.

TABLE A6　Quality Control Measurements of Diameter 1 (Chapter 8).

TABLE A7　Quality Control Measurements of Diameter 2 (Chapter 8).

TABLE A8　Monthly Averages of the Weekly Total Investment in the U.S. Treasury Securities by the Large Commercial Banks in New York City (Chapter 9).

TABLE A9　Monthly Averages of Money Market Rates from Weekly Market Yields of 3-Month Bills of U.S. Government Securities (Chapter 9).

TABLE A10　University of Wisconsin Hospital In-patient Census April 21 Through July 29, 1974 (Chapter 9).

TABLE A11　Monthly U.S. Consumer Price Index from 1953 Through 1970 (Chapter 9).

TABLE A12　Monthly U.S. Wholesale Price Index from 1953 Through 1970 (Chapter 9).

TABLE A13　International Airline Passengers: Monthly Totals (Thousands of Passengers) January 1949 through December 1960. (Chapter 10).

TABLE A14　Coded Crack Length Data (Chapter 10).

TABLE A15　Wood Surface Profile (Chapter 10).

TABLE A16　Papermaking Machine Step Input Test Data (Chapter 10).

TABLE A17   Chemical Concentration Data (Chapter 10).

TABLE A18   Quarterly Gross National Product of the U.S. From 1946 Through 1970 (Chapter 10).

TABLE A19   Quarterly Expenditures on Producers' Durables in the U.S. from 1946 Through 1970 (Chapter 10).

TABLE A20   Normal Operating Record of a Papermaking Machine (Chapter 11).

TABLE A21   Hourly Pulping Digester Operation Data (Chapter 11).

Table A1 Observed Values of Adjustments in the Gate Opening in a Papermaking Process Taken Every Twenty Minutes

33.5	34.0	33.5	33.5	33.5	33.0	33.0	33.5	33.5	33.3
33.3	33.5	33.0	32.5	32.5	32.8	32.8	32.5	32.5	32.2
32.5	32.5	32.5	32.3	32.0	32.5	32.5	32.5	34.0	33.3
33.0	32.5	32.5	32.5	32.8	32.8	32.8	32.2	32.7	33.0
33.0	33.0	32.5	32.0	32.0	31.8	31.8	31.8	32.5	32.7
32.7	32.0	32.0	32.6	32.4	32.4	32.5	32.5	32.5	32.5
32.8	32.5	32.5	32.4	32.4	32.4	32.4	32.0	32.1	32.1
32.1	31.6	31.6	31.5	31.5	31.2	31.2	31.4	31.4	31.4
31.4	31.6	31.2	31.8	32.2	32.0	32.3	32.0	31.5	31.2
31.2	31.2	31.2	31.2	31.2	31.0	31.0	31.2	31.2	32.2
32.2	32.2	31.5	31.5	31.5	31.6	31.6	31.8	31.6	31.4
31.4	31.4	31.8	32.0	31.8	31.8	31.6	31.6	31.8	31.8
31.8	31.8	32.5	32.3	32.0	32.0	32.0	31.5	31.5	31.5
31.5	31.5	32.0	32.0	32.3	32.3	32.3	32.0	32.0	31.2
31.2	31.0	30.7	30.7	30.7	30.7	30.7	30.7	31.0	31.0
30.7	30.7	31.0	31.0	31.3	31.3	32.0	31.3	31.5	31.5

Total 160 observations (read row-wise)

Table A2 Yearly Average Sunspot Numbers from 1749 Through 1924

80.9	83.4	47.7	47.8	30.7	12.2	9.6	10.2	32.4	47.6
54.0	62.9	85.9	61.2	45.1	36.4	20.9	11.4	37.8	69.8
106.1	100.8	81.6	66.5	34.8	30.6	7.0	19.8	92.5	154.4
125.9	84.8	68.1	38.5	22.8	10.2	24.1	82.9	132.0	130.9
118.1	89.9	66.6	60.0	46.9	41.0	21.3	16.0	6.4	4.1
6.8	14.5	34.0	45.0	43.1	47.5	42.2	28.1	10.1	8.1
2.5	0.0	1.4	5.0	12.2	13.9	35.4	45.8	41.1	30.4
23.9	15.7	6.6	4.0	1.8	8.5	16.6	36.3	49.7	62.5
67.0	71.0	47.8	27.5	8.5	13.2	56.9	121.5	138.3	103.2
85.8	63.2	36.8	24.2	10.7	15.0	40.1	61.5	98.5	124.3
95.9	66.5	64.5	54.2	39.0	20.6	6.7	4.3	22.8	54.8
93.8	95.7	77.2	59.1	44.0	47.0	30.5	16.3	7.3	37.3
73.9	139.1	111.2	101.7	66.3	44.7	17.1	11.3	12.3	3.4
6.0	32.3	54.3	59.7	63.7	63.5	52.2	25.4	13.1	6.8
6.3	7.1	35.6	73.0	84.9	78.0	64.0	41.8	26.2	26.7
12.1	9.5	2.7	5.0	24.4	42.0	63.5	53.8	62.0	48.5
43.9	18.6	5.7	3.6	1.4	9.6	47.4	57.1	103.9	80.6
63.6	37.6	26.1	14.2	5.8	16.7				

Total 176 observations (read row-wise)

Table A3 IBM Stock Prices (daily) from May 17th, 1961 Through November 2, 1962

460	457	452	459	462	459	463	479	493	490
492	498	499	497	496	490	489	478	487	491
487	482	479	478	479	477	479	475	479	476
476	478	479	477	476	475	475	473	474	474
474	465	466	467	471	471	467	473	481	488
490	489	489	485	491	492	494	499	498	500
497	494	495	500	504	513	511	514	510	509
515	519	523	519	523	531	547	551	547	541
545	549	545	549	547	543	540	539	532	517
527	540	542	538	541	541	547	553	559	557
557	560	571	571	569	575	580	584	585	590
599	603	599	596	585	587	585	581	583	592
592	596	596	595	598	598	595	595	592	588
582	576	578	589	585	580	579	584	581	581
577	577	578	580	586	583	581	576	571	575
575	573	577	582	584	579	572	577	571	560
549	556	557	563	564	567	561	559	553	553
553	547	550	544	541	532	525	542	555	558
551	551	552	553	557	557	548	547	545	545
539	539	535	537	535	536	537	543	548	546
547	548	549	553	553	552	551	550	553	554
551	551	545	547	547	537	539	538	533	525
513	510	521	521	521	523	516	511	518	517
520	519	519	519	518	513	499	485	454	462
473	482	486	475	459	451	453	446	455	452
457	449	450	435	415	398	399	361	383	393
385	360	364	365	370	374	359	335	323	306
333	330	336	328	316	320	332	320	333	344
339	350	351	350	345	350	359	375	379	376
382	370	365	367	372	373	363	371	369	376
387	387	376	385	385	380	373	382	377	376
379	386	387	386	389	394	393	409	411	409
408	393	391	388	396	387	383	388	382	384
382	383	383	388	395	392	386	383	377	364
369	355	350	353	340	350	349	358	360	360
366	359	356	355	367	357	361	355	348	343
330	340	339	331	345	352	346	352	357	

Total 369 observations (read row-wise)

Table A4 Response Data of a Mechanical Vibratory System of Mass, Spring, and Dashpot (sampling interval $\Delta = 0.02$ sec)

30.0	28.0	25.0	24.0	23.0	21.0	20.0	22.0	24.0	27.0
30.0	31.0	34.0	37.0	33.0	28.0	25.0	23.0	21.0	19.0
18.0	17.0	16.0	17.0	18.5	22.0	29.0	32.0	32.0	30.0
25.0	20.0	17.0	14.0	13.0	17.0	22.0	27.0	33.0	30.0
21.0	15.0	12.0	10.0	9.0	6.0	6.0	8.0	10.0	12.0
15.0	16.0	16.5	18.0	21.0	15.0	7.0	4.0	3.0	7.0
15.5	22.0	30.0	40.0	40.0	39.0	38.0	35.0	30.0	25.0
20.0	18.0	20.0	22.5	27.0	32.0	32.5	33.0	32.0	30.0
25.5	23.3	23.3	24.0	27.0	31.5	35.0	36.0	34.0	30.5
29.0	25.0	20.0	19.0	21.0	23.5	28.0	33.5	36.0	37.5
38.0	36.0	33.0	29.5	28.0	28.0	30.0	30.5	30.0	30.0
28.0	25.0	23.0	24.5	27.0	31.0	34.0	33.0	25.0	16.0
13.0	14.0	17.0	22.0	29.0	32.0	30.0	26.0	24.0	24.0

Total 130 observations (read row-wise)

Source: Data from H.J. Stedudel, "A Time Series Approach to Modeling Second Order Mechanical Systems," M.S. Thesis, Univ. of Wisconsin-Milwaukee, 1971.

Table A5 Grinding Wheel Profile Data ($\times$ 10^{-3} in.) (Sampling Interval $\Delta = 0.002$ in.)

13.5	4.0	4.0	4.5	3.0	3.0	10.0	10.2	9.0	10.0
8.5	7.0	10.5	7.5	7.0	10.5	9.5	7.0	12.0	13.5
12.5	15.0	13.0	11.0	9.0	10.5	10.5	11.0	10.5	9.0
8.2	8.5	9.2	8.5	10.0	14.5	13.0	2.0	6.0	6.0
11.0	9.5	12.5	13.8	12.0	12.0	12.0	13.0	12.0	14.0
14.5	13.5	12.3	7.0	7.0	7.0	6.5	12.5	15.0	12.5
11.6	11.0	10.0	8.5	3.0	11.5	11.5	11.5	11.0	9.0
2.5	7.0	6.0	6.6	14.0	11.0	9.0	6.5	4.0	6.0
12.0	11.0	12.0	12.5	12.5	13.6	13.0	8.0	6.5	6.8
6.0	7.2	10.2	8.0	7.5	11.0	11.8	11.8	6.5	8.0
9.0	8.0	8.0	9.0	9.5	10.0	9.0	12.0	13.5	13.8
15.0	12.5	11.0	11.5	14.5	11.5	11.8	13.0	15.0	14.5
13.0	9.0	11.0	9.0	10.0	14.0	13.5	3.0	2.2	6.0
8.0	9.0	9.0	9.0	7.0	6.0	6.5	7.0	7.5	8.5
9.0	9.5	10.0	11.5	11.2	12.5	11.6	8.0	7.0	6.0
6.0	6.0	9.0	12.0	13.5	13.0	3.5	1.8	1.6	7.5
8.0	7.9	11.6	12.5	10.5	8.0	9.0	11.6	11.8	12.6
10.2	10.0	5.0	7.0	-1.0	0.0	0.0	3.0	11.0	12.0
12.2	11.0	8.0	7.0	5.5	10.0	11.5	7.0	4.0	7.0
7.0	10.0	9.0	8.0	10.0	13.0	10.0	6.5	11.0	13.0
13.0	14.0	13.0	12.5	12.0	9.0	8.5	7.0	8.5	10.0
8.0	4.0	3.0	10.0	13.0	13.0	13.0	12.5	11.0	11.0
11.0	14.5	14.0	14.0	13.5	10.0	9.5	10.0	12.5	10.0
9.0	9.0	4.0	3.0	6.0	5.0	7.0	6.0	5.0	8.5
10.5	11.1	11.0	10.0	11.2	8.0	2.5	5.0	13.2	14.0

Total 250 observations (read row-wise)

Table A6   Quality Control Measurements of Diameter 1: Last Three Digits of Micrometer Readings (Chapter 8)

384	394	384	391	393	392	393	394	395	396	394	393
393	393	390	395	385	393	392	385	396	395	396	392
394	397	398	396	387	382	399	402	393	394	399	387
397	398	394	396	395	385	397	397	395	396	391	394
396	398	397	394	399	400	394	396	394	398	397	397
395	388	378	397	398	399	397	401	399	397	398	398
388	393	398	402	401	398	395	399	398	402	394	397
396	400	400	399	401	400	397	392	399	397	398	398
399	400	401	401	400	401	391	403	394	402	404	405
404	398	403	403	402	405	403	408	406	397	404	403
402	394	398	402	397	398	374	392	401	403	398	403
390	399	339	396	395	397	396	395	404	402	398	401
399	402	402	401	400	402	403	388	397	404	389	401
404	392	398	393	399	401	403	398	401	401	392	399
403	403	404	400	404	398	400	401	405	401	400	405
400	398	403	405	402	403	400	401	400	401	398	401
399	403	400	400	402	398	403	400	401	399	400	401
399	398	399	404	401	401	403	394	398	400	402	405
390	403	386	387	404	403	402	404	408	407	399	397
402	402	405	402	404	403	403	397	396	392	400	400
402	401	404	407	399	402	402	401	404	406	401	400
403	398	397	405	403	393	404	398	405	399	401	403
401	396	403	404	402	403	403	395	407	391	390	406
404	400	393	406	399	408	398	407	402	403	403	396
409	381	406	404	404	406	405	407	406	407	410	407
406	400	396	406	402	405	401	403	403	400	407	403
407	410	401	407	403	398	406	407	407	406	409	409
408	407	407	403	408	406	402	406	388	407	398	405
398	406	402	405	409	401	407	404	402	403	406	408
407	398	403	400	407	392	397	399	392	407	411	407
403	400	403	404	403	407	407	400	399	404	400	404
403	407	405	408	407	409	409	409	404	406	407	407
402	399	410	409	407	407	404	407	409	405	409	410
406	392	403	407	404	407	405	407	382	408	410	406
404	407	405	407	403	408	409	368	393	405	401	402
408	406	408	408	404	389	405	383	408	409	407	391
406	400	409	405	404	407	406	406	408	400	405	403
407	409	411	410	389	408	384	401	394	409	409	410
390	409	410	410	394	406	411	394	411	401	406	403
408	410	412	408	402	400	410	400	401	405	408	407
408	401	406	410	410	411	411	402	402	411	400	410

Table A6—*Continued*

410	405	409	414	402	414	409	410	411	405	411	408
410	411	401	399	411	410	399	411	403	409	412	404
412	402	412	402	411	408	411	397	411	412	408	410
411	396	409	399	414	404	397	400	410	406	410	414
408	412	406	408	410	411	410	412	408	405	411	410
410	395	411	400	405	410	410	406	412	402	410	403
409	410	410	412	409	397	407	402	405	411	411	408
401	406	411	411	405	412	403	411	408	408	409	387
397	409	400	410	409	411	409	406	411	396	407	411

Total 600 observations (read row-wise)

Table A7   Quality Control Measurements of Diameter 2: Last Three Digits of Micrometer Readings (Chapter 8)

391	400	368	377	470	386	397	388	374	392	387	367
383	389	411	375	370	409	401	381	366	381	418	407
410	411	381	382	392	388	413	384	383	413	407	405
410	381	402	404	416	382	383	405	413	419	374	409
411	409	373	402	414	399	409	408	411	382	413	409
414	404	403	387	413	405	406	410	386	408	411	408
410	413	413	417	417	412	365	415	410	411	419	418
419	384	411	396	414	393	413	387	403	380	414	411
389	373	407	403	378	390	414	383	409	411	391	408
409	398	413	389	395	378	384	406	394	415	414	414
412	400	414	404	406	411	406	406	404	407	355	401
407	401	406	399	409	404	407	403	397	409	410	405
407	408	404	375	403	405	406	406	406	398	406	406
406	407	406	405	410	409	408	403	412	409	389	406
408	405	412	412	414	413	408	410	409	406	408	404
413	410	403	406	365	405	403	400	364	408	385	405
400	404	407	404	400	400	401	404	398	404	398	399
402	399	403	397	398	399	402	400	407	404	408	408
404	406	402	405	368	408	399	402	400	398	401	405
402	399	402	403	392	401	394	397	373	398	397	401
395	365	390	398	384	383	372	393	388	393	366	359
394	395	371	397	392	386	399	373	389	395	370	387
406	401	398	397	377	370	398	379	382	383	407	390
398	371	404	401	367	379	400	371	398	391	367	395
405	376	399	403	381	396	376	399	395	400	398	400

Total 600 observations (read row-wise) *continued next page*

Table A7   Quality Control Measurements of Diameter 2: Last Three Digits of Micrometer Readings (Chapter 8)—*Continued*

402	383	389	399	366	385	403	400	384	407	370	391
397	393	389	391	403	400	400	400	390	397	359	374
374	399	405	367	401	390	394	398	402	377	404	406
397	387	399	377	401	374	400	404	412	409	405	402
389	403	385	393	406	401	405	396	404	395	404	399
408	412	410	412	408	410	413	397	404	408	413	415
411	414	417	418	423	425	416	418	423	418	415	431
436	435	428	429	430	427	427	405	420	420	419	427
423	424	426	424	397	415	421	415	423	420	419	426
422	418	422	421	424	419	419	423	422	423	421	412
424	416	422	424	426	420	427	422	431	425	419	426
424	428	427	421	421	423	421	423	418	422	424	422
420	417	427	423	421	423	427	416	384	419	417	415
424	418	423	423	418	421	419	420	420	415	417	418
424	399	420	419	416	420	419	413	420	420	406	420
386	407	396	412	387	401	415	414	414	416	424	412
415	419	400	399	389	421	401	421	411	412	424	420
402	419	395	419	407	401	418	397	394	398	406	416
391	422	396	420	421	389	412	402	422	395	401	397
391	419	422	389	410	430	396	419	415	392	391	387
417	422	419	398	419	421	422	420	418	419	394	418
410	418	386	417	390	419	423	395	409	385	417	378
423	424	426	424	418	426	423	434	406	414	421	403
414	400	391	420	416	400	418	405	419	419	417	421
416	406	421	417	390	411	415	419	411	417	418	417

Total 600 observations (read row-wise)

Table A8   Monthly Averages of the Weekly Total Investment (Millions of Dollars) in the U.S. Treasury Securities by Large Commercial Banks in New York City from 1965 Through 1975 (Chapter 9)

Year	Jan.	Feb.	March	April	May	June	July	Aug.	Sept.	Oct.	Nov.	Dec.
1965	5446	4846	4671	4528	4418	4461	4644	4461	4260	4628	4612	4760
66	4639	4329	3944	4270	3992	3840	3901	3908	4232	3915	3946	4366
67	4538	4860	4970	4666	4743	4867	5116	4899	5287	5794	5833	5598
68	5313	5193	4807	4756	4631	4728	4773	5299	5910	5648	5645	5965
69	5165	4581	4145	4509	3990	4030	4255	4292	4324	4164	4616	4962
70	4550	4136	4232	4937	4507	4330	4500	5169	4993	4880	5095	5460
71	5545	5343	5106	5439	4864	4657	4989	4427	4563	4528	5331	5264
72	5112	5031	5552	5134	5005	4901	4602	4746	5283	4715	4983	5040
73	4931	4363	4138	4316	3761	4247	3832	3769	4154	4258	4868	5788
74	5435	5117	4983	4781	3793	3097	3373	4118	4679	4144	4625	4932
75	4719	4881	5624	5908	6284	6930	7049	6845	7452	7531	8332	8369

Table A9 Monthly Averages of Money Market Rates from Weekly Market Yields of 3-Month Bills of U.S. Government Securities from 1965 Through 1975 (Chapter 9)

Year	Jan.	Feb.	March	April	May	June	July	Aug.	Sept.	Oct.	Nov.	Dec.
1965	3.81	3.93	3.93	3.93	3.89	3.81	3.83	3.84	3.90	4.02	4.08	4.33
66	4.55	4.64	4.60	4.59	4.63	4.52	4.72	4.94	5.29	5.35	5.33	4.98
67	4.74	4.56	4.31	3.89	3.61	3.47	4.11	4.22	4.38	4.55	4.67	4.95
68	5.01	4.95	5.14	5.36	5.63	5.56	5.33	5.10	5.20	5.31	5.46	5.93
69	6.13	6.13	6.03	6.13	6.02	6.44	6.90	6.99	7.08	6.99	7.19	7.78
70	7.90	7.12	6.63	6.42	6.83	6.70	6.46	6.44	6.19	5.90	5.32	4.87
71	4.52	3.69	3.35	3.79	4.09	4.66	5.36	4.98	4.65	4.49	4.21	4.11
72	3.45	3.18	3.68	3.73	3.68	3.88	3.99	3.92	4.63	4.74	4.76	5.04
73	5.35	5.57	6.06	6.26	6.27	7.14	7.98	8.65	8.37	7.19	7.83	7.47
74	7.77	7.18	7.86	8.29	8.33	7.93	7.53	8.80	8.22	7.30	7.56	7.15
75	6.52	5.55	5.49	5.61	5.32	6.09	6.41	6.41	6.41	6.11	5.50	5.47

Table A10 University of Wisconsin Hospital In-Patient Census April 21 Through July 19, 1974 (Chapter 9)

397	462	486	483	477	438	407	421	480	484
486	479	415	400	419	477	510	503	500	435
408	417	478	497	500	512	450	421	423	471
496	478	463	413	396	366	375	444	469	480
439	402	442	492	507	518	493	439	399	428
476	499	488	460	419	380	406	472	502	495
490	443	398	417	490	505	499	484	430	384
392	452	455	426	414	405	379	410	485	514
525	511	461	436	444	488	494	510	493	429
392	420	466	476	494	484	423	388	411	472

Total 100 observations (read row-wise)

Table A11   Monthly U.S. Consumer Price Index (1957–1959 = 100)
(Chapter 9)

Year	Jan.	Feb.	March	April	May	June	July	Aug.	Sept.	Oct.	Nov.	Dec.
1953	92.8	92.4	92.6	92.7	92.9	93.3	93.5	93.7	93.9	94.0	93.7	93.6
1954	93.9	93.7	93.6	93.4	93.7	93.8	93.9	93.7	93.5	93.3	93.4	93.1
1955	93.1	93.1	93.1	93.1	93.1	93.2	93.5	93.3	93.6	93.6	93.7	93.5
1956	93.4	93.4	93.5	93.6	94.0	94.7	95.3	95.2	95.4	95.9	96.0	96.2
1957	96.3	96.7	96.9	97.2	97.5	98.0	98.4	98.6	98.7	98.7	99.1	99.1
1958	99.7	99.8	100.5	100.6	100.7	100.8	101.0	100.8	100.8	100.8	101.0	100.8
1959	100.9	100.8	100.8	101.0	101.0	101.5	101.8	101.7	102.0	102.3	102.4	102.3
1960	102.2	102.4	102.4	102.8	102.9	103.1	103.2	103.2	103.3	103.7	103.7	103.8
1961	103.8	103.9	103.9	103.9	103.8	104.0	104.4	104.3	104.6	104.6	104.6	104.5
1962	104.5	104.8	105.0	105.2	105.2	105.3	105.5	105.5	106.1	106.0	106.0	105.8
1963	106.0	106.1	106.2	106.2	106.6	107.1	107.1	107.1	107.1	107.2	107.4	107.6
1964	107.7	107.6	107.7	107.8	107.8	108.0	108.3	108.2	108.4	108.5	108.7	108.8
1965	108.9	108.9	109.0	109.3	109.6	110.1	110.2	110.0	110.2	110.4	110.6	111.0
1966	111.0	111.6	112.0	112.5	112.6	112.9	113.3	113.8	114.1	114.5	114.6	114.7
1967	114.7	114.8	115.0	115.3	115.6	116.0	116.5	116.9	117.1	117.5	117.8	118.2
1968	118.6	119.0	119.5	119.9	120.3	120.9	121.5	121.9	122.2	122.9	123.4	123.7
1969	124.1	124.6	125.6	126.4	126.8	127.6	128.2	128.7	129.3	129.8	130.5	131.3
1970	131.8	132.5	133.2	134.0	134.6	135.2	135.7	136.0	136.6	137.4	137.8	138.5

Table A12   Monthly U.S. Wholesale Price Index (1957–1959 = 100)
(Chapter 9)

Year	Jan.	Feb.	March	Apr.	May	June	July	Aug.	Sept.	Oct.	Nov.	Dec.
1953	92.7	92.4	92.7	92.2	92.6	92.3	93.5	93.2	93.6	92.9	92.6	92.8
1954	93.5	93.2	93.2	93.6	93.5	92.7	93.1	93.2	92.7	92.5	92.7	92.3
1955	92.8	93.1	92.7	93.2	92.7	93.0	93.2	93.5	94.2	94.1	93.7	93.8
1956	94.3	94.8	95.1	95.8	96.4	96.3	96.1	96.7	97.4	97.5	97.7	98.0
1957	98.6	98.6	98.6	98.8	98.7	99.0	99.7	99.8	99.5	99.3	99.6	99.9
1958	100.2	100.3	100.9	100.6	100.7	100.5	100.5	100.4	100.4	100.3	100.5	100.5
1959	100.7	100.7	100.8	101.2	101.1	100.9	100.7	100.4	100.9	100.4	100.2	100.2
1960	100.6	100.6	101.2	101.2	100.9	100.7	100.9	100.5	100.5	100.8	100.8	100.7
1961	101.0	101.0	101.0	100.5	100.0	99.5	99.9	100.1	100.0	100.0	100.0	100.4
1962	100.8	100.7	100.7	100.4	100.2	100.0	100.4	100.5	101.2	100.6	100.7	100.4
1963	100.5	100.2	99.9	99.7	100.0	100.3	100.6	100.4	100.3	100.5	100.7	100.3
1964	101.0	100.5	100.4	100.3	100.1	100.0	100.4	100.3	100.7	100.8	100.7	100.7
1965	101.0	101.2	101.3	101.7	102.1	102.8	102.9	102.9	103.0	103.1	103.5	104.1
1966	104.6	105.4	105.4	105.5	105.6	105.7	106.4	106.8	106.8	106.2	105.9	105.9
1967	106.2	106.0	105.7	105.3	105.8	106.3	106.5	106.1	106.2	106.1	106.2	106.8
1968	107.2	108.0	108.2	108.3	108.5	108.7	109.1	108.7	109.1	109.1	109.6	109.8
1969	110.7	111.1	111.7	111.9	112.8	113.2	113.3	113.4	113.6	114.0	114.7	115.1
1970	116.0	116.4	116.6	116.6	116.8	117.0	117.7	117.2	117.8	117.8	117.7	117.8

Table A13   International Airline Passengers: Monthly Totals (Thousands of Passengers) January 1949 Through December 1960

Year	Jan.	Feb.	March	Apr.	May	June	July	Aug.	Sept.	Oct.	Nov.	Dec.
1949	112	118	132	129	121	135	148	148	136	119	104	118
1950	115	126	141	135	125	149	170	170	158	133	114	140
1951	145	150	178	163	172	178	199	199	184	162	146	166
1952	171	180	193	181	183	218	230	242	209	191	172	194
1953	196	196	236	235	229	243	264	272	237	211	180	201
1954	204	188	235	227	234	264	302	293	259	229	203	229
1955	242	233	267	269	270	315	364	347	312	274	237	278
1956	284	277	317	313	318	374	413	405	355	306	271	306
1957	315	301	356	348	355	422	465	467	404	347	305	336
1958	340	318	362	348	363	435	491	505	404	359	310	337
1959	360	342	406	396	420	472	548	559	463	407	362	405
1960	417	391	419	461	472	535	622	606	508	461	390	432

Total 144 observations

Table A14   Coded Crack Length Data (Chapter 10)

10.9800	11.0800	11.2400	11.3300	11.4600
11.5100	11.5800	11.6500	11.7700	11.8600
11.9900	12.0900	12.1900	12.3701	12.5401
12.6301	12.8101	12.9801	13.1301	13.2801
13.3901	13.5401	13.6401	13.7401	13.8301
14.0201	14.1101	14.1801	14.2601	14.4001
14.4501	14.5101	14.6102	14.6602	14.7502
14.8602	14.9902	15.1202	15.2802	15.4802
15.5702	15.7002	15.7502	15.8602	15.9402
15.9702	16.1202	16.0902	16.4102	16.5102
16.6102	16.7102	16.8103	16.9103	17.0103
17.1103	17.1903	17.2803	17.3703	17.4603
17.5603	17.6603	17.6603	17.8103	17.9103
18.0103	18.1103	18.2103	18.3103	18.4103
18.5103	18.6103	18.7104	18.8104	18.9104
19.0604	19.2104	19.3104	19.4004	19.4604
19.5804	19.7104	19.8104	19.9104	20.0104
20.1104	20.2604	20.3104	20.4104	20.5104

Total 90 observations (read row-wise)

Table A15 Wood Surface Profile (Chapter 10)

0.106	0.111	0.111	0.107	0.105	0.107	0.110	0.108	0.111	0.119
0.117	0.107	0.105	0.107	0.109	0.105	0.104	0.102	0.108	0.113
0.113	0.107	0.103	0.103	0.098	0.102	0.103	0.104	0.105	0.105
0.105	0.101	0.103	0.107	0.109	0.104	0.100	0.103	0.100	0.105
0.102	0.105	0.106	0.107	0.104	0.107	0.109	0.108	0.111	0.107
0.107	0.106	0.107	0.102	0.102	0.101	0.103	0.103	0.103	0.100
0.101	0.101	0.100	0.102	0.101	0.096	0.096	0.098	0.104	0.107
0.107	0.102	0.105	0.101	0.105	0.110	0.111	0.111	0.100	0.102
0.102	0.107	0.112	0.114	0.113	0.108	0.106	0.103	0.103	0.101
0.103	0.106	0.107	0.106	0.107	0.107	0.104	0.111	0.117	0.118
0.115	0.107	0.110	0.117	0.121	0.122	0.123	0.119	0.117	0.118
0.115	0.111	0.108	0.107	0.105	0.105	0.105	0.103	0.105	0.107
0.109	0.110	0.111	0.108	0.107	0.106	0.108	0.107	0.105	0.102
0.101	0.102	0.101	0.097	0.100	0.105	0.108	0.108	0.105	0.103
0.103	0.100	0.103	0.106	0.107	0.097	0.098	0.100	0.101	0.097
0.099	0.101	0.104	0.107	0.109	0.111	0.109	0.103	0.105	0.102
0.108	0.113	0.113	0.108	0.107	0.102	0.106	0.106	0.106	0.103
0.097	0.103	0.107	0.102	0.107	0.111	0.110	0.107	0.103	0.099
0.097	0.099	0.100	0.099	0.100	0.099	0.100	0.099	0.099	0.098
0.100	0.102	0.102	0.106	0.112	0.113	0.109	0.107	0.105	0.097
0.105	0.110	0.113	0.108	0.101	0.095	0.099	0.100	0.097	0.092
0.098	0.101	0.103	0.101	0.092	0.095	0.091	0.086	0.086	0.087
0.093	0.097	0.095	0.091	0.086	0.087	0.088	0.088	0.089	0.087
0.090	0.088	0.087	0.089	0.090	0.090	0.087	0.086	0.088	0.083
0.085	0.085	0.087	0.091	0.093	0.096	0.095	0.089	0.089	0.085
0.088	0.089	0.092	0.095	0.091	0.087	0.083	0.083	0.082	0.081
0.081	0.080	0.081	0.082	0.080	0.076	0.072	0.073	0.075	0.077
0.075	0.080	0.081	0.081	0.081	0.081	0.081	0.084	0.086	0.087
0.088	0.086	0.084	0.082	0.080	0.079	0.082	0.082	0.076	0.081
0.083	0.082	0.081	0.075	0.078	0.078	0.078	0.079	0.082	0.082
0.084	0.082	0.077	0.077	0.077	0.075	0.077	0.073	0.075	0.076
0.080	0.077	0.068	0.071	0.071	0.068	0.067	0.069	0.072	0.082

Total 320 observations (read row-wise)

Table A16 Papermaking Machine Step Input Test Data (Chapter 10)
(sampling interval $\Delta = 1$ sec)

Stock Flow (gal/min)	Measured Basis Weight (lb/3300 sq ft)	Stock Flow (gal/min)	Measured Basis Weight (lb/3300 sq ft)
18.738	39.120	18.787	40.683
18.930	39.149	18.843	40.674
18.581	39.186	18.749	40.576
18.670	39.115	18.786	40.487
18.805	39.106	18.783	40.469
18.798	39.151	18.764	40.680

Total 100 observations (read downward)

Table A16—*Continued*

Stock Flow (gal/min)	Measured Basis Weight (lb/3300 sq ft)	Stock Flow (gal/min)	Measured Basis Weight (lb/3300 sq ft)
18.843	39.171	18.813	40.367
18.869	39.240	18.764	40.222
18.824	39.271	18.820	40.223
18.832	39.359	18.854	40.195
18.824	39.375	18.787	40.264
18.876	39.360	18.907	40.436
18.869	39.275	18.685	40.439
18.816	39.282	18.799	40.448
18.787	39.226	18.787	40.605
18.820	39.361	18.801	40.611
18.843	39.366	18.786	40.713
18.801	39.404	18.843	40.604
18.843	39.652	18.798	40.410
18.835	39.475	18.817	40.486
18.824	39.608	18.772	40.475
18.892	39.697	18.805	40.527
18.765	39.736	18.824	40.702
18.805	39.854	18.831	40.608
18.858	39.559	18.779	40.912
18.869	39.649	18.848	40.807
18.820	40.064	18.757	40.865
18.910	40.066	18.828	40.874
18.797	40.265	18.818	40.774
18.708	40.470	18.835	40.522
18.772	40.439	18.794	40.205
18.862	40.433	18.850	40.206
18.787	40.364	18.779	40.231
18.854	40.335	18.869	40.228
18.723	40.339	18.835	40.521
18.828	40.109	18.813	40.438
18.862	40.194	18.820	40.439
18.824	40.181	18.779	40.562
18.861	40.196	18.783	40.556
18.790	40.211	18.805	40.753
18.802	40.516	18.768	40.539
18.854	40.367	18.849	40.382
18.801	40.379	18.798	40.481
18.809	40.572	18.843	40.476
18.846	40.623	18.764	40.457
18.805	40.366	18.738	40.482
18.865	40.311	18.847	40.636
18.745	40.310	18.775	40.970
18.850	40.406	18.817	40.891
18.813	40.592	18.783	40.819

Total 100 observations (read downward)

Table A17 Chemical Concentration Data (Chapter 10) (sampling interval Δ = 0.02 sec)

56	55	54	52	50	49	49	48	47	47	46	45	45	45	45	45	45	44	44	42
42	42	42	42	42	42	41	41	40	41	40	40	40	40	39	39	39	39	39	38
37	38	39	37	37	37	37	37	37	36	36	36	36	36	36	35	35	35	35	35
35	35	35	35	35	35	35	35	34	34	34	33	33	34	34	33	34	34	33	33
33	33	33	32	33	33	33	32	32	32	32	32	33	32	32	33	32	32	32	32

Total 100 observations (read row-wise)

Table A18 Quarterly Gross National Product of the U.S. (1946 to 1970)

	Quarterly			
Year	I Q	II Q	III Q	IV Q
200, Gross National Product in Current Dollars (Annual Rate, Billions of Dollars)				
1946	196.5	204.0	214.2	219.2
1947	223.6	227.6	231.8	242.1
1948	248.0	255.6	262.5	263.9
1949	258.5	255.2	257.1	255.0
1950	266.0	275.4	293.1	304.5
1951	318.0	325.8	332.8	336.9
1952	339.5	339.1	345.6	357.7
1953	364.2	367.5	365.8	360.8
1954	360.7	360.4	364.7	373.4
1955	386.2	394.4	402.5	408.8
1956	410.6	416.2	420.6	429.5
1957	436.9	439.9	446.3	441.5
1958	434.7	438.3	451.4	464.4
1959	474.0	486.9	484.0	490.5
1960	503.0	504.7	504.2	503.3
1961	503.6	514.9	524.2	537.7
1962	547.8	557.2	564.4	572.0
1963	577.4	584.2	594.7	605.8
1964	617.7	628.0	638.9	645.1
1965	662.8	675.7	691.1	710.0
1966	729.5	743.3	755.9	770.7
1967	774.4	784.5	800.9	815.9
1968	834.0	857.4	875.2	890.2
1969	907.0	923.5	941.7	948.9
1970	958.0	971.7	986.3	989.7

Table A19 Quarterly Expenditures on Producers' Durables in the U.S. from 1946 to 1970.

Year	Quarterly			
	I Q	II Q	III Q	IV Q
243. Gross Private Domestic Fixed Investment, Producers' Durable Equipment (Annual Rate, Billions of Dollars)				
1946	7.5	8.9	11.1	13.4
1947	15.5	15.7	15.6	16.7
1948	18.0	17.4	17.9	18.8
1949	17.6	17.0	16.1	15.7
1950	15.9	17.9	20.3	20.4
1951	20.2	20.5	20.9	20.9
1952	21.1	21.4	18.2	20.1
1953	21.4	21.3	21.9	21.3
1954	20.4	20.4	20.7	20.7
1955	20.9	23.0	24.9	26.5
1956	25.6	26.1	27.0	27.2
1957	28.1	28.0	29.1	28.3
1958	25.7	24.5	24.4	25.5
1959	27.0	28.7	29.1	29.0
1960	29.6	31.2	30.6	29.8
1961	27.6	27.7	29.0	30.3
1962	31.0	32.1	33.5	33.2
1963	33.2	33.8	35.5	36.8
1964	37.9	39.0	41.0	41.6
1965	43.7	44.4	46.6	48.3
1966	50.2	52.1	54.0	56.0
1967	53.9	55.6	55.4	56.2
1968	57.9	57.3	58.8	60.1
1969	63.1	63.5	64.8	65.7
1970	64.8	65.6	67.2	62.1

Table A20   Normal Operating Record of a Papermaking Machine
(Chapter 11) (Sampling Interval $\Delta = 1$ sec.)

Stock Flow (gal/min)	Measured Basis Weight (lb/3300 sq ft)
18.299	38.808
18.218	38.832
18.324	38.832
18.336	38.834
18.258	38.914
18.255	38.962
18.276	38.962
18.302	38.927
18.291	38.971
18.291	38.909
18.250	38.864
18.284	39.051
18.259	39.083
18.376	39.036
18.255	39.176
18.328	39.133
18.259	39.259
18.358	39.215
18.277	39.159
18.365	39.099
18.313	39.167
18.343	38.793
18.358	38.839
18.240	38.846
18.270	38.851
18.321	38.916
18.325	38.941
18.336	39.004
18.284	38.991
18.365	38.950
18.336	39.019
18.270	38.942
18.409	38.741
18.294	38.838
18.250	38.851
18.264	38.934
18.244	38.901
18.399	39.017
18.269	38.970
18.276	38.968
18.321	38.943
18.295	38.922
18.258	38.887

Total 50 observations (read downward)

Table A20—*Continued*

Stock Flow (gal/min)	Measured Basis Weight (lb/3300 sq ft)
18.261	38.957
18.196	39.018
18.300	38.981
18.377	38.999
18.340	39.102
18.392	39.054
18.347	39.075
Total 50 observations (read downward)	

Table A21   Hourly Pulping Digester Operation Data (Chapter 11)

Temperature (°F)	Blow-to-Feed Ratio (gpm/rpm)	K number	Temperature (°F)	Blow-to-Feed Ratio (gpm/rpm)	K number
339.4	57.1	18.90	339.4	53.3	17.30
339.5	55.2	20.50	339.4	52.5	17.80
339.7	57.9	20.60	339.2	52.9	16.10
343.9	53.4	27.60	338.9	52.8	16.10
343.0	56.8	24.00	339.1	53.1	18.60
343.2	59.5	24.60	338.9	52.8	16.60
343.1	57.9	23.20	338.7	52.9	16.80
343.1	57.9	21.30	338.5	52.9	16.50
343.5	55.0	23.50	338.4	52.8	16.90
343.7	52.8	21.60	338.5	55.4	18.30
340.8	44.6	19.90	338.4	57.9	17.50
341.5	52.5	24.90	338.6	57.4	19.60
341.3	52.6	18.60	338.6	56.2	19.30
340.8	56.7	17.20	338.6	55.9	19.00
340.7	57.3	16.90	340.4	55.2	18.40
340.9	54.2	17.00	340.1	54.3	17.90
338.9	55.9	17.90	340.1	53.1	18.50
338.5	52.9	15.29	340.4	53.0	20.50
337.9	52.9	14.10	340.8	53.0	22.20
337.5	53.2	13.50	340.7	52.7	20.70
337.5	52.8	14.80	340.4	52.6	20.00
337.6	53.0	15.40	341.0	·51.8	20.00
337.6	52.9	16.60	341.0	50.5	19.80
337.3	52.8	16.00	341.0	53.2	19.10
337.1	53.1	16.80	344.4	51.9	15.60
337.0	52.8	16.90	344.3	50.0	15.70
337.1	53.1	18.20	344.1	49.9	14.20
342.6	53.0	14.60	344.3	49.4	15.80
Total 110 observations (read downward) *continued next page*					

Table A21 Hourly Pulping Digester Operation Data (Chapter 11)
—*Continued*

Temperature (°F)	Blow-to-Feed Ratio (gpm/rpm)	K number	Temperature (°F)	Blow-to-Feed Ratio (gpm/rpm)	K number
344.5	56.0	16.10	342.1	56.8	19.60
345.8	53.5	14.60	342.5	55.4	19.60
344.6	60.4	12.50	342.2	53.5	18.70
342.9	57.6	13.30	343.0	56.7	20.60
339.4	55.6	13.10	340.2	57.2	21.40
340.0	55.3	18.90	341.5	51.7	20.70
339.9	58.1	15.90	341.2	55.7	19.30
339.5	55.2	17.00	340.9	50.1	18.00
339.9	56.3	21.00	341.2	52.8	20.60
340.4	55.7	22.70	341.5	52.0	21.40
340.9	57.4	25.10	341.4	52.6	19.40
341.9	56.9	22.20	341.1	53.0	19.30
341.8	58.6	22.80	342.3	51.9	18.00
341.8	55.6	22.40	345.0	53.4	17.20
342.1	55.9	23.00	343.2	53.2	16.30
342.2	56.7	22.50	344.5	55.6	16.10
342.7	54.8	25.80	343.5	54.4	16.00
345.9	51.5	23.60	343.3	54.0	15.70
339.6	55.1	21.40	343.2	50.5	15.80
340.4	53.9	17.80	343.3	51.3	13.20
343.0	53.2	17.20	343.2	51.0	14.90
341.6	53.1	19.20	341.9	51.0	14.50
341.4	56.4	16.80	341.8	51.2	15.20
341.1	56.7	16.60	341.6	50.9	15.00
340.7	57.9	15.20	341.3	51.0	14.50
340.7	59.1	16.40	340.0	51.0	15.50
341.7	55.8	13.80	339.8	52.4	15.60

Total 110 observations (read downward)

NORMAL, t, χ^2, and F TABLES

Table A Cumulative Normal Distribution[a]

$$\Phi(z) = \frac{1}{\sqrt{2\pi}} \int_{-\infty}^{z} e^{-u^2/2}\, du$$

z	0.00	0.01	0.02	0.03	0.04	0.05	0.06	0.07	0.08	0.09
0.0	0.50000	0.50399	0.50798	0.51197	0.51595	0.51994	0.52392	0.52790	0.53188	0.53586
0.1	0.53983	0.54380	0.54776	0.55172	0.55567	0.55962	0.56356	0.56749	0.57142	0.57535
0.2	0.57926	0.58317	0.58706	0.59095	0.59483	0.59871	0.60257	0.60642	0.61026	0.61409
0.3	0.61791	0.62172	0.62552	0.62930	0.63307	0.63683	0.64058	0.64431	0.64803	0.65173
0.4	0.65542	0.65910	0.66276	0.66640	0.67003	0.67364	0.67724	0.68082	0.68439	0.68793
0.5	0.69146	0.69497	0.69847	0.70194	0.70540	0.70884	0.71226	0.71566	0.71904	0.72240
0.6	0.72575	0.72907	0.73237	0.73565	0.73891	0.74215	0.74537	0.74857	0.75175	0.75490
0.7	0.75804	0.76115	0.76424	0.76730	0.77035	0.77337	0.77637	0.77935	0.78230	0.78524
0.8	0.78814	0.79103	0.79389	0.79673	0.79955	0.80234	0.80511	0.80785	0.81057	0.81327
0.9	0.81594	0.81859	0.82121	0.82381	0.82639	0.82894	0.83147	0.83398	0.83646	0.83891
1.0	0.84134	0.84375	0.84614	0.84850	0.85083	0.85314	0.85543	0.85769	0.85993	0.86214
1.1	0.86433	0.86650	0.86864	0.87076	0.87286	0.87493	0.87698	0.87900	0.88100	0.88298
1.2	0.88493	0.88686	0.88877	0.89065	0.89251	0.89435	0.89617	0.89796	0.89973	0.90147
1.3	0.90320	0.90490	0.90658	0.90824	0.90988	0.91149	0.91309	0.91466	0.91621	0.91774
1.4	0.91924	0.92073	0.92220	0.92364	0.92507	0.92647	0.92786	0.92922	0.93056	0.93189
1.5	0.93319	0.93448	0.93574	0.93699	0.93822	0.93943	0.94062	0.94179	0.94295	0.94408
1.6	0.94520	0.94630	0.94738	0.94845	0.94950	0.95053	0.95154	0.95254	0.95352	0.95449
1.7	0.95543	0.95637	0.95728	0.95818	0.95907	0.95994	0.96080	0.96164	0.96246	0.96327
1.8	0.96407	0.96485	0.96562	0.96638	0.96712	0.96784	0.96856	0.96926	0.96995	0.97062
1.9	0.97128	0.97193	0.97257	0.97320	0.97381	0.97441	0.97500	0.97558	0.97615	0.97670
2.0	0.97725	0.97778	0.97831	0.97882	0.97932	0.97982	0.98030	0.98077	0.98124	0.98169
2.1	0.98214	0.98257	0.98300	0.98341	0.98382	0.98422	0.98461	0.98500	0.98537	0.98574
2.2	0.98610	0.98645	0.98679	0.98713	0.98745	0.98778	0.98809	0.98840	0.98870	0.98899
2.3	0.98928	0.98956	0.98983	0.99010	0.99036	0.99061	0.99086	0.99111	0.99134	0.99158
2.4	0.99180	0.99202	0.99224	0.99245	0.99266	0.99286	0.99305	0.99324	0.99343	0.99361
2.5	0.99379	0.99396	0.99413	0.99430	0.99446	0.99461	0.99477	0.99492	0.99506	0.99520
2.6	0.99534	0.99547	0.99560	0.99573	0.99585	0.99598	0.99609	0.99621	0.99632	0.99643
2.7	0.99653	0.99664	0.99674	0.99683	0.99693	0.99702	0.99711	0.99720	0.99728	0.99736
2.8	0.99744	0.99752	0.99760	0.99767	0.99774	0.99781	0.99788	0.99795	0.99801	0.99807
2.9	0.99813	0.99819	0.99825	0.99831	0.99836	0.99841	0.99846	0.99851	0.99856	0.99861
3.0	0.99865	0.99869	0.99874	0.99878	0.99882	0.99886	0.99889	0.99893	0.99897	0.99900
3.1	0.99903	0.99906	0.99910	0.99913	0.99916	0.99918	0.99921	0.99924	0.99926	0.99929
3.2	0.99931	0.99934	0.99936	0.99938	0.99940	0.99942	0.99944	0.99946	0.99948	0.99950
3.3	0.99952	0.99953	0.99957	0.99957	0.99958	0.99960	0.99961	0.99962	0.99964	0.99965
3.4	0.99966	0.99968	0.99969	0.99970	0.99971	0.99972	0.99973	0.99974	0.99975	0.99976
3.5	0.99977	0.99978	0.99978	0.99979	0.99980	0.99981	0.99981	0.99982	0.99983	0.99983
3.6	0.99984	0.99985	0.99985	0.99986	0.99986	0.99987	0.99987	0.99988	0.99988	0.99989
3.7	0.99989	0.99990	0.99990	0.99990	0.99991	0.99991	0.99992	0.99992	0.99992	0.99992
3.8	0.99993	0.99993	0.99993	0.99994	0.99994	0.99994	0.99994	0.99995	0.99995	0.99995
3.9	0.99995	0.99995	0.99996	0.99996	0.99996	0.99996	0.99996	0.99996	0.99997	0.99997

[a] Reproduced with permission from Pearson and Hartley, *Biometrika Tables for Statisticians*, Vol. 1 (1958), pp. 104–108.

Table B Percentage Points of the χ^2 Distribution[a]
Values of $\chi^2_{\nu, P}$ such that

$$P = \frac{1}{2^{\nu/2}\Gamma(\nu/2)} \int_0^{\chi^2_{\nu,P}} y^{\nu/2-1} e^{-y/2} \, dy$$

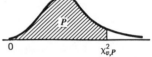

ν \ P	0.005	0.010	0.025	0.050	0.100	0.250	0.500
1	0.00004	0.00016	0.00098	0.00393	0.01579	0.1015	0.4549
2	0.0100	0.0201	0.0506	0.1026	0.2107	0.5754	1.386
3	0.0717	0.1148	0.2158	0.3518	0.5844	1.213	2.366
4	0.2070	0.2971	0.4844	0.7107	1.064	1.923	3.357
5	0.4117	0.5543	0.8312	1.145	1.610	2.675	4.351
6	0.6757	0.8721	1.2373	1.635	2.204	3.455	5.348
7	0.9893	1.239	1.690	2.167	2.833	4.255	6.346
8	1.344	1.646	2.180	2.733	3.490	5.071	7.344
9	1.735	2.088	2.700	3.325	4.168	5.899	8.343
10	2.156	2.558	3.247	3.940	4.865	6.737	9.342
11	2.603	3.053	3.816	4.575	5.578	7.584	10.34
12	3.074	3.571	4.404	5.226	6.304	8.438	11.34
13	3.565	4.107	5.009	5.892	7.041	9.299	12.34
14	4.075	4.660	5.629	6.571	7.790	10.17	13.34
15	4.601	5.229	6.262	7.261	8.547	11.04	14.34
16	5.142	5.812	6.908	7.962	9.312	11.91	15.34
17	5.697	6.408	7.564	8.672	10.09	12.79	16.34
18	6.265	7.015	8.231	9.390	10.86	13.68	17.34
19	6.844	7.633	8.907	10.12	11.65	14.56	18.34
20	7.434	8.260	9.591	10.85	12.44	15.45	19.34
21	8.034	8.897	10.28	11.59	13.24	16.34	20.34
22	8.643	9.542	10.98	12.34	14.04	17.24	21.34
23	9.260	10.20	11.69	13.09	14.85	18.14	22.34
24	9.886	10.86	12.40	13.85	15.66	19.04	23.34
25	10.52	11.52	13.12	14.61	16.47	19.94	24.34
26	11.16	12.20	13.84	15.38	17.29	20.84	25.34
27	11.81	12.88	14.57	16.15	18.11	21.75	26.34
28	12.46	13.56	15.31	16.93	18.94	22.66	27.34
29	13.12	14.26	16.05	17.71	19.77	23.57	28.34
30	13.79	14.95	16.79	18.49	20.60	24.48	29.34
40	20.71	22.16	24.43	26.51	29.05	33.66	39.34
50	27.99	29.71	32.36	34.76	37.69	42.94	49.33
60	35.53	37.48	40.48	43.19	46.46	52.29	59.33
70	43.28	45.44	48.76	51.74	55.33	61.70	69.33
80	51.17	53.54	57.15	60.39	64.28	71.14	79.33
90	59.20	61.75	65.65	69.13	73.29	80.62	89.33
100	67.33	70.06	74.22	77.93	82.36	90.13	99.33

[a] Reproduced with permission from Pearson and Hartley, *Biometrika Tables for Statisticians*, Vol. 1, pp. 130–131 (1958).

Table B (continued)

0.750	0.900	0.950	0.975	0.990	0.995	0.999
1.323	2.706	3.841	5.024	6.635	7.879	10.83
2.773	4.605	5.991	7.378	9.210	10.60	13.82
4.108	6.251	7.815	9.348	11.34	12.84	16.27
5.385	7.779	9.488	11.14	13.28	14.86	18.47
6.626	9.236	11.07	12.83	15.09	16.75	20.52
7.841	10.64	12.59	14.45	16.81	18.55	22.46
9.037	12.02	14.07	16.01	18.48	20.28	24.32
10.22	13.36	15.51	17.53	20.09	21.96	26.12
11.39	14.68	16.92	19.02	21.67	23.59	27.88
12.55	15.99	18.31	20.48	23.21	25.19	29.59
13.70	17.28	19.68	21.92	24.72	26.76	31.26
14.85	18.55	21.03	23.34	26.22	28.30	32.91
15.98	19.81	22.36	24.74	27.69	29.82	34.53
17.12	21.06	23.68	26.12	29.14	31.32	36.12
18.25	22.31	25.00	27.49	30.58	32.80	37.70
19.37	23.54	26.30	28.85	32.00	34.27	39.25
20.49	24.77	27.59	30.19	33.41	35.72	40.79
21.60	25.99	28.87	31.53	34.81	37.16	42.31
22.72	27.20	30.14	32.85	36.19	38.58	43.82
23.83	28.41	31.41	34.17	37.57	40.00	45.32
24.93	29.62	32.67	35.48	38.93	41.40	46.80
26.04	30.81	33.92	36.78	40.29	42.80	48.27
27.14	32.01	35.17	38.08	41.64	44.18	49.73
28.24	33.20	36.42	39.36	42.98	45.56	51.18
29.34	34.38	37.65	40.65	44.31	46.93	52.62
30.43	35.56	38.89	41.92	45.64	48.29	54.05
31.53	36.74	40.11	43.19	46.96	49.64	55.48
32.62	37.92	41.34	44.46	48.28	50.99	56.89
33.71	39.09	42.56	45.72	49.59	52.34	58.30
34.80	40.26	43.77	46.98	50.89	53.67	59.70
45.62	51.80	55.76	59.34	63.69	66.77	73.40
56.33	63.17	67.50	71.42	76.15	79.49	86.66
66.98	74.40	79.08	83.30	88.38	91.95	99.61
77.58	85.53	90.53	95.02	100.4	104.2	112.3
88.13	96.58	101.9	106.6	112.3	116.3	124.8
98.65	107.6	113.1	118.1	124.1	128.3	137.2
109.1	118.5	124.3	129.6	135.8	140.2	149.4

Table C Percentage Points of Student's t-Distribution[a]
Values of $t_{v,P}$ such that

$$P = \int_{-\infty}^{t_{v,P}} \frac{1}{\sqrt{v\pi}} \frac{\Gamma\left(\frac{v+1}{2}\right)}{\Gamma\left(\frac{v}{2}\right)} \left(1 + \frac{t^2}{v}\right)^{-(v+1)/2} dt$$

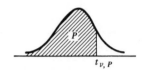

v \ P	0.750	0.900	0.950	0.975	0.990	0.995	0.999	0.9995
1	1.000	3.078	6.314	12.706	31.821	63.657	318.31	636.62
2	0.816	1.886	2.920	4.303	6.965	9.925	22.326	31.598
3	0.765	1.638	2.353	3.182	4.541	5.841	10.213	12.924
4	0.741	1.533	2.132	2.776	3.747	4.604	7.173	8.610
5	0.727	1.476	2.015	2.571	3.365	4.032	5.893	6.869
6	0.718	1.440	1.943	2.447	3.143	3.707	5.208	5.959
7	0.711	1.415	1.895	2.365	2.998	3.499	4.785	5.408
8	0.706	1.397	1.860	2.306	2.896	3.355	4.501	5.041
9	0.703	1.383	1.833	2.262	2.821	3.250	4.297	4.781
10	0.700	1.372	1.812	2.228	2.764	3.169	4.144	4.587
11	0.697	1.363	1.796	2.201	2.718	3.106	4.025	4.437
12	0.695	1.356	1.782	2.179	2.681	3.055	3.930	4.318
13	0.694	1.350	1.771	2.160	2.650	3.012	3.852	4.221
14	0.692	1.345	1.761	2.145	2.624	2.977	3.787	4.140
15	0.691	1.341	1.753	2.131	2.602	2.947	3.733	4.073
16	0.690	1.337	1.746	2.120	2.583	2.921	3.686	4.015
17	0.689	1.333	1.740	2.110	2.567	2.898	3.646	3.965
18	0.688	1.330	1.734	2.101	2.552	2.878	3.610	3.922
19	0.688	1.328	1.729	2.093	2.539	2.861	3.579	3.883

Table C (continued) Percentage Points of Student's t-Distribution[a]

Values of $t_{v,P}$ such that

$$P = \int_{-\infty}^{t_{v,P}} \frac{1}{\sqrt{v\pi}} \frac{\Gamma\left(\dfrac{v+1}{2}\right)}{\Gamma\left(\dfrac{v}{2}\right)} \left(1 + \frac{t^2}{v}\right)^{-(v+1)/2} dt$$

P / v	0.750	0.900	0.950	0.975	0.990	0.995	0.999	0.9995
20	0.687	1.325	1.725	2.086	2.528	2.845	3.552	3.850
21	0.686	1.323	1.721	2.080	2.518	2.831	3.527	3.819
22	0.686	1.321	1.717	2.074	2.508	2.819	3.505	3.792
23	0.685	1.319	1.714	2.069	2.500	2.807	3.485	3.767
24	0.685	1.318	1.711	2.064	2.492	2.797	3.467	3.745
25	0.684	1.316	1.708	2.060	2.485	2.787	3.450	3.725
26	0.684	1.315	1.706	2.056	2.479	2.779	3.435	3.707
27	0.684	1.314	1.703	2.052	2.473	2.771	3.421	3.690
28	0.683	1.313	1.701	2.048	2.467	2.763	3.408	3.674
29	0.683	1.311	1.699	2.045	2.462	2.756	3.396	3.659
30	0.683	1.310	1.697	2.042	2.457	2.750	3.385	3.646
40	0.681	1.303	1.684	2.021	2.423	2.704	3.307	3.551
60	0.679	1.296	1.671	2.000	2.390	2.660	3.232	3.460
120	0.677	1.289	1.658	1.980	2.358	2.617	3.160	3.373
∞	0.674	1.282	1.645	1.960	2.326	2.576	3.090	3.291

[a] Reproduced from Pearson and Hartley, *Biometrika Tables for Statisticians*, Vol. 1 (1958), p. 138, and from Table III, Fisher and Yates, *Statistical Tables for Biological, Agricultural and Medical Research*, published by Longman Group Ltd., London (previously published by Oliver and Boyd, Edinburgh, 1953) and by permission of authors and publishers.

Table D Percentage Points of the F-Distribution[a]
Values of $F_{v_1,v_2,P}$ such that

$$P = \frac{1}{B(v_1/2,\, v_2/2)} \int_0^{v_1 F_{v_1,v_2,P}/v_2} g^{v_1/2-1}(1+g)^{-(v_1+v_2)/2}\, dg$$

$P = 0.95$

v_2 \ v_1	1	2	3	4	5	6	7	8	9
1	161.4	199.5	215.7	224.6	230.2	234.0	236.8	238.9	240.5
2	18.51	19.00	19.16	19.25	19.30	19.33	19.35	19.37	19.38
3	10.13	9.55	9.28	9.12	9.01	8.94	8.89	8.85	8.81
4	7.71	6.94	6.59	6.39	6.26	6.16	6.09	6.04	6.00
5	6.61	5.79	5.41	5.19	5.05	4.95	4.88	4.82	4.77
6	5.99	5.14	4.76	4.53	4.39	4.28	4.21	4.15	4.10
7	5.59	4.74	4.35	4.12	3.97	3.87	3.79	3.73	3.68
8	5.32	4.46	4.07	3.84	3.69	3.58	3.50	3.44	3.39
9	5.12	4.26	3.86	3.63	3.48	3.37	3.29	3.23	3.18
10	4.96	4.10	3.71	3.48	3.33	3.22	3.14	3.07	3.02
11	4.84	3.98	3.59	3.36	3.20	3.09	3.01	2.95	2.90
12	4.75	3.89	3.49	3.26	3.11	3.00	2.91	2.85	2.80
13	4.67	3.81	3.41	3.18	3.03	2.92	2.83	2.77	2.71
14	4.60	3.74	3.34	3.11	2.96	2.85	2.76	2.70	2.65
15	4.54	3.68	3.29	3.06	2.90	2.79	2.71	2.64	2.59
16	4.49	3.63	3.24	3.01	2.85	2.74	2.66	2.59	2.54
17	4.45	3.59	3.20	2.96	2.81	2.70	2.61	2.55	2.49
18	4.41	3.55	3.16	2.93	2.77	2.66	2.58	2.51	2.46
19	4.38	3.52	3.13	2.90	2.74	2.63	2.54	2.48	2.42
20	4.35	3.49	3.10	2.87	2.71	2.60	2.51	2.45	2.39
21	4.32	3.47	3.07	2.84	2.68	2.57	2.49	2.42	2.37
22	4.30	3.44	3.05	2.82	2.66	2.55	2.46	2.40	2.34
23	4.28	3.42	3.03	2.80	2.64	2.53	2.44	2.37	2.32
24	4.26	3.40	3.01	2.78	2.62	2.51	2.42	2.36	2.30
25	4.24	3.39	2.99	2.76	2.60	2.49	2.40	2.34	2.28
26	4.23	3.37	2.98	2.74	2.59	2.47	2.39	2.32	2.27
27	4.21	3.35	2.96	2.73	2.57	2.46	2.37	2.31	2.25
28	4.20	3.34	2.95	2.71	2.56	2.45	2.36	2.29	2.24
29	4.18	3.33	2.93	2.70	2.55	2.43	2.35	2.28	2.22
30	4.17	3.32	2.92	2.69	2.53	2.42	2.33	2.27	2.21
40	4.08	3.23	2.84	2.61	2.45	2.34	2.25	2.18	2.12
60	4.00	3.15	2.76	2.53	2.37	2.25	2.17	2.10	2.04
120	3.92	3.07	2.68	2.45	2.29	2.17	2.09	2.02	1.96
∞	3.84	3.00	2.60	2.37	2.21	2.10	2.01	1.94	1.88

[a] Reproduced with permission from Pearson and Hartley, *Biometrika Tables for Statisticians*, Vol. 1 (1958), pp. 159–163.

Table D (continued)

10	12	15	20	24	30	40	60	120	∞
241.9	243.9	245.9	248.0	249.1	250.1	251.1	252.2	253.3	254.3
19.40	19.41	19.43	19.45	19.45	19.46	19.47	19.48	19.49	19.50
8.79	8.74	8.70	8.66	8.64	8.62	8.59	8.57	8.55	8.53
5.96	5.91	5.86	5.80	5.77	5.75	5.72	5.69	5.66	5.63
4.74	4.68	4.62	4.56	4.53	4.50	4.46	4.43	4.40	4.36
4.06	4.00	3.94	3.87	3.84	3.81	3.77	3.74	3.70	3.67
3.64	3.57	3.51	3.44	3.41	3.38	3.34	3.30	3.27	3.23
3.35	3.28	3.22	3.15	3.12	3.08	3.04	3.01	2.97	2.93
3.14	3.07	3.01	2.94	2.90	2.86	2.83	2.79	2.75	2.71
2.98	2.91	2.85	2.77	2.74	2.70	2.66	2.62	2.58	2.54
2.85	2.79	2.72	2.65	2.61	2.57	2.53	2.49	2.45	2.40
2.75	2.69	2.62	2.54	2.51	2.47	2.43	2.38	2.34	2.30
2.67	2.60	2.53	2.46	2.42	2.38	2.34	2.30	2.25	2.21
2.60	2.53	2.46	2.39	2.35	2.31	2.27	2.22	2.18	2.13
2.54	2.48	2.40	2.33	2.29	2.25	2.20	2.16	2.11	2.07
2.49	2.42	2.35	2.28	2.24	2.19	2.15	2.11	2.06	2.01
2.45	2.38	2.31	2.23	2.19	2.15	2.10	2.06	2.01	1.96
2.41	2.34	2.27	2.19	2.15	2.11	2.06	2.02	1.97	1.92
2.38	2.31	2.23	2.16	2.11	2.07	2.03	1.98	1.93	1.88
2.35	2.28	2.20	2.12	2.08	2.04	1.99	1.95	1.90	1.84
2.32	2.25	2.18	2.10	2.05	2.01	1.96	1.92	1.87	1.81
2.30	2.23	2.15	2.07	2.03	1.98	1.94	1.89	1.84	1.78
2.27	2.20	2.13	2.05	2.01	1.96	1.91	1.86	1.81	1.76
2.25	2.18	2.11	2.03	1.98	1.94	1.89	1.84	1.79	1.73
2.24	2.16	2.09	2.01	1.96	1.92	1.87	1.82	1.77	1.71
2.22	2.15	2.07	1.99	1.95	1.90	1.85	1.80	1.75	1.69
2.20	2.13	2.06	1.97	1.93	1.88	1.84	1.79	1.73	1.67
2.19	2.12	2.04	1.96	1.91	1.87	1.82	1.77	1.71	1.65
2.18	2.10	2.03	1.94	1.90	1.85	1.81	1.75	1.70	1.64
2.16	2.09	2.01	1.93	1.89	1.84	1.79	1.74	1.68	1.62
2.08	2.00	1.92	1.84	1.79	1.74	1.69	1.64	1.58	1.51
1.99	1.92	1.84	1.75	1.70	1.65	1.59	1.53	1.47	1.39
1.91	1.83	1.75	1.66	1.61	1.55	1.50	1.43	1.35	1.25
1.83	1.75	1.67	1.57	1.52	1.46	1.39	1.32	1.22	1.00

Table D (continued)

Values of $F_{v1.v2,P}$ such that

$$P = \frac{1}{B(v_1/2,\ v_2/2)} \int_0^{v_1\ F_{v_1,v_2,P/v_2}} g^{v_1/2-1}(1+g)^{-(v_1+v_2)/2}\ dg$$

$P = 0.99$

v_2 \ v_1	1	2	3	4	5	6	7	8	9
1	4052	4999.5	5403	5625	5764	5859	5928	5982	6022
2	98.50	99.00	99.17	99.25	99.30	99.33	99.36	99.37	99.39
3	34.12	30.82	29.46	28.71	28.24	27.91	27.67	27.49	27.35
4	21.20	18.00	16.69	15.98	15.52	15.21	14.98	14.80	14.66
5	16.26	13.27	12.06	11.39	10.97	10.67	10.46	10.29	10.16
6	13.75	10.92	9.78	9.15	8.75	8.47	8.26	8.10	7.98
7	12.25	9.55	8.45	7.85	7.46	7.19	6.99	6.84	6.72
8	11.26	8.65	7.59	7.01	6.63	6.37	6.18	6.03	5.91
9	10.56	8.02	6.99	6.42	6.06	5.80	5.61	5.47	5.35
10	10.04	7.56	6.55	5.99	5.64	5.39	5.20	5.06	4.94
11	9.65	7.21	6.22	5.67	5.32	5.07	4.89	4.74	4.63
12	9.33	6.93	5.95	5.41	5.06	4.82	4.64	4.50	4.39
13	9.07	6.70	5.74	5.21	4.86	4.62	4.44	4.30	4.19
14	8.86	6.51	5.56	5.04	4.69	4.46	4.28	4.14	4.03
15	8.68	6.36	5.42	4.89	4.56	4.32	4.14	4.00	3.89
16	8.53	6.23	5.29	4.78	4.44	4.20	4.03	3.89	3.78
17	8.40	6.11	5.18	4.67	4.34	4.10	3.93	3.79	3.68
18	8.29	6.01	5.09	4.58	4.25	4.01	3.84	3.71	3.60
19	8.18	5.93	5.01	4.50	4.17	3.94	3.77	3.63	3.52
20	8.10	5.85	4.94	4.43	4.10	3.87	3.70	3.56	3.46
21	8.02	5.78	4.87	4.37	4.04	3.81	3.64	3.51	3.40
22	7.95	5.72	4.82	4.31	3.99	3.76	3.59	3.45	3.35
23	7.88	5.66	4.76	4.26	3.94	3.71	3.54	3.41	3.30
24	7.82	5.61	4.72	4.22	3.90	3.67	3.50	3.36	3.26
25	7.77	5.57	4.68	4.18	3.85	3.63	3.46	3.32	3.22
26	7.72	5.53	4.64	4.14	3.82	3.59	3.42	3.29	3.18
27	7.68	5.49	4.60	4.11	3.78	3.56	3.39	3.26	3.15
28	7.64	5.45	4.57	4.07	3.75	3.53	3.36	3.23	3.12
29	7.60	5.42	4.54	4.04	3.73	3.50	3.33	3.20	3.09
30	7.56	5.39	4.51	4.02	3.70	3.47	3.30	3.17	3.07
40	7.31	5.18	4.31	3.83	3.51	3.29	3.12	2.99	2.89
60	7.08	4.98	4.13	3.65	3.34	3.12	2.95	2.82	2.72
120	6.85	4.79	3.95	3.48	3.17	2.96	2.79	2.66	2.56
∞	6.63	4.61	3.78	3.32	3.02	2.80	2.64	2.51	2.41

Table D (continued)

10	12	15	20	24	30	40	60	120	∞
6056	6106	6157	6209	6235	6261	6287	6313	6339	6366
99.40	99.42	99.43	99.45	99.46	99.47	99.47	99.48	99.49	99.50
27.23	27.05	26.87	26.69	26.60	26.50	26.41	26.32	26.22	26.13
14.55	14.37	14.20	14.02	13.93	13.84	13.75	13.65	13.56	13.46
10.05	9.89	9.72	9.55	9.47	9.38	9.29	9.20	9.11	9.02
7.87	7.72	7.56	7.40	7.31	7.23	7.14	7.06	6.97	6.88
6.62	6.47	6.31	6.16	6.07	5.99	5.91	5.82	5.74	5.65
5.81	5.67	5.52	5.36	5.28	5.20	5.12	5.03	4.95	4.86
5.26	5.11	4.96	4.81	4.73	4.65	4.57	4.48	4.40	4.31
4.85	4.71	4.56	4.41	4.33	4.25	4.17	4.08	4.00	3.91
4.54	4.40	4.25	4.10	4.02	3.94	3.86	3.78	3.69	3.60
4.30	4.16	4.01	3.86	3.78	3.70	3.62	3.54	3.45	3.36
4.10	3.96	3.82	3.66	3.59	3.51	3.43	3.34	3.25	3.17
3.94	3.80	3.66	3.51	3.43	3.35	3.27	3.18	3.09	3.00
3.80	3.67	3.52	3.37	3.29	3.21	3.13	3.05	2.96	2.87
3.69	3.55	3.41	3.26	3.18	3.10	3.02	2.93	2.84	2.75
3.59	3.46	3.31	3.16	3.08	3.00	2.92	2.83	2.75	2.65
3.51	3.37	3.23	3.08	3.00	2.92	2.84	2.75	2.66	2.57
3.43	3.30	3.15	3.00	2.92	2.84	2.76	2.67	2.58	2.49
3.37	3.23	3.09	2.94	2.86	2.78	2.69	2.61	2.52	2.42
3.31	3.17	3.03	2.88	2.80	2.72	2.64	2.55	2.46	2.36
3.26	3.12	2.98	2.83	2.75	2.67	2.58	2.50	2.40	2.31
3.21	3.07	2.93	2.78	2.70	2.62	2.54	2.45	2.35	2.26
3.17	3.03	2.89	2.74	2.66	2.58	2.49	2.40	2.31	2.21
3.13	2.99	2.85	2.70	2.62	2.54	2.45	2.36	2.27	2.17
3.09	2.96	2.81	2.66	2.58	2.50	2.42	2.33	2.23	2.13
3.06	2.93	2.78	2.63	2.55	2.47	2.38	2.29	2.20	2.10
3.03	2.90	2.75	2.60	2.52	2.44	2.35	2.26	2.17	2.06
3.00	2.87	2.73	2.57	2.49	2.41	2.33	2.23	2.14	2.03
2.98	2.84	2.70	2.55	2.47	2.39	2.30	2.21	2.11	2.01
2.80	2.66	2.52	2.37	2.29	2.20	2.11	2.02	1.92	1.80
2.63	2.50	2.35	2.20	2.12	2.03	1.94	1.84	1.73	1.60
2.47	2.34	2.19	2.03	1.95	1.86	1.76	1.66	1.53	1.38
2.32	2.18	2.04	1.88	1.79	1.70	1.59	1.47	1.32	1.00

Table D (continued)

Values of $F_{v_1,v_2,P}$ such that

$$P = \frac{1}{B(v_1/2, \, v_2/2)} \int^{v_1 \, F_{v_1,v_2,P/v_2}} g^{v_1/2-1}(1+g)^{-(v_1+v_2)/2} \, dg$$

$P = 0.999$

v_1 / v_2	1	2	3	4	5	6	7	8	9
1	4051*	5000*	5404*	5625*	5764*	5859*	5929*	5981*	6023*
2	998.5	999.0	999.2	999.2	999.3	999.3	999.4	999.4	999.4
3	167.0	148.5	141.1	137.1	134.6	132.8	131.6	130.6	129.9
4	74.14	61.25	56.18	53.44	51.71	50.53	49.66	49.00	48.47
5	47.18	37.12	33.20	31.09	29.75	28.84	28.16	27.64	27.24
6	35.51	27.00	23.70	21.92	20.81	20.03	19.46	19.03	18.69
7	29.25	21.69	18.77	17.19	16.21	15.52	15.02	14.63	14.33
8	25.42	18.49	15.83	14.39	13.49	12.86	12.40	12.04	11.77
9	22.86	16.39	13.90	12.56	11.71	11.13	10.70	10.37	10.11
10	21.04	14.91	12.55	11.28	10.48	9.92	9.52	9.20	8.96
11	19.69	13.81	11.56	10.35	9.58	9.05	8.66	8.35	8.12
12	18.64	12.97	10.80	9.63	8.89	8.38	8.00	7.71	7.48
13	17.81	12.31	10.21	9.07	8.35	7.86	7.49	7.21	6.98
14	17.14	11.78	9.73	8.62	7.92	7.43	7.08	6.80	6.58
15	16.59	11.34	9.34	8.25	7.57	7.09	6.74	6.47	6.26
16	16.12	10.97	9.00	7.94	7.27	6.81	6.46	6.19	5.98
17	15.72	10.66	8.73	7.68	7.02	6.56	6.22	5.96	5.75
18	15.38	10.39	8.49	7.46	6.81	6.35	6.02	5.76	5.56
19	15.08	10.16	8.28	7.26	6.62	6.18	5.85	5.59	5.39
20	14.82	9.95	8.10	7.10	6.46	6.02	5.69	5.44	5.24
21	14.59	9.77	7.94	6.95	6.32	5.88	5.56	5.31	5.11
22	14.38	9.61	7.80	6.81	6.19	5.76	5.44	5.19	4.99
23	14.19	9.47	7.67	6.69	6.08	5.65	5.33	5.09	4.89
24	14.03	9.34	7.55	6.59	5.98	5.55	5.23	4.99	4.80
25	13.88	9.22	7.45	6.49	5.88	5.46	5.15	4.91	4.71
26	13.74	9.12	7.36	6.41	5.80	5.38	5.07	4.83	4.64
27	13.61	9.02	7.27	6.33	5.73	5.31	5.00	4.76	4.57
28	13.50	8.93	7.19	6.25	5.66	5.25	4.93	4.69	4.50
29	13.39	8.85	7.12	6.19	5.59	5.18	4.87	4.64	4.45
30	13.29	8.77	7.05	6.12	5.53	5.12	4.82	4.58	4.39
40	12.61	8.25	6.60	5.70	5.13	4.73	4.44	4.21	4.02
60	11.97	7.76	6.17	5.31	4.76	4.37	4.09	3.87	3.69
120	11.38	7.32	5.79	4.95	4.42	4.04	3.77	3.55	3.38
∞	10.83	6.91	5.42	4.62	4.10	3.74	3.47	3.27	3.10

* Multiply these entries by 100.

Table D (continued)

10	12	15	20	24	30	40	60	120	∞
6056*	6107*	6158*	6209*	6235*	6261*	6287*	6313*	6340*	6366*
999.4	999.4	999.4	999.4	999.5	999.5	999.5	999.5	999.5	999.5
129.2	128.3	127.4	126.4	125.9	125.4	125.0	124.5	124.0	123.5
48.05	47.41	46.76	46.10	45.77	45.43	45.09	44.75	44.40	44.05
26.92	26.42	25.91	25.39	25.14	24.87	24.60	24.33	24.06	23.79
18.41	17.99	17.56	17.12	16.89	16.67	16.44	16.21	15.99	15.75
14.08	13.71	13.32	12.93	12.73	12.53	12.33	12.12	11.91	11.70
11.54	11.19	10.84	10.48	10.30	10.11	9.92	9.73	9.53	9.33
9.89	9.57	9.24	8.90	8.72	8.55	8.37	8.19	8.00	7.81
8.75	8.45	8.13	7.80	7.64	7.47	7.30	7.12	6.94	6.76
7.92	7.63	7.32	7.01	6.85	6.68	6.52	6.35	6.17	6.00
7.29	7.00	6.71	6.40	6.25	6.09	5.93	5.76	5.59	5.42
6.80	6.52	6.23	5.93	5.78	5.63	5.47	5.30	5.14	4.97
6.40	6.13	5.85	5.56	5.41	5.25	5.10	4.94	4.77	4.60
6.08	5.81	5.54	5.25	5.10	4.95	4.80	4.64	4.47	4.31
5.81	5.55	5.27	4.99	4.85	4.70	4.54	4.39	4.23	4.06
5.58	5.32	5.05	4.78	4.63	4.48	4.33	4.18	4.02	3.85
5.39	5.13	4.87	4.59	4.45	4.30	4.15	4.00	3.84	3.67
5.22	4.97	4.70	4.43	4.29	4.14	3.99	3.84	3.68	3.51
5.08	4.82	4.56	4.29	4.15	4.00	3.86	3.70	3.54	3.38
4.95	4.70	4.44	4.17	4.03	3.88	3.74	3.58	3.42	3.26
4.83	4.58	4.33	4.06	3.92	3.78	3.63	3.48	3.32	3.15
4.73	4.48	4.23	3.96	3.82	3.68	3.53	3.38	3.22	3.05
4.64	4.39	4.14	3.87	3.74	3.59	3.45	3.29	3.14	2.97
4.56	4.31	4.06	3.79	3.66	3.52	3.37	3.22	3.06	2.89
4.48	4.24	3.99	3.72	3.59	3.44	3.30	3.15	2.99	2.82
4.41	4.17	3.92	3.66	3.52	3.38	3.23	3.08	2.92	2.75
4.35	4.11	3.86	3.60	3.46	3.32	3.18	3.02	2.86	2.69
4.29	4.05	3.80	3.54	3.41	3.27	3.12	2.97	2.81	2.64
4.24	4.00	3.75	3.49	3.36	3.22	3.07	2.92	2.76	2.59
3.87	3.64	3.40	3.15	3.01	2.87	2.73	2.57	2.41	2.23
3.54	3.31	3.08	2.83	2.69	2.55	2.41	2.25	2.08	1.89
3.24	3.02	2.78	2.53	2.40	2.26	2.11	1.95	1.76	1.54
2.96	2.74	2.51	2.27	2.13	1.99	1.84	1.66	1.45	1.00

COMPUTER PROGRAMS

DESCRIPTION OF PROGRAM INPUT

1. TITLE : ONE CARD WITH A MAXIMUM OF 40 CHARACTERS.

2. NOB,NSER,DELTA,NMOD,INCPH,INCTH

   NOB: NUMBER OF OBSERVATIONS PER SERIES.

   NSER: NUMBER OF SERIES TO BE MODELED SIMULTANEOUSLY.
   (1 FOR ARMA MODELS, >1 FOR EXTENDED ARMA MODELS)

   DELTA: SAMPLING INTERVAL. (TIME BETWEEN DATA POINTS)

   NMOD: NUMBER OF MODELS TO FIT. IF NMOD IS POSITVE, THE
   PROGRAM FITS NMOD MODELS AND GENERATES ITS OWN INITIAL
   VALUES. IF NMOD IS NEGATIVE, THE PROGRAM FITS NMOD MODELS
   WITH THE INITIAL VALUES SUPPLIED BY THE USER.

   INCPH: INCREMENT IN AUTOREGRESSIVE ORDER BETWEEN MODELS.

   INCTH: INCREMENT IN MOVING AVERAGE ORDER BETWEEN MODELS.

3. [(NAR(I,J), I=1,NSER), NMA(J), J=1,NSER]

   NAR: INITIAL AUTOREGRESSIVE ORDER(S) OF THE MODEL.

   NMA: INITIAL MOVING AVERAGE ORDER OF THE MODEL.

   NOTE: FOR EACH SERIES MODELED THERE ARE NSER NAR'S AND 1 NMA.
   IE. FOR NSER=3, NAR=2, AND NMA=1, THIS CARD READS:
   2,2,2,1, 2,2,2,1, 2,2,2,1
   THE NAR'S DO NOT HAVE TO BE EQUAL.

4. STEP,CONTOL,ITMAX,IFREF

   STEP: CONTROLS STEP SIZE IN NON-LINEAR ESTIMATION ROUTINE.

   CONTOL: CONVERGENCE CRITERION FOR NON-LINEAR ESTIMATION.

   ITMAX: MAXIMUM NUMBER OF ITERATIONS ALLOWED.

   IFREF: AUTO-REFINEMENT FLAG. IF IFREF > 0, THE VALUES OF STEP
   AND CONTOL ARE DECREASED BY $10^{**-IFREF}$ AND THE ESTIMATION
   IS REDONE.

   NOTE: IN GENERAL, 1.E-3,1.E-5,20,0 GIVES SATISFACTORY
   RESULTS FOR THIS CARD. FURTHER ADJUSTMENTS SHOULD
   BE MADE BASED ON THE RESULTS OF THE MODELING.

5. IFMEAN,IFAT,IFDIF,IFLAG,IFSEAS,IFSER,IFPRT

   IFMEAN: =0 THE AVERAGE OF THE DATA IS TAKEN AS THE MEAN.

   >0 THE MEAN OF THE DATA IS ESTIMATED BY THE PROGRAM.

   =1 THE AVERAGE OF THE DATA IS SUBTRACTED FOR
   INITIAL VALUE ESTIMATION.
   =N X(N-1) IS SUBTRACTED FOR INITIAL ESTIMATION.

   <0 X(-IFMEAN) IS SUBTRACTED FROM THE DATA BEFORE MODELING,
   AND THE MEAN IS NOT ESTIMATED.

   IFAT: =0 THE FIRST NAR RESIDUALS ARE SET EQUAL TO C.
   (RECOMMENDED FOR STABLE/STATIONARY DATA.)

   >0 A(1)=X(1),A(2)=X(2)-PHI(1)*X(1)+THETA(1)*A(1), ETC.
   IE. ASSUME A(-NAR),...,A(C) = 0, AND X(-NAR),...,
   X(0) = 0.   (RECOMMENDED FOR UNSTABLE/NON-
   STATIONARY DATA.)

   <0 THE INITIAL RESIDUALS ARE ESTIMATED.
   (FOR MAXIMUM LIKELIHOOD ESTIMATES)

IFDIF: =0 NUMERICAL DERIVATIVES ARE USED IN ESTIMATION.

#0 ANALYTICAL DERIVATIVES ARE USED IN ESTIMATION.

NOTE: IFDIF MUST = 0 IF IFSEAS > 0.   (# = 'NOT EQUAL')

IFLAG: =0 NO SPECIAL LAGS BETWEEN SERIES.

#0 LAGS BETWEEN SERIES ARE READ IN.

IFSEAS: =0 NO DETERMINISTIC COMPONENT IS ESTIMATED.

=1 ESTIMATE DETERMINISTIC PARAMETERS ONLY.

=2 ESTIMATE STOCHASTIC PARAMETERS ONLY AFTER DETERMINISTIC PORTION IS SUBTRACTED FROM THE DATA. (USE THE RESULTS FROM ISEAS=1 FOR THE DETERMINISTIC PARAMETERS.)

=3 ESTIMATE BOTH DETERMINISTIC AND STOCHASTIC PARAMETERS. (USING THE RESULTS FROM ISEAS=1 AND 2 AS INITIAL VALUES)

IFSER: =0 MODELS FOR ALL SERIES ARE ESTIMATED.

=N A MODEL IS ESTIMATED FOR SERIES N ONLY.

IFPRT: =0 EXCESS PRINTOUT IS SUPPRESSED.

>0 FULL PRINT OUT OF PROGRAM.

6.   IF IFLAG # 0, THE LAGS BETWEEN SERIES ARE READ IN.

NSER CARDS   (REFERENCE CHAPTER 11.2.3)
LAG(1,2),LAG(1,3),....,LAG(1,NSER)
LAG(2,1),LAG(2,3),....,LAG(2,NSER)
.
.
LAG(NSER,1),LAG(NSER,2),....,LAG(NSER,NSER-1)

THE LAG BETWEEN A SERIES AND ITSELF IS SET EQUAL TO 1 AND IS NOT INPUT.

7.   IF IFSEAS >0, THE DETERMINISTIC PARAMETERS AND THEIR CONSTRAINT FLAGS ARE READ.

I.   NPOLY,NEXP,NSIN
THESE CORRESPOND TO THE NUMBER OF POLYNOMIAL, EXPONENTIAL, AND SINUSOIDAL TRENDS TO BE ESTIMATED ACCORDING TO THE FORM:

$$X_{DET} = B(0) + \sum_{I=1}^{NPOLY} B(I)*T**I + \sum_{J=I+1}^{NEXP+I} B(J)*E^{B(J+1)*T} + $$

$$\sum_{K=J+1}^{NSIN+J} B(K)*E^{B(K+1)*T} [B(K+2)*SIN(B(K+3)*T) + \sqrt{1-B(K+2)**2} *COS(B(K+3)*T)]$$

WHERE T IS TIME, FREQUENCIES IN CYCLES/UNIT TIME.

II.   [(B(I),IC(I), I=0,NPOLY+2*NEXP+4*NSIN]
WHERE THE B'S ARE INITIAL VALUE INPUTS CORRESPONDING TO THE PREVIOUS EXPRESSION, AND THE IC'S INDICATE WHETHER THE CORRESPONDING PARAMETER IS TO BE ESTIMATED (IC=0), OR FIXED AT THE INPUT VALUE (IC#0). THE MAXIMUM NUMBER OF SEASONALITY PARAMETERS IS 30.
WARNING THE INITIAL VALUES OF ESTIMATED PARAMETERS MUST NOT BE ZERO.

8.   DATA: FREE FORMAT   SERIES 1, SERIES 2, ETC.

9.   IF NMOD < 0,   READ INITIAL VALUES - ONE SET FOR EACH SERIES
                      AND/OR MODEL.

    I.   IF IFMEAN >0, INITIAL VALUES OF MEANS, SERIES 1,...

    II.   INITIAL VALUES OF THE PARAMETERS... PHIQ'S FIRST, THEN
           PHI'S IN ORDER OF INCREASING LAG, SERIES 1 FIRST,...
           THEN THETA'S IN ORDER OF INCREASING LAG.

    III.   IF IFAT < 1, INITIAL VALUES OF THE AT'S.

        WARNING THE INITIAL VALUES MUST NOT BE ZERO.

EXAMPLE RUNSTREAM: CHAPTER 2, SUNSPOT DATA, 176 OBSERVATIONS.
FIT AR(1) AND AR(2) MODELS.

```
@RUN DDS,BME466012/USER-ID#,,3,25
@PASSWD MYPASS
@XQT DDS*ABS.DISCRETE
  WOLFER SUNSPOT NUMBER
176,1,1,,2,1,0
1,0
1,E-3,1.E-5,20,0
0,0,0,0,0,0,0
@ADD,P DDS*DATA.A2
@FIN
```

EXAMPLE OUTPUT:

@XQT DDS*ABS.DISCRETE

                    DATA DEPENDENT SYSTEMS MODELING ROUTINE - MTU 1982

          THIS PROGRAM FITS UNIVARIATE (ARMA) AND EXTENDED (EARMA)
          MODELS TO TIME SERIES DATA, WITH OPTIONAL DETERMINISTIC TREND
          ESTIMATION FOR UNIVARIATE MODELS. IT IS DESIGNED TO BE USED
          WITH THE BOOK:
               "TIME SERIES AND SYSTEM ANALYSIS WITH APPLICATIONS"
                    BY S. M. PANDIT AND S. M. WU

               THIS CODE IS INTENDED FOR INSTRUCTIONAL USE ONLY!
     ****

                    WOLFER SUNSPOT NUMBER
     1 SERIES WITH 176 OBSERVATIONS, DELTA = .1000+001

     INPUT STARTING MODEL ORDERS:

     SERIES 1: ARMA(1, 0)

                    PROGRAM CONTROL FLAGS
     IFMEAN: 0   IFAT: 0   IFDIF: 0   NMOD: 2 IFPRT: 0
     IFLAG: 0   IFSEAS: 0   IFSER: 0 INCPH: 1 INCTH: 0
     STEP= .1000-002   CONVERGENCE TOLERANCE= .1000-004
          20 ITERATIONS MAX   IFREF: 0

@ADD,P DDS*DATA.A2

     CHARACTERISTICS OF THE DATA

     SERIES 1   AVERAGE= .4476+002 VARIANCE= .1209+004

     ****

     FINAL RESIDUAL SUM OF SQUARES= .717085+005
          SERIES 1: ARMA(1, 0)

BEST PARAMETER VALUES AND 95% CONFIDENCE LIMITS ESTIMATED
BY LINEARIZATION FOR THE INDIVIDUAL PARAMETERS ARE AS FOLLOWS:

        AT LAG =   1

  PHI(1 1 LAG) = .81076+000 +/-.86419-001

STANDARD ERROR OF RESIDUALS = .20243+002 ESTIMATED WITH 176
RESIDUALS AND 175 DEGREES OF FREEDOM.
****

                        THE AVERAGE OF THE RESIDUALS FOR SERIES   1 =   -.334594+000

                          AUTO CORRELATIONS OF SERIES 1 RESIDUALS
                          LAG   CORRELATION   UNIFIED
                           1   .5248608   6.9630056
                           2   .1141603   1.1852306
                           3   -.2462459   -2.6025300
                           4   -.3857844   -3.9285961
                           5   -.4505766   -4.2325241
                           6   -.3699828   -3.1879301
                           7   -.2209815   -1.7894226
                           8   .0452122   .3662790
                           9   .2663355   2.1207699
                          10   .3943859   2.8300913
                          11   .3715501   2.7627445
                          12   .2269073   1.6189169
                          13   .0187266   .1316278
                          14   -.1333012   -.9368710
                          15   -.1906902   -1.3335796
                          16   -.1964389   -1.3739558
                          17   -.1161134   -.7934620
                          18   -.1294555   -.8836937
                          19   -.0774514   -.5263708
                          20   -.0204506   -.1387667
                          21   .0772002   .5242323
                          22   .2107952   1.4279604
                          23   .2198956   1.4726442
                          24   .1610760   1.0656775
                          25   -.0366325   -.2408137
                          26   -.1036402   -.6810758
                          27   -.1855892   -1.2164068
                          28   -.1495060   -.9717791
                          29   -.1689072   -1.0920317
                          30   -.1584025   -1.0172469

                        THE SUM OF SQUARES OF THE AUTO CORRELATIONS =   .1659+001

****

****

FINAL RESIDUAL SUM OF SQUARES = .413141+005
    SERIES 1: ARMA(2, 0)

```
BEST PARAMETER VALUES AND 95% CONFIDENCE LIMITS ESTIMATED
BY LINEARIZATION FOR THE INDIVIDUAL PARAMETERS ARE AS FOLLOWS:

            AT LAG  =      1                      2
  PHI( 1  1 LAG) = .13364+001 +/-.11279+000   -.65049+000 +/-.11285+000

STANDARD ERROR OF RESIDUALS =   .15409+002 ESTIMATED WITH  176
RESIDUALS AND  174 DEGREES OF FREEDOM.
****************************************************************************


F-TEST WITH LAST MODEL FOR THIS SERIES =   128.010


                        THE AVERAGE OF THE RESIDUALS FOR SERIES    1  =   -.115806+000

                        AUTO CORRELATIONS OF SERIES  1 RESIDUALS
                   LAG      CORRELATION                     UNIFIED
                    1      -.0688014                      -.9127543
                    2      -.0518805                      -.9756481
                    3      -.0525490                      -.6901415
                    4      -.1037547                      1.3589691
                    5      -.0231692                      -.3003265
                    6      -.0380923                      -.4935215
                    7      -.0761653                      -.9854389
                    8       .0808726                      1.0406091
                    9       .1283488                      1.6414287
                   10       .0852920                      1.0744592
                   11       .0062530                       .0062530
                   12       .1603002                      1.3529733
                   13       .0110476                       .0731846
                   14      -.0060416                      -.3421987
                   15      -.0282503                      -.0215659
                   16      -.0017816                     -1.0236878
                   17      -.0845671                      1.1400909
                   18      -.0864323                     -1.1465762
                   19      -.0958400                      -.2595814
                   20      -.0218650                      -.5119601
                   21      -.0431398                      -.3806534
                   22       .0574399                      1.1461237
                   23       .0969750                       .2759853
                   24       .0235251                       .2163591
                   25      -.0184505                       .0207885
                   26      -.0017733                      -.6676467
                   27      -.0956446                     -2.0520763
                   28      -.1969713                       .0271143
                   29      -.0023801                      -.8969065
                   30      -.0787313                     -1.2232018
                   30      -.1078635

                   THE SUM OF SQUARES OF THE AUTO CORRELATIONS =    .1959+000



****************************************************************************
```

```
EXAMPLE RUNSTREAM: CHAPTER 4, SUNSPOT DATA, 176 OBSERVATIONS.
                   FIT ARMA(2,1) THROUGH ARMA(4,3) MODELS.

     @RUN DDS,BME466012/USER-ID#,,3,25
     @PASSWD MYPASS
     @XQT DDS*ABS.DISCRETE
        WOLFER SUNSPOT NUMBER
     176,1,1,,2,2,2
     2,1
     1.E-3,1.E-5,20,0
     1,0,0,0,0,0,1
     @ADD,P DDS*DATA.A2
     @FIN
```


```
        EXAMPLE OUTPUT:

@XQT DDS*ABS.DISCRETE

                              DATA DEPENDENT SYSTEMS MODELING ROUTINE - MTU 1982

                      THIS PROGRAM FITS UNIVARIATE (ARMA) AND EXTENDED (EARMA)
                      MODELS TO TIME SERIES DATA, WITH OPTIONAL DETERMINISTIC TREND
                      ESTIMATION FOR UNIVARIATE MODELS. IT IS DESIGNED TO BE USED
                      WITH THE BOOK:
                          "TIME SERIES AND SYSTEM ANALYSIS WITH APPLICATIONS"
                                   BY S. M. PANDIT AND S. M. WU
                ********************THIS CODE IS INTENDED FOR INSTRUCTIONAL USE ONLY!
                ***************************************************************************

                  WOLFER SUNSPOT NUMBER
    1 SERIES WITH  176 OBSERVATIONS, DELTA = .1000+001

    INPUT STARTING MODEL ORDERS:

    SERIES   1: ARMA( 2, 1 )

                      PROGRAM CONTROL FLAGS
        IFMEAN:  1    IFAT:  0   IFDIF:  0   NMOD:  2   IFPRT:  1
        IFLAG:   0   IFSEAS:  0  IFSER:  0   INCPH:  2   INCTH:  2
        STEP= .1000-002    CONVERGENCE TOLERANCE= .1000-004
             20 ITERATIONS MAX      IFREF:  0

@ADD,P DDS*DATA.A2

    CHARACTERISTICS OF THE DATA
        SERIES  1    AVERAGE= .4476+002  VARIANCE=  .1209+004



        SERIES   1: ARMA( 2, 1 )

THE INVERSE FUNCTION COEFFICIENTS ARE:
  .126765+001  -.500592+000  -.110781+000

INITIAL PARAMETERS =  .44765+002  .14890+001 -.78112+000  .22130+000

   VERSION 4R1 OF LS, 09-03-82, WITH   4 PARAMETERS
START ITERATION NO.  1 NO. OF CALLS TO MODEL   0
```

```
INITIAL SUM OF SQUARES =   .4108+005
PARAMETER   4 LIMITS THE CORRECTIONS TO     .6095+000 TIMES THE GAUSS-NEWTON VALUES.
SEARCH CONVERGED AFTER  5 CYCLES, WITH LAMBDA =   .10071+001 AND SSQ =   .40880+005

START ITERATION NO.  2 NO. OF CALLS TO MODEL  11
SEARCH CONVERGED AFTER  1 CYCLES, WITH LAMBDA =   .10000+001 AND SSQ =   .40879+005


*************************************************************************

FINAL RESIDUAL SUM OF SQUARES =  .408790+005
    SERIES   1: ARMA( 2,  1 )



BEST PARAMETER VALUES AND 95% CONFIDENCE LIMITS ESTIMATED
BY LINEARIZATION FOR THE INDIVIDUAL PARAMETERS ARE AS FOLLOWS:

   SERIES  1 MEAN =  .44520+002 +/-.65797+001

                                                 2
          AT LAG =         1
   PHI( 1  1 LAG) =  .14239+001 +/-.15640+000   -.72126+000 +/-.13798+000

          AT LAG =         1
 THETA( 1  1 LAG) =  .15112+000 +/-.22345+000

STANDARD ERROR OF RESIDUALS =   .15417+002 ESTIMATED WITH  176
RESIDUALS AND  172 DEGREES OF FREEDOM.
*************************************************************************

NORMALIZED CORRELATION MATRIX OF PARAMETER ESTIMATES
    1.0000000
     .0206142   1.0000000
    -.0262964   -.9060250   1.0000000
     .0125369    .7483103  -.6989811   1.0000000


                THE AVERAGE OF THE RESIDUALS FOR SERIES   1 =   -.192675-001

                   AUTO CORRELATIONS OF SERIES 1 RESIDUALS
                   LAG      CORRELATION            UNIFIED
                    1       -.0039282            -.0521132
                    2       -.0298837             .3964465
                    3       -.0370061             .5262833
                    4        .1361095            1.8012263
                    5        .0418883             .5433996
                    6        .0742174             .9629297
                    7       -.0426891            -.5509713
                    8        .0916866            1.1813258
                    9        .1338671            1.7112806
                   10        .0882124            1.1093500
                   11        .1568677            1.9590999
                   12        .1076795            1.3163947
                   13        .0007837             .0094881
                   14       -.0277835            -.3353608
                   15        .0064459             .0779868
                   16       -.0539886            -.6531706
                   17        .1001541            1.2087673
                   18       -.0744680            -.8913908
                   19       -.0303305            -.3621431
                   20        .0605606            .7211320
                   21       -.0701699            -.8330975
                   22        .0802914             .9495285
                   23        .0190162             .2237434
```

```
24     .0083902              -.0989002
25    -.0146082              -.0955802
26    -.0088903              -.0434902
27    -.0023-003             -.2602302
28    -.0231022              -.0775712
29     .0209551             -1.3715868
30
```

THE SUM OF SQUARES OF THE AUTO CORRELATIONS =   .1990+000

*********************MODEL CHARACTERISTICS*********************

GREEN'S FUNCTION OF SERIES 1 WITH ITS RESIDUALS

G(J) = .15706+001*(.84927+000)**J*COS(2*PI* .91780-001*DELTA*J+(-.88058+000))

FOR J GREATER THAN OR EQUAL TO 0

Frequency should be multiplied by the sample interval before computing G_J.

CHARACTERISTICS OF THE AUTOREGRESSIVE OPERATOR

DISCRETE COMPLEX ROOTS		NATURAL FREQUENCY (HZ)	DAMPING RATIO
REAL	IMAG		
.7119	+/- .4631	.9939-001	.2726+000

CHARACTERISTICS OF THE MOVING AVERAGE OPERATOR

DISCRETE COMPLEX ROOTS		NATURAL FREQUENCY (HZ)	DAMPING RATIO
REAL	IMAG		
.1511		.3007+000	

****

SERIES 1: ARMA(4, 3)

****

THE INVERSE FUNCTION COEFFICIENTS ARE:

.128051+001 -.505942+000 -.173896+000 .214256+000 -.178991+000 -.322598-001 .134005+000

INITIAL PARAMETERS = .44765+002 .79218+000 WITH 8 PARAMETERS .28381+000 -.38377+000 -.48833+000 -.34343+000 .26502+000

VERSION 4R1 OF LS, 09-03-82, WITH 8 PARAMETERS

START ITERATION NO. 1 NO. OF CALLS TO MODEL 0

INITIAL SUM OF SQUARES = .4061+005

PARAMETER 4 LIMITS THE CORRECTIONS TO .1657+000 TIMES THE GAUSS-NEWTON VALUES.
SEARCH CONVERGED AFTER 7 CYCLES, WITH LAMBDA = .95769+000 AND SSQ = .39223+005

START ITERATION NO. 2 NO. OF CALLS TO MODEL 17

```
FINAL RESIDUAL SUM OF SQUARES = .390159+005

   SERIES 1: ARMA( 4, 3 )


BEST PARAMETER VALUES AND 95% CONFIDENCE LIMITS ESTIMATED
BY LINEARIZATION FOR THE INDIVIDUAL PARAMETERS ARE AS FOLLOWS:

   SERIES 1 MEAN = .44837+002 +/-.65790+001

        AT LAG =     1                2                    3                    4

   PHI( 1  1 LAG) = .42009+000 +/-.10017+001   .28267+000 +/-.74809+000   -.83264-001 +/-.49276+000   -.32596+000 +/-.50150+000

        AT LAG =     1                2                    3

   THETA( 1  1 LAG) = -.88901+000 +/-.10043+001   -.30748+000 +/-.76385+000   .14092+000 +/-.23338+000

STANDARD ERROR OF RESIDUALS = .15239+002 ESTIMATED WITH 176
RESIDUALS AND 168 DEGREES OF FREEDOM.
*******************************************************

NORMALIZED CORRELATION MATRIX OF PARAMETER ESTIMATES

 1.0000000
 -.0000890   1.0000000
 -.7489290   .0000834   1.0000000
  .7518273  -.8078196  -.7080920   1.0000000
  .0345409  -.8648225  -.0824160   .9889097   1.0000000
 -.6448170  .8208700   .4736949  -.8240859   -.8323852   1.0000000
 -.6147778  -.0050781   .3569949   -.3361660  -.3678854   .3687813   1.0000000
  .0070965  -.3235531   -.2056645   .0250233   .0240160   .0240160   .4593789   1.0000000

F-TEST WITH LAST MODEL FOR THIS SERIES = 2.006


       THE AVERAGE OF THE RESIDUALS FOR SERIES   1 =  -.559762-001


          AUTO CORRELATIONS OF SERIES  1 RESIDUALS

          LAG      CORRELATION      UNIFIED
           1        -.0076686        .1017354
           2         .0022004       -.2945704
           3        -.0084908       -.1125782
           4         .0068405       -.0124267
           5         .0063513        .0839152
           6         .0748563        .9836683
```

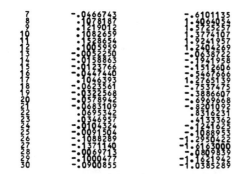

7	-.0466743	-.6101135
8	.1078187	1.4064054
9	.1219012	1.5725257
10	.1082659	1.3774107
11	.1528654	1.9241957
12	.1665959	2.2404269
13	-.0053850	-.0638722
14	-.0158863	-.1921958
15	-.0123766	-.1512606
16	-.0447440	-.5467666
17	.1046393	1.2765139
18	-.0623561	-.7537475
19	-.0322568	-.3886607
20	.0578942	.6969668
21	-.0683109	-.8201092
22	.0695342	.8316231
23	-.0346957	-.4133362
24	.0104334	.1241625
25	-.0091504	-.1088953
26	.1088289	1.2950422
27	-.1371140	-1.6163000
28	-.0069713	-.0809839
29	.1000477	1.1621242
30	-.0900855	-1.0385289

THE SUM OF SQUARES OF THE AUTO CORRELATIONS = .1703+000

********************MODEL CHARACTERISTICS********************

GREEN'S FUNCTION OF SERIES 1 WITH ITS RESIDUALS

G(J) = .16057+001*(.84922+000)**J*COS[2*PI* .89873-001*DELTA*J+(-.84896+000)]

+ .92555-001*(.67230+000)**J*COS[2*PI* .38609+000*DELTA*J+(.22903+001)]

FOR J GREATER THAN OR EQUAL TO 0

Frequency should be multiplied by the sample interval before computing G_j.

CHARACTERISTICS OF THE AUTOREGRESSIVE OPERATOR

| DISCRETE COMPLEX ROOTS | | NATURAL FREQUENCY | DAMPING |
REAL	IMAG	(HZ)	RATIO
-.7174 +/-	.4545	.9356-001	.2780+000
-.5073 +/-	.4411	.3912+000	.1615+000

CHARACTERISTICS OF THE MOVING AVERAGE OPERATOR

| DISCRETE COMPLEX ROOTS | | NATURAL FREQUENCY | DAMPING |
REAL	IMAG	(HZ)	RATIO
-.2423		.2256+000	
-.5657 +/-	.5115	.3854+000	.1119+000

****

```
EXAMPLE RUNSTREAMS: CHAPTER 10, CRACK LENGTH DATA. FIT AN ARMA
               MODEL WITH A TREND (THREE RUNSTREAMS).

STEP 1:
    @RUN DDS,BME466C12/USER-ID#,,3,25
    @PASSWD MYPASS
    @XQT DDS*ABS.DISCRETE
    CRACK LENGTH DATA
    90,1,1.,1,0,0
    1,0
    0,E-3,1,E-5,20,0
    0,1,0,0,1,0,0
    1,0,0
    11.02783,0,0,10658,0
    @ADD,P DDS*DATA.A14
    @FIN
```

```
EXAMPLE OUTPUT:

@XQT DDS*ABS.DISCRETE
                        DATA DEPENDENT SYSTEMS MODELING ROUTINE - MTU 1982

                        THIS PROGRAM FITS UNIVARIATE (ARMA) AND EXTENDED (EARMA)
                        MODELS TO TIME SERIES DATA, WITH OPTIONAL DETERMINISTIC TREND
                        ESTIMATION FOR UNIVARIATE MODELS. IT IS DESIGNED TO BE USED
                        WITH THE BOOK:
                            "TIME SERIES AND SYSTEM ANALYSIS WITH APPLICATIONS"
                              BY S. M. PANDIT AND S. M. WU

                        THIS CODE IS INTENDED FOR INSTRUCTIONAL USE ONLY!
    *************************************************************************

          CRACK LENGTH DATA
    1 SERIES WITH   90 OBSERVATIONS, DELTA = .1000+001

    INPUT STARTING MODEL ORDERS:

    SERIES  1: ARMA( 1,  0 )

              PROGRAM CONTROL FLAGS
    IFMEAN:  0    IFAT:  1   IFDIF:  0  NMOD:  1  IFPRT:  0
    IFLAG:   0  IFSEAS:  1  IFSER:  0  INCPH:  0   INCTH:  0
    STEP= .1000-002    CONVERGENCE TOLERANCE= .1000-004
       20 ITERATIONS MAX    IFREF:  C

    SEASONALITY: 1 POLYNOMIAL,  0 EXPONENTIAL, AND  0 SINUSOIDAL TRENDS

    BETA( 1)= .1103+002    BETA( 2)= .1066+000

    ESTIMATION FLAGS:
    IC( 1)=  C              IC( 2)=  0
@ADD,P DDS*DATA.A14

    CHARACTERISTICS OF THE DATA

    SERIES  1    AVERAGE= .1588+002  VARIANCE= .7768+001

    *************************************************************************

FINAL RESIDUAL SUM OF SQUARES =  .132626+001
    SERIES  1: ARMA( 1,  0 )
```

```
BEST PARAMETER VALUES AND 95% CONFIDENCE LIMITS ESTIMATED
BY LINEARIZATION FOR THE INDIVIDUAL PARAMETERS ARE AS FOLLOWS:

        BETA(1) =  .11028+002 +/-.51152-001   .10658+000 +/-.97630-003

STANDARD ERROR OF RESIDUALS =  .12276+000 ESTIMATED WITH   90
RESIDUALS AND   88 DEGREES OF FREEDOM.
```

****

```
              THE AVERAGE OF THE RESIDUALS FOR SERIES    1  =   .225173-007

                      AUTO CORRELATIONS OF SERIES  1 RESIDUALS
                  LAG         CORRELATION              UNIFIED
                   1          .9214596             8.7417330
                   2          .8625183             4.9814352
                   3          .7822337             3.6288749
                   4          .7049872             2.8751577
                   5          .6236659             2.3378223
                   6          .5307743             1.8787919
                   7          .4530126             1.5441194
                   8          .3719423             1.2354785
                   9          .3133038             1.0234857
                  10          .2440521              .7881370
                  11          .1765017              .5660973
                  12          .1133692              .3623233
                  13          .0444242              .1417715
                  14         -.0062143             -.0198273
                  15         -.0350768             -.1119154
                  16         -.0784633             -.2503089
                  17         -.1014276             -.3233434
                  18         -.1209491             -.3841737
                  19         -.1205895             -.3833557
                  20         -.1184217             -.3758509
                  21         -.1127811             -.3573880
                  22         -.1053340             -.3333165
                  23         -.0980260             -.3098091
                  24         -.1099852             -.3472358
                  25         -.1146343             -.3614397
                  26         -.1235607             -.3890094
                  27         -.1287365             -.4046249
                  28         -.1598241             -.5014035
                  29         -.1831136             -.5728922
                  30         -.1998211             -.6228961

              THE SUM OF SQUARES OF THE AUTO CORRELATIONS =   .4171+001
```

****

```
      STEP 2:
          @RUN DDS,BME466012/USER-ID#,,3,25
          @PASSWD MYPASS
          @XQT DDS*ABS.DISCRETE
             CRACK LENGTH DATA
          90,1,1.,1,0,0
          1,0
          1.E-3,1.E-5,20,0
          0,1,0,2,0,0
          1,0,0
          11.027811,0,0,.10658025,0
          @ADD,P DDS*DATA.A14
          @FIN
```

EXAMPLE OUTPUT:

@XQT DDS*ABS.DISCRETE

DATA DEPENDENT SYSTEMS MODELING ROUTINE - MTU 1982

THIS PROGRAM FITS UNIVARIATE (ARMA) AND EXTENDED (EARMA)
MODELS TO TIME SERIES DATA, WITH OPTIONAL DETERMINISTIC TREND
ESTIMATION FOR UNIVARIATE MODELS. IT IS DESIGNED TO BE USED
WITH THE BOOK:
    "TIME SERIES AND SYSTEM ANALYSIS WITH APPLICATIONS"
        BY S. M. PANDIT AND S. M. WU

THIS CODE IS INTENDED FOR INSTRUCTIONAL USE ONLY!
****

             CRACK LENGTH DATA
   1 SERIES WITH   90 OBSERVATIONS, DELTA = .1000+001

   INPUT STARTING MODEL ORDERS:

   SERIES 1: ARMA(1, 0)

             PROGRAM CONTROL FLAGS
   IFMEAN: 0   IFAT: 1   IFDIF: C   NMOD: 1   IFPRT: 0
    IFLAG: 0   IFSEAS: 2   IFSER: 0   INCPH: 0   INCTH: 0
   STEP= .1000-002   CONVERGENCE TOLERANCE= .1000-004
      20 ITERATIONS MAX   IFREF: C

   SEASONALITY: 1 POLYNOMIAL, 0 EXPONENTIAL, AND 0 SINUSOIDAL TRENDS

   BETA(1)= .1103+002   BETA(2)= .1066+000

   ESTIMATION FLAGS:
     IC(1)=   0   IC(2)=   0
@ADD,P DDS*DATA.A1.

   CHARACTERISTICS OF THE DATA

   SERIES 1   AVERAGE= .1588+002 VARIANCE= .7768+001

****

FINAL RESIDUAL SUM OF SQUARES = .189849+000
    SERIES 1: ARMA(1, 0)

BEST PARAMETER VALUES AND 95% CONFIDENCE LIMITS ESTIMATED
BY LINEARIZATION FOR THE INDIVIDUAL PARAMETERS ARE AS FOLLOWS:

        AT LAG =   1
  PHI(1 1 LAG) = .92989+000 +/-.78964-001

STANDARD ERROR OF RESIDUALS = .46186-001 ESTIMATED WITH   90
RESIDUALS AND   89 DEGREES OF FREEDOM.
****

```
THE AVERAGE OF THE RESIDUALS FOR SERIES    1 =   -.113275-002

              AUTO CORRELATIONS OF SERIES  1 RESIDUALS
              LAG        CORRELATION             UNIFIED
               1        -.0878709              -.8336163
               2         .1734646             1.6330685
               3         .0447540              .4093766
               4         .0285416              .2605925
               5        -.1035528             -.9447531
               6        -.0728904             -.6585092
               7         .0443130              .3984193
               8        -.1255737            -1.1270504
               9        -.0717495             -.6350662
              10        -.0046073             -.0405983
              11         .0061403              .0541058
              12         .0539273              .4751700
              13        -.1105621             -.9717623
              14        -.1605372            -1.3973011
              15         .0752776              .6410422
              16        -.1264117            -1.0716038
              17        -.0194196             -.1625610
              18        -.1216105            -1.0176983
              19         .0104524              .0864813
              20        -.0197157             -.1631111
              21        -.0089540             -.0740560
              22         .0034412              .0284596
              23         .1277271             1.0563188
              24         .0455520              .3721355
              25         .0293983              .2397997
              26        -.0238116             -.1941051
              27         .1433580             1.1673090
              28        -.0004972             -.0039916
              29        -.0462417             -.3712133
              30         .0130233              .1043867

       THE SUM OF SQUARES OF THE AUTO CORRELATIONS =   .2006+000
```

****

STEP 3:

```
@RUN DDS,BME466C12/USER-ID#,,3,25
@PASSWD MYPASS
@XQT DDS*ABS.DISCRETE
     CRACK LENGTH DATA
90,1,1.,-1,0,0
1,0
1.E-3,1.E-5,20,0
0,1,0,0,3,0,0
1,0,0
11.027811,0,0.10658025,0
@ADD,P DDS*DATA.A14
0.929887
@FIN
```

EXAMPLE OUTPUT:

```
@XQT DDS*ABS.DISCRETE

                        DATA DEPENDENT SYSTEMS MODELING ROUTINE - MTU 1982
                        THIS PROGRAM FITS UNIVARIATE (ARMA) AND EXTENDED (EARMA)
                        MODELS TO TIME SERIES DATA, WITH OPTIONAL DETERMINISTIC TREND
                        ESTIMATION FOR UNIVARIATE MODELS. IT IS DESIGNED TO BE USED
                        WITH THE BOOK:
                           "TIME SERIES AND SYSTEM ANALYSIS WITH APPLICATIONS"
                               BY S. M. PANDIT AND S. M. WU
                        THIS CODE IS INTENDED FOR INSTRUCTIONAL USE ONLY!
        ****************************************************************

                   CRACK LENGTH DATA
     1 SERIES WITH   90 OBSERVATIONS, DELTA = .1000+001

     INPUT STARTING MODEL ORDERS:
     SERIES  1: ARMA( 1, 0 )

                   PROGRAM CONTROL FLAGS
     IFMEAN:  0    IFAT:  1  IFDIF:  0  NMOD: -1  IFPRT:  0
     IFLAG:   0    IFSEAS: 3  IFSER:  0  INCPH:  0  INCTH:  0
     STEP= .1000-002   CONVERGENCE TOLERANCE= .1000-004
        20 ITERATIONS MAX     IFREF:  0

     SEASONALITY: 1 POLYNOMIAL, 0 EXPONENTIAL, AND  0 SINUSOIDAL TRENDS

     BETA( 1)= .1103+002     BETA( 2)= .1066+000

     ESTIMATION FLAGS:
       IC( 1)=    0              IC( 2)=   0
@ADD,P DDS*DATA.A14

     CHARACTERISTICS OF THE DATA
     SERIES 1    AVERAGE= .1588+002  VARIANCE= .7768+001




****************************************************************

FINAL RESIDUAL SUM OF SQUARES = .169309+000
     SERIES  1: ARMA( 1, 0 )



BEST PARAMETER VALUES AND 95% CONFIDENCE LIMITS ESTIMATED
BY LINEARIZATION FOR THE INDIVIDUAL PARAMETERS ARE AS FOLLOWS:
              I =       1                      2
           BETA(I) = .10881+002 +/-.86894-001    .10799+000 +/-.35135-002

           AT LAG =    1
     PHI( 1 1 LAG) = .96002+000 +/-.67320-001

STANDARD ERROR OF RESIDUALS = .44114-001 ESTIMATED WITH   90
RESIDUALS AND   87 DEGREES OF FREEDOM.
****************************************************************
```

```
THE AVERAGE OF THE RESIDUALS FOR SERIES   1  =    .234146-002

         AUTO CORRELATIONS OF SERIES  1 RESIDUALS
         LAG          CORRELATION              UNIFIED
          1          -.1314367            -1.2469179
          2           .2222449             2.0728942
          3           .0182721              .1628280
          4           .0345164              .3074963
          5           .0575075              .5117789
          6          -.1293068            -1.1474119
          7           .0073468              .0642588
          8          -.1466567            -1.2826791
          9           .0588203              .5052958
         10           .0050877              .0435822
         11          -.0052478             -.0449533
         12           .0514718              .4409006
         13          -.0707135             -.6044183
         14          -.1335347            -1.1367721
         15          -.0756577             -.6350157
         16          -.0824377             -.6888423
         17          -.0355646             -.2956203
         18          -.0973063             -.8080462
         19          -.0323495             -.2667070
         20          -.0107360             -.0884437
         21           .0346927              .2857754
         22           .0082110              .0675758
         23           .1514662             1.2464829
         24          -.0575412             -.4655632
         25           .1118278              .9026224
         26          -.0306357             -.2450683
         27           .1457993             1.1655364
         28          -.0120386             -.0948177
         29          -.0165239             -.1301314
         30          -.0215654             -.1698023

     THE SUM OF SQUARES OF THE AUTO CORRELATIONS =   .2263+000


  *****************************************************************
```

```
EXAMPLE RUNSTREAM: CHAPTER 11, PAPERMAKING DATA, 2 SERIES, 50
                   OBSERVATIONS. FIT EARMA(1 1,0) AND EARMA(2 2,1).

@RUN DDS,BME466012/USER-ID#,,3,25
@PASSWD MYPASS
@XQT DDS*ABS.DISCRETE
   PAPERMAKING DATA
50,2,1,2,1,1
1,1,0,1,1,0
0,E-0,1,E-5,20,0
0,0,0,0,0,0,0
@ADD DDS*DATA.A20
@FIN



EXAMPLE OUTPUT:

@XQT DDS*ABS.DISCRETE
                          DATA DEPENDENT SYSTEMS MODELING ROUTINE - MTU 1982

                 THIS PROGRAM FITS UNIVARIATE (ARMA) AND EXTENDED (EARMA)
                 MODELS TO TIME SERIES DATA, WITH OPTIONAL DETERMINISTIC TREND
                 ESTIMATION FOR UNIVARIATE MODELS. IT IS DESIGNED TO BE USED
                 WITH THE BOOK:
                    "TIME SERIES AND SYSTEM ANALYSIS WITH APPLICATIONS"
                        BY S. M. PANDIT AND S. M. WU

                    THIS CODE IS INTENDED FOR INSTRUCTIONAL USE ONLY!
        **************************************************************************


                  PAPERMAKING DATA
     2 SERIES WITH   50 OBSERVATIONS, DELTA = .1000+001

   INPUT STARTING MODEL ORDERS:
   SERIES   1: ARMA(  1  1,  0 )

   SERIES   2: ARMA(  1  1,  0 )


                   PROGRAM CONTROL FLAGS
     IFMEAN:  0     IFAT:   0    IFDIF:   0   NMOD:   2  IFPRT:   0
     IFLAG:   0     IFSEAS:  0    IFSER:   0   INCPH:   0   INCTH:   1
     STEP= .1000-002    CONVERGENCE TOLERANCE= .1000-004
        20 ITERATIONS MAX     IFREF:   0

     SERIES   1: LAGS=     1,    0,
     SERIES   2: LAGS=     1,    0,

   CHARACTERISTICS OF THE DATA

     SERIES  1    AVERAGE= .1830+002  VARIANCE= .2399-002
     SERIES  2    AVERAGE= .3897+002  VARIANCE= .1344-001




   ****************************************************************************

   FINAL RESIDUAL SUM OF SQUARES =  .105134+000
        SERIES   1: ARMA(  1  1,  0 )
```

```
BEST PARAMETER VALUES AND 95% CONFIDENCE LIMITS ESTIMATED
BY LINEARIZATION FOR THE INDIVIDUAL PARAMETERS ARE AS FOLLOWS:

          AT LAG =     1
  PHI( 1  1 LAG) = .24747-002 +/-.27639+000

  PHI( 1  2  0) = .10360+000 +/-.16759+000

          AT LAG =     1
  PHI( 1  2 LAG) = .19116+000 +/-.16414+000

STANDARD ERROR OF RESIDUALS =  .47296-001 ESTIMATED WITH   50
RESIDUALS AND   47 DEGREES OF FREEDOM.
********************************************************************
```

```
          THE AVERAGE OF THE RESIDUALS FOR SERIES   1 =   .165075-001

               AUTO CORRELATIONS OF SERIES  1 RESIDUALS
               LAG       CORRELATION              UNIFIED
                1       -.1132625               -.8008866
                2        .0279155                .1948868
                3        .0194640                .1357960
                4       -.0926749               -.6463329
                5        .0849466                .5875454
                6        .0928545                .6378559
                7       -.0704801               -.4805625
                8        .1284055                .8709670
                9       -.0844052               -.5640225
               10        .0517245                .3434612
               11        .0509955                .3378242
               12       -.1951204              -1.2896528
               13       -.0078246               -.0500777
               14       -.1314355               -.8411528
               15        .0129134                .0814973
               16       -.0125863               -.0794224
               17       -.0178814               -.1128213
               18       -.0046473               -.0293145
               19        .0474237                .2991349
               20        .0569253                .3327691
               21       -.0519105               -.3260410
               22       -.0079646               -.0499182
               23        .0762706               .4780039
               24       -.1282351               -.8000293
               25       -.1442101               -.8883903
               26        .1037756                .6415603
               27        .0393756                .2368868
               28       -.0374707               -.2251746
               29        .0468837                .2814554
               30        .0893801                .5357244

          THE SUM OF SQUARES OF THE AUTO CORRELATIONS =   .2039+000


          *********************MODEL CHARACTERISTICS*********************

               GREEN'S FUNCTION OF SERIES  1 WITH ITS RESIDUALS
          G(J) =   .10000+001*( .24747-002)**J
               FOR J GREATER THAN OR EQUAL TO  0
```

CHARACTERISTICS OF THE AUTOREGRESSIVE OPERATOR

DISCRETE COMPLEX ROOTS REAL IMAG	NATURAL FREQUENCY (HZ)	DAMPING RATIO
.0025	.9552+000	

****

****

FINAL RESIDUAL SUM OF SQUARES = .305954+000

   SERIES   2: ARMA(1 1, 0)

BEST PARAMETER VALUES AND 95% CONFIDENCE LIMITS ESTIMATED
BY LINEARIZATION FOR THE INDIVIDUAL PARAMETERS ARE AS FOLLOWS:

   AT LAG =   1

   PHI(2 1 LAG) = .24663+000 +/-.46123+000

   AT LAG =   1

   PHI(2 2 LAG) = .69612+000 +/-.19492+000

STANDARD ERROR OF RESIDUALS = .79838-001 ESTIMATED WITH   50
RESIDUALS AND 48 DEGREES OF FREEDOM.

****

THE AVERAGE OF THE RESIDUALS FOR SERIES   2 =   .424149-001

CORRELATIONS OF SERIES   1 RESIDUALS WITH SERIES   2 RESIDUALS

LAG	UNIFIED	LAG	CORRELATION

CORRELATIONS OF SERIES   1 RESIDUALS

LAG	CORRELATION

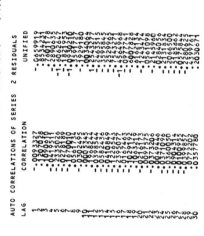

THE SUM OF SQUARES OF THE CROSS CORRELATIONS = .9751+000

AUTO CORRELATIONS OF SERIES 2 RESIDUALS

LAG   CORRELATION   UNIFIED

THE SUM OF SQUARES OF THE AUTO CORRELATIONS = .2773+000

*********************MODEL CHARACTERISTICS*********************

IMP(J) =   IMPULSE RESPONSE FUNCTION OF SERIES 2 TO SERIES 1

    = .24663+000*(.69612+C00)**J

    FOR J GREATER THAN OR EQUAL TO 1

G(J) =   GREEN'S FUNCTION OF SERIES 2 WITH ITS RESIDUALS

    = .10000+0C1*(.69612+C00)**J

    FOR J GREATER THAN OR EQUAL TO 0

```
                                     CHARACTERISTICS OF THE AUTOREGRESSIVE OPERATOR
                                     DISCRETE              NATURAL
                                   COMPLEX ROOTS          FREQUENCY              DAMPING
                                   REAL     IMAG            (HZ)                  RATIO
                                 ------------------------------------------------------------
                                   .6961                  .5765-001
```

****

THE INVERSE FUNCTION COEFFICIENTS ARE:
  -.194578-001   .423847-001 -.150570+000   .279856-001   .309320+000 -.107257+000   .129352+000

INITIAL PARAMETERS = .12935+000 -.75254+000 .28121-001 -.66840-001 .32984+000 -.73308+000

****

FINAL RESIDUAL SUM OF SQUARES = .846373-001
   SERIES 1: ARMA(2 2, 1)

BEST PARAMETER VALUES AND 95% CONFIDENCE LIMITS ESTIMATED
BY LINEARIZATION FOR THE INDIVIDUAL PARAMETERS ARE AS FOLLOWS:

         AT LAG =   1   2
  PHI(1 1 LAG) = -.28989+000 +/-.56955+000   .23399-001 +/-.27661+000

  PHI(1 2 0) = .12081+000 +/-.16389+000

         AT LAG =   1   2
  PHI(1 2 LAG) = .62335-002 +/-.17593+000   .25990+000 +/-.17273+000

         AT LAG =   1
THETA(1 1 LAG) = -.22179+000 +/-.64337+000

STANDARD ERROR OF RESIDUALS = .43859-001 ESTIMATED WITH   50
RESIDUALS AND   44 DEGREES OF FREEDOM.
****

F-TEST WITH LAST MODEL FOR THIS SERIES =   3.552

                        THE AVERAGE OF THE RESIDUALS FOR SERIES   1 = .569268-001

                            AUTO CORRELATIONS OF SERIES 1 RESIDUALS
                            LAG   CORRELATION   UNIFIED
```
```

```
 1      .0270287              .1911217
 2      .1255924              .8874245
 3     -.0679315             -.4726114
 4     -.0655294             -.4558768
 5     -.0485607             -.3349693
 6     -.0499276             -.3436278
 7     -.2076150            -1.4255512
 8      .1108159              .7317381
 9     -.1206563             -.7883187
10      .0836621              .5399440
11      .0003828              .0024565
12     -.0632339             -.4057443
13     -.0989524             -.6328534
14      .0650069              .4124633
15     -.0688414             -.4353142
16     -.0596521             -.3757849
17     -.0269189             -.1691018
18      .0173407              .1088704
19     -.0428829             -.2691678
20      .0990495              .6208159
21      .0606383              .3771689
22      .0172409              .1069343
23      .0190170              .1179232
24      .0783263              .4855611
25     -.0308421             -.1903013
26      .0900144              .5550036
27     -.0423591             -.2595803
28     -.0142535             -.0872291
29      .0577425              .3533214
30     -.0432886             -.2642202
```

THE SUM OF SQUARES OF THE AUTO CORRELATIONS =   .1729+000

*********************MODEL CHARACTERISTICS*********************

        GREEN'S FUNCTION OF SERIES 1 WITH ITS RESIDUALS

G(J) =   .68233+000*(.65786-001)**J

     +   .31767+000*(-.35568+000)**J

        FOR J GREATER THAN OR EQUAL TO 0

        CHARACTERISTICS OF THE AUTOREGRESSIVE OPERATOR

| DISCRETE COMPLEX ROOTS | | NATURAL FREQUENCY (HZ) | DAMPING RATIO |
REAL	IMAG		
.0658		.4331+000	
-.3557		.5000+000	

        CHARACTERISTICS OF THE MOVING AVERAGE OPERATOR

| DISCRETE COMPLEX ROOTS | | NATURAL FREQUENCY (HZ) | DAMPING RATIO |
REAL	IMAG		
-.2218		.5000+000	

****

THE INVERSE FUNCTION COEFFICIENTS ARE:

.209165+000 -.502717+000 -.701819-001 .671098+000 .889039-001 -.498996-001

INITIAL PARAMETERS = .20916+000 -.56346+000 .96150+000 -.10598+000 .29040+000

****

FINAL RESIDUAL SUM OF SQUARES = .197465+000
    SERIES 2: ARMA(2 2, 1)

BEST PARAMETER VALUES AND 95% CONFIDENCE LIMITS ESTIMATED
BY LINEARIZATION FOR THE INDIVIDUAL PARAMETERS ARE AS FOLLOWS:

                    AT LAG =   1   2

PHI(2 1 LAG) = .25096+000 +/-.45368+000   -.57845+000 +/-.46265+000

                    AT LAG =   1   2

PHI(2 2 LAG) = .14931+001 +/-.25643+000   -.54477+000 +/-.24619+000

                    AT LAG =   1

THETA(2 2 LAG) = .11301+001 +/-.27954+000

STANDARD ERROR OF RESIDUALS = .6519-001 ESTIMATED WITH 50
RESIDUALS AND 46 DEGREES OF FREEDOM.
****

F-TEST WITH LAST MODEL FOR THIS SERIES =   8.241

THE AVERAGE OF THE RESIDUALS FOR SERIES   2 =   .267475-002

CORRELATIONS OF SERIES 1 RESIDUALS WITH SERIES 2 RESIDUALS

THE SUM OF SQUARES OF THE CROSS CORRELATIONS = .9916+000

AUTO CORRELATIONS OF SERIES 2 RESIDUALS

LAG   CORRELATION   UNIFIED

THE SUM OF SQUARES OF THE AUTO CORRELATIONS = .2644+000

*********************MODEL CHARACTERISTICS*********************

IMPULSE RESPONSE FUNCTION OF SERIES 2 TO SERIES 1

$IMP(J) = -.32897+001*(.63444+000)**J$
$+ .35407+001*(.85867+000)**J$

FOR J GREATER THAN OR EQUAL TO 1

GREEN'S FUNCTION OF SERIES 2 WITH ITS RESIDUALS

$G(J) = -.22103+001*(.63444+000)**J$
$+ -.12103+001*(.85867+000)**J$

FOR J GREATER THAN OR EQUAL TO 0

```
CHARACTERISTICS OF THE AUTOREGRESSIVE OPERATOR
       DISCRETE              NATURAL
    COMPLEX ROOTS           FREQUENCY           DAMPING
   REAL      IMAG             (HZ)               RATIO
-------------------------------------------------------
   .6344                    .7242-001
   .8587                    .2425-001


CHARACTERISTICS OF THE MOVING AVERAGE OPERATOR
       DISCRETE              NATURAL
    COMPLEX ROOTS           FREQUENCY           DAMPING
   REAL      IMAG             (HZ)               RATIO
-------------------------------------------------------
 1.1301                    -.1946-001


*********************************************************
```

```
  1:*****************************************************************
  2:*                                                               *
  3:*    THIS PROGRAM FITS UNIVARIATE (ARMA) AND EXTENDED (EARMA)    *
  4:*    MODELS TO TIME SERIES DATA, WITH OPTIONAL DETERMINISTIC TREND*
  5:*    ESTIMATION FOR UNIVARIATE MODELS. IT IS DESIGNED TO BE USED *
  6:*    WITH THE BOOK:                                              *
  7:*         "TIME SERIES AND SYSTEM ANALYSIS WITH APPLICATIONS"    *
  8:*                 BY S. M. PANDIT AND S. M. WU                   *
  9:*                                                               *
 10:*          THIS CODE WRITTEN BY WILLIAM WITTIG                   *
 11:*                            AT                                 *
 12:*              MICHIGAN TECHNOLOGICAL UNIVERSITY                 *
 13:*                          1982                                 *
 14:*                                                               *
 15:*        SUBROUTINE LS ADAPTED FROM CODE WRITTEN AT             *
 16:*              UNIVERSITY OF WISCONSIN-MADISON                  *
 17:*                                                               *
 18:*      THIS CODE IS INTENDED FOR INSTRUCTIONAL USE ONLY!        *
 19:*                                                               *
 20:****WARNING*** THE FORTRAN 77 PRETEST DO LOOP FEATURE           *
 21:*              IS USED EXTENSIVLY!                              *
 22:*                                                               *
 23:*   PARAMETER STORAGE IN THE PAR ARRAY IS AS FOLLOWS:           *
 24:*   1. ESTIMATED SEASONALITY PARAMETERS, 2. MEANS, 3. PHIO TERMS,*
 25:*   4. PHI TERMS: SERIES 1, SERIES 2, ETC., 5. THETA TERMS      *
 26:*                                                               *
 27:*   ZERO LAG INVERSE FUNCTION COEFFICIENTS ARE STORED AT THE    *
 28:*   END OF THE LAST COLUMN OF THE XTX ARRAY. (SUBROUTINE EAR)   *
 29:*****************************************************************
 30:      PARAMETER MXSER=3,MXSR2=MXSER**2,MXNOB=369,MXPAR=45,MXINV=50,
 31:     >   MXNOB1=MXNOB+1,MXINV1=MXINV+1
 32:      DOUBLE PRECISION XTX(MXINV,MXINV1)
 33:      COMMON /INDEX/ INDEX,MAXLAG,NOB,NSER,NPHIO,NPHI,NSEAS
 34:      COMMON /LAG/ LAG(MXSER,MXSER),NAR(MXSER,MXSER),NMA(MXSER)
 35:      COMMON /FLAG/ IFMEAN,IFAT,IFSEAS
 36:      COMMON /SEASON/ NPOLY,NEXP,NSIN,ISEAS(30),SEAS(30),DELTA
 37:      COMMON /DATA/ Y(MXNOB,MXSER),X(MXNOB,MXSER)
 38:      COMMON /BLOK1/ STEP,CONTOL,ITMAX,IFREF
 39:      DIMENSION AT(MXNOB1,MXSER),PAR(MXPAR),AVG(MXSER),VAR(MXSER)
 40:     >   ,OSSQ(MXSER),SCA(MXSER)
 41:      CHARACTER TITLE*40,FORMT*50
 42:      DATA LAG,IPAR,INDEX,IFINV,NPHIO,NPHI,NSEASP,NSEAS/MXSR2*1,2*1,5*0/
 43:C***READ INPUT DATA
 44:      READ 10,TITLE
 45:10    FORMAT (A)
 46:      READ*,NOB,NSER,DELTA,NMOD,INCPH,INCTH
 47:      READ*,((NAR(J,I),J=1,NSER),NMA(I),I=1,NSER)
 48:      READ*,STEP,CONTOL,ITMAX,IFREF
 49:      READ*,IFMEAN,IFAT,IFDIF,IFLAG,IFSEAS,IFSER,IFPRT
 50:C***ECHO CHECK INPUT
 51:      WRITE(FORMT,20) NSER
 52:20    FORMAT (25H(5X,'SERIES',I4,': ARMA(',I3,16HI3,',',I3,' )'/))
 53:      PRINT 30,TITLE
 54:30    FORMAT (36X,'DATA DEPENDENT SYSTEMS MODELING ROUTINE - MTU 1982'
 55:     >//36X,'THIS PROGRAM FITS UNIVARIATE (ARMA) AND EXTENDED (EARMA)'/3
 56:     >6X,'MODELS TO TIME SERIES DATA, WITH OPTIONAL DETERMINISTIC TREND'
 57:     >/36X,'ESTIMATION FOR UNIVARIATE MODELS. IT IS DESIGNED TO BE USED'
 58:     >/36X,'WITH THE BOOK:'
 59:     >/36X,'     "TIME SERIES AND SYSTEM ANALYSIS WITH APPLICATIONS"'
 60:     >//36X,'         BY S. M. PANDIT AND S. M. WU'
 61:     >//36X,'   THIS CODE IS INTENDED FOR INSTRUCTIONAL USE ONLY!'
 62:     >/25X,80('*')////15X,A)
 63:      PRINT 40,NSER,NOB,DELTA
 64:40    FORMAT (I6,' SERIES WITH',I5,' OBSERVATIONS, DELTA =',E10.4///
 65:     >          ' INPUT STARTING MODEL ORDERS:'/)
 66:      DO 50 I=1,NSER
 67:50    PRINT FORMT,I,(NAR(J,I),J=1,NSER),NMA(I)
 68:      PRINT 60,IFMEAN,IFAT,IFDIF,NMOD,IFPRT,IFLAG,IFSEAS,IFSER,
 69:     >INCPH,INCTH
 70:60    FORMAT (/15X,'PROGRAM CONTROL FLAGS'/5X,'IFMEAN:',I3,4X,
 71:     >'IFAT:',I3,' IFDIF:',I3,' NMOD:',I3,' IFPRT:',I3/6X,'IFLAG:',
 72:     >I3,' IFSEAS:',I3,' IFSER:',I3,' INCPH:',I3,' INCTH:',I3)
 73:      PRINT 70,STEP,CONTOL,ITMAX,IFREF
 74:70    FORMAT (5X,'STEP=',E10.4,' CONVERGENCE TOLERANCE=',E10.4
 75:     >/5X,I5,' ITERATIONS MAX    IFREF:',I4//)
 76:C***CHECK FOR ERRORS IN INPUT
 77:      IF (NSER.GT.MXSER) STOP'TOO MANY DATA SETS, MXSER=3'
 78:      IF (NOB.GT.MXNOB) STOP'TOO MANY OBSERVATIONS, MXNOB=369'
 79:      IF (IFSEAS.GT.0.AND.IFDIF.NE.0)
 80:     >   STOP'NUMERICAL DERIVATIVES MUST BE USED WITH SEASONALITY'
 81:      IF (IFSEAS.GT.0.AND.NSER.GT.1)
 82:     >   STOP'ONLY ONE SERIES MAY BE USED WITH SEASONALITY'
 83:      IF (IFSEAS.EQ.1.AND.NMOD.LT.0)
```

```
 84:          >     STOP'NO STOCHASTIC INITIAL VALUES FOR IFSEAS = 1'
 85:          IF ((IFSEAS.EQ.1.OR.IFSEAS.EQ.3).AND.NMOD.GT.1)
 86:          >     STOP'ONLY ONE MODEL AT A TIME FOR IFSEAS=1 OR 3'
 87:C***DETERMINE LAGS FOR MULTI-VARIATE MODELS
 88:          IF (NSER.GT.1) THEN
 89:            DO 100 I=1,NSER
 90:              IF (IFLAG.EQ.0) THEN
 91:                DO 80 J=I+1,NSER
 92:80              LAG(J,I)=0.0
 93:              ELSE
 94:                IF (I.EQ.1) THEN
 95:                  READ*,(LAG(J,I),J=2,NSER)
 96:                ELSEIF (I.EQ.NSER) THEN
 97:                  READ*,(LAG(J,I),J=1,I-1)
 98:                ELSE
 99:                  READ*,(LAG(J,I),J=1,I-1),(LAG(J,I),J=I+1,NSER)
100:                ENDIF
101:              ENDIF
102:              PRINT 90,I,(LAG(J,I),J=1,NSER)
103:90            FORMAT (5X,'SERIES'I4,': LAGS='10(I5,',')/)
104:100         CONTINUE
105:          ENDIF
106:C***READ SEASONALITY INPUT IF SEASONALITY FLAG IS SET
107:          IF (IFSEAS.GT.0) THEN
108:            READ*,NPOLY,NEXP,NSIN
109:            NSEASP=NPOLY+2*NEXP+4*NSIN+1
110:            IF (NSEASP.GT.30)
111:          >     STOP'TOO MANY SEASONALITY PARAMETERS, MAX=30'
112:            READ*,(SEAS(I),ISEAS(I),I=1,NSEASP)
113:            PRINT 110,NPOLY,NEXP,NSIN,(I,SEAS(I),I=1,NSEASP)
114:110         FORMAT(5X,'SEASONALITY:'I3,' POLYNOMIAL,'I3,' EXPONENTIAL,'
115:          > ' AND'I3,' SINUSOIDAL TRENDS'//6(5X,5('BETA('I2,')=',
116:          > E11.4,5X)/))
117:            PRINT 115,(I,ISEAS(I),I=1,NSEASP)
118:115         FORMAT (/5X,'ESTIMATION FLAGS:'/6(7X,5('IC('I2,')=',I4,14X)/)//)
119:C***MODIFY SEASONALITY PARAMETERS FOR SAMPLING INTERVAL
120:            N=NPOLY+1
121:            DO 120 I=1,NEXP
122:              N=N+2
123:120           SEAS(N)=SEAS(N)*DELTA
124:            DO 130 I=1,NSIN
125:              N=N+2
126:              SEAS(N)=SEAS(N)*DELTA
127:              N=N+2
128:130           SEAS(N)=SEAS(N)*DELTA*6.2831853
129:C***LOAD SEASONALITY PARAMETERS INTO PARAMETER ARRAY
130:            IF (IFSEAS.EQ.1.OR.IFSEAS.EQ.3) THEN
131:              DO 140 I=1,NSEASP
132:                IF (ISEAS(I).EQ.0) THEN
133:                  PAR(IPAR)=SEAS(I)
134:                  ISEAS(IPAR)=I
135:                ENDIF
136:                IPAR=IPAR+
137:140           CONTINUE
138:              NSEAS=IPAR-1
139:            ENDIF
140:          ENDIF
141:C***READ IN DATA SERIES
142:          READ*,((Y(I,J),I=1,NOB),J=1,NSER)
143:C***DETREND DATA TO ESTIMATE STOCHASTIC PART OF COMBINED MODELS
144:          IF (IFSEAS.EQ.2)     CALL DETPAR(NOB)
145:C***FIND AVERAGE AND VARIANCE OF EACH SERIES AND PRINT
146:          PRINT 145
147:145       FORMAT (//'  CHARACTERISTICS OF THE DATA'/)
148:          DO 190 I=1,NSER
149:            SUM=0.0
150:            VAR(I)=0.0
151:            DO 150 J=1,NOB
152:150           SUM=SUM+Y(J,I)
153:            AVG(I)=SUM/NOB
154:            DO 160 J=1,NOB
155:160           VAR(I)=VAR(I)+(Y(J,I)-AVG(I))**2
156:            VAR(I)=VAR(I)/(NOB-1)
157:            PRINT 170,I,AVG(I),VAR(I)
158:170         FORMAT (5X,'SERIES',I3,4X,'AVERAGE=',E11.4,'  VARIANCE=',
159:          > E11.4)
160:C***SCALE DATA FOR EARMA MODELS
161:            IF (NSER.GT.1) THEN
162:              SCA(I)=VAR(I)
163:              IF (IFMEAN.GT.0) SCA(I)=SCA(I)+AVG(I)**2
164:              SCA(I)=SQRT(SCA(I))
165:              DO 185 J=1,NOB
166:185             Y(J,I)=Y(J,I)/SCA(I)
```

```
167:                        AVG(I)=AVG(I)/SCA(I)
168:                    ENDIF
169:C***SUBTRACT AVERAGE FROM DATA IF MEAN IS NOT ESTIMATED
170:                    IF (IFMEAN.LE.0.AND.IFSEAS.EQ.0) THEN
171:                        IF (IFMEAN.LT.0) AVG(I)=X(-IFMEAN,I)
172:                        DO 180 J=1,NOB
173:                            Y(J,I)=Y(J,I)-AVG(I)
174:180                         X(J,I)=Y(J,I)
175:                    ENDIF
176:190         CONTINUE
177:C***CHECK FOR INITIAL VALUE FLAG
178:            IF (NMOD.LT.0) THEN
179:                IFINV=1
180:                NMOD=-NMOD
181:            ENDIF
182:C***LOOP ON NUMBER OF MODELS
183:            DO 290 MOD=1,NMOD
184:C***SET INITIAL VALUES OF MEANS IF ESTIMATED
185:            IF (IFMEAN.GT.0.AND.IFSEAS.EQ.0) THEN
186:                IPAR=NSEAS+1
187:                DO 210 I=1,NSER
188:                    PAR(IPAR)=AVG(I)
189:                    IF (IFMEAN.GT.1) PAR(IPAR)=Y(IFMEAN-1,I)
190:                    IPAR=IPAR+1
191:C***SUBTRACT AVERAGE TO RESET DATA FOR INITIAL VALUES OF NEW MODEL
192:                    DO 210 J=1,NOB
193:210                     X(J,I)=Y(J,I)-AVG(I)
194:            ENDIF
195:C***SKIP STOCHASTIC ESTIMATION FOR DETERMINISTIC MODELS
196:            NPAR=NSEAS
197:            IF (IFSEAS.EQ.1) GO TO 250
198:C***CHECK SERIES INDEX FLAG
199:            IF (IFSER.GT.0.AND.IFSER.LE.NSER) THEN
200:                INDEX=IFSER
201:                GO TO 220
202:            ENDIF
203:C***LOOP ON NUMBER OF DATA SERIES
204:            DO 280 INDEX=1,NSER
205:C***FIND NUMBER OF PHIO'S AND PHI'S, AND MAXLAG FOR THIS SERIES
206:220         NPHIO=0
207:            NPHI=0
208:            MAXLAG=1
209:            DO 230 I=INDEX+1,NSER
210:230             IF (LAG(I,INDEX).EQ.0)NPHIO=NPHIO+1
211:            DO 235 I=1,NSER
212:                ML=LAG(I,INDEX)+NAR(I,INDEX)-1
213:                IF(ML.GT.MAXLAG) MAXLAG=ML
214:235             NPHI=NPHI+NAR(I,INDEX)
215:            MAXLAG=MAX(MAXLAG,NMA(INDEX))
216:            NPAR=NPHIO+NPHI+NMA(INDEX)+IPAR-1
217:C***SET INITIAL VALUES OF INITIAL AT'S IF ESTIMATED
218:            IF (IFAT.LT.0) THEN
219:                DO 240 I=1,MAXLAG
220:240                 PAR(NPAR+I)=0.1
221:                NPAR=NPAR+MAXLAG
222:            ENDIF
223:            IF (NPAR.GT.MXPAR) STOP'TOO MANY PARAMETERS, MXPAR=45'
224:            WRITE(6,'(/////)')
225:            IF (IFPRT.GT.0)
226:        >       PRINT FORMT, INDEX,(NAR(J,INDEX),J=1,NSER),NMA(INDEX)
227:C***PURE AR MODELS
228:            IF (NMA(INDEX).EQ.0.AND.IFINV.NE.1) THEN
229:                ISIZ=NPHI+NPHIO
230:                ISIZ1=ISIZ+1
231:                CALL EAR(XTX,NAR(1,INDEX),LAG(1,INDEX),PAR,ISIZ)
232:                CALL GJR(XTX,ISIZ,4,MXINV,MXINV1)
233:                IPAR=IPAR-1
234:C***LOAD PHIO'S INTO PAR
235:                DO 245 I=1,NPHIO
236:245                 PAR(I+IPAR)=XTX(NPHI+I,ISIZ1)
237:                IPAR=IPAR+NPHIO
238:C***LOAD PHI'S INTO PAR
239:                DO 246 I=1,NPHI
240:246                 PAR(I+IPAR)=XTX(I,ISIZ1)
241:                IPAR=IPAR-NPHIO+1
242:                GO TO 250
243:            ENDIF
244:C***READ INITIAL VALUES IF IFINV = 1 (NMOD LT 0)
245:            IF (IFINV.EQ.1) THEN
246:                IF (IFMEAN.GT.0.AND.IFSEAS.EQ.0)
247:        >           READ*,(PAR(I),I=IPAR-NSER,IPAR-1)
248:                READ*,(PAR(I),I=IPAR,NPAR)
249:            ELSE
```

```
250:C***FIND INITIAL VALUES BY INVERSE FUNCTION METHOD
251:                              CALL INVAL(XTX,PAR,SCA,NPAR)
252:                        ENDIF
253:C***SCALE INITIAL PARAMETERS FOR EARMA MODELS
254:                        IF (NSER.GT.1) THEN
255:                           N=NSEAS+1
256:                           IF (IFMEAN.GE.1) N=N+NSER
257:C***SCALE ZERO LAG TERMS
258:                           DO 247 I=INDEX+1,NSER
259:                              IF (LAG(I,INDEX).EQ.0) THEN
260:                                 PAR(N)=PAR(N)*SCA(I)/SCA(INDEX)
261:                                 N=N+1
262:                              ENDIF
263:247                        CONTINUE
264:C***SCALE REST OF PHI TERMS
265:                           DO 249 I=1,NSER
266:                              IF (I.NE.INDEX) THEN
267:                                 NN=N
268:                                 SCALE=SCA(I)/SCA(INDEX)
269:                                 DO 248 J=1,NAR(I,INDEX)
270:                                    PAR(NN)=PAR(NN)*SCALE
271:248                                 NN=NN+1
272:                              ENDIF
273:249                           N=N+NAR(I,INDEX)
274:                        ENDIF
275:C***DO NON-LINEAR LEAST SQUARES SEARCH TO FIND BEST PARAMETERS
276:250                     CALL LS(NPAR,PAR,AT(1,INDEX),FORMT,IFDIF,SCA,IFPRT)
277:C***CHECK FOR AUTO-REFINEMENT OF PARAMETERS
278:                        IF (IFREF.GT.0) THEN
279:                           IFREF=-IFREF
280:                           STEP=STEP*10.**IFREF
281:                           CONTOL=CONTOL*10.**IFREF
282:                           IF (IFPRT.GT.0) PRINT 255,STEP,CONTOL
283:255                        FORMAT(///'    NEW CONVERGENCE PARAMETERS:',5X,'STEP='
284:     >                     ,E10.4,5X,'CONVERGENCE TOLERANCE=',E10.4)
285:C***REDO ESTIMATION AND RESTORE ESTIMATION PARAMETERS
286:                           CALL LS(NPAR,PAR,AT(1,INDEX),FORMT,IFDIF,SCA,IFPRT)
287:                           IFREF=-IFREF
288:                           STEP=STEP*10.**IFREF
289:                           CONTOL=CONTOL*10.**IFREF
290:                        ENDIF
291:C***FIND AND PRINT F-TEST VALUE
292:                        IF (NMOD.GT.1) THEN
293:                           IF (MOD.GT.1) THEN
294:                              FTST=(OSSQ(INDEX)/AT(NOB+1,INDEX)-1)*
295:     >                        (NOB-NPAR)/(NSER*INCPH+INCTH)
296:                              PRINT 265,FTST
297:265                           FORMAT(//' F-TEST WITH LAST MODEL FOR THIS SERIES =',F10.3)
298:                           ENDIF
299:                           OSSQ(INDEX)=AT(NOB+1,INDEX)
300:                        ENDIF
301:C***FIND CORRELATIONS OF RESIDUALS
302:                        CALL ATSCOR(AT,NOB,INDEX,IFSER)
303:C***FIND ROOTS, FREQUENCIES, DAMPING, AND GREEN'S FUNCTION COEFFICIENTS
304:                        IF (IFSEAS.NE.1.AND.IFPRT.GT.0) CALL ANALYS(PAR,DELTA)
305:C***UPDATE MODEL ORDERS FOR NEXT MODEL
306:                        DO 270 I=1,NSER
307:270                        NAR(I,INDEX)=NAR(I,INDEX)+INCPH
308:                        NMA(INDEX)=NMA(INDEX)+INCTH
309:280                     PRINT 285
310:285                     FORMAT (///1X,131('*')//)
311:290                  CONTINUE
312:         STOP
313:         END
314:C#=#=#=#=#=#=#=#=#=#=#=#=#=#=#=#=#=#=#=#=#=#=#=#=#=#=#=#=#=#=#=#=#=#=#
315:         SUBROUTINE ANALYS(PAR,DELTA)
316:C***THIS SUBROUTINE FINDS THE ROOTS,FREQUENCIES,AND DAMPING, AND THE
317:C***GREEN'S FUNCTION AND IMPULSE RESPONSE COEFFICIENTS OF THE MODEL
318:         COMMON /INDEX/ INDEX,MAXLAG,NOB,NSER,NPHIO,NPHI,NSEAS
319:         COMMON /LAG/ LAG(MXSER,MXSER),NAR(MXSER,MXSER),NMA(MXSER)
320:         COMMON /FLAG/ IFMEAN,IFAT,IFSEAS
321:         COMPLEX ROOT(MXPAR)
322:         DIMENSION PAR(MXPAR)
323:C***FIND ROOTS OF AUTOREGRESSIVE SIDE
324:         IF (NAR(INDEX,INDEX).EQ.0) RETURN
325:         N=NPHIO+NSEAS+1
326:         IF (IFMEAN.GE.1.AND.IFSEAS.EQ.0) N=N+NSER
327:         DO 10 I=1,INDEX-1
328:10          N=N+NAR(I,INDEX)
329:         NR=NAR(INDEX,INDEX)
330:         CALL EIGEN(PAR(N),NR,ROOT)
331:         PRINT 30
```

```
333:30       FORMAT (////37X,20('*'),'MODEL CHARACTERISTICS',20('*')////)
334:C***FIND EXPLICIT GREEN'S FUNCTION AND IMPULSE RESPONSE
335:         N=NSEAS+1
336:         IF (IFMEAN.GE.1.AND.IFSEAS.EQ.0) N=N+NSER
337:         N1=N
338:         N=N+NPHIO
339:         M=N+NPHI
340:         DO 63 I=1,NSER
341:C***SET ZERO LAG TERMS
342:             IF (LAG(I,INDEX).EQ.0) THEN
343:                 PHIO=PAR(N1)
344:                 N1=N1+1
345:                 N2=N
346:                 NN=NAR(I,INDEX)
347:             ELSE
348:                 PHIO=PAR(N)
349:                 N2=N+1
350:                 NN=NAR(I,INDEX)-1
351:             ENDIF
352:C***FIND FUNCTION FOR SERIES I
353:             IF (I.NE.INDEX.AND.NR.GT.NN) THEN
354:                 PRINT 61,INDEX,I
355:61               FORMAT (44X,'IMPULSE RESPONSE FUNCTION OF SERIES',
356:     >           I3,' TO SERIES',I3/)
357:                 CALL GREEN(PHIO,PAR(N2),ROOT,NR,NN,LAG(I,INDEX),'IMP(j) =',
358:     >           DELTA)
359:             ELSEIF (I.EQ.INDEX.AND.NR.GT.NMA(INDEX)) THEN
360:                 PRINT 62,INDEX
361:62               FORMAT (44X,'GREEN''S FUNCTION OF SERIES',I3,
362:     >           ' WITH ITS RESIDUALS'/)
363:                 CALL GREEN(1.0,PAR(M),ROOT,NR,NMA(INDEX),0,'  G(j) =',
364:     >           DELTA)
365:             ENDIF
366:63       N=N+NAR(I,INDEX)
367:C***FIND FREQENCIES AND DAMPING OF AUTOREGRESSIVE OPERATOR
368:         PRINT 64
369:64       FORMAT (////45X,'CHARACTERISTICS OF THE AUTOREGRESSIVE OPERATOR')
370:         CALL FREQ(NR,ROOT,DELTA)
371:C***FIND ROOTS OF MOVING AVERAGE SIDE
372:         NR=NMA(INDEX)
373:         IF (NR.GT.0) THEN
374:             N=NSEAS+NPHIO+NPHI+1
375:             IF (IFMEAN.GE.1.AND.IFSEAS.EQ.0) N=N+NSER
376:             CALL EIGEN(PAR(N),NR,ROOT)
377:C***FIND FREQENCIES AND DAMPING OF MOVING AVERAGE OPERATOR
378:             PRINT 80
379:80           FORMAT(/////45X,'CHARACTERISTICS OF THE MOVING',
380:     >       ' AVERAGE OPERATOR')
381:             CALL FREQ(NR,ROOT,DELTA)
382:         ENDIF
383:         RETURN
384:         END
385:C#=#=#=#=#=#=#=#=#=#=#=#=#=#=#=#=#=#=#=#=#=#=#=#=#=#=#=#=#=#=#=#=#=#=#=#
386:         SUBROUTINE ATSCOR(AT,NOB,INDEX,IFSER)
387:C***THIS SUBROUTINE CALCULATES CORRELATIONS OF THE RESIDUALS FOR 30 LAGS
388:         PARAMETER MXSER=3,MXSR2=MXSER**2,MXNOB=369,MXNOB1=MXNOB+1
389:         DOUBLE PRECISION SUM,SUM1,SUM2
390:         DIMENSION AT(MXNOB1,MXSER)
391:         A=2./NOB
392:C***FIND AND PRINT THE AVERAGE OF THE RESIDUALS
393:         AVG=0.0
394:         DO 10 I=1,NOB
395:10       AVG=AVG+AT(I,INDEX)
396:         AVG=AVG/NOB
397:         PRINT 20,INDEX,AVG
398:20       FORMAT (////36X,'THE AVERAGE OF THE RESIDUALS FOR SERIES',I5,
399:     >   ' =',E15.6)
400:C***MEAN SUBTRACT THE AT'S FOR THIS SERIES
401:         DO 25 I=1,NOB
402:25       AT(I,INDEX)=AT(I,INDEX)-AVG
403:C***CALCULATE THE CORRELATIONS
404:         DO 110 I=1,INDEX
405:             IF (I.NE.INDEX.AND.IFSER.EQ.0) THEN
406:C***FIND CROSS CORRELATIONS BETWEEN THE SERIES
407:                 SSQ=0.0
408:                 Q=SQRT(AT(NOB+1,I)*AT(NOB+1,INDEX))
409:                 PRINT 30,I,INDEX
410:30               FORMAT (//37X,'CORRELATIONS OF SERIES',I3,' RESIDUALS ',
411:     >           'WITH SERIES',I3,' RESIDUALS'//13X,'LAG',7X,'CORRELATION',
412:     >           13X,'UNIFIED',22X,'LAG',7X,'CORRELATION',13X,'UNIFIED'/)
413:                 DO 55 K=0,30
414:                     SUM1,SUM2=0.0
415:                     DO 40 J=K+1,NOB
```

```
416:                                SUM1=SUM1+AT(J,I)*AT(J-K,INDEX)
417:40                              SUM2=SUM2+AT(J-K,I)*AT(J,INDEX)
418:                        R1=SUM1/Q
419:                        R2=SUM2/Q
420:                        SE=SQRT(NOB-K)
421:                        S1=R1*SE
422:                        S2=R2*SE
423:                        SSQ=SSQ+R1**2+R2**2
424:55                      PRINT 60,K,R1,S1,-K,R2,S2
425:60                      FORMAT(12X,I3,5X,F12.7,10X,F12.7,21X,I3,5X,F12.7,10X
426:       >                       ,F12.7)
427:                        PRINT 65,SSQ
428:65                      FORMAT(//39X,'THE SUM OF SQUARES OF THE CROSS',
429:       >                  ' CORRELATIONS =',E12.4)
430:                     ELSEIF (I.EQ.INDEX) THEN
431:C***FIND THE AUTOCORRELATIONS OF THE INDEX SERIES
432:                        SSQ=0.0
433:                        VAR=1./NOB
434:                        Q=AT(NOB+1,INDEX)
435:                        PRINT 70,INDEX
436:70                      FORMAT(//45X,'AUTO CORRELATIONS OF SERIES',I3,' RESIDUALS'
437:       >                //45X,'LAG',7X,'CORRELATION',13X,'UNIFIED'//)
438:                        DO 90 K=1,30
439:                           SUM=0.0
440:                           DO 80 J=1,NOB-K
441:80                            SUM=SUM+AT(J,INDEX)*AT(J+K,INDEX)
442:                           R=SUM/Q
443:                           S=R/SQRT(VAR)
444:                           SSQ=SSQ+R**2
445:                           VAR=VAR+A*R**2
446:90                         PRINT 100,K,R,S
447:100                     FORMAT (44X,I3,5X,F12.7,10X,F12.7)
448:                        PRINT 105,SSQ
449:105                     FORMAT(//39X,'THE SUM OF SQUARES OF THE AUTO',
450:       >                  ' CORRELATIONS =',E12.4)
451:                     ENDIF
452:110               CONTINUE
453:            RETURN
454:            END
455:C#=#=#=#=#=#=#=#=#=#=#=#=#=#=#=#=#=#=#=#=#=#=#=#=#=#=#=#=#=#=#=#=#=#=#
456:            SUBROUTINE DETPAR(NOB)
457:C***THIS SUBROUTINE SUBTRACTS THE DETERMINISTIC TRENDS FROM THE DATA
458:            PARAMETER MXSER=3,MXNOB=369,MXPAR=45
459:            COMMON /DATA/ Y(MXNOB,MXSER),X(MXNOB,MXSER)
460:            COMMON /SEASON/ NPOLY,NEXP,NSIN,ISEAS(30),SEAS(30),DELTA
461:            DO 40 IT=1,NOB
462:               N=2
463:               S=SEAS(1)
464:               DO 10 I=1,NPOLY
465:                  S=S+SEAS(N)*(DELTA*IT)**I
466:10                N=N+1
467:               DO 20 I=1,NEXP
468:                  S=S+SEAS(N)*EXP(SEAS(N+1)*IT)
469:20                N=N+2
470:               DO 30 I=1,NSIN
471:                  S=S+SEAS(N)*EXP(SEAS(N+1)*IT)*
472:       >(SEAS(N+2)*SIN(SEAS(N+3)*IT)+SQRT(1-SEAS(N+2)**2)*
473:       >COS(SEAS(N+3)*IT))
474:30                N=N+4
475:40             X(IT,1)=Y(IT,1)-S
476:            RETURN
477:            END
478:C#=#=#=#=#=#=#=#=#=#=#=#=#=#=#=#=#=#=#=#=#=#=#=#=#=#=#=#=#=#=#=#=#=#=#
479:            SUBROUTINE DIF(PAR,Z,AT,IS,IFMEAN)
480:C***THIS SUBROUTINE FINDS THE ANALYTICAL DERIVATIVES OF THE
481:C***RESIDUALS WITH RESPECT TO THE STOCHASTIC PARAMETERS
482:            PARAMETER MXSER=3,MXNOB=369,MXNOB1=MXNOB+1,MXPAR=45,MXPAR1=MXPAR+1
483:            DOUBLE PRECISION S
484:            COMMON /INDEX/ INDEX,MAXLAG,NOB,NSER,NPHIQ,NPHI,NSEAS
485:            COMMON /LAG/ LAG(MXSER,MXSER),NAR(MXSER),NMA(MXSER)
486:            COMMON /DATA/ Y(MXNOB,MXSER),X(MXNOB,MXSER)
487:            DIMENSION PAR(MXPAR),AT(MXNOB1),Z(MXNOB,MXPAR1)
488:            IZ=1
489:            KPAR=NPHIQ+NPHI+1
490:C***FIND DERIVATIVES WITH RESPECT TO MEANS IF ESTIMATED
491:            IF(IFMEAN.GE.1)THEN
492:C***RESTORE ORIGINAL DATA VALUES
493:               DO 5 I=1,NSER
494:                  DO 5 J=1,NOB
495:5                    X(J,I)=Y(J,I)
496:               KPAR=KPAR+NSER
497:               IPAR=NSER+1
498:               JPAR=IPAR+NPHIQ
```

```
499:                    DO 30 IZ=1,NSER
500:                    SUM=0.0
501:                    IF(IZ.EQ.INDEX)SUM=-1.0
502:                       IF(IZ.GT.INDEX.AND.LAG(IZ,INDEX).EQ.0)THEN
503:                          SUM=-PAR(IPAR)
504:                          IPAR=IPAR+1
505:                       ENDIF
506:                       DO 10 J=1,NAR(IZ,INDEX)
507:                          SUM=SUM+PAR(JPAR)
508:10                        JPAR=JPAR+1
509:                       DO 30 IT=1,NOB
510:                       S=SUM
511:                       K=KPAR
512:                          DO 20 J=1,MIN(IT-1,NMA(INDEX))
513:                          S=S+PAR(K)*Z(IT-J,IZ)
514:20                        K=K+1
515:30                     Z(IT,IZ)=S
516:                    ENDIF
517:C***FIND DERIVATIVES WITH RESPECT TO PHI0'S
518:                    DO 60 I=1+INDEX,NSER
519:                    IF(LAG(I,INDEX).EQ.0)THEN
520:                       DO 50 IT=1,NOB
521:                       K=KPAR
522:                       S=X(IT,I)
523:                          DO 40 J=1,MIN(IT-1,NMA(INDEX))
524:                          S=S+PAR(K)*Z(IT-J,IZ)
525:40                        K=K+1
526:50                     Z(IT,IZ)=S
527:                       IZ=IZ+1
528:                    ENDIF
529:60                 CONTINUE
530:C***FIND DERIVATIVES WITH RESPECT TO PHI'S - PHI1 FIRST
531:                    DO 140 I=1,NSER
532:                    LAGT=LAG(I,INDEX)
533:                    IF(LAGT.EQ.0)LAGT=1
534:                    N=1
535:65                  DO 80 IT=1,LAGT
536:                    S=0.0
537:                       IF(IFMEAN.GE.1)S=PAR(I)
538:                    K=KPAR
539:                       DO 70 J=1,MIN(IT-1,NMA(INDEX))
540:                       S=S+PAR(K)*Z(IT-J,IZ)
541:70                     K=K+1
542:80                  Z(IT,IZ)=S
543:                    DO 100 IT=LAGT+1,NOB
544:                    S=0.0
545:                       IF(IFMEAN.GE.1)S=PAR(I)
546:                    K=KPAR
547:                       DO 90 J=1,MIN(IT-1,NMA(INDEX))
548:                       S=S+PAR(K)*Z(IT-J,IZ)
549:90                     K=K+1
550:100                 Z(IT,IZ)=S-X(IT-LAGT,I)
551:C***DO REST OF PHI'S FOR THIS SERIES
552:                    IZ=IZ+1
553:                    IF(IFMEAN.GE.1)THEN
554:                       N=N+1
555:                       LAGT=LAGT+1
556:                       IF(N.LE.NAR(I,INDEX))GO TO 65
557:                    ELSE
558:                       DO 130 J=2,NAR(I,INDEX)
559:                          DO 110 IT=I,J
560:110                       Z(IT,IZ)=0.0
561:                          DO 120 IT=J+1,NOB
562:120                       Z(IT,IZ)=Z(IT-1,IZ-1)
563:130                       IZ=IZ+1
564:                    ENDIF
565:140                 CONTINUE
566:C***FIND DERIVATIVES WITH RESPECT TO THETA - THETA1 FIRST
567:                    Z(1,IZ)=0.0
568:                    DO 160 IT=2,NOB
569:                    S=0.0
570:                    K=KPAR
571:                       DO 150 J=1,MIN(IT-1,NMA(INDEX))
572:                       S=S+PAR(K)*Z(IT-J,IZ)
573:150                    K=K+1
574:160                 Z(IT,IZ)=S+AT(IT-1)
575:C***DO REST OF THETA TERMS
576:                    DO 180 I=2,NMA(INDEX)
577:                    IZ=IZ+1
578:                       DO 170 IT=1,I
579:170                    Z(IT,IZ)=0.0
580:                       DO 180 IT=I+1,NOB
581:180                    Z(IT,IZ)=Z(IT-1,IZ-1)
```

```
582:        IS=MAXLAG
583:        RETURN
584:        END
585:C#=#=#=#=#=#=#=#=#=#=#=#=#=#=#=#=#=#=#=#=#=#=#=#=#=#=#=#=#=#=#=#
586:        SUBROUTINE EAR(XTX,NAR,LAG,PAR,ISIZ)
587:C***THIS SUBROUTINE FORMS THE XTX AND XTY MATRICES FOR EAR MODELS
588:        PARAMETER MXSER=3,MXNOB=369,MXPAR=45,MXINV=50,MXINV1=MXINV+1
589:        DOUBLE PRECISION XTX,SUM
590:        COMMON /INDEX/ INDEX,MAXLAG,NOB,NSER,NPHIO,NPHI,NSEAS
591:        COMMON /FLAG/ IFMEAN,IFAT,IFSEAS
592:        COMMON /DATA/ Y(MXNOB,MXSER),X(MXNOB,MXSER)
593:        DIMENSION XTX(MXINV,MXINV1),LAG(MXSER),LAGT(MXSER)
594:        DIMENSION PAR(MXPAR),NAR(MXSER)
595:C***SUBTRACT MEAN AND SEASONALITY FOR INITIAL VALUES
596:        IF (IFSEAS.EQ.3) CALL MODEL(PAR,AT,0)
597:C***SET TEMPORARY LAGS
598:        DO 10 I=1,NSER
599:        LAGT(I)=LAG(I)
600:10      IF(LAGT(I).EQ.0) LAGT(I)=1
601:C***FORM INVARIANT PART OF XTX BY BLOCK COLUMNS
602:        IROW=1
603:        ICOL=1
604:        DO 30 I=1,NSER
605:C***FORM DIAGONAL BLOCK
606:        CALL XX(X(2-LAGT(I),I),XTX,NAR(I),IROW,ICOL,MAXLAG,NOB)
607:C***FORM REST OF COLUMN
608:        DO 20 J=I+1,NSER
609:        IROW=IROW+NAR(J-1)
610:20      CALL XA(X(2-LAGT(I),I),X(2-LAGT(J),J),XTX,NAR(I),
611:     >  NAR(J),IROW,ICOL,MAXLAG,NOB)
612:        ICOL=ICOL+NAR(I)
613:30      IROW=ICOL
614:C***FORM IMMEDIATE BLOCKS - LOOP DOWN THE ROWS
615:        DO 50 I=INDEX+1,NSER
616:        IF(LAG(I).EQ.0) THEN
617:        ICOL=1
618:C***LOOP ACROSS THE BLOCKS
619:        DO 40 J=1,NSER
620:C***LOOP ON ELEMENTS PER BLOCK
621:        DO 40 K=1,NAR(J)
622:        SUM=0.0
623:        DO 35 L=MAXLAG+1,NOB
624:35      SUM=SUM-X(L-K-LAGT(J)+1,J)*X(L,I)
625:        XTX(IROW,ICOL)=SUM
626:40      ICOL=ICOL+1
627:C***FORM IMMEDIATE DIAGONAL BLOCK
628:        DO 70 J=INDEX+1,I
629:        IF(LAG(J).EQ.0) THEN
630:        SUM=0.0
631:        DO 60 K=MAXLAG+1,NOB
632:60      SUM=SUM+X(K,I)*X(K,J)
633:        XTX(IROW,ICOL)=SUM
634:        ICOL=ICOL+1
635:        ENDIF
636:70      CONTINUE
637:        IROW=IROW+1
638:        ENDIF
639:50      CONTINUE
640:C***FORM XTY - SECTION CORRESPONDING TO INVARIANT PART
641:        IROW=1
642:        DO 90 I=1,NSER
643:        DO 90 J=1,NAR(I)
644:        SUM=0.0
645:        DO 80 K=MAXLAG+1,NOB
646:80      SUM=SUM+X(K-J-LAGT(I)+1,I)*X(K,INDEX)
647:        XTX(IROW,ISIZ+1)=SUM
648:90      IROW=IROW+1
649:C***IMMEDIATE PART
650:        DO 110 I=INDEX+1,NSER
651:        IF(LAG(I).EQ.0) THEN
652:        SUM=0.0
653:        DO 100 J=MAXLAG+1,NOB
654:100     SUM=SUM-X(J,I)*X(J,INDEX)
655:        XTX(IROW,ISIZ+1)=SUM
656:        IROW=IROW+1
657:        ENDIF
658:110     CONTINUE
659:C***DEFINE UPPER TRIANGLE OF XTX
660:        DO 120 I=1,ISIZ-1
661:        DO 120 J=I+1,ISIZ
662:120     XTX(I,J)=XTX(J,I)
663:        RETURN
664:        END
```

```
665:C#=#=#=#=#=#=#=#=#=#=#=#=#=#=#=#=#=#=#=#=#=#=#=#=#=#=#=#=#=#=#=#=#=#
666:      SUBROUTINE XX(X,XTX,N,IR,IC,MAXLAG,NOB)
667:C***THIS SUBROUTINE FORMS THE LOWER TRIANGLE OF THE DIAGONAL BLOCKS
668:C***IR AND IC ARE THE POSITION OF THE UPPER LEFT CORNER
669:      PARAMETER MXSER=3,MXNOB=369,MXPAR=45,MXINV=50,MXINV1=MXINV+1
670:      DOUBLE PRECISION XTX,SUM
671:      DIMENSION X(MXNOB),XTX(MXINV,MXINV1)
672:      K=NOB+1
673:      L=N+1
674:      DO 20 I=0,N-1
675:      SUM=0.0
676:C***FORM INITIAL WINDOW
677:      DO 10 J=MAXLAG,NOB-1
678:10    SUM=SUM+X(J-I)*X(J)
679:      XTX(IR+I,IC)=SUM
680:C***WINDOW REST OF DIAGONAL
681:      DO 20 J=1,N-I-1
682:      IF(J.EQ.0) GO TO 20
683:      SUM=SUM+X(N-J)*X(N-J-I)-X(NOB-J)*X(NOB-J-I)
684:20    XTX(IR+I+J,IC+J)=SUM
685:      RETURN
686:      END
687:C#=#=#=#=#=#=#=#=#=#=#=#=#=#=#=#=#=#=#=#=#=#=#=#=#=#=#=#=#=#=#=#=#=#
688:      SUBROUTINE XA(X1,X2,XTX,N,M,IR,IC,MAXLAG,NOB)
689:C***THIS SUBROUTINE FORMS THE OFF DIAGONAL BLOCKS
690:C***IR AND IC ARE THE POSITION OF THE UPPER LEFT ELEMENT
691:      PARAMETER MXSER=3,MXNOB=369,MXPAR=45,MXINV=50,MXINV1=MXINV+1
692:      DOUBLE PRECISION XTX,SUM
693:      DIMENSION X1(MXNOB),X2(MXNOB),XTX(MXINV,MXINV1)
694:      NOBM1=NOB-1
695:      NP1=N+1
696:C***FORM UPPER RIGHT HALF
697:C***GET INITIAL WINDOW
698:      DO 20 I=N-1,1,-1
699:      SUM=0.0
700:      DO 10 J=MAXLAG,NOBM1
701:10    SUM=SUM+X1(J-I)*X2(J)
702:      XTX(IR,IC+I)=SUM
703:C***WINDOW REST OF DIAGONAL
704:      DO 20 J=1,MIN(M-1,N-I-1)
705:      SUM=SUM+X1(N-J-I)*X2(N-J)-X1(NOB-J-I)*X2(NOB-J)
706:20    XTX(IR+J,IC+I+J)=SUM
707:C***FORM LOWER LEFT HALF
708:C***GET INITIAL WINDOW
709:      DO 40 I=0,M-1
710:      SUM=0.0
711:      DO 30 J=MAXLAG,NOBM1
712:30    SUM=SUM+X1(J)*X2(J-I)
713:      XTX(IR+I,IC)=SUM
714:C***WINDOW REST OF DIAGONAL
715:      DO 40 J=1,M-I-1
716:      SUM=SUM+X1(M-J)*X2(M-J-I)-X1(NOB-J)*X2(NOB-J-I)
717:40    XTX(IR+I+J,IC+J)=SUM
718:      RETURN
719:      END
720:C#=#=#=#=#=#=#=#=#=#=#=#=#=#=#=#=#=#=#=#=#=#=#=#=#=#=#=#=#=#=#=#=#=#
721:      SUBROUTINE EIGEN(PAR,N,ROOT)
722:C***THIS SUBROUTINE USES A MODIFIED BAIRSTOW METHOD FOR ROCT EXTRACTION
723:      PARAMETER MXPAR=45,MP=6*MXPAR/5,MP1=MP+1,MXPAR1=MXPAR+1
724:      IMPLICIT REAL*8(A-H,O-Z)
725:      REAL*4 PAR(1),A(MXPAR1)
726:      COMPLEX ROOT(MXPAR)
727:      DIMENSION H(MP),B(MP1),C(MP1),U(MXPAR),V(MXPAR)
728:      N1=N+1
729:      A(1)=-1.
730:      A(N1)=1./PAR(N)
731:      DO 5 I=2,N
732:      II=N1-I
733:5     A(I)=-PAR(II)*A(N1)
734:      IREV=1
735:      NC=N+1
736:C***THIS LOOP STORES THE COEFFICIENTS OF THE ORIGINAL POLYNOMIAL INTO H
737:      DO 10 I=1,NC
738:10    H(I)=A(I)
739:      P=0.
740:      Q=0.
741:      R=0.
742:C***THE FOLLOWING SECTION CHECKS TO SEE IF 0 IS A ROOT
743:30    IF(H(1).NE.0)GO TO 40
744:      NC=NC-1
745:      V(NC)=0.
746:      U(NC)=0.
747:C***THIS LOOP STORES THE COEFFICIENTS OF THE DEPRESSED POLYNOMIAL INTO H
```

```
748:      DO 20 I=1,NC
749:   20 H(I)=H(I+1)
750:      GO TO 30
751:C***GENERATE FIRST OR SECOND ORDER ROOTS
752:   40 IF(NC.EQ.1)GO TO 320
753:      IF(NC.NE.2)GO TO 50
754:      R=-H(1)/H(2)
755:      GO TO 200
756:   50 IF(NC.NE.3)GO TO 60
757:      P=H(2)/H(3)
758:      Q=H(1)/H(3)
759:      GO TO 240
760:C***REVERSE COEFFICIENT ORDER IF WILL SPEED CONVERGENCE OF P,Q,& R
761:   60 IF(DABS(H(NC-1)/H(NC)).GE.DABS(H(2)/H(1)))GO TO 100
762:C***REVERSE COEFFICIENTS AND SET IREV FOR INTERCHANGE
763:      IREV=-IREV
764:      M=NC/2
765:      DO 70 I=1,M
766:      NL=NC+1-I
767:      TEMP=H(NL)
768:      H(NL)=H(I)
769:   70 H(I)=TEMP
770:C***INITIAL GUESS OF P,Q,& R AFTER FIRST FACTOR AND ITS ROOTS FOUND
771:      IF(Q.EQ.0)GO TO 80
772:      P=P/Q
773:      Q=1./Q
774:      GO TO 90
775:   80 P=0.
776:   90 IF(R.EQ.0)GO TO 100
777:      R=1./R
778:C***SET ERROR CRITERION AND FIRST 2 B'S AND C'S
779:  100 E=5.E-10
780:      B(NC)=H(NC)
781:      C(NC)=H(NC)
782:      B(NC+1)=0.
783:      C(NC+1)=0.
784:      NP=NC-1
785:C***LOOP TO FIND THE LINEAR AND QUADRATIC FACTORS OF THE GIVEN POLYNOMIAL
786:  190 DO 110 J=1,1000
787:C***THIS SECTION FINDS THE LINEAR FACTORS
788:      DO 120 K=1,NP
789:      I=NC-K
790:  120 B(I)=H(I)+R*B(I+1)
791:      C(I)=B(I)+R*C(I+1)
792:C***CHECK IF R IS CONSTANT TERM OF FACTOR - ELSE NEW R
793:      IF(DABS(B(1)/H(1)).LE.E)GO TO 200
794:      IF(C(2).EQ.0.)GO TO 130
795:      R=R-B(1)/C(2)
796:      GO TO 140
797:  130 R=R+1
798:C***THIS SECTION FINDS THE QUADRATIC FACTORS
799:  140 DO 150 K=1,NP
800:      I=NC-K
801:      B(I)=H(I)-P*B(I+1)-Q*B(I+2)
802:  150 C(I)=B(I)-P*C(I+1)-Q*C(I+2)
803:C***CHECK IF P&Q ARE COEFFICIENTS OF QUADRATIC FACTOR - ELSE NEW P&Q
804:      IF(H(2).EQ.0.) IF(DABS(B(2)/H(1))-E)160,160,170
805:      IF(DABS(B(2)/H(2))-E)160,160,170
806:  160 IF(DABS(B(1)/H(1))-E)240,240,170
807:  170 DENOM=C(3)**2-(C(2)-B(2))*C(4)
808:      IF(DENOM.EQ.0)GO TO 180
809:      P=P+(B(2)*C(3)-B(1)*C(4))/DENOM
810:      Q=Q+(-B(2)*(C(2)-B(2))+B(1)*C(3))/DENOM
811:      GO TO 110
812:  180 P=P-2.
813:      Q=Q*(Q+1.)
814:  110 CONTINUE
815:C***CHANGE E IF NOT CONVERGED AFTER 1000 ITERATIONS AND TRY AGAIN
816:      E=E*10.
817:      GO TO 190
818:C***THIS SECTION FINDS THE ROOTS OF THE LINEAR FACTORS
819:  200 NC=NC-1
820:      V(NC)=0.
821:      IF(IREV.EQ.-1)GO TO 210
822:      U(NC)=R
823:      GO TO 220
824:  210 U(NC)=1./R
825:C***STORE DEPRESSED POLYNOMIAL IN H TO FIND NEXT FACTOR
826:  220 DO 230 I=1,NC
827:  230 H(I)=B(I+1)
828:      GO TO 40
829:C***THIS SECTION FINDS THE ROOTS OF THE QUADRATIC FACTORS
830:  240 NC=NC-2
```

```
831:            IF(IREV.EQ.-1)GO TO 250
832:            QP=Q
833:            PP=P/2.
834:            GO TO 260
835:     250    QP=1./Q
836:            PP=P/(Q*2.)
837:     260    DISCRM=PP**2-QP
838:C***DETERMINE IF QUADRATIC ROOT IS REAL OR IMAGINARY
839:            IF(DISCRM.GE.0.)GO TO 270
840:            U(NC+1)=-PP
841:            U(NC)=-PP
842:            V(NC+1)=DSQRT(-DISCRM)
843:            V(NC)=-V(NC+1)
844:            GO TO 280
845:     270    IF(PP.EQ.0.)GO TO 290
846:            U(NC+1)=-(PP/DABS(PP))*(DABS(PP)+DSQRT(DISCRM))
847:            GO TO 300
848:     290    U(NC+1)=-DSQRT(DISCRM)
849:     300    V(NC+1)=0.
850:            U(NC)=QP/U(NC+1)
851:            V(NC)=0.
852:C***STORE DEPRESSED POLYNOMIAL IN H TO FIND NEXT FACTOR
853:     280    DO 310 I=1,NC
854:     310    H(I)=B(I+2)
855:            GO TO 40
856:     320    CONTINUE
857:            DO 330 I=1,N
858:     330    ROOT(I)=CMPLX(SNGL(U(I)),SNGL(V(I)))
859:            RETURN
860:            END
861:C#=#=#=#=#=#=#=#=#=#=#=#=#=#=#=#=#=#=#=#=#=#=#=#=#=#=#=#=#=#
862:            SUBROUTINE FREQ(NR,ROOT,DELTA)
863:C***THIS SUBROUTINE CALCULATES FREQUENCY AND DAMPING FROM THE ROOTS
864:            COMPLEX ROOT(1),CROOT
865:            PRINT 10
866:10          FORMAT (/46X,'DISCRETE',14X,'NATURAL'/43X,'COMPLEX ROOTS',11X,
867:           >'FREQUENCY',10X,'DAMPING'/43X,'REAL',6X,'IMAG',12X,'(HZ)',
868:           >14X,'RATIO'/41X,52('-'))
869:            DO 40 I=1,NR
870:C***FIND FREQUENCY AND DAMPING FOR 2ND ORDER ROOT
871:            AIROOT=AIMAG(ROOT(I))
872:            IF (AIROOT.GT.0) THEN
873:               CROOT=CLOG(ROOT(I))/DELTA
874:               ACROOT=CABS(CROOT)
875:               W=ACROOT/6.2831853
876:               Z=-REAL(CROOT)/ACROOT
877:               PRINT 20,ROOT(I),W,Z
878:20             FORMAT (40X,F7.4,' +/-',F7.4,2(7X,E11.4))
879:C***FIND BREAK FREQUENCY FOR REAL ROOTS
880:            ELSEIF (AIROOT.EQ.0) THEN
881:               RROOT=REAL(ROOT(I))
882:               IF (RROOT.LE.0) THEN
883:C***NEGATIVE ROOT - SET BREAK FREQUENCY AT NYQUIST FREQUENCY
884:                  W=0.5/DELTA
885:               ELSE
886:C***POSITIVE ROOT - CALCULATE BREAK FREQUENCY
887:                  W=-ALOG(RROOT)/(DELTA*6.2831853)
888:               ENDIF
889:               PRINT 30,REAL(ROOT(I)),W
890:30             FORMAT (40X,F7.4,18X,E11.4)
891:            ENDIF
892:40          CONTINUE
893:            RETURN
894:            END
895:C#=#=#=#=#=#=#=#=#=#=#=#=#=#=#=#=#=#=#=#=#=#=#=#=#=#=#=#=#=#
896:            SUBROUTINE GJR (A,N,IW,ISIZ,ISIZ1)
897:C***THIS SUBROUTINE DOES A GAUSS-JORDAN REDUCTION
898:            PARAMETER MXINV=50
899:            DOUBLE PRECISION A
900:            DIMENSION A(ISIZ,ISIZ1),JC(MXINV)
901:C***IW=: 1- FIND INVERSE, 4- SOLVE AX=B, 5- BOTH
902:C***JC IS THE PERMUTATION VECTOR
903:C***KI IS THE OPTION KEY FOR MATRIX INVERSION
904:C***L IS THE COLUMN CONTROL FOR AX=B
905:C***M IS THE COLUMN CONTOL FOR MATRIX INVERSION
906:            K=0
907:            M=1
908:            L=N+(IW/4)
909:            KI=2-MOD(IW,2)
910:            GO TO (10,30), KI
911:C***INITIALIZE JC FOR INVERSION
912:10          DO 20 I=1,N
913:20          JC(I)=I
```

```
914:C***SEARCH FOR PIVOT ROW
915:30         DO 120 I=1,N
916:           GO TO (50,40), KI
917:40         M=I
918:50         IF (I.EQ.N) GO TO 95
919:           X=-1.
920:           DO 60 J=I,N
921:           IF (X.GT.ABS(A(J,I))) GO TO 60
922:           X=ABS(A(J,I))
923:           K=J
924:60         CONTINUE
925:           IF (K.EQ.I) GO TO 95
926:           GO TO (70,80), KI
927:70         MU=JC(I)
928:           JC(I)=JC(K)
929:           JC(K)=MU
930:C***INTERCHANGE ROW I AND ROW K
931:80         DO 90 J=M,L
932:           X=A(I,J)
933:           A(I,J)=A(K,J)
934:90         A(K,J)=X
935:C***TEST FOR SINGULARITY
936:95         IF (ABS(A(I,I)).EQ.0.) THEN
937:C***MATRIX IS SINGULAR
938:           PRINT*,'***ERROR*** IN GJR. MATRIX IS SINGULAR'
939:           STOP
940:           ENDIF
941:           X=A(I,I)
942:           A(I,I)=1.
943:C***REDUCTION OF THE I-TH ROW
944:           DO 100 J=M,L
945:100        A(I,J)=A(I,J)/X
946:C***REDUCTION OF ALL REMAINING ROWS
947:           DO 120 K=1,N
948:           IF (K.EQ.I) GO TO 120
949:           X=A(K,I)
950:           A(K,I)=0.
951:           DO 110 J=M,L
952:           A(K,J)=A(K,J)-X*A(I,J)
953:110        CONTINUE
954:120        CONTINUE
955:C***AX=B AND DET.(A) ARE NOW COMPUTED
956:           GO TO (130,180), KI
956:C***PERMUTATION OF THE COLUMNS FOR MATRIX INVERSION
958:130        DO 170 J=1,N
959:           IF (JC(J).EQ.J) GO TO 170
960:           JJ=J+1
961:           DO 140 I=JJ,N
962:           IF (JC(I).EQ.J) GO TO 150
963:140        CONTINUE
964:150        JC(I)=JC(J)
965:           DO 160 K=1,N
966:           X=A(K,I)
967:           A(K,I)=A(K,J)
968:160        A(K,J)=X
969:170        CONTINUE
970:180        JC(1)=N
971:           RETURN
972:           END
973:C#=#=#=#=#=#=#=#=#=#=#=#=#=#=#=#=#=#=#=#=#=#=#=#=#=#=#=#=#=#=#=#=#
974:           SUBROUTINE GREEN(THTO,THETA,ROOT,N,M,LAG,TITLE,DELTA)
975:C***THIS SUBROUTINE FINDS THE SMALL G'S OF THE GREEN'S FUNCTION
976:           CHARACTER*8 TITLE,TTL
977:           COMPLEX ROOT,CNUM,DEN,G
978:           DIMENSION ROOT(1),THETA(1)
979:           TTL=TITLE
980:           DO 50 I=1,N
981:           IF (AIMAG(ROOT(I)).GE.0.0) THEN
982:           CNUM=THTO*ROOT(I)**(N-1)
983:           DO 10 J=1,M
984:10         CNUM=CNUM-THETA(J)*ROOT(I)**(N-1-J)
985:           DEN=CMPLX(1.0,0.0)
986:           DO 20 J=1,N
987:20         IF (I.NE.J) DEN=DEN*(ROOT(I)-ROOT(J))
988:           G=CNUM/DEN
989:C***PRINT TERM CORRESPONDING TO THIS G
990:           IF (AIMAG(ROOT(I)).EQ.0) THEN
991:           PRINT 30,TTL,REAL(G),REAL(ROOT(I))
992:30         FORMAT(35X,A,3X,E11.5,'*(',E11.5,')**j'/)
993:           ELSE
994:           CAR=CABS(ROOT(I))
995:           PRINT 40,TTL,2.*CABS(G),CAR,ACOS(REAL(ROOT(I))/CAR)
996:     >     /(DELTA*6.2831853),ATAN2(AIMAG(G),REAL(G))
```

```
997:40                          FORMAT (35X,A,3X,E11.5,'*(',E11.5,')**]*',
998:    >                       'CoS[2*PI*',E11.5,'*DELTA*j+(',E11.5,'']]*'/)
999:                     ENDIF         +'
1000:                  TTL=''
1001:              ENDIF
1002:50        CONTINUE
1003:         PRINT 60,LAG
1004:60       FORMAT(45X,'FOR J GREATER THAN OR EQUAL TO',I3///)
1005:         RETURN
1006:         END
1007:C#=#=#=#=#=#=#=#=#=#=#=#=#=#=#=#=#=#=#=#=#=#=#=#=#=#=#=#=#=#=#=#
1008:         SUBROUTINE INVAL(XTX,PAR,SCA,NPAR)
1009:C***THIS SUBROUTINE FINDS THE INITIAL VALUES OF THE STOCHASTIC PARAMETERS
1010:C***USING THE INVERSE FUNCTION METHOD
1011:         PARAMETER MXSER=3,MXNOB=369,MXPAR=45,MXINV=50,MXINV1=MXINV+1
1012:         DOUBLE PRECISION XTX(MXINV,MXINV1)
1013:         COMMON /INDEX/ INDEX,MAXLAG,NOB,NSER,NPHIO,NPHI,NSEAS
1014:         COMMON /LAG/ LAG(MXSER,MXSER),NAR(MXSER,MXSER),NMA(MXSER)
1015:         COMMON /FLAG/ IFMEAN,IFAT,IFSEAS
1016:         DIMENSION PAR(MXPAR),N(MXSER),THETA(MXPAR),SCA(MXSER)
1017:C***FIND MAXIMUM AUTOREGRESSIVE ORDER AND INVERSE FUNCTION ORDER
1018:         NM=1
1019:         DO 10 I=1,NSER
1020:10           IF (NAR(I,INDEX).GT.NM) NM=NAR(I,INDEX)
1021:C***SET INDEXES
1022:         M=NMA(INDEX)
1023:         NP=0
1024:         IF (NAR(INDEX,INDEX).EQ.0.OR.NM.EQ.1.AND.M.EQ.1) NP=1
1025:         NM=MAX(NM,M)+M+NP
1026:         ISIZ=NSER*NM+NPHIO
1027:         NXTY=ISIZ+1
1028:         OFFSET=NSEAS
1029:         IF (IFMEAN.GT.0.AND.IFSEAS.EQ.0) OFFSET=OFFSET+NSER
1030:         NN=OFFSET+NPHIO+NPHI
1031:         MXLAG=MAXLAG
1032:         MAXLAG=1
1033:C***SET AR ORDERS AND FIND NEW MAXIMUM LAG
1034:         DO 20 I=1,NSER
1035:             N(I)=NM
1036:             ML=NM+LAG(I,INDEX)-1
1037:20           IF(ML.GT.MAXLAG) MAXLAG=ML
1038:C***FIT EAR(NM) MODEL TO GET INVERSE FUNCTION COEFFICIENTS
1039:         CALL EAR(XTX,N,LAG(1,INDEX),PAR,ISIZ)
1040:         CALL GJR(XTX,ISIZ,4,MXINV,MXINV1)
1041:C***UNSCALE INVERSE FUNCTION COEFFICIENTS
1042:         K=1
1043:         DO 22 I=1,NSER
1044:             IF (I.NE.INDEX) THEN
1045:                 SCALE=SCA(INDEX)/SCA(I)
1046:                 KK=K
1047:                 DO 21 J=1,N(I)
1048:                     XTX(KK,NXTY)=XTX(KK,NXTY)*SCALE
1049:21                   KK=KK+1
1050:             ENDIF
1051:22           K=K+N(I)
1052:C***UNSCALE ZERO LAG TERMS
1053:         DO 23 I=INDEX+1,NSER
1054:             IF (LAG(I,INDEX).EQ.0) THEN
1055:                 XTX(K,NXTY)=XTX(K,NXTY)*SCA(INDEX)/SCA(I)
1056:                 K=K+1
1057:             ENDIF
1058:23       CONTINUE
1059:         PRINT 25,(XTX(I,NXTY),I=1,ISIZ)
1060:25       FORMAT (//' THE INVERSE FUNCTION COEFFICIENTS ARE:'/9(/8E14.6))
1061:C***FIND INITIAL THETA'S
1062:         MP1=M+1
1063:         DO 40 J=1,M
1064:             DO 40 K=1,MP1
1065:                 SUM=0.0
1066:                 DO 30 I=0,(NSER-1)*NM,NM
1067:30                   SUM=SUM+XTX(I+J+K,NXTY)
1068:40               XTX(J,K)=SUM
1069:         CALL GJR(XTX,M,4,MXINV,MXINV1)
1070:C***CHECK FOR INVERTIBILITY AND FIX IF UNSTABLE
1071:         DO 50 I=1,M
1072:             II=MP1-I
1073:50           THETA(I)=XTX(II,MP1)
1074:         CALL STBLIZ(THETA,M)
1075:C***LOAD THETAS INTO PAR
1076:         DO 60 I=1,M
1077:60           PAR(NN+I)=THETA(I)
1078:C***FIND INITIAL PHI'S
1079:         JPAR=OFFSET+NPHIO+1
```

```
1080:         L=OFFSET+1
1081:         DO 80 I=1,NSER
1082:            IPAR=(I-1)*NM+1
1083:C***FIND PHIO'S FOR SERIES I
1084:            PHIO=0.0
1085:            IF (I.EQ.INDEX) PHIO=1.0
1086:            IF (LAG(I,INDEX).EQ.0) THEN
1087:               PAR(L)=XTX(NSER*NM+L-OFFSET,NXTY)
1088:               PHIO=PAR(L)
1089:               L=L+1
1090:            ENDIF
1091:C***FIND REST OF PHI'S FOR SERIES I
1092:            DO 80 J=1,NAR(I,INDEX)
1093:               SUM=0.0
1094:               IF(J.LE.M) SUM=THETA(J)*PHIO
1095:               DO 70 K=1,MIN(J-1,M)
=SUM-THETA(K)*XTX(IPAR-K,NXTY)
1096:               PAR(JPAR)=XTX(IPAR,NXTY)+SUM
1097:               JPAR=JPAR+1
1098:               IPAR=IPAR+1
1099:80
1100:C***PRINT INITIAL PARAMETER VALUES
1101:         WRITE (6,90) (PAR(I),I=1,NPAR)
1102:90       FORMAT (//' INITIAL PARAMETERS =',9E12.5)
1103:         MAXLAG=MXLAG
1104:         RETURN
1105:         END
1106:C#=#=#=#=#=#=#=#=#=#=#=#=#=#=#=#=#=#=#=#=#=#=#=#=#=#=#=#=#=#=#=#=#
1107:         SUBROUTINE LS (NPAR,PAR,F,FORMT,IFDIF,SCA,IFPRT)
1108:C***THIS SUBROUTINE DOES A MARQUART-LEVENBERG TYPE NON-LINEAR SEARCH
1109:         PARAMETER MXSER=3,MXNOB=369,MXNOB1=MXNOB+1,MXPAR=45,MXPAR1=MXPAR+1
1110:         LOGICAL LG,FLAG
1111:         CHARACTER FORMT*50
1112:         DOUBLE PRECISION PIVOT,CMULT,DENOM,SSRED,TRED
1113:         DOUBLE PRECISION A(MXPAR1,MXPAR1),PARB(MXPAR),XX,FLR
1114:         COMMON /INDEX/ INDEX,MAXLAG,NOB,NSER,NPHIO,NPHI,NSEAS
1115:         COMMON /LAG/ LAG(MXSER,MXSER),NAR(MXSER,MXSER),NMA(MXSER)
1116:         COMMON /FLAG/ IFMEAN,IFAT,IFSEAS
1117:         COMMON /SEASON/ NPOLY,NEXP,NSIN,ISEAS(30),SEAS(30),DELTA
1118:         COMMON /BLOK1/ STEP,CONTOL,ITMAX,IFREF
1119:         DIMENSION PAR(MXPAR),CHMAX(MXPAR),Z(MXNOB,MXPAR1),F(MXNOB1),
1120:         DIMENSION SS(3),FL(3),FD(4),SD(4),LSTP(MXPAR1),SPDA(MXPAR1),
1121:        >         SCA(MXSER)
1122:         FINF=1.E19
1123:         IF (IFPRT.GT.0) WRITE (6,10) NPAR
1124:10       FORMAT (//'  VERSION 4R1 OF LS, 09-03-82, WITH',I4,
1125:        >' PARAMETERS'/)
1126:         DEL=-.001
1127:         BNDLW=-FINF
1128:         BNDUP=FINF
1129:         DO 20 I=1,NPAR
1130:20          CHMAX(I)=.2*ABS(PAR(I))
1131:         DO 60 I=1,NPAR
1132:            IF (PAR(I).EQ.0.0) STOP 'A PARAMETER IS EQUAL TO ZERO'
1133:            IF (PAR(I).LT.BNDLW.OR.PAR(I).GT.BNDUP)
1134:        1      STOP 'A PARAMETER IS OUTSIDE ITS BOUNDS'
1135:60       CONTINUE
1136:C***MAIN ITERATION LOOP
1137:         ITNO=1
1138:         NPAR1=NPAR+1
1139:         NFUNC=0
1140:70       IF (IFPRT.GT.0) WRITE (6,80) ITNO,NFUNC
1141:80       FORMAT ('0START ITERATION NO.',I3,' NO. OF CALLS TO MODEL',I4)
1142:         FLAG=.TRUE.
1143:C***FIND AT'S OF ORIGINAL PARAMETERS
1144:85       CALL MODEL (PAR,F,1)
1145:         NFUNC=NFUNC+1
1146:C***STORE AT'S IN LAST COLUMN OF Z
1147:         DO 90 IOB=1,NOB
1148:90          Z(IOB,NPAR1)=-F(IOB)
1149:         ITNO=ITNO+1
1150:C***FIND DERIVATIVES WITH RESPECT TO EACH PARAMETER
1151:         IF (IFDIF.NE.0) THEN
1152:C***ANALYTICAL
1153:            CALL DIF (PAR,Z,F,IS,IFMEAN)
1154:         ELSE
1155:C***NUMERICAL
1156:            DO 130 IPAR=1,NPAR
1157:               IF (ABS(PAR(IPAR)).LE.1.E-25) THEN
1158:                  WRITE (6,100) IPAR
1159:100      1         FORMAT ('0THE VALUE OF PAR(',I3,') IS TOO SMALL FOR',
1160:        1         'DETERMINING THE DERIVATIVE')
1161:                  STOP
1162:               ENDIF
```

```
1163:                PARD=PAR(IPAR)
1164:                DPAR=ABS(PAR(IPAR)*DEL)
1165:                S1=BNDUP-PARD-DPAR
1166:                S2=PARD-DPAR-BNDLW
1167:                IF (S1.LT.0..AND.S2.GT.S1) THEN
1168:                   PAR(IPAR)=AMAX1(PARD-DPAR,BNDLW)
1169:                   DENOM=PARD-PAR(I)
1170:                   CALL MODEL (PAR,F,1)
1171:                   NFUNC=NFUNC+1
1172:                   DO 110 IOB=1,NOB
1173:110                   Z(IOB,IPAR)=-(Z(IOB,NPAR1)+F(IOB))/DENOM
1174:                ELSE
1175:                   PAR(IPAR)=AMIN1(PARD+DPAR,BNDUP)
1176:                   DENOM=PAR(IPAR)-PARD
1177:                   CALL MODEL (PAR,F,1)
1178:                   NFUNC=NFUNC+1
1179:                   DO 120 IOB=1,NOB
1180:120                   Z(IOB,IPAR)=(F(IOB)+Z(IOB,NPAR1))/DENOM
1181:                ENDIF
1182:                PAR(IPAR)=PARD
1183:130             CONTINUE
1184:                IS=1
1185:             ENDIF
1186:C***FORM XTX AND XTY MATRIX
1187:             DO 150 IPAR=1,NPAR1
1188:             DO 150 JPAR=1,IPAR
1189:                XX=0.0
1190:                DO 140 IOB=IS,NOB
1191:140                XX=XX+1.0D0*Z(IOB,IPAR)*Z(IOB,JPAR)
1192:                A(IPAR,JPAR)=XX
1193:150             A(JPAR,IPAR)=XX
1194:             IF (ITNO.EQ.2.AND.IFPRT.GT.0) WRITE (6,160) A(NPAR1,NPAR1)
1195:160          FORMAT (//' INITIAL SUM OF SQUARES =',D12.4/)
1196:C***FIND GAUSS-NEWTON CORRECTIONS (XX'XX)**-1(XX'Y)
1197:             NES=0
1198:             NTRANS=0
1199:             SSB=A(NPAR1,NPAR1)
1200:             DO 170 I=1,NPAR
1201:             LSTP(I)=0
1202:             PARB(I)=PAR(I)
1203:170          SPDA(I)=STEP*A(I,I)
1204:180          SSRED=0
1205:             NPIV=0
1206:             DO 190 I=1,NPAR
1207:             IF (LSTP(I).NE.0.OR.A(I,I).LE.SPDA(I).OR.ABS(CHMAX(I))
1208:           1    .LT.1./FINF) GO TO 190
1209:             TRED=A(I,NPAR1)**2/A(I,I)
1210:             IF (TRED.GE.SSRED) THEN
1211:                SSRED=TRED
1212:                NPIV=I
1213:             ENDIF
1214:190          CONTINUE
1215:             IF (NPIV.EQ.0) GO TO 225
1216:             NTRANS=NTRANS+1
1217:             NES=NES+1
1218:             LSTP(NPIV)=NPIV
1219:C***DO GAUSS JORDAN MODIFICATION FOR THIS ROW AND COLUMN
1220:             PIVOT=A(NPIV,NPIV)
1221:             A(NPIV,NPIV)=1.D0
1222:             DO 200 J=1,NPAR1
1223:200          A(NPIV,J)=A(NPIV,J)/PIVOT
1224:             DO 220 I=1,NPAR1
1225:             IF (I.NE.NPIV) THEN
1226:                CMULT=A(I,NPIV)
1227:                DO 210 J=1,NPAR1
1228:210                IF(J.NE.NPIV)A(I,J)=A(I,J)-CMULT*A(NPIV,J)
1229:                A(I,NPIV)=-A(I,NPIV)/PIVOT
1230:             ENDIF
1231:220          CONTINUE
1232:             GO TO 180
1233:225          IF (NTRANS.LE.0) STOP'NO PARAMETER CHANGES POSSIBLE. INSPECT BOUND
1234:           1S & CHMAX ARRAY'
1235:             SSE1=A(NPAR1,NPAR1)
1236:             DO 230 I=1,NPAR
1237:             IF(LSTP(I).EQ.0)A(I,NPAR1)=0.D0
1238:230          CONTINUE
1239:C***FIND MAXIMUM ALLOWABLE LAMBDA
1240:             ILAM=0
1241:             FLAM=1
1242:             ILMAX=0
1243:             FLMAX=FINF
1244:             QMAX=FINF
1245:             DO 240 I=1,NPAR
```

```
1246:                   ABSA=DABS(A(I,NPAR1))
1247:                      IF (ABSA.GE.1./FINF) THEN
1248:                      QLAM=ABS(CHMAX(I))
1249:                      IF(CHMAX(I).LE.0.0)QLAM=QLAM*DABS(PARB(I))
1250:                         IF (FLAM*ABSA.GT.QLAM) THEN
1251:                         ILAM=I
1252:                         FLAM=QLAM/ABSA
1253:                         ENDIF
1254:                      IF(A(I,NPAR1).GT.0.0)QMAX=BNDUP-PARB(I)
1255:                      IF(A(I,NPAR1).LT.0.0)QMAX=PARB(I)-BNDLW
1256:                         IF (QMAX.LT.FLMAX*ABSA) THEN
1257:                         ILMAX=I
1258:                         FLMAX=QMAX/ABSA
1259:                         ENDIF
1260:                      ENDIF
1261:240                CONTINUE
1262:       IF (ILAM.NE.0.AND.IFPRT.GT.0) WRITE (6,250) ILAM,FLAM
1263:250    FORMAT (' PARAMETER',I4,' LIMITS THE CORRECTIONS TO ',E12.4,
1264:       >' TIMES THE GAUSS-NEWTON VALUES.')
1265:       IF (FLMAX.LT.1..AND.ILMAX.NE.ILAM.AND.IFPRT.GT.0)
1266:       >WRITE (6,250) ILMAX,FLMAX
1267:       IF (FLAM.LT.1./FINF) STOP'NO PARAMETER CHANGES POSSIBLE. INSPECT B
1268:       >OUNDS & CHMAX ARRAY'
1269:       SBEST=SSB
1270:       FBEST=0
1271:       FLR=2*FINF
1272:       SSP=SSE1
1273:       SS(1)=SSB
1274:       FL(1)=0
1275:       SS(2)=1.01*FINF
1276:       FL(2)=1.01*FLMAX
1277:       SS(3)=1.02*FINF
1278:       FL(3)=1.02*FLMAX
1279:       FLT=FLAM
1280:       KEY=0
1281:       LG=.TRUE.
1282:C***ADJUST LAMBDA UNTIL BEST SSQ FOUND FOR THIS ITERATION
1283:       DO 390 IGRID=1,ITMAX
1284:260       DO 270 I=1,NPAR
1285:          PAR(I)=PARB(I)+FLT*A(I,NPAR1)
1286:          IF(PAR(I).GT.BNDUP)PAR(I)=BNDUP
1287:270       IF(PAR(I).LT.BNDLW)PAR(I)=BNDLW
1288:C***FIND SSQ OF AT'S FOR NEW PARAMETERS
1289:       CALL MODEL (PAR,F,1)
1290:       NFUNC=NFUNC+1
1291:       SST=0.
1292:       DO 290 IOB=1,NOB
1293:          DF=ABS(F(IOB))
1294:C***CHECK FOR UNSTABLE MODEL
1295:          IF (DF.GT.1.E15) THEN
1296:          WRITE (6,280) IOB,F(IOB)
1297:280       FORMAT ('0AT(',I3,') =',E10.3,' IS TOO LARGE')
1298:             IF (FLAG) THEN
1299:             PRINT*,'UNSTABLE MODEL- FIX AND RESTART ITERATION'
1300:             FLAG=.FALSE.
1301:             N=NSEAS+NPHIO+NPHI+1
1302:             IF (IFMEAN.GT.0.AND.IFSEAS.EQ.0) N=N+NSER
1303:             CALL STBLIZ(PAR(N),NMA(INDEX))
1304:             GO TO 85
1305:             ELSE
1306:             STOP'UNSTABLE MODEL - SECOND TRY'
1307:             ENDIF
1308:          ENDIF
1309:290       SST=SST+DF**2
1310:       SSR=SST
1311:       LG=LG.AND.SST.GT.SSB
1312:       IF (KEY.EQ.1) GO TO 300
1313:C***CHECK FOR CONVERGENCE
1314:       IF ((ABS(FLT-1.).LE.CONTOL.OR.ABS(FLT-FLMAX).LE.CONTOL).AND.
1315:       >  ABS(SST-SSE1).LE.ABS(SSE1)*CONTOL.AND..NOT.LG) THEN
1316:       FLR=FLT
1317:300    IF (IFPRT.GT.0) WRITE (6,310) IGRID,FLR,SSR
1318:310    FORMAT(' SEARCH CONVERGED AFTER',I3,' CYCLES, WITH LAMBDA ='
1319:       >  ,D13.5,' AND SSQ =',E13.5/)
1320:       GO TO 410
1321:       ENDIF
1322:       INS=0
1323:       K=0
1324:       DO 320 I=1,3
1325:          IF(FL(I).GT.FLT.AND.INS.EQ.0)INS=I
1326:          IF(INS.GT.0)K=1
1327:          IK=I+K
1328:          FD(IK)=FL(I)
```

```
1329:320         SD(IK)=SS(I)
1330:          IF(INS.EQ.0)INS=4
1331:          FD(INS)=FLT
1332:          SD(INS)=SST
1333:          K=0
1334:          IF((SD(2).GT.SD(3).OR.INS.EQ.4).AND.IGRID.GT.2)K=1
1335:          IF(SD(1).LE.SD(2))K=0
1336:          DO 330 I=1,3
1337:          IK=I+K
1338:          FL(I)=FD(IK)
1339:330       SS(I)=SD(IK)
1340:C***STORE BEST VALUES OF SSQ AND LAMBDA
1341:          IF (SST.LT.SBEST) THEN
1342:          SBEST=SST
1343:          FBEST=FLT
1344:          ENDIF
1345:C***FIND NEW LAMBDA
1346:          IF (FL(3).LE.FLMAX) GO TO 340
1347:          IF (SS(1).LE.SS(2)) GO TO 350
1348:          FLT=0.1*FL(1)+0.9*FL(2)
1349:          GO TO 390
1350:340       DENOM=(FL(3)-FL(1))*(SS(2)-SS(1))+(FL(1)-FL(2))*(SS(3)-SS(1))
1351:          IF (DENOM.LE.-1./FINF.AND.FL(3).LT.FINF) GO TO 370
1352:          SSP=FINF
1353:          IF (SS(1).GT.SS(2)) GO TO 360
1354:350       FLT=0.9*FL(1)+0.1*FL(2)
1355:          GO TO 390
1356:360       FLT=FLMAX
1357:          IF(FL(3).GE.0.98*FLMAX)FLT=0.1*FL(2)+0.9*FL(3)
1358:          IF(FL(3).LT.0.49*FLMAX)FLT=2.*FL(3)
1359:          GO TO 390
1360:370       FOLD=FLR
1361:          FLR=((FL(3)**2-FL(1)**2)*(SS(2)-SS(1))+(FL(1)**2-FL(2)**2)*
1362:         1(SS(3)-SS(1)))/2./DENOM
1363:          IF(FLR.GE.FLMAX)FLR=FLMAX
1364:          IF(FLR.LE.FL(1))FLR=FL(1)
1365:          SSR=SS(1)+(SS(2)-SS(1))*(FLR-FL(1))*(FLR-FL(3))/(FL(2)-FL(1))/(FL(
1366:         12)-FL(3))+(SS(3)-SS(1))*(FLR-FL(1))*(FLR-FL(2))/(FL(3)-FL(1))/(FL(
1367:         23)-FL(2))
1368:          IF (ABS(SSR-SSP).GT.ABS(CONTOL*SSP).AND.DABS(FOLD-FLR).GT.
1369:         >DABS(CONTOL*FLR)) GO TO 380
1370:          IF (SSR.LT.0..OR.FLR.LE.FL(1).OR.FLR.GT.FL(3).OR.LG) GO TO 380
1371:          FLT=FLR
1372:          KEY=1
1373:          GO TO 260
1374:380       SSP=SSR
1375:          FLT=0.9*FL(1)+0.1*FL(2)
1376:          IF(FLR.GT.FLT)FLT=FLR
1377:          FT=0.1*FL(1)+0.9*FL(2)
1378:          IF(FLR.GT.FT)FLT=FT
1379:          FT=0.9*FL(2)+0.1*FL(3)
1380:          IF(FLR.LT.FT)FLT=FT
1381:          IF(FLR.GE.FT)FLT=FLR
1382:          FT=0.1*FL(2)+0.9*FL(3)
1383:          IF(FLR.GT.FT)FLT=FT
1384:          IF (FLR.GT.FL(3)) GO TO 360
1385:390       CONTINUE
1386:C***
1387:          IGRID=ITMAX+1
1388:          IF (IFPRT.GT.0) WRITE (6,400) ITMAX,FBEST,SBEST
1389:400       FORMAT (' SEARCH TOOK THE FULL',I4,' CYCLES. BEST TRIAL POINT,'
1390:         1'LAMBDA =',E13.5,' SSQ =',E13.5)
1391:          FLR=FBEST
1392:          SSR=SBEST
1393:C***MODIFY PARAMETERS FOR BEST CASE
1394:410       DO 420 I=1,NPAR
1395:          PAR(I)=PARB(I)+A(I,NPAR1)*FLR
1396:          IF(PAR(I).GT.BNDUP)PAR(I)=BNDUP
1397:          IF(PAR(I).LT.BNDLW)PAR(I)=BNDLW
1398:420       CONTINUE
1399:          IF (SSB.LT.SSR.AND.IFPRT.GT.0) WRITE (6,430) SSR,SSB
1400:430       FORMAT (' OCURRENT SUM OF SQUARES',E15.8,' EXCEEDS RESULT',E15.8,
1401:         1' OF PREVIOUS ITERATION')
1402:          IF (ITNO.LE.ITMAX.AND.ABS((SSR-SSB)/CONTOL).GT.SSB.AND.IGRID.GT.1)
1403:         1 GO TO 70
1404:          IF (ABS((SSR-SSB)/CONTOL).GT.SSB.AND.IGRID.GT.1) WRITE (6,440)
1405:440       FORMAT (//' *******CONVERGENCE CRITERION IS NOT SATISFIED,',
1406:         1' MAXIMUM NUMBER OF ITERATIONS WAS REACHED ********')
1407:          IF (IFREF.GT.0) RETURN
1408:C***FIND CONFIDENCE BOUNDS
1409:          NDF=NOB-NES
1410:          SEXT=0
1411:          IF(NDF.GT.0)SEXT=SQRT(SSR/FLOAT(NDF))
```

```
1412:C***REFORM XTX MATRIX AND INVERT IT
1413:          DO 460 I=1,NPAR
1414:          DO 460 J=1,I
1415:             XX=0.0
1416:             DO 450 K=IS,NOB
1417:450             XX=XX+Z(K,I)*Z(K,J)
1418:             A(I,J)=XX
1419:460          A(J,I)=XX
1420:          CALL GJR (A,NPAR,1,MXPAR1,MXPAR1)
1421:          DO 470 I=1,NPAR
1422:             IF (A(I,I).LT.0.0) THEN
1423:                A(I,I)=-A(I,I)
1424:                IF (IFPRT.GT.0) PRINT*,'***WARNING***',I,
1425:       >          ' TH DIAGONAL OF COVARIANCE MATRIX IS NEGATIVE'
1426:             ENDIF
1427:             A(I,NPAR1)=DSQRT(A(I,I))
1428:             PARB(I)=A(I,NPAR1)*SEXT*1.96
1429:470          CONTINUE
1430:C***FIND FINAL AT'S AND PRINT FINAL SSQ
1431:          CALL MODEL (PAR,F,1)
1432:          NFUNC=NFUNC+1
1433:          XX=0.0
1434:          DO 630 I=1,NOB
1435:630          XX=XX+F(I)*F(I)
1436:          F(NOB+1)=SNGL(XX)
1437:C***UNSCALE RESIDUAL SSQ FOR EARMA MODELS
1438:          IF (NSER.GT.1) XX=XX*SCA(INDEX)**2
1439:          WRITE (6,640) XX
1440:640       FORMAT(//80('*')///1X,'FINAL RESIDUAL SUM OF SQUARES = ',D12.6/)
1441:C***UNSCALE DETERMINISTIC PARAMETERS
1442:          IF (IFSEAS.EQ.1.OR.IFSEAS.EQ.3) THEN
1443:             N=1
1444:             DO 471 I=1,NEXP
1445:                N=N+2
1446:                PAR(N)=PAR(N)/DELTA
1447:471             PARB(N)=PARB(N)/DELTA
1448:             DO 472 I=1,NSIN
1449:                N=N+2
1450:                PAR(N)=PAR(N)/DELTA
1451:                PARB(N)=PARB(N)/DELTA
1452:                N=N+2
1453:                PAR(N)=PAR(N)/DELTA/6.2831853
1454:472             PARB(N)=PARB(N)/DELTA/6.2831853
1455:          ENDIF
1456:C***UNSCALE EARMA PARAMETERS AND CONFIDENCE BOUNDS
1457:          IF (NSER.GT.1) THEN
1458:             N=NSEAS+1
1459:             IF (IFMEAN.GE.1) THEN
1460:                DO 476 I=1,NSER
1461:                   PAR(N)=PAR(N)*SCA(I)
1462:                   PARB(N)=PARB(N)*SCA(I)
1463:476                N=N+1
1464:             ENDIF
1465:C***UNSCALE ZERO LAG TERMS
1466:             DO 473 I=INDEX+1,NSER
1467:                IF (LAG(I,INDEX).EQ.0) THEN
1468:                   PAR(N)=PAR(N)*SCA(INDEX)/SCA(I)
1469:                   PARB(N)=PARB(N)*SCA(INDEX)/SCA(I)
1470:                   N=N+1
1471:                ENDIF
1472:473             CONTINUE
1473:C***SCALE REST OF PHI TERMS
1474:             DO 475 I=1,NSER
1475:                IF (I.NE.INDEX) THEN
1476:                   NN=N
1477:                   SCALE=SCA(INDEX)/SCA(I)
1478:                   DO 474 J=1,NAR(I,INDEX)
1479:                      PAR(NN)=PAR(NN)*SCALE
1480:                      PARB(NN)=PARB(NN)*SCALE
1481:474                   NN=NN+1
1482:                ENDIF
1483:475             N=N+NAR(I,INDEX)
1484:          ENDIF
1485:C***PRINT PARAMETERS AND CONFIDENCE BOUNDS
1486:          PRINT FORMT, INDEX, (NAR(J,INDEX),J=1,NSER),NMA(INDEX)
1487:          WRITE (6,480)
1488:480       FORMAT (/////'*'///' BEST PARAMETER VALUES AND 95% CONFIDENCE LIMITS ',
1489:       >'ESTIMATED'/' BY LINEARIZATION FOR THE INDIVIDUAL PARAMETERS ARE',
1490:       >' AS FOLLOWS:')
1491:          NP=1
1492:C***PRINT SEASONALITY PARAMETERS
1493:          IF (IFSEAS.EQ.1.OR.IFSEAS.EQ.3) THEN
1494:             J1=(NSEAS+3)/4
```

```
1495:              DO 490 J2=1,J1
1496:                 I1=(J2-1)*4+1
1497:                 I2=MINO(NSEAS,J2*4)
1498:                 WRITE (6,485) (ISEAS(J),J=I1,I2)
1499:485              FORMAT (/16X,'I  = ',3X,I3,3(25X,I3))
1500:490              WRITE(6,495) (PAR(J),PARB(J),J=I1,I2)
1501:495              FORMAT (11X,'BETA(I) = ',4(E11.5,' +/-',E10.5,3X))
1502:              NP=NSEAS+1
1503:           ENDIF
1504:C***PRINT MEANS
1505:           IF (IFMEAN.GT.0.AND.IFSEAS.EQ.0) THEN
1506:              WRITE (6,500) (I,PAR(I),PARB(I),I=NP,NP+NSER-1)
1507:500           FORMAT (/4X,'SERIES',I3,' MEAN = ',E11.5,' +/-',E10.5)
1508:              NP=NP+NSER
1509:           ENDIF
1510:C***PRINT PHI'S
1511:           IF (IFSEAS.NE.1) THEN
1512:              L=NP
1513:              NP=NP+NPHIO-1
1514:              DO 550 I=1,NSER
1515:                 IF (LAG(I,INDEX).EQ.0) THEN
1516:                    WRITE (6,510) INDEX,I,PAR(L),PARB(L)
1517:510                 FORMAT (//,3X,'PHI(',2I3,'  0) = ',E11.5,' +/-',E10.5)
1518:                    LAGT=1
1519:                    L=L+1
1520:                 ELSE
1521:                    LAGT=LAG(I,INDEX)
1522:                 ENDIF
1523:                 J1=(NAR(I,INDEX)+3)/4
1524:                 DO 530 J2=1,J1
1525:                    I1=(J2-1)*4+1
1526:                    I2=MINO(NAR(I,INDEX),J2*4)
1527:                    WRITE (6,520) (LAGT+J-1,J=I1,I2)
1528:520                 FORMAT (//11X,'AT LAG  =',4X,I3,3(25X,I3))
1529:530                 WRITE(6,540) INDEX,I,(PAR(J+NP),PARB(J+NP),J=I1,I2)
1530:540              FORMAT (//'   PHI(',2I3,' LAG) = ',4(E11.5,' +/-',E10.5,3X))
1531:550              NP=NP+NAR(I,INDEX)
1532:C***PRINT THETAS
1533:              J1=(NMA(INDEX)+3)/4
1534:              DO 560 J2=1,J1
1535:                 I1=(J2-1)*4+1
1536:                 I2=MINO(NMA(INDEX),J2*4)
1537:                 WRITE (6,520) (J,J=I1,I2)
1538:560              WRITE (6,570) INDEX,INDEX,(PAR(J+NP),PARB(J+NP),J=I1,I2)
1539:570           FORMAT (//' THETA(',2I3,' LAG) = ',4(E11.5,' +/-',E10.5,3X))
1540:              NP=NP+NMA(INDEX)
1541:              IF (IFAT.LT.0)
1542:     >         PRINT 575,(PAR(I),PARB(I),I=NP+1,MAXLAG+NP)
1543:575           FORMAT ('         INITIAL AT''S = ',4(E11.5,' +/-',E10.5,3X))
1544:           ENDIF
1545:        IF (NSER.GT.1) SEXT=SEXT*SCA(INDEX)
1546:        WRITE (6,580) SEXT,NOB,NDF
1547:580     FORMAT (///' STANDARD ERROR OF RESIDUALS =',E13.5,' ESTIMATED',
1548:     >  ' WITH',I5,' RESIDUALS AND',I5,' DEGREES OF FREEDOM.'//80('*')//)
1549:C***FIND AND PRINT CORRELATION MATRIX
1550:        DO 590 I=1,NPAR
1551:           DO 590 J=1,I
1552:590           A(I,J)=A(I,J)/A(I,NPAR1)/A(J,NPAR1)
1553:        IF (IFPRT.GT.0) WRITE (6,600)
1554:600     FORMAT ('0NORMALIZED CORRELATION MATRIX OF PARAMETER ESTIMATES'/)
1555:        J1=(NPAR+9)/10
1556:        DO 610 J2=1,J1
1557:           I1=(J2-1)*10+1
1558:           I2=MINO(NPAR,J2*10)
1559:           DO 610 I=I1,NPAR
1560:              II=MINO(I,I2)
1561:610           IF (IFPRT.GT.0) WRITE (6,620) (A(I,J),J=I1,II)
1562:620     FORMAT (6X,10(F10.7,2X))
1563:        RETURN
1564:        END
1565:C#=#=#=#=#=#=#=#=#=#=#=#=#=#=#=#=#=#=#=#=#=#=#=#=#=#=#=#=#=#=#=#=#=#=#
1566:        SUBROUTINE MODEL(PAR,AT,IFIRST)
1567:C***THIS SUBROUTINE FINDS THE RESIDUALS (AT'S) OF THE MODEL
1568:        PARAMETER MXSER=3,MXNOB=369,MXNOB1=MXNOB+1,MXPAR=45
1569:        DOUBLE PRECISION SUM
1570:        COMMON /INDEX/ INDEX,MAXLAG,NOB,NSER,NPHIO,NPHI,NSEAS
1571:        COMMON /LAG/ LAG(MXSER,MXSER),NAR(MXSER,MXSER),NMA(MXSER)
1572:        COMMON /FLAG/ IFMEAN,IFAT,IFSEAS
1573:        COMMON /SEASON/ NPOLY,NEXP,NSIN,ISEAS(30),SEAS(30),DELTA
1574:        COMMON /DATA/ Y(MXNOB,MXSER),X(MXNOB,MXSER)
1575:        DIMENSION PAR(MXPAR),AT(MXNOB)
1576:        IPAR=1
1577:C***SEASONALITY PARAMETERS
```

```
1578:          IF (IFSEAS.EQ.1.OR.IFSEAS.EQ.3) THEN
1579:C***LOAD SEASONALITY ARRAY
1580:          N=1
1581:          DO 10 I=1,NSEAS
1582:          SEAS(ISEAS(I))=PAR(IPAR)
1583:          IPAR=IPAR+1
1584:10        CONTINUE
1585:C***SUBTRACT SEASONALITY FROM DATA
1586:          CALL DETPAR(NOB)
1587:          IF (IFIRST.EQ.0) RETURN
1588:C***SUBTRACT MEAN IF ESTIMATED
1589:          ELSEIF (IFMEAN.GE.1.AND.IFSEAS.EQ.0) THEN
1590:          DO 40 I=1,NSER
1591:          DO 30 J=1,NOB
1592:30        X(J,I)=Y(J,I)-PAR(IPAR)
1593:40        IPAR=IPAR+1
1594:          ENDIF
1595:          IF (IFSEAS.EQ.1) THEN
1596:          DO 20 I=1,NOB
1597:20        AT(I)=X(I,1)
1598:          RETURN
1599:          ENDIF
1600:C***FIND FIRST MAXLAG AT'S
1601:          JPAR=IPAR
1602:          IF (IFAT.EQ.0) THEN
1603:          DO 50 I=1,MAXLAG
1604:50        AT(I)=0.0
1605:          ELSEIF (IFAT.GT.0) THEN
1606:          DO 100 IT=1,MAXLAG
1607:          IPAR=JPAR
1608:          SUM=0.0
1609:C***FIND PHI(0) TERMS
1610:          DO 60 I=INDEX+1,NSER
1611:          IF (LAG(I,INDEX).EQ.0) THEN
1612:          SUM=SUM+PAR(IPAR)*X(IT,I)
1613:          IPAR=IPAR+1
1614:          ENDIF
1615:60        CONTINUE
1616:C***FIND REST OF PHI TERMS
1617:          KPAR=IPAR
1618:          DO 80 I=1,NSER
1619:          LAGT=LAG(I,INDEX)
1620:          IF(LAGT.EQ.0) LAGT=1
1621:          IPAR=KPAR
1622:          DO 70 J=LAGT,MIN(IT,LAGT+NAR(I,INDEX))-1
1623:          SUM=SUM-PAR(IPAR)*X(IT-J,I)
1624:70        IPAR=IPAR+1
1625:80        KPAR=KPAR+NAR(I,INDEX)
1626:C***DO THETA TERMS
1627:          DO 90 I=1,MIN(IT-1,NMA(INDEX))
1628:          SUM=SUM+PAR(KPAR)*AT(IT-I)
1629:90        KPAR=KPAR+1
1630:100       AT(IT)=X(IT,INDEX)+SUM
1631:          ELSE
1632:          NP=IPAR+NPHI0+NPHI+NMA(INDEX)-1
1633:          DO 1 I=1,MAXLAG
1634:1         AT(I)=PAR(NP+I)
1635:          ENDIF
1636:C***FIND REST OF AT'S
1637:          DO 140 IT=MAXLAG+1,NOB
1638:          IPAR=JPAR
1639:          SUM=0.0
1640:          DO 110 I=INDEX+1,NSER
1641:          IF (LAG(I,INDEX).EQ.0) THEN
1642:          SUM=SUM+PAR(IPAR)*X(IT,I)
1643:          IPAR=IPAR+1
1644:          ENDIF
1645:110       CONTINUE
1646:          DO 120 I=1,NSER
1647:          LAGT=LAG(I,INDEX)
1648:          IF(LAGT.EQ.0) LAGT=1
1649:          DO 120 J=LAGT,LAGT+NAR(I,INDEX)-1
1650:          SUM=SUM-PAR(IPAR)*X(IT-J,I)
1651:120       IPAR=IPAR+1
1652:          DO 130 I=1,NMA(INDEX)
1653:          SUM=SUM+PAR(IPAR)*AT(IT-I)
1654:130       IPAR=IPAR+1
1655:140       AT(IT)=X(IT,INDEX)+SUM
1656:          RETURN
1657:          END
1658:C#=#=#=#=#=#=#=#=#=#=#=#=#=#=#=#=#=#=#=#=#=#=#=#=#=#=#=#=#=#=#=#=#
1659:          SUBROUTINE STBLIZ(THETA,N)
1660:C***THIS SUBROUTINE CHECKS FOR STABILITY OF THETA AND FIXES IF NECESSARY
```

```
1661:        PARAMETER MXPAR=45,MXPAR1=MXPAR+1
1662:        COMPLEX ROOT(MXPAR),CTHETA(MXPAR),UNSRT,RPRIME,F,FPRIME
1663:        DIMENSION THETA(MXPAR),NSTABL(MXPAR)
1664:C***COMPUTE DISCRETE ROOTS
1665:        CALL EIGEN(THETA,N,ROOT(1))
1666:C***CHECK FOR UNSTABLE MODEL
1667:        ISTABL=0
1668:        DO 20 I=1,N
1669:          IF(CABS(ROOT(I)).GT.1.0) THEN
1670:            ISTABL=ISTABL+1
1671:            NSTABL(ISTABL)=I
1672:          END IF
1673:20      CONTINUE
1674:C***STABILIZE ANY UNSTABLE ROOTS - IF NECESSARY
1675:        IF(ISTABL.GT.0) THEN
1676:          DO 30 I=1,N
1677:30          CTHETA(I)=CMPLX(THETA(I),0.)
1678:          DO 50 I=1,ISTABL
1679:            UNSRT=ROOT(NSTABL(I))
1680:            RPRIME=1./UNSRT
1681:C***PARTIAL REVERSE ROUTINE - UPDATE PARAMETERS FOR ONE ALTERED ROOT
1682:            F=CTHETA(1)-UNSRT
1683:            CTHETA(1)=F+RPRIME
1684:            DO 40 J=2,N
1685:              FPRIME=F*RPRIME
1686:              F=CTHETA(J)+F*UNSRT
1687:40            CTHETA(J)=F-FPRIME
1688:50        CONTINUE
1689:          DO 60 I=1,N
1690:60          THETA(I)=REAL(CTHETA(I))
1691:        END IF
1692:        RETURN
1693:        END
```

REFERENCES

Anderson, T. W. (1971), *The Statistical Analysis of Time Series*, Wiley, New York.

Bartlett, M. S. (1946), "On the Theoretical Specification of Sampling Properties of Autocorrelated Time Series," *J. Royal Stat. Soc.*, B8, p. 27.

Box, G. E. P., and Jenkins, G. M. (1970), *Time Series Analysis Forecasting and Control*, Rev. Ed. 1976, Holden-Day, San Francisco.

Brown, R. G. (1962), *Smoothing, Forecasting and Prediction of Discrete Time Series*, Prentice-Hall, Englewood Cliffs, New Jersey.

Chu, B. B. (1972), "Parametric Modeling of Linear Dynamic System and Applications," Ph.D. Thesis, University of Wisconsin-Madison.

Coddington, E. A., and N. Levinson (1955), *Theory of Ordinary Differential Equations*, McGraw Hill, New York.

Cox, D. R. (1961), "Prediction by Exponentially Weighted Moving Averages and Related Methods," *J. Royal Stat. Soc.*, B23, p. 414.

Den Hartog, J. P. (1956), *Mechanical Vibrations*, 4th Ed., McGraw Hill, New York.

Doob, J. L. (1953), *Stochastic Processes*, Wiley, New York.

Dornfeld, D. A. (1976), "Investigation of the Fundamentals of Mechanical Pulping," Ph.D. Thesis, University of Wisconsin-Madison.

Durbin, J. (1960), "The Fitting of Time Series Models," *Rev. Int. Inst. Stat.*, 28, p. 233.

Durbin, J. (1970), "Testing for Serial Correlation in Least Square Regression When Some of the Regressors Are Lagged Dependent Variables," *Econometrica*, 38, p. 410.

Goh, T. N. (1973), "Extensions and Engineering Applications of Time Series Techniques," Ph.D. Thesis, University of Wisconsin-Madison.

Jury, E. I. (1964), *Theory and Applications of the Z-Transform Method*, Wiley, New York.

Hannan, E. J. (1970), *Multiple Time Series*, Wiley, New York.

Kalman, R. E. (1960), "A New Approach to Linear Filtering and Prediction Problems," *Trans. ASME J. of Basic Eng.*, D82, p. 35.

Kendall, M. G. (1945), "On the Analysis of Oscillatory Time Series," *J. Royal Stat. Soc.*, Vol. 108, p. 93.

Khintchin, A. Y. (1934), "Korrelations Theorie der Stationaren Stochastischen Prozesse," *Math. Ann.*, 109, p. 604.

Kolmogorov, A. N. (1931), "Uber die Analytischen Methoden in der Wahrscheinlichkeitsrechnung," *Math. Ann.*, 104, p. 415.

Kolmogorov, A. N. (1933), "Grundbergriffe der Wahrscheinlichkeitsrechnung," *Eng. Mat.*, 2, No. 3.

Kolmogorov, A. N. (1939), "Sur L'Interpolation et L'Extrapolation des Suites Stationnaires," *C. R. Acad. Sci. Paris*, 208, p. 2043.

Kolmogorov, A. N. (1941a), "Stationary Sequences in Hilbert Space," *Bull. Moscow Univ.*, 2, No. 6.

Kolmogorov, A. N. (1941b), "Interpolation and Extrapolation von Stationaren Zufulligen Folgen," *Bull. Acad. Nauk U.R.S.S.*, Ser Math, 5, p. 3.

Kuo, B. C. (1970), *Discrete Data Control Systems*, Prentice Hall, Englewood Cliffs, New Jersey.

Marquardt, D. W. (1963), "An Algorithm for Least Squares Estimation of Nonlinear Parameters," *J. of Soc. Ind. Appl. Math.*, 11, p. 431.

Miller, K. S. (1968) *Linear Difference Equations*, W. A. Benjamin, New York.

Muth, J. F. (1960), "Optimal Properties of Exponentially Weighted Forecasts," *J. Am. Stat. Assoc.*, Vol. 55, p. 299.

Nelson, C. R. (1973), *Applied Time Series Analysis for Managerial Forecasting*, Holden-Day, San Francisco.

Pandit, S. M. (1973), "Data Dependent Systems: Modeling Analysis and Optimal Control Via Time Series," Ph.D. Thesis, University of Wisconsin-Madison.

Pandit, S. M., Goh, T. N., and Wu, S. M. (1976), "Modeling of System Dynamics and Disturbance from Papermaking Process Data," *Trans. ASME, J. of Dynamic Systems, Measurement and Control*, Vol. 98G, p. 197.

Pandit, S. M., Subramanian, T. L., and Wu, S. M. (1975), "Stability of Random Vibrations with Special Reference to Machine Tool Chatter," *Trans. ASME, J. of Engineering for Industry*, Vol. 97B, p. 211.

Pandit, S. M., and Wu, S. M. (1973), "Characterization of Abrasive Tools by Continuous Time Series," *Trans. ASME, J. of Engineering for Industry*, Vol. 95B, p. 821.

Pandit, S. M., and Wu, S. M. (1974), "Exponential Smoothing as a Special Case of a Linear Stochastic System," *Operations Research*, Vol. 22, p. 868.

Pandit, S. M., and Wu, S. M. (1975), "Unique Estimates of the Parameters of a Continuous Stationary Stochastic Process," *Biometrika*, Vol. 62, p. 497.

Pandit, S. M., and Wu, S. M. (1977), "Modeling and Analysis of Closed-Loop Systems from Operating Data," *Technometrics*, Vol. 19, p. 477.

Pugachev, V. S. (1957), *Theory of Random Functions and Its Applications to Automatic Control Problems*, (Russian) English Trans. of 3rd Ed. by O. M. Blum, Pargamon Press, New York, 1965.

Rao, C. R. (1965), *Linear Statistical Inference and Its Applications*, 2nd Ed., 1973, Wiley, New York.

Slutsky, E. (1927), "The Summation of Random Causes as the Source of Cyclic Processes," (Russian) *Problems of Economic Conditions*, 3, p. 1, English Trans. in *Econometrica*, 5, 1937, p. 105.

Stralkowski, C. M., Wu, S. M., and DeVor, R. E. (1969), "Characterization of Grinding Wheel Profiles by Autoregressive-Moving Average Models," *Int. J. Mach. Tool Des., Res.*, Vol. 9, p. 145.

Whittle, P. (1963), *Prediction and Regulation*, D. Van Nostrand, Princeton, New Jersey.

Wiener, N. (1949), *Extrapolation, Interpolation and Smoothing of Stationary Time Series*, Wiley, New York.

Wold, H. O. (1938), *A Study in the Analysis of Stationary Time Series*, Almquist and Wicksell, Uppsala, (2nd Ed. 1954).

Wu, S. M., and Dalal, J. G. (1971), "Stochastic Model for Machining Processes: Optimal Decision-Making and Control," *Trans. ASME, J. of Engineering for Industry*, Vol. 93B, p. 593.

Yaglom, A. M. (1952), *An Introduction to the Theory of Stationary Random Functions* (Russian), *Uspekhi Matematicheskikh Nauk*, Vol. 7, No. 5, English Trans. by R. A. Silverman (1962) Prentice-Hall, Dover edition 1973.

Yaglom, A. M. (1955), "The Correlation Theory of Processes Whose nth Difference Constitute a Stationary Process," *Matem. Sb.*, 37, p. 141; English Trans. in *Amer. Math Soc. Trans.*, Series 2, 8, p. 87.

Yule, G. U. (1927), "On a Method of Investigating Periodicities in Disturbed Series, with Special Reference to Wolfer's Sunspot Numbers," *Phil. Trans.*, A226, p. 267.

APPLICATION BIBLIOGRAPHY

The methodology outlined in this work has been extensively applied in real life to a variety of fields. However, each of these applications requires a considerable in-depth knowledge of the field for understanding and appreciating it. This precludes the inclusion of these applications in the text. However, readers already familiar with the field would not only benefit professionally by reading these applications, but also would better understand and appreciate the methodology itself, which is the aim of the book. Therefore, for the readers' convenience, a number of major application references explicitly employing the methods in the book are listed below, with some early related work included as background material.

IRON AND STEELMAKING

1 Pandit, S. M., Clum, J. A., and Wu, S. M. (1975), "Modeling, Prediction and Control of Blast Furnace Operation from Observed Data by Multivariate Time Series," *Proc. AIME Ironmaking Conference*, Vol. 34, pp. 403–412

2 Pandit, S. M. (1975), "Data Dependent Systems: A New Approach to Prediction and Control in Steelmaking," *Proc. CAPC Symposium on the Control of Basic Oxygen Steelmaking*, C.I.M., Toronto, Canada, Sept. 2–4, 1975, pp. 5.2–26.

3 Nayeb Hashemi, A. A., Clum, J. A., and Wu, S. M. (1977), "Blast Furnace Modeling and Control by the DDS Method," *Proc. 36th Ironmaking Conference*, Pittsburgh, Vol. 36, pp. 283–297.

MANUFACTURING—GENERAL

1 Wu, S. M., and Pandit, S. M. (1974), "A New Approach to Manufacturing Systems," *Proc. NAMRC II*, pp. 1–13.

2   Pandit, S. M., and Wu, S. M. (1974), "Data Dependent Systems: A New Approach to Manufacturing Systems Analysis," *Proc. Int. Conference on Production Engineering*, Tokyo, Japan, pp. 82–87.

3   Pandit, S. M. (1977), "Data Dependent Systems: A Quantitative Key to Productivity," *Proc. Int. Conference on Production Engineering*, New Dehli, pp. 166–177.

4   Wu, S. M., and Kapoor, S. G. (1979), "Further Application of Dynamic Data System (DDS) with Application to Manufacturing Processes," *SME Technical Paper No. M579-171*.

MANUFACTURING—ABRASIVES AND GRINDING

1   Deutsch, S. J., and Wu, S. M. (1973), "Relationship Between the Parameters of an Autoregressive Model and Grinding Wheel Constituents," *Trans. ASME*, Vol. 95, Series B, No. 4, pp. 979–982.

2   Law, S. S., Joglekar, A. M., and Wu, S. M. (1973), "On Building Models for the Grinding Process," *Trans ASME*, Vol. 95, Series B, No. 4, pp. 983–991.

3. Pandit, S. M., and Wu, S. M. (1973), "Characterization of Abrasive Tools by Continuous Time Series," *Trans. ASME, J. of Engineering for Industry*, Vol. 95B, pp. 821–826.

4   Deutsch, S. J., and Wu, S. M. (1974), "Analysis of Mechanical Wear During Grinding by Empirical-Stochastic Models," *Wear*, 29, pp. 247–257.

5   Phadke, M. S., Burney, F. A., and Wu, S. M. (1975), "Evaluation of Coated Abrasive Grain Geometry and Wear via Continuous Time Series Models," *Wear*, 31, pp. 29–38.

6   Burney, F. A., and Wu, S. M. (1976), "Evaluation of Coated Abrasive Wear via Statistical Parameters," *Wear*, Vol. 36, No. 2, pp. 225–234.

7   Dornfeld, D., and Wu, S. M. (1977), "An Investigation of Groundwood Surfaces as Related to Pulp and Stone," *Wear*, Vol. 42, No. 1, pp. 141–153.

8   Pandit, S. M. (1977), "Quantitative Evaluation of Abrasive Tool and Machined Surface Topography," *1977 Symposium on New Developments in Tool Materials and Applications*, IIT, Chicago, pp. 107–114.

9   DeVries, W. R., Dornfeld, D., and Wu, S. M. (1978), "Bivariate Time Series Analysis of the Effective Force Vibration and Friction

Coefficient Distributions in Wood Grinding," *Trans. ASME*, Series B, Vol. 100, No. 2, pp. 181–185.

10 Wu, S. M., Kapoor, S. G., and DeVries, M. F. (1978), "Characterization of Coated Abrasive Wear by Statistical Invariants," *Proc. NAMRC VI*, pp. 346–350.

11 Pandit, S. M., and Sathyanarayanan, G. (1981), "A New Approach to Wheel-Workpiece Interaction in Surface Grinding," *Proc. NAMRC IX*, pp. 275–282.

12 Pandit, S. M., and Sathyanarayanan, G. (1982), "A Model for Surface Grinding Based on Abrasive Geometry and Elasticity," *Trans. ASME, J. of Engineering for Industry*, Vol. 104, pp. 349–357.

MANUFACTURING—MACHINE TOOL AND DYNAMICS

1 Pandit, S. M., Subramanian, T. L., and Wu, S. M. (1975), "Modeling Machine Tool Chatter by Time Series," *Trans. ASME, J. of Engineering for Industry*, Vol. 97B, pp. 211–215.

2 Burney, F. A., Pandit, S. M., and Wu, S. M. (1976), "A Stochastic Approach to Characterization of Machine Tool System Dynamics Under Actual Working Conditions," *Trans. ASME, J. of Engineering for Industry*, Vol. 99B, pp. 585–590.

3 Garcia-Gardea, E., Burney, F. A., and Wu, S. M. (1976), "Evaluation of a Minimum Vibration Face Milling Cutter by Time Series Methods," *Proc. NAMRC IV*, pp. 437–444.

4 Subramanian, T. L., DeVries, M. F., and Wu, S. M. (1976), "An Investigation of Computer Control of Machinery Chatter," *Trans. ASME, J. of Engineering for Industry*, Vol. 98B, pp. 1209–1214.

5 Burney, F. A. Pandit, S. M., and Wu, S. M. (1977), "A New Approach to the Analysis of Machine Tool System Stability Under Working Conditions," *Trans. ASME, J. of Engineering for Industry*, Vol. 99B, pp. 585–590.

6 Wu, S. M., DeVries, M. F., and DeVries, W.R. (1977), "Analysis of Machining Operations by the Dynamic Data System Approach," *Proc. NAMRC V*, pp. 219–223.

7 Burney, F. A., Kapoor, S. G., and Wu, S. M. (1977), "Effect of the Minimum Vibration Face Milling Cutter on the Surface Finish of the Workpieces," *Proc. NAMRC V*, pp. 316–321.

8 Garcia-Gardea, E., Burney, F. A., and Wu, S. M. (1979), "Determination of True Cutting Signal by Separation of the Instrumen-

tation Dynamics from Measured Response," *Trans. ASME*, Series B, Vol. 101, No. 3, pp. 264–268.

9 Eman, K., and Wu, S. M. (1980), "A Comparative Study of Classical Techniques and the Dynamic Data System (DDS) Approach for Machine Tool Structure Identification," *Proc. NAMRC VIII*, pp. 401–404.

10 Pandit, S. M., and Revach, S. (1981), "Data Dependent Systems Approach to Dynamics of Surface Generation in Turning," *Trans. ASME, J. of Engineering for Industry*, Vol. 103B, pp. 437–445.

11 Pandit, S. M., and Balkuvar, A. (1982), "Frequency Decomposition of Cutting Forces in End Milling," *Proc. NAMRC X*, pp. 393–400.

MANUFACTURING—SURFACE CHARACTERIZATION

1 DeVor, R. E., and Wu, S. M. (1972), "Surface Profile Characterization by Autoregressive Moving Average Models," *Trans. ASME*, Vol. 94, Series B, No. 3, pp. 825–832.

2 Pandit, S. M. Suratkar, P. T., and Wu, S. M. (1976), "Mathematical Model of a Ground Surface Profile with the Grinding Process as a Feedback System," *Wear*, Vol. 39, pp. 205–217.

3 Suratkar, P. T., Pandit, S. M., and Wu, S. M. (1976), "A Stochastic Approach to the Mode of Deformation and Contact Between Rough Surfaces," *Wear*, Vol. 39, pp. 239–250.

4 Pandit, S. M. Nassipour, F., and Wu, S. M. (1977), "Stochastic Geometry of Anisotropic Random Surfaces with Application of Coated Abrasives," *Trans. ASME, J. of Engineering for Industry*, Vol. 99B, pp. 218–224.

5 Nassirpour, F., and Wu, S. M. (1977), "Statistical Evaluation of Surface Finish and its Relationship to Cutting Parameters in Turning," *Int. J. of Machine Tool Design and Research*, Vol. 17, pp. 197–208.

6 Pandit, S. M., and Rajurkar, K. P. (1978), "A Mathematical Model for Electro-Discharge Machined Surface Roughness," *Proc. NAMRC VI*, pp. 339–345.

7 Nassipour, F., Kapoor, S.G., and Wu, S. M. (1978), "Runway Roughness Characterization by DDS Approach," *Trans. J. of ASCE*, Vol. 104, No. TE2, pp. 213–226.

8 Shuiab, A. R., and Wu, S. M. (1979), "Computer Evaluation of Sur-

face Topography (CEST) of Sheet Steel," *Proc. NAMRC VII,* Ann Arbor, Mich.

9 Pandit, S. M., and Rajurkar, K. P. (1980), "Crater Geometry and Volume from Electro-Discharge Machined Surface Profiles by Data Dependent Systems," *Trans. ASME, J. of Engineering for Industry,* Vol. 102B, pp. 289–295.

10 Pandit, S. M., and Revach, S. (1980), "Wavelength Decomposition of Surface Roughness in Turning," *Proc. NAMRC VIII,* pp. 358–365.

MANUFACTURING—PROCESS MODELING

1 Pandit, S. M., and Rajurkar, K. P. (1980), "Data Dependent Systems Approach to EDM Process Modeling from Surface Roughness Profiles," *Annals of the CIRP,* Vol. 29/1, pp. 107–112.

2 Pandit, S. M., and Rajurkar, K. P. (1981), "Analysis of Electro Discharge Machining of Cemented Carbides," *Annals of the CIRP,* Vol. 30/1, pp. 111–116.

3 Rajurkar, K. P., and Pandit, S. M. (1982), "Prediction of Metal Removal Rate and Surface Roughness in Electrical Discharge Machining," *Proc. NAMRC X,* pp. 444–450.

MANUFACTURING—ON-LINE MONITORING

1 Pandit, S. M., and Kashou, S. (1982), "A Data Dependent Systems Approach to On-Line Tool Wear Sensing," *Trans. ASME, J. of Engineering for Industry,* Vol. 104B, pp. 217–223.

MANUFACTURING—TOOL LIFE

1 Pandit, S. M. (1978), "Data Dependent System Approach to Stochastic Tool Life and Reliability," *Trans. ASME, J. of Engineering for Industry,* Vol. 100, pp. 318–322.

2 Pandit, S. M., and Kahng, C. H. (1978), "Reliability and Life Distribution of Ceramic Tools by Data Dependent Systems," *Annals of CIRP,* Vol. 27/1, pp. 23–27.

METHODOLOGY AND SYSTEM IDENTIFICATION

1 Stralkowski, C. M., DeVor, R. E., and Wu, S. M. (1970), "Charts for the Interpretation and Estimation of the Second Order Autoregressive Models," *Technometrics,* Vol. 12, No. 3, pp. 669–685.

2 Stralkowski, C. M., DeVor, R. E., and Wu, S. M. (1974), "Charts for the Interpretation and Estimation of the Second Order Moving Average and Mixed First Order Autoregressive Moving Average Models," *Technometrics,* Vol. 16, No. 2, pp. 275–285.

3. Pandit, S. M., and Wu, S. M. (1974), "Exponential Smoothing as a Special Case of a Linear Stochastic System," *Operations Research* Vol. 22, pp. 868–879.

4 Phadke, M. S., and Wu, S. M. (1974), "Modeling of Continuous Stochastic Processes from Discrete Observations with Applications to Sunspot Data," *JASA,* Vol. 69, No. 346, pp. 326–329.

5 Phadke, M. S., and Wu, S. M. (1974), "Identification of Multi-Input Multi-Output Transfer Function and Noise Model of a Blast Furnace from Closed Loop Data," *IEEE Trans. on Automatic Control,* Vol. AC-19, No. 6, pp. 944–951.

6 Chu, B. B., Wu, S. M., and Goh, T. N. (1975), "A Note on Parametric Time-Series Modeling of Second Order Systems," *Trans. ASME,* Series G, Vol. 97, No. 3, pp. 309.

7 Pandit, S. M., and Wu, S. M. (1977), "Modeling and Analysis of Closed Loop Systems from Operating Data," *Technometrics,* Vol. 19, pp. 477–485.

8 Wu, S. M. (1977), "Dynamic Data System—A New Modeling Technique," *Trans. ASME,* Series B, Vol. 99, No. 3, pp. 708–714.

9 Pandit, S. M. (1977), "Data Dependent Systems: A Quantitative Key to Productivity," *Proc. Int. Conference on Production Engineering,* New Delhi, pp. 166–177.

10 Pandit, S. M. (1977), "Stochastic Linearization by Data Dependent Systems," *Trans. ASME, J. of Dynamic Systems, Measurement, and Control,* Vol. 99G, pp. 221–226.

11 Huang, H. P., and Wu, S. M. (1978), "Simulation Studies of System Modeling and Identification via DDS Approach," *Proc. 9th Annual Pittsburgh Conference,* Vol. 9, pp. 1155–1160.

12 Pandit, S. M. (1981), "Characteristic Shapes and Wavelength Decomposition of Surfaces in Machining," *Annals of the CIRP,* Vol. 30/1, pp. 487–492.

13 Pandit, S. M., and King, D. R. (1981), "Data Dependent Systems Linearization of Nonlinear Pendulums," ASME Paper No. 81-WA/DSC-6.

14 Pandit, S. M., and Revach, S. (1981), "Data Dependent Systems Approach to Dynamics of Surface Generation in Turning," *Trans. ASME, J. of Engineering for Industry,* Vol. 103B, pp. 437–445.

BUSINESS, ECONOMICS, AND MANAGEMENT

1 Steudel, H. J., and Wu, S. M. (1977), "A Time Series Approach to Queing Systems with Applications for Modeling Job-shop In-process Inventories," *Management Science,* Vol. 23, No. 7, pp. 745–755.

2 Steudel, H. J., Pandit, S. M., and Wu, S. M. (1977), "Multiple Time Series Approach to Modeling The Manufacturing Job-shop as a Network of Queues," *Management Science,* Vol. 24, pp. 456–463.

3 Steudel, H. J., Pandit, S. M., and Wu, S. M. (1978), "Interpretation of Dispatching Policies on Queue Behavior via Simulation and Time Series Analysis," *AIIE Trans.,* Vol. 10, No. 3, pp. 292–297.

4 Pandit, S. M. (1980), "Data Dependent Systems and Exponential Smoothing," *Analyzing Time Series,* Ed. O. D. Anderson, North-Holland, pp. 217–238.

5 Pandit, S. M. (1982), "Data Dependent Systems Approach to Trend and Seasonality," *Time Series Analysis—Theory and Practice I,* Ed. O. D. Anderson, North-Holland, pp. 515–526.

NUCLEAR POWER PLANT SURVEILLANCE

1 Chow, M. C., Wu, S. M., and Cain, D. G. (1977), "Malfunction Detection of Nuclear Reactor by Dynamic Data System Methodology," *Proc. SMORN II,* pp. 595–602.

2 Chow, M. C., and Wu, S. M. (1977), "System Identification of Nuclear Reactor by Dynamic Data System Methodology," *Proc. IEEE, Control Conference.*

3 Chow, M. C., Wu, S. M., and Ermer, D. (1979), "A Time Series Control Chart for a Nuclear Reactor," *Proc. 1979 Annual Reliability and Machinability Symposium.*

4 Wu, S. M., Hsu, M. C., and Chow, M. C. (1979), "The Determination of Time Constants of Reactor Pressure and Temperature Sensors: The DDS Method," *Nuclear Science and Engineering,* Vol. 27, pp. 89–96.

SOLAR ENERGY

1 Pandit, S. M., and Rajurkar, K. P. (1981), "Data Dependent Systems Approach to Solar Energy Simulation Inputs," *Proc. ASME/SSEA Conference, Solar Engineering—1981*, pp. 389–398.

HEAT TRANSFER

1 Pandit, S. M., and Rajurkar, K. P. (1981), "A New Approach to Thermal Modeling Applied to Electro Discharge Machining," ASME Paper No. 81-WA/HT-3, to appear in *ASME Journal of Heat Transfer*.

FRICTION AND WEAR

1. Pandit, S. M., and Kashou, S. (1983), "Variation of Friction Coefficient with Tool Wear" *Wear*, Vol. 84, pp. 65–79.

PULP AND PAPERMAKING

1 Tee, L. H., and Wu, S. M. (1972), "An Application of Stochastic and Dynamic Models for the Control of a Papermaking Process," *Technometrics,* Vol. 14, No. 2, pp. 481–496.

2 Phadke, M. S., and Wu, S. M. (1976), "Identification of Papermaking Process From Closed Loop Data by Bivariate Time Series Analysis," *Trans. ASME, Series G,* Vol. 98, No. 3, pp. 291–295.

3 DeVries, W. R., Pandit, S. M., and Wu, S. M. (1977), "Evaluation of the Stability of Paper Basis Weight Profiles Using Multivariate Time Series," *IEEE Trans. on Automatic Control,* Vol. AC-22, pp. 590–954.

4 DeVries, W. R., and Wu, S. M. (1977), "Characterization of the Formation of Base Sheet Paper Using Spectral Moments Determined From Time Series Models," *J. of Fluid Mechanics,* Vol. 82, p. 261.

5 Kapoor, S. G., and Wu, S. M. (1978), "A Stochastic Approach to Paper Surface Characterization and Printability Criteria," *J. of Physics D: Applied Physics,* Vol. 11, pp. 83–97.

6 DeVries, S. G., and Wu, S. M. (1978), "Evaluation of Process Control Effectiveness and Diagnosis of Variation in Paper Basis Weight via Multivariate Time Series Analysis," *IEEE Trans. on Automatic Control,* Vol. AC-23, No. 4, pp. 702–708.

7   Kapoor, S. G., Wu, S. M.. and Pandit, S. M. (1978), "A New Method for Evaluating Printing Smoothness of Coated Papers," *TAPPI,* Vol. 61, No. 6, pp. 71–74.

8   Goh, T. N., and Wu, S. M. (1978), "Use of Secondary Input for Process Control—An Application to a Digester Pulping Process," *Trans. ASME, Series G, J. of Dynamic Systems, Measurement and Controls,* Vol. 100, No. 1, pp. 88–91.

9   Kapoor, S. G., and Wu, S. M. (1979), "Predicting Printability of Gravure and Embossed Papers by Time Series Analysis," *J. of Physics D,* Vol. 12, No. 12.

VIBRATIONS

1   Pandit, S. M. (1977), "Analysis of Vibration Records by Data Dependent Systems," *Shock and Vibration Bulletin,* No. 47, Part 4, pp. 161–174.

2   Eman, K., and Wu, S. M. (1980), "A Feasibility Study of On-Line Identification of Chatter in Tuning Operations," *Trans. ASME, J. of Engineering for Industry,* Vol. 102, pp. 315–321.

3   Eman, K., and Wu, S. M. (1980), "Forecasting Control of Machining Chatter," *Computer Applications in Manufacturing Systems, presented at the 1980 Winter Meeting of ASME,* Chicago, Illinois, pp. 37–52.

SIGNATURE ANALYSIS

1   Pandit, S. M., Suzuki, H., and Kahng, C. H. (1980), "Application of Data Dependent Systems to Diagnostic Vibration Analysis," *Trans. ASME, J. of Mechanical Design,* Vol. 102, pp. 233–241.

2   Wu, S. M., Tobin, T., and Chow, M. C. (1980), "Signature Analysis for Mechanical Systems via Dynamic Data System (DDS) Monitoring Technique," *Trans. ASME, J. of Mechanical Design,* April, Vol. 102, pp. 217–221.

BIOMEDICAL ENGINEERING

1.   Ambardar, A., Walworth, M. E., Pandit, S. M., and Ray, G. (1983), "Data Dependent Systems Analysis of M-Mode Echocardiograms," *Proc. 1st IASTED Symposium on Applied Informatics.*

MISCELLANEOUS

1 Phadke, M., Joglekar, A. M., and Wu, S. M. (1977), "Investigation of Spiking in Electron Beam Welding Using Stationary Stochastic Models," *Trans. ASME, Series B,* Vol. 9, No. 2, pp. 323–326.

2 Ermer, D., and Wu, S. M. (1979), "Improved Hydrologic Forecasting for Regulation of the Great Lakes," *10th Annual Pittsburgh Conference,* Vol. 10.

3 Kapoor, S. G., Kunpanitchakit, C., and Wu, S. M. (1979), "Modeling and Forecasting of a Maneuvering Tank for Gunfire Control," *10th Annual Pittsburgh Conference,* Vol. 10.

4 Adeyemi, S. O., Wu, S. M., and Berthouex, P. M. (1979), "Modeling and Control of a Phosphorus Removal Process by Multivariate Time Series Method," *Water Research,* Vol. 13, pp. 105–112.

INDEX

A(1) continuous first order autoregressive
model, 220
response generation, 222
uniformly sampled discrete, *see* Covariance
equivalent AR(1)
variance, 227
A(2) continuous second order autoregressive
model, 254
advantage in spectral estimation, 279
response generation, 262
uniformly sampled discrete, *see* Covariance
equivalent ARMA(2,1)
variance, 261
Adequacy checks, 27, 31, 33, 34, 37
as hypothesis testing, 160–161
on residual autocorrelations, 163
see also F-criterion
Airline passenger ticket sales series:
comparison of forecasts, 390–391
logged series (stochastic) trends and periods,
347–350
multiplicative model, 348
original series (deterministic) trends and
periods, 378–389
AM(2,1), continuous second order
autoregressive first order moving
average model, 295–296
feedback block diagram, 303
uniformly sampled model, *see* Covariance
equivalent ARMA(2,1) from AM(2,1)
variance, 297, 311
Amplitude, *see* Phase angle
Analysis of variance, 66
Anderson, 42, 59
Application:
areas of, 7
bibliography, 567
AR(1) discrete first order autoregressive
model, 3, 23
conditional distribution, 30

confidence interval on Φ_1, 53, 76
estimation, 25–27
forecasting or prediction, 29–30
response generation, 82–87
stability, 85–87
structure, 25
time constant, 230–231, 246
variance, 76, 88, 115
variance of $\hat{\Phi}$, 53, 76
AR(1) to ARMA(2,1):
via autocovariance function, 125
via Green's function, 98
via regression, 34–35
via sum of two, 141
AR(2), 36
confidence interval on Φ_1, Φ_2, 74, 77
correlation between $\hat{\Phi}_1$, $\hat{\Phi}_2$, 77
estimation of, 36, 73
as special case of ARMA(2,1), 36, 42–44,
74, 100, 139, 141
variance, 73, 123, 140
variance-covariance of, $\hat{\Phi}_1$, $\hat{\Phi}_2$, 73, 77
ARIMA, 44, 105
as special cases of ARMA, 31, 44, 74, 105,
109, 344, 345
see also Integrated moving average;
Integrated random walk; Random walk
AR(n):
as modeling approach, 164
estimation, 153–154
AR(p) for initial values, 156
ARMA, discrete autoregressive moving
average models, 13
characteristics of, 78
with constant added to a_t, 409–411
difference equation form, 79
extended (for multiple series) *see* Extended
ARMA
genesis of, 43, 205
Green's function form, 80, 139

inverse function form, 106
related to exponential smoothing, 204
systems and dynamics, 78
transfer function for, 111
ARMA(1,1):
as special case of ARMA(2,1), 40, 141
variance, 125
ARMA(2,1), 34
importance in modeling, 145, 152–153
related to continuous systems, *see*
Covariance equivalent ARMA(2,1)
response generation, 93–104
variance, 120, 121
variance decomposition, 121–122, 127, 139
ARMA(2,1) from AR(2), 141
ARMA(2,1) to ARMA(4,3)
via sum of two, 141
ARMA(2,1) to ARMA(3,2), via regression
41
ARMA(3,2), 41, 146
ARMA (n,m), 42
with $m \neq n - 1$, 44, 145–146
ARMA(n,n):
generating mechanism, 141
sequence for modeling, 165
ARMA(n, n−1), 5, 41, 43
adequacy of, 44
analogy with n-degrees-of-freedom system,
151
for stochastic part in deterministic trend,
364
from a sum of n AR(1), 44
ARMA(2n, 2n−1), 151, 158, 161–163, 165
ARMAV, Autoregressive Moving Average
Vector models simplified, special form
of, 418–421. *See also* Extended
ARMA
a_t distribution and properties, *see* Residual
properties
a_t's, computation of, 154
three options, 166–167
a_t's too large, 111
Autocorrelation, 2, 27, 76, 112
of ARMA/AM models, *see* Autocovariance
Autocorrelation, residual (a_t), 28, 114, 425
distribution of, 62
standard error of, 54, 163
unified, 163
variance of, 53, 114
Autocorrelation sample (estimated), 26, 113
variance of, 53

Autocovariance, 76, 78, 112
A(1), 227
A(2), 261
AR(1) explicit, 116
AR(1) implicit, 115
AR(2), 123
ARMA(1.1), 124–125
ARMA(2.1) explicit, 121
ARMA(2.1) implicit, 119–120
ARMA(n, m), 129
contrast with Green's function, 122
difference equation for, 116, 120, 129
MA(1), 118–119
MA(2), 129
relation to Green's function, 128, 226
Autocovariance sample, 113
Autoregressive model, 3
continuous, *see* A(1), A(2)
discrete, *see* AR(1), AR(2), AR(n)
Autoregressive Integrated Moving Average,
see ARIMA
Autoregressive Moving Average model:
continuous, *see* AM
discrete, *see* ARMA
Autospectrum:
A(1) model, 228
A(2) model, 263
ARMA models, 132
extrapolation beyond Nyquist frequency, 279

B (Backward shift) operator, 80
Bachelier, 33, 236
Bachelier-Wiener Process, 236, 239
Bartlett, 163
Basis weight step response series, 365
first order dynamics, 366–371
plot, 365
second order dynamics, 407
Bivariate time series, 418–424
inverse function for, 475–479
Box and Jenkins, 42, 106, 163, 179, 279, 355
Brown, 199, 300
Brownian motion, 236, 263

Calculus, 12
Calibration of measuring instruments, 363
Central limit, 58
Characteristic equation, roots (continuous), 244
first order, 213
relation to discrete, 264, 272
second order, 250–251

Characteristic roots (discrete), 91
  for seasonality, 325
  for trends, 314
Chapterwise summary, 7–10
Chatter, 272
Chemical relaxation series, 371
  first order models, 372–375
  plot, 371
Chi-square,
  degrees of freedom of, 60
  in defining F-distribution, 67
  distribution, 59
  mean of, 60
  probability table, 504–505
  test on residuals, 62, 163–164
Coddington and Levinson, 224
Conditional distribution, 29–30, 185–186
Conditional expectation, 29, 179
  as orthogonal projection, 184
  rules for, 184
Conditional least squares, 53, 142, 154
Conditional linear regression, 3, 25, 53
  of ARMA models, 39
  of ARMA(2,1) model, 35, 39
  of ARMA$(n, n-1)$ model, 41, 42
Confidence intervals, 39, 50
  on parameters of ARMA models, 73–74,
    155
Consumer and wholesale price index series,
    341
  growth trend by Green's function, 354, 396
  instability of, 345, 354, 401
  modeling tables—deterministic trend,
    397–402
  modeling tables—stochastic trend, 343–
    345
  stochastic versus deterministic, 402–403
Continuous systems (models), 211
  first order, see A(1)
  second order, see A(2) or AM(2,1)
Convergence criterion, 161
Convergence difficulties, 155, 167
Convergence tolerance, 154
Convolution, 128, 221–222, 259
Correlation, 3, 47
  of forecast errors:
    AR(1), 186, 188
    ARMA(n,m), 190–191
    graphical representation, 191
  reduced to zero, 231–238
  sample, estimate, 49

Course text:
  one-semester, 12
  two-quarter, 12
Covariance, 46, 55
  sample, estimate, 49
Covariance equivalent, AR(1), 229–231
  effect of α_0, 232–235
  effect of sampling interval, 231–232
  parametric relations, 230
Covariance equivalent ARMA(2,1) from A(2):
  advantage in spectral estimation, 278–279
  effect of damping ratio, 286–289
  effect of sampling interval and natural
    frequency, 280–286
  estimation, 278
  experimental, verification, 292
  parametric, relations, 265–267, 294
  reduced, stability region, 269
Covariance equivalent ARMA(2,1) from
    AM(2,1), 298
  effect of natural frequency and sampling
    interval, 305–306
  limiting case of exponential smoothing, 300
  parametric relations, 298–299
Cox, 301
Crack propagation series, 357
  linear deterministic trend plus stochastic
    model, 361
  physical interpretation, 362–363
Cross correlation, 425

Damped (natural) frequency, 252
Damping:
  critical, 250, 252
  force, 249
  ratio ζ, 250
  in stock prices, 290–291
Dashpot constant, 249
Dead time or delay, 417–418
  effect on optimal control, 451–452
  estimation in modeling, 436–439
Degenerate random variable, 51
Degrees of freedom:
  of chi-square, 60
  of vibration system, 151, 250
Delta, Dirac δ (t), 217–219, 245–246
  Kronecker δ_k, 112, 128, 217
Den-Hartog, 270
Dependence, 2, 13
  static versus dynamic, 78
Deseasonalizing, see Seasonal adjustment

Detrending, *see* Trend removal
Difference equation, 4, 79
  form of ARMA models, 79
  for Green's function, 90, 92
Differencing operators *after* modeling,
    advantages of, 74, 313
Differencing operators *before* modeling,
    dangers of, 324–326
Differential equation, first order:
  formulation of homogeneous, 213
  formulation of nonhomogeneous, 220
  review, 241–242
  solution of homogeneous, 213
  solution of nonhomogeneous, 224
  stability, 215–216
Differential equation, second order
  formulation of homogeneous, 249
  formulation of nonhomogeneous, 255
  review, 243
  solution of homogeneous, 250–251
  solution of nonhomogeneous, 256–260
  stability, 251–253
Dirac, *see* Delta
Doob, 236
Durbin, 114, 131
Dynamic modes, *see* Modes
Dynamics of ARMA models, 3, 78
  via autocovariance function, 125, 139
  via Green's function, 98, 139
  in limiting cases, 237
Dynamics same for system and noise,
    370–371, 374–375, 408–411

Eigenvalues, *see* Characteristic roots
Estimation:
  of AR models, 153
  of ARMA models, 154–155
Estimators, 49
  consistency of, 50
  unbiased, 49
Eventual forecasts, 197–198
  variance of, 209
EWMA, *see* Exponential smoothing
Expected value, 46, 54
Exponential function, 242
  differentiation, 243
  integration, 244
  useful values, 244
Expenditures on producers' durables (EDP)
    series, 411

Exponential smoothing, 110, 170
  choice of smoothing constant, 301, 312
  constant or parameter, 201
  feedback interpretation, 301–304
  interpretation and advantages, 201–204
  as a random walk with added white noise,
    210
  related to ARMA, 204–208
  as a special case of updating, 199
  as a sum of two components, 209
  vs. ARMA model for forecasting, 206–207
  from weighted moving averages, 201–202
Exponential smoothing via AM(2,1), 295
  advantages, 295–296
  robustness explained, 301
  versatility explained, 303–304
Extended ARMA models (for multiple series),
    424
  advantages for transfer function modeling,
    432
  forecasting, 464–475
  modeling, 432
  modeling examples, 426–441
  modeling procedure, 425–426
  optimal control, 452–464
  as special form of ARMAV, 421–424
  transfer function form, 415–416
  transformation to ARMAV, 432–433, 443
Extrapolation of time series, 177, 178

F-criterion for model adequacy:
  application for data independence, 69
  for ARMA models, 161–162
  derivation from linear regression, 68
  for desired models, 163
  as stopping criterion, 162, 165
F-distribution, 67
  probability table, 508–513
Feedback:
  block diagram for AM(2,1), 303
  in business, economics and quality control,
    304
  effect on variance, 311
  negative, instability, 311
  presence in ARMAV models, 419–420
  time constant, 303–304
  unintentional, 310
Forcing function, 79, 220, 249
Forecasts:
  AR(1), 29–30, 184–188
  ARMA(1,1), 192–195, 198–199

ARMA(2,1), 195–197
ARMA(n,m), 188–192
  effect of transformation and differencing, 390–391
  eventual, 197
  in terms of present and past values, 209
  of multiple series, 464–475
  of series with deterministic trends, 407
  of series with stochastic trends, 354
  updating, 198–199
Forecast error, *see* Forecasts
F-test, *see* F-criterion
Fundamental theorem, 149

Gain, first order, 366, 431
  any system, 431
Gauss method, 154
Geometric series, 12
  sum of, 408
Green's function G_j, 4, 78, 79
  A(1), 226
  A(2), 256–257
  AM(2,1), 297
  AR(1), 79
  AR(2), 95–97
  ARMA(1,1), 96–97
  ARMA (2,1) explicit, 91–92
  ARMA(2,1) implicit, 89
  with complex roots, 94, 98, 134–135
  component computation, 352–354
  components illustrated, 97, 332–333
  determining seasonality, 320–326
  determining trends, 314–319, 345, 354
  with equal roots, 103
  general ARMA(n,m), 105
  implicit method, 89–90
  MA(2), 128
  multiple, 456, 460, 474
  operator (transfer function) for, 108, 111
  physical interpretation, 80
  relation to inverse function, 209
  in response generation, 82–104, 222, 262
  as a sum of exponentials, 97, 257
Grinding wheel profile, 13
  continuous A(2) model, 293
  discrete AR(2) model, 172
  modeling table, 173
Gross national product (GNP) series, 410–411

Hannan, 42
Harmonics, 339, 379
Hilbert space, 44, 148
Holt, 199
Homogeneous difference equation, 92
  differential equation, 241
Hospital patient census series, 334
  modeling table, 336
  plot, 335
  seasonality analysis, 337–341

θ_i from roots v_i, *see* Φ

IBM stock prices series, 13
  continuous first order model, 239
  continuous second order model, 290
  discrete AR(1) model, 30
  discrete ARMA(1,1) model, 170
  modeling table, 171
  nearly random walk, 31, 117, 118, 239, 290
  physical interpretation, 291
  plot of, 15
  sample autocorrelation, 118
  stochastic constant trend, 315
Impulse, unit, 217–218
Impulse response, 4, 364. *See also* Green's function
Independence of residuals, 27, 31, 54, 59
Independent increment process, 235
Independent observations, 2, 13, 49–50, 54, 59
  sampling interval for, 232, 282
  time constant τ or α_0 for, 233–235
Independent random variables, 46, 55
  variance of their sum, 48
Initial values, 39
  from autocorrelation, 175–176
  bivariate illustration, 475–479
  in deterministic trends, 367, 378
  from inverse function, 155–159, 174
  invertibility of , 157, 159, 174
Instructor guidelines, 11. *See also* Course
Integrated models (processes), 44, 105, 279
Integrated moving average (IMA):
  as a limit of sampled A(2), 281
  as a limit of sampled AM(2,1), 299–300
  as special case of ARMA(1,1), 109, 170
  as stochastic linear trend, 316
  in quality control, 309
Integrated random walk, 105
  as stochastic linear trend, 316

Inverse function, I_j, 78, 106
  AR(1), 107
  ARMA(1,1), 109
  ARMA(1,2) explicit, 109
  ARMA(1,2) implicit, 108
  ARMA(n,m), 108–110
  duality with Green's function, 108
  general implicit relations, 156
  MA(1), 107
  MA(2), 109
  multiple, 475–479
  operator for, 108, 111
Invertibility:
  computational reasons for, 111
  of initial values, 157
  see also Inverse function
Invertibility conditions (region):
  ARMA($n,2$), 101, 109
  ARMA($n,3$), 140
  general ARMA(n,m) from parameters,
    136–138
  general ARMA(n,m) from roots, 110
Investment and money market rate series, 327
  modeling tables, 330–331
  model interpretation, 330–334
  plots of data, 328–329
  plots of Green's functions and components,
    332–333

Jury, 136

Kalman, 179
Kalman filtering, 412
Kendall, 114
Khintchin, 132, 178
Kolmogorov, 177–178
Kolmogorov-Wiener prediction theory, 179
Kuo, 479

Lag, 26. *See also* Dead time; Delay
Laplace transform, 256
  generalized, 395
Leading indicator(s) for forecasting:
  examples, 464–469
  multiple, 473–475
  single, 471–473
  usefulness, 469–471
Least squares, double linear, 157. *See also*
    regression
Least squares estimates, 18, 22
  derivation of, 45, 70

Linear regression multiple:
  confidence intervals on parameters, 72
  estimates, 22, 70
  example of, 71
  mean of estimators, 72
  prediction, 22
  review, 21–23
  variance of estimators, 72
Linear regression, simple:
  conditional, *see* Conditional linear
    regression
  confidence intervals on parameters, 58, 76
  confidence interval on regression line,
    65–66, 76
  estimates, 17–18, 45–46, 76
  example of, 20–21
  F-ratio for hypothesis testing, 67
  mean and variance of estimators, 52, 76
  prediction, 19
  review of, 17
Linearization:
  of nonlinear models, 39
  of nonlinear systems, 6
Local linear hypothesis, 155

MA(1), first order moving average model, 40
  variance, 119
MA(2), second order moving average model,
    128
  variance, 129
Markov processes, 177
Mathematical level of contents, 11
Maximum likelihood, 154, 167
Mean, 46, 55
  confidence interval on, 58, 63
  estimated in, ARMA model, 155
  estimate of, 48–49
Measuring instrument dynamics, 363
Mechanical vibratory displacement series, 13
  continuous A(2) model, 291–292
  discrete ARMA(2,1), 173–174
  experimental conditions for, 291
  modeling table, 173
  plot of, 15
Memory, 3, 13, 78
  in limiting cases, 237
Microprocessor implementation, 157
Miller, 79
Minimum mean squared error control, *see*
    Optimal control

Modal analysis, *see* Modal (power) decomposition

Modal decomposition, 105. *See also* Variance decomposition

Modal power decomposition, 130

Modeling as decomposition, 393–395

Modeling procedure:
ARMA, 165
ARMA flow chart, 166
extended ARMA (simplified ARMAV), 425–426

Modeling strategy, 40–45, 114
deterministic trends, 364
increment in AR order, 151–153
moving average order, 150

Modeling, system approach to:
adequacy of, 148–149
via autocovariance, 149–150
via dynamics, 144–147

Model order, 42

Modes, 105, 401
in vibration data, 173

Moments, *see* Probability moments

Moving average, 200
exponentially weighted, 201
part of ARMA, 35

Moving average order $(n-1)$, 150

Multi-input-one-output optimal control, 459–464

Multiple regression, *see* Linear regression, multiple

Multiple series, 412
modeling, *see* Extended ARMA models

Multivariate time series, 2, 413

Muth, 199, 209

Natural frequency ω_n, 250
for independent data, 282
multiplicity and its resolution, 273–277

NID, normally independently distributed, 18

Nonhomogeneous difference equation, 79
differential equation, 241

Nonlinear least squares, 39, 154

Nonlinear regression:
of ARMA(2,1), 39
of ARMA$(n, n-1)$, 42
see also Nonlinear least squares

Nonlinear systems, 6

Nonstationary time series, 26, 75, 106, 355
general model for, 392
practical aspects in modeling, 395–396

Normal distribution, 18
of linear forms, 55
multivariate, 112
of observation with forecast as mean, 19
probability density function, 54

Normal probability:
computation, 56
limits, 57
table of, 503

Notation, 11

Nyquist frequency, 274
autospectrum beyond, 278–279

One-input-one-output system:
modeling, 426–429
optimal control, 445–454

Optimal (minimum variance) control:
development procedure, 445, 452
efficiency, 446, 447, 449, 457, 458
general, multi-input-one-output, 459–464
general, one-input-one-output, 452–454
illustrative examples, 445–452, 454–458
output after control, 444–445
strategy, 444

Orthogonal decomposition, 3, 25, 87–88, 181, 221–224
of A(1), 226
of A(2), 260
of AM(2,1), 297
of AR(1), 25, 80
of ARMA(2,1), 91

Orthogonal projection, 179–182

Overdamped system, 250, 267

Papermaking-input gate opening series, 13
continuous A(1) model, 230–231
discrete AR(1) model, 25–27
discrete ARMA(2,1) model, 171
modeling table, 172
plot, 14
sample autocorrelation, 118
stochastic constant trend, 315

Papermaking input-output series, 426
modeling, 426–429
optimal control, 445–449
plot, 426

Papermaking process, 413–415

Paper pulping digester series:
dead time determination, 436–439
modeling, 439–441

optimal control, 458
plot, 436
Parsimonious models, 165, 328
Partial autocorrelation, 130–131
Partial fractions, 91, 105, 132
Phase angle:
  in deterministic trend, 377, 388–389
  in Green's function, 94–95, 353
  in vibration response, 252
PID, proportional plus integral plus derivative,
    controller, 446, 481–482
Prediction as orthogonal projection, 180
Prediction error, *see* Forecasts
Preprocessing, 313
Probability density, 54
  joint, 55
Probability distribution, 3, 54
Probability limits on forecasts:
  AR(1), 29, 185–188
  AR(1), using t-distribution, 19, 21, 65, 77
  ARMA(1,1), 192–193
  ARMA(2,1), 196–197
  ARMA(n,m), 189
  extended ARMA, 467–469
  multiple regression, 22
  multiple series, 467–469
  simple regression, 19, 57, 64, 76
Probability moments, 3, 6
Pugachev, 106
Pulping digester operation, 434–435
Pulse, unit, 217–218

Quality control, 295, 304
  feedback loop for, 303
  out-of-control, instability, 310–311
Quality control measurements series, 306
  interpretation, 310–311
  modeling, 307–309

Ramp function, 219
Random vibration, 248
Random walk:
  autocorrelation of, 117
  continuous time, 234, 239, 247
  illustrated by IBM series, 30–31, 170
  integrated, 105
  justification from estimate, 77
  as a limit of AR(1), 30–33
  as a limit of sampled A(1), 232–235
  as a limit of sampled A(2), 288–289
  as a limit of sampled AM(2,1), 305

physical interpretation, 237–238
  as stochastic constant trend, 314–315
  variance of, 76, 89, 355
Regression, 3. *See also* Linear regression;
    Nonlinear regression
Relaxation process, 371
  time constant, 372
Residual autocorrelation:
  sample or estimated, 114, 163
  theoretical, *see* Residual properties
Residual crosscorrelation: 425
Residual properties:
  in ARMA models a_t, 24, 29, 119
  in linear regression ε_t, 18, 22
  in multiple series a_{it}, 421, 424
Residual variance estimate:
  ARMA, 26
  autoregression, 25–26, 36, 154
  multiple regression, 22
  simple regression, 18
  unbiased estimate, 18, 26
Resonance curves, 264
Restoring (spring) force, 248–249
  in stock prices, 291
Retail trading, 401
Roots of unity, 335, 337, 346, 349, 350
Routh-Hurwitz criterion, 136

Seasonal adjustment, 339–340
Seasonal component contribution, 341
Seasonal models, deterministic, 376–392
Seasonal models, stochastic:
  modeling examples, 327–352
  simulated examples, 320–323
Seasonal operators, 325
Seasonal operators *before* modeling, dangers
    of, 324–326
Seasonalities:
  arbitrary, 323, 325
  quarterly, 321
  yearly, 320
Self-excited vibrations, 272
Self-study guidelines, *see* Instructor
Slutsky, 177, 325
Spectrum, *see* Autospectrum
Spring-mass-dashpot system, 248–249, 253
  for business, economics, quality control, 295
Stability:
  asymptotic, 85, 216, 251–254, 257–258
  of A(1), 215–216, 227
  of A(2), AM(2,1), 251–253

of AR(1), 26, 76, 85
of ARMA(2,1), 100–105
of ARMA(3,m), 140
with equal characteristic roots, 102–104, 253
of general ARMA from parameters, 136–138
of general ARMA from roots, 106
related to eventual forecasts, 197–198
static and dynamic, 270–272
Stability region:
  ARMA(2,m), 101
  ARMA(3,m), 140
Stable but not asymptotically, 85, 102, 252
  effect on eventual forecast, 198
  illustrated by papermaking data, 171
  as a limit of sampled A(2), 286–287
  in stochastic constant trend, 314–315
Standard deviation (error), 46
Standard normal distribution, 56
  table of, 503
  variables, 59
Starred sections, 11
State space, 412–413
Static and dynamic stability, 270–272
Stationarity region, *see* Stability region
Stationary time series, 3, 26, 76
  data characteristics, 13
  strictly, 3, 112
  wide sense, 3, 112
Steepest descent, 39, 154–155
Step function, 218
Step response, 364
  first order system, 366–370
  second order system, 259–260, 407
Stochastic difference equation, 1, 5
  first order, *see* AR(1)
  nth order, *see* ARMA($n,n-1$), ARMA(n,m)
  second order, *see* AR(2), ARMA(2,1)
Stochastic differential equation, 1, 5
  first order, *see* A(1)
  second order, *see* A(2) or AM(2,1)
Stochastic process, 1, 112
  stationary, 3, 112
Stock prices:
  random walk of, 33
  restoring and damping force for, 290–291
  see also IBM
Summary of chapters, 7–10
Sunspot series, 13
  AR(1) model, 33, 170
  AR(2) model, 37, 169, 170

modeling table, 170
  plot of, 14
Surface characterization, 293
System identification, 364–365

t-distribution, 63
  probability table, 506–507
Time constant estimation:
  of feedback systems, 303–304
  by impulse response, 214–215, 371–375
  by step response, 214–215
Time constant τ, 214–215
  from AR(1), 230–231, 246, 363, 373
  for independence, 232
  for random walk, 235, 238–239
  of relaxation, 372–375
  4 τ, 215, 244
Time series:
  definition, 1, 2
  multiple, 2, 412
  multivariate, 2, 413
  nonstationary, 26, 355
  univariate, 2
Transfer function—continuous to discrete, 479–481
Transfer function from deterministic input response, *vs.* operating data, 429–432
Transfer function models, *see* Extended ARMA models
Transfer function *vs.* state variable, 416–417
Trend analysis, 314–319
Trend component contribution, *see* Seasonal component contribution
Trend, deterministic, 355
  in ARMA models, 408–411
  combined, 376–377
  exponential, 363–375
  general, 392
  growth and decay, 376, 378
  known, circular form, 403–406
  linear, 356–363
  periodic (seasonal), 375–392
Trend, deterministic *vs.* stochastic, 402–403, 406–407
Trend indication from model, 74, 313
Trend removal, *see* Seasonal adjustment
Trend, stochastic:
  constant, 314–316
  linear, 316–318

polynomial, 318–319
quadratic, 319
Two-input-one-output system:
    modeling, 433–441
    optimal control, 458

Unbiased estimate:
    of mean, 49
    of variance, 61
    of σ_a^2, 18, 26
    of σ_ε^2, 62
Unconditional least squares, 53
Unconditional regression:
    AR(2) and ARMA(2,1) models, 39
    and dynamics, 78
Underdamped system, 250, 268
Unified autocorrelation, 163, 165
Uniqueness of A(2) model parameters,
    275–278
Univariate time series, 2
Unstable system, 85–86
    A(1), 216
    A(2), AM(2,1), 253
    AR(1), 85–86
    ARMA(2,1), 102–104
    different representations illustrated, 345,
        354, 400–403
    in quality control, 310–311
    in stochastic linear and polynomial trend,
        316–319
Unstable time series, 26, 75
    eventual forecast for, 198
    illustrated by airline data, 390
    illustrated by simulated data, 317, 319
Updating forecasts, 198–199
    related to exponential smoothing, 199
User guidelines, *see* instructor

Φ_i (θ_i) from roots λ_i (ν_i):
    fourth order, 152
    nth order, 147, 157
    second order, 91–109
    third order, 146

Variance. 46
    of linear forms, 47–48
    of X, 50
    sample, estimate, 48
Variance components, *see* Variance
        decomposition
Variance decomposition:
    of ARMA(2,1), 127
    of ARMA($n,n-1$), 130
Variance of forecast (error):
    ARMA models, 182, 189
    deterministic and stochastic trends
        compared, 390
    from log transformed series, 391
    multiple series, 474
    two series, 472
Vibration system, mechanical, 249–254
    data, *see* Mechanical vibratory displacement

White noise, 2
    continuous, 217
    discrete, 217
    variance of, 224
    variance of the integral of, 224
Whittle, 179, 279
Wiener, 106, 178–179, 236
Wold, 88, 174, 178
Wold's decomposition, 87–89. *See also*
        Orthogonal decomposition
Wolfer, *see* Sunspot series
Wood surface profile series, 403
    characterization by estimates, 406
    circular trend modeling tables, 405–406

$\bar{X}$, 48
    distribution of, 58
    mean of, 49
    standard deviation of, 50
    variance of, 50
$\dot{X}_t$, 16, 113, 155

Yaglom, 132, 179, 279, 318
Yule, 3, 177
Yule-Walker equations, 131

Z-transform, 412–430